RCL23

HOW TO FIND CHEMICAL INFORMATION

HOW TO FIND CHEMICAL INFORMATION

A Guide for Practicing Chemists, Educators, and Students

THIRD EDITION

Robert E. Maizell
Director, Technology Information Consultants
Science Park
New Haven, Connecticut

Formerly Manager
Business and Scientific Intelligence Centers
Olin Corporation

A WILEY-INTERSCIENCE PUBLICATION
JOHN WILEY & SONS, INC.
New York • Chichester • Weinheim • Brisbane • Singapore • Toronto

This book is printed on acid-free paper. ♾

Published simultaneously in Canada.

Library of Congress Cataloging-in-Publication Data

Maizell, Robert E. (Robert Edward), 1924–
How to find chemical information : a guide for practicing chemists, educators, and students / Robert E. Maizell.—3rd ed.
p. cm.
"A Wiley-Interscience publication."
Includes bibliographical references and index.
ISBN 0-471-12579-2 (cloth : acid-free paper)
1. Chemical literature. I. Title.
QD8.5.M34 1998
540′.7—dc21 97-29120

Printed in the United States of America.

10 9 8 7 6 5 4 3

CONTENTS

LIST OF TABLES

PREFACE

Chemical information tools are more crucial than ever to help the chemist and chemical engineer save valuable time, control costs, and achieve the most effective work. For many activities in chemistry and chemical engineering, optimum use of chemical information sources is key to success or failure.

More chemical information tools are available today than ever before, and many are more powerful than ever imagined before. Chemists and engineers are bombarded with advertisements promoting new information products, many of which are, unfortunately, very expensive. There are also some excellent low-cost sources, most notably the *Merck Index*, and no-cost sources, such as the U.S. Patent and Trademark Office and the International Business Machines Corporation Internet databases of U.S. patents. Competition between new information products is intense. The availability of the Internet, whose files are "free" in many cases (but by no means all), has added a significant and exciting new ingredient to the mix. The future of chemical information tools is bright and promising. However, these tools can offer a daunting challenge in their complexity and diversity. This volume is an organized review and guide that describes and discusses how to evaluate and select the most appropriate chemical information tools and how to utilize them to best advantage.

As far as is known, this is the first book of its kind that discusses, explains, and places into perspective for the reader all the major types of sources of chemical information: printed materials of all types, traditional online tools, compact disks, the Internet, and other sources including those of an informal nature.

The author hopes this book will be helpful for desktop reference, for daily use, and as a text (in the classroom or for independent study). The primary audiences for whom the book is written include practicing chemists and engineers, educators, and students. Information professionals, including librarians and chemical information specialists, should also find this volume helpful. The approach is practical so as to enhance the value of the book in the daily activities of the reader.

In order to reflect the many changes that have taken place since the second edition, this edition has been thoroughly revised and largely rewritten, and a great deal of new material has been added. All the formats in which chemical information is disseminated and can be searched are included, as for example, printed, electronic [online, Internet, CD-ROM (compact disk–read-only memory), and diskette], and informal means.

One of the strengths of this edition is that it describes and discusses the more enduring principles, strategies, and approaches that the chemist and engineer can utilize in the selection, evaluation, and use of chemical information tools. These principles, strategies, and approaches can be utilized with existing tools as well as with the new and improved tools that will become available in the future. Thus, a particular emphasis of this book is methods of evaluation that the chemist can apply to new chemical information tools.

In addition to its coverage of methods of evaluation and selection, this new edition describes the principal tools and methods of chemical information today. Pros and cons of the most important sources are given whenever feasible.

Revisions and updating have been carefully introduced throughout. For example, the chapter on online tools (Chapter 10) has been expanded, revised, and improved to include the latest information about online databases, host systems, the Internet, compact disks, and other electronic products. (Electronic tools are also discussed in most of the other chapters in the book.) This chapter includes valuable information about search strategies. In Chapter 12, the section on the major reference books (Section I) has been revised and updated, and an important new section about journals in chemistry (Section II) has been added. The two full chapters about Chemical Abstracts Service and its products (Chapters 6 and 7) have been updated and revised to include the most important recent information.

Another area of emphasis is the new and revised material related to environmental and safety information sources and tools (Chapter 14). The chapter on physical properties (Chapter 15) has been updated and revised, with special emphasis on electronic sources. The chapter on the business and marketing aspects of chemistry (Chapter 16) has also been expanded and improved.

Every effort has been made to avoid jargon or, if it was used, to define it at the outset.

In addition to descriptions and evaluations of key tools and methods, this book contains useful facts and figures including statistical data, names of key people, addresses, phone numbers, Internet addresses, and prices. Of course, many of these data change with time. Thus, for example, prices usually increase with time; they also are subject to variations depending on format, such as printed or electronic; quantity or volume discounts are frequently offered; there are often academic, industrial, and personal rate schedules; and there can be rate variation with number of users, networking capabilities, and geographic location of the user. In addition, products change in scope from time to time. The recommended way to get the most accurate current price and current product capabilities is to contact the producer of the information tool or product. Nevertheless, the data in this book will usually give the reader an excellent idea of what to expect. Tracking products and their producers on the Internet can also be very helpful.

In addition to the types of changes noted above, product and service ownership change often. Thus, VCH Publishers and Van Nostrand Reinhold are now owned by John Wiley & Sons, Inc.; according to reports, Elsevier apparently plans an interest in Beilstein; and the DIALOG Corporation is now part of M.A.I.D. Further, new Internet and other products are continuously introduced, while others change scope, address, and name, and still others are discontinued. At the same time, downsizing among some major information providers has occurred. Although landmark changes are a dominant theme in chemical information sources and tools, they can be "managed" by the astute user.

The perspective brought to this revision is that of over 12 years of intensive experience as a chemical information consultant working with chemists, chemical engineers, and other scientists and managers at large, medium-sized, and small chemical and advanced materials companies, as well as with universities and government. Further prior background includes several decades with Olin Corporation as Manager of Business and Scientific Intelligence Centers, and as Consulting Scientist for New Technologies. The author knows first hand the excitement of actively participating in some of the more important developments described. Personal experience includes, additionally, extensive meetings, telephone contacts, or correspondence with many of those involved at the cutting edge in their respective areas of specialization. This experience also includes instructional and training activities.

The author wishes to express sincere thanks to those who were kind enough to either review chapters or sections of the text, supply materials and data used in the writing of the book, or share their thoughts and perspectives on the latest developments. These include the following persons:

Dr. Harry Agahigian
Mr. Harry Allcock
Mr. Robert Bachman
Dr. Steven Barbee
Mr. Craig Bryson
Dr. John Buckingham
Ms. Susan Carino
Ms. Gill Cockhead
Dr. Andrew Cowan
Mr. James Dawson
Dr. Stephen DeVito
Ms. Carol Frischmann
Dr. Mildred Green
Dr. Stephen Heller
Dr. Martin Hicks
Dr. David Hughes
Dr. Stuart Kaback
Dr. Deborah Kahn
Dr. Roger Kolb
Mrs. Faye Kramer
Dr. Nancy Lambert
Mr. Brian Lewis
Dr. David Lide
Mr. Jorge Manrique
Mr. Robert Massie
Dr. William Mayer
Dr. W. Val Metanomski
Dr. Ralf Michaelis
Mrs. Jane Myers
Dr. Michael O'Hara
Dr. Louis Okorn
Dr. Shelly Rahman
Mrs. Patricia Rosso
Dr. John Rumble
Mrs. Marie Scandone
Mrs. Yvonne Schickel
Mr. Theodore Selover, Jr.
Mr. Eric Shively
Dr. Edelyn Simmons
Mrs. Darlene Slaughter
Dr. Arleen Somerville
Mr. Charles Sullivan
Dr. George Thomson
Dr. David Trurip
Dr. Patricia Turley
Dr. Jörn Von Jouanne
Dr. Donald Walter
Dr. David Weisgerber
Mr. Jack Westbrook

Mr. Eric Shively and Dr. W. Val Metanomski, noted above, were particularly helpful. The author may have inadvertently omitted the names of some who contributed, and for that he extends his apologies.

Thanks for their able assistance go also to Wiley Editor Betty Sun, Assistant Managing Editor Christine Punzo, and Copy Editor Cathy Hertz.

Capable graphic arts technical assistance and advice were supplied by Mrs. Linda Battalene (of the firm The Perfect Image, North Haven, CT), Mrs. Mary Ann Bisaccia, and Mrs. Sheila Klein, and the author is grateful for their work. To Mona Maizell go sincere thanks for understanding, patience, and enthusiastic moral support and encouragement at every stage. Ms. Liz Maizell assisted as well, and this is appreciated.

ROBERT E. MAIZELL

Science Park
New Haven, CT

LIST OF PRODUCTS AND/OR SERVICES REQUIRING TRADEMARK (™), REGISTERED (®), OR SERVICEMARK (℠) SYMBOLS*

ASM International®
Benchtop/PBM™
BIOSIS Previews®
Caplus™
CAS BioTech Updates®
CA Selects®
CA Selects Plus®
CASurveyor®
CASurveyor™
CCR®
ChemAdvisor®
ChemFinder®
Chemical Abstracts®
Chemical Market Reporter™
Chemical Titles®
Chemscape™
CLAIMS®
CrossFire™
CrossFireplusReactions™
CS Catalyst™
Current Chemical Reactions®
Current Contents™
Current Contents Connect™
Custom DIALOG™
Customized Research Alerts®

DataStar℠
Design Institute for Physical Property Data®
DIALINDEX®
DIALOG®
DIALOG *Ei ChemDisc*™
DIALOG *Ei Compendex®*
DialogLink™
DIALOG *Medline®*
DIALOG *Metadex®*
DIALOG OnDisc™
DIALOG *Petroleum Abstracts*™
DIALOG *QuickStart*℠
DIALOG *ScienceBase®*
DIALOG *Select*℠
DIPPR®
Document Detective Service℠
Document Solution™
EEE®
Engineering Information Village™
ESTOC®
EventLine®
EXCEL®
Galenet®
The Genuine Article®
HAZARDLINE®
HYDROPROP®
HYSYS Concept®
IBM®
IC®
Index Chemicus (IC)®
Institute for Scientific Information®
International Plastics Selector™
Internet Database Service®
ISI®
ISI Document Solution℠
ISIS™
KR SourceOne℠
LEXIS®-NEXIS®
Macintosh®

Marpat®
Microsoft®
Microsoft Internet Explorer®
Microsoft® Windows®
NetFire™
Netscape Navigator®
One Search®
OneSearch®
Oracle®
PatentWEB™
Permuterm®
Plastics D/I/G/E/S/T/®
QPAT-US™
Questel·Orbit®
Reaction Citation Index®
Request-A-Print®
Research ALERT®
Research Alert Direct®
RetroChem®
ROVER™
SCI®
SciFinder™
Selects®
SelectsPlus™
SourceOne℠
STN™
STN Easy℠
STN Express®
STN International℠
Trademark Checker™
Ulrich's™ *International Periodicals Dictionary*
UNIX®
U.S. PatentImages®
VAX®
Web Translator™
Westlaw®
Windows®

*Symbols do not appear in running text. The author apologizes for any inadvertent omissions. Omission of a tradename or mark from the above list does not mean that the name is not a tradename or trademark.

1 Basic Concepts

Why make regular use of the journal literature, patents, computerized databases, and other chemical information sources? A premise of this book is that the effective, regular use of appropriate printed and electronic chemical information sources is an essential key to achieving success in research and development and other functions in the profession of chemistry and chemical engineering.

Effective use of chemical information helps avoid duplicating previous work. This achieves savings in time (1) and funds and avoids infringing on the proprietary rights of others. In addition, even if there is no directly related previous work, the chemist who makes effective use of information can plan and act on a solid foundation of background data. Further, as a source of ideas, or for idea development, chemical information sources are invaluable fountains of inspiration and serendipity.

Many outstanding chemists are noted for their extensive use of the literature. For example, a recent biography (2) of Wallace H. Carothers (the inventor of nylon based on his work at the DuPont Company) makes frequent mention of his extensive use of the literature. Another famous DuPont chemist, William E. Hanford (coinventor of U.S. Patent 2,284,896 recognized by the National Inventors Hall of Fame in 1991 as the basis for polyurethane manufacture) was also an avid user of the literature and patents. In his later career as Vice President for Research at Olin Corporation, Hanford instilled his zeal for the literature in his staff, including this author, so that it became an integral part of the daily research effort for everyone. Furthermore, at least one research study shows that creative chemists use the literature more than less creative chemists, although literature use does not necessarily make a person creative (3).

Printed and electronic sources of chemical information are not always the first or best choices when looking for needed data. Thus, for example, it may sometimes be easier and more efficient to ask a colleague in an adjacent laboratory, especially if that person is believed to be both knowledgeable and reliable in the field of interest and has the desired information readily at hand.

Nevertheless, the chemist who knows how to utilize chemical information sources quickly and efficiently, and who has the required patience, energy, and perceptiveness, will usually have a clear advantage over the chemist who either lacks these skills and qualities or is too lazy to utilize them.

Although use of chemical information sources is not easy, the often significant benefits usually make the effort well worth it.

1.1. THE INFORMATION PROFESSIONAL AND LITERATURE USE

A good working knowledge—and regular use—of the appropriate specialized journal and patent literature and other so-called *primary* (original) chemical information sources (frequently available in both printed and electronic forms) is now virtually imperative for all practicing chemists, especially for those engaged in research and development efforts. As these primary sources continue to increase in volume, important changes are also taking place in the size and sophistication of *secondary* chemical information services (such as computerized online abstracting and indexing services and other related electronic and printed tools) intended to provide quicker and more effective access to primary chemical sources.

In addition to these trends, an increasing number of significant electronic tools are available that are specifically designed for direct use by the laboratory chemist and chemical engineer as the end user without the assistance of an information professional as described below. These include certain thoughtfully designed online products, mostly on the Internet, and compact disk products (CD-ROMs), and are discussed more fully in Chapters 6, 7, and 10.

With the important exception of tools specifically designed for the end user, the secondary sources of chemical information are now so important, extensive, complex, and expensive that they are usually most effectively utilized by chemists with specialized training as an information professional. These individuals are designated by such job titles as *chemistry librarians, chemical information specialists, information scientists,* and *literature chemists.*

Recognition is given to this field within the framework of the American Chemical Society by activities of the Division of Chemical Information to which these specialties are a prime interest and by publication of the *Journal of Chemical Information and Computer Sciences,* edited by George W. A. Milne, National Institutes of Health. For laboratory chemists, these persons are surrogates who can be of considerable help in planning or performance of literature and database studies, as appropriate, and who, in addition, can sometimes advise and guide as to research project strategies.

Although laboratory chemists can find that information professionals are invaluable colleagues, there is no substitute for firsthand familiarity with chemical information sources. It is a mistake to rely too heavily on a surrogate for information gathering needs because only by *firsthand* use of the literature can the chemist keep fully up-to-date and professionally stimulated.

Another consideration is capability of the surrogate in the field of chemistry under study. Some information chemists are highly skilled not only in computer or information sciences but also in specific fields of chemistry. Others, especially those who serve a diversified clientele in several disciplines, have only a generalist's knowledge of chemistry and may not have either time or opportunity to develop the highly specialized expertise that many laboratory chemists possess. In addition, the surrogate may concentrate on a specific topic as directed by the laboratory chemist. The result could be that the surrogate may not have the opportunity to recognize peripheral, but significant, data that could be identified by a person who specializes in the field and that could be crucial to success of this or related projects. Furthermore, information professionals are frequently not as readily available as might be desired, as a result of significant reductions in certain types of funding and staff reductions at many industrial organizations in recent years. Universities have experienced fewer staff reductions; however, Somerville (University of Rochester) and others have noted that some librarians in college and university chemistry departments may not have sufficient background in chemistry or related sciences to give optimal assistance. This is another good reason why laboratory chemists should attempt to obtain at least a basic working knowledge of the most significant information sources and tools in chemistry, not only those that are intended specifically for the end user, but the other important tools as well.

Even in the case of the increasing number of electronic products that are intended specifically for the end user, the information professional can still play a significant role that can benefit the laboratory scientist. For example, with regard to the landmark end-user product *SciFinder,* discussed in Chapters 6 and 7 of this book, fruitful collaboration between the laboratory chemist and the information professional has been described by Williams on the basis of experience at the Monsanto Company (4). Among other topics, Williams writes about the role of the information professional in explaining nuances, in pointing out advantages and limitations, and in offering support in use of *SciFinder* by laboratory scientists.

The best approach is a full partnership in which the laboratory chemist shares project goals, progress, and information about problems with the information professional, and in which the two work closely together to achieve desired objectives in the most cost-efficient manner. One way to achieve this is to require that qualified information scientists be full members of all major project teams.

In the ideal situation, laboratory chemist and information chemist develop search strategy (see Chapters 3 and 10) together. If computer-based, online searching (see Chapter 10) is to be done, the laboratory and information chemist preferably sit together, and both participate as search results begin to appear so that strategy can be modified if needed. If this is not possible, as may frequently be the case, the two should consult closely as soon as possible to review initial results and plan next steps. This kind of professional partnership can ordinarily yield results that are more than satisfactory. The partnership described can be especially helpful when crucial matters are at stake. These include such issues as safety and toxicity, for example, or when significant technology or financial decisions are involved.

1.2. THE ADMINISTRATOR AND INFORMATION

If a chemical organization or department is to flourish and prosper to the fullest, it is important that the administrators appreciate the value of literature and other information sources and of information professionals as mentioned in the previous section. Literature and information use and searching are probably the most economical forms of chemical activities and can be the most productive. There is significant infusion of pertinent new information and ideas into the organization, and other important benefits can be achieved, such as avoiding unnecessary repetition of work done by others. Support by management of the chemical information center or library is vital to the fruitful work of chemists or engineers in any organization, whether they are professors, industrial or governmental employees, or students.

REFERENCES

1. "ACS Report Rates Information System Efficiency," *Chem. Eng. News* **47**(31), 45–46 (1969). This study by the American Chemical Society Corporation Associates is one of the better studies on the specific value of information tools.
2. M. E. Hermes, *Enough for One Lifetime—Wallace Carothers, Inventor of Nylon,* American Chemical Society, Washington, DC, 1996.
3. R. E. Maizell, "Information Gathering Patterns and Creativity," *Am. Doc.,* **11,** 9–17 (1960).
4. J. Williams, "Information at the Desktop for Scientists—*SciFinder*" *Online,* 61–66 (July/Aug. 1995).

"The Chemical Information Instructor" Quarterly Column in *Journal of Chemical Education;* Other Instructional Sources

The "Chemical Information Instructor" is a column that appears periodically in the *Journal of Chemical Education.* It is edited by Dr. Arleen N. Somerville, Carlson Library, University of Rochester, Rochester NY 14627-0236 (phone 716-275-4465; fax 716-473-1712; e-mail address ansv@dbv.cc.rochester.edu).*

This column provides instructors with practical information on a wide range of topics related to teaching information searching skills to undergraduates, graduate students, and other researchers. While oriented toward academic situations, all chemists would benefit from this material. Information is provided in print, on the Web (World Wide Web; http://jchemed.chem.wisc.edu) via *JCE Online,* and via a discussion forum on the Internet. Topics include, but are not limited to, courses—semester, short courses; workshops; integration of information instruction into one or more courses; integration of WWW sources into instruction; specific types of information, including inorganic and bioorganic chemistry, organic reactions and polymers; specific types of materials, such as patents and journal articles; specific sources and databases, such as *Science Citation Index, Chemical Abstracts;* specific types of searches, including structure, reaction, and citation indexing; ways to stay current on a topic; and teaching techniques, such as interactive teaching.

Contributions aim to provide information needed by readers to replicate similar experiences in their institutions. Information includes description of the instruction; staffing; costs involved and how resolved; logistics—hardware, software, and scheduling; practice questions and exam questions; how this instruction contributed to the overall instruction program; and any other information the author thinks other instructors would find valuable.

*Note: All information below is provided through the courtesy of Arleen N. Somerville.

The editor welcomes contributions by potential authors and by all readers about topics that they would like to see discussed in the column and especially in the Internet discussion forum. Potential authors are urged to contact the editor with an early draft version before completing the paper.

The contents to date, in chronological order, is as follows:

A. N. Somerville, "Information Sources for Organic Chemistry, 1: Type of Reaction and Name Reactions," *J. Chem. Educ.,* **68,** 553–561 (1991).

A. N. Somerville, "Information Sources for Organic Chemistry, 2: Functional Group Chemistry," *J. Chem. Educ.,* **68,** 842–853 (1991).

H. I. Abrash, "A Course in Chemical Information Retrieval," *J. Chem. Educ.,* **69,** 143 (1992).

J. A. Jenkins, "Undergraduate Instruction in Online Searching of *Chemical Abstracts,*" *J. Chem. Educ.,* **69,** 639–643 (1992).

A. N. Somerville, "Information Sources for Organic Chemistry, 3: Reagents and Solvents," *J. Chem. Educ.,* **69,** 379–386 (1992).

C. Carr, "Teaching and Using Chemical Information; An Updated Bibliography," *J. Chem. Educ.,* **70,** 719–726 (1993).

A. N. Somerville, "Subject Search of *Chemical Abstracts* Online," *J. Chem. Educ.,* **70,** 200–203 (1993).

R. C. Cooke, "Undergraduate Online Chemistry Literature Searching: An Open Ended Course Segment Approach," *J. Chem. Educ.,* **71,** 867–871 (1994).

S. J. Penhale and W. J. Stratton, "Online Searching Assignments in a Chemistry Course for Nonscience Majors," *J. Chem. Educ.,* **71,** 227–229 (1994).

L. M. Wier, "*Chemical Abstracts:* Switching from Hard Copy to Electronic Access," *J. Chem. Educ.,* **71,** 578 (1994).

G. Baysinger, "Identifying Unknowns: Library Support in an Undergraduate Organic Chemistry Course," *J. Chem. Educ.,* **72,** 1107–1111 (1995).

P. O'Neill and E. Goetz, "Electronic Users Group: A Forum for Experienced Searchers," *J. Chem. Educ.,* **72,** 604–605 (1995).

P. A. Thompson, J. Jenkins, and D. R. Buhler, "Teaching Toxicology Graduate Students Online Data Systems," *J. Chem. Educ.,* **72,** 324–326 (1995).

C. Holmes and J. Warden, "CIStudio—a World Wide Web-based Interactive Chemical Information Course," *J. Chem. Educ.,* **73,** 325–331 (1996).

F. J. Matthews, "Chemical Literature: A Course Composed of Traditional and Online Teaching," *J. Chem. Educ.,* **14,** 1011–1014 (1997).

A. Smith, "Teaching Citation Searching" (in press).

G. Baysinger, "Instructional Examples Using *Dictionary of Organic Compounds* CD-ROM" (*J. Chem. Educ.,* in press).

A. N. Somerville, "Teaching Aids Available for Information Instructors" (*J. Chem. Educ.,* in press).

In addition to the above papers, the American Chemical Society's Division of Chemical Information Education Committee has published teaching modules relating to: computer searching of *Chemical Abstracts* on DIALOG and on STN International; citation searching; and patent searching on STN. Requests for these should be directed to Mrs. Arleen N. Somerville, University of Rochester, Rochester, NY 14627, phone 716-275-4465.

Further, the American Chemical Society and Special Libraries Association maintain a clearinghouse for chemical instructional materials. Requests for these should be directed to Dr. Gary D. Wiggins, Chemistry Library, Indiana University, Bloomington, IN 47405, phone 812-855-9452.

2 Information Flow and Communication Patterns in Chemistry

The chemist should be aware of the broad framework of information flow. For example, many chemists and other scientists obtain much of their information from colleagues by such informal means as face-to-face conversations, telephone calls, e-mail (electronic mail), and written correspondence. Informal contacts of this type have been called "invisible colleges" or networks. In addition, chemists acquire much information at national, international, regional, and local technical meetings sponsored by professional societies and trade associations.

Primary journals, patents, trade and review journals, manufacturer trade literature, reports, books, and abstracting and indexing services are all highly significant sources of information. Computerized online databases and CD-ROM (compact disk–read-only memory) products permit the chemist to search retrospectively, to keep with new developments by such means as current alerting methodologies, and also offer other capabilities, including, for example, viewing and searching of the full text of documents and ordering of document copies. In addition, the Internet has emerged as a major factor in communication among chemists. All of these modes are described in this book.

The sequence of communications between chemists often follows this approximate sequence of overlapping principal steps (many of these steps can be conducted using a computer and results or output can be stored on a computer):

1. Conducting search of the literature and other information sources, including contacts with colleagues, to identify previous work, to build on a foundation of facts, and to avoid replication of what others have already done.
2. Performing laboratory work.
3. Entering data in laboratory notebooks. These are internally used, proprietary documents. Notebooks are especially important in obtaining patent protection. Entries are best made in accordance with requirements of the organization with which the chemist is affiliated; hence exact mode of entry and type of data required is variable. Systems for electronic laboratory notebooks are available, but security and legal issues must be properly addressed in any organization making use of such a system. Retention of notebooks is usually governed by special records retention procedures as determined by the chemist's organization.

4. Writing letters for in-house use. These are important to the inner workings of most organizations. Some locations find centralized correspondence centers useful. Others encourage personal filing systems. Many organizations utilize internal e-mail systems. Letters are not a substitute for reports.
5. Writing research reports and related documents for in-house use. These are proprietary internal documents in most cases. An exception would be, for example, reports based on government sponsored research that, if unclassified, may be available to the public.
6. Filing patent applications. In almost all major industrialized nations, patent applications are published 18 months after filing, and, for at least one major patent issuing authority, also become available online at about this time. Patent applications must be examined by patent examiners before they are granted.
7. Informally exchanging results with colleagues in other organizations (face-to-face meetings, telephone calls, Internet [including e-mail] communications, written [paper] correspondence). Excluded from such an exchange would be information that is confidential, proprietary, or that would interfere with the obtaining of a patent by the investigator or organization (requirements vary with country).
8. Presenting results at professional society, trade association, and other technical meetings. Writing and submitting paper to journal editors for possible publication. In many organizations, these actions require clearance by research directors, public relations people, and legal staff to avoid premature release of information. For almost all primary research journals, submitted papers are subject to review by anonymous referees prior to publication.
9. Publication of accepted papers and/or issuing of granted patents. The full texts of papers for an increasing number of important journals and of U.S. patents are now available at nearly the same time in both printed and online forms. The full text of European Patent Office patent documents is also available online. A handful of journals in chemistry appear in electronic form only, but this is not a major trend. In an increasing number of cases, advance copies of titles and abstracts of papers in chemistry may appear several weeks in advance of the printed versions in printed or electronic form. Although advance full text (preprints) of some journal papers in some fields of science, especially in physics, may appear on the Internet or in other electronic forms some weeks before the printed version is received by subscribers, this is seldom the case in chemistry* because of the priority date concerns related to patents (see Chapter 13). However, in some cases, chemists have begun to post "preprints" on the Internet prior to journal publication, and this is a matter of concern to journal editors. (For example, see the editorial by W. H. Glaze that appeared in *Environmental Science and Technology,* July 1996, on page 273a). The

*But, beginning in 1998, for all ACS journals, all papers that have been refereed, copyedited, and author-checked and accepted for publication are to be on the Internet before appearing in the printed journal.

Journal of the American Chemical Society will not accept manuscripts that have been previously posted on the Internet.

10. Announcement of publications and patents by current awareness and alerting services. Initial listing in some abstracting and indexing services preliminary to full abstracting and indexing.
11. Abstracting and indexing of published papers and patent documents by one or more of the major abstracting and indexing services. Concurrent input of abstracting and indexing data for both online and printed availability. (*Online* is meant to imply both "commercial" online services and Internet availability. Most online services are becoming available not only commercially but also on the Internet, perhaps in modified form.)
12. Printing of corresponding abstracts and indexes in full-size, microfilm, or microfiche form, and online availability. In some cases, compact disk versions are also available.
13. Use of abstracts and indexes, both printed and electronic versions.
14. Summarization and evaluation in review articles, monographs, encyclopedias, and information analysis centers; citations in other literature.

Some of the newer technologies can significantly telescope or otherwise shorten some of these steps by electronic capture of data in initial steps and subsequent reuse in later steps. For example, recording of data by use of a computer permits not only initial capture of data but also subsequent analysis and use in writing of letters, reports, patent applications, and papers, as well as exchange of these data with colleagues through electronic mail. This capability is readily available today to those chemists who have the required hardware and software.

It is important to emphasize also that communication frequently may not follow the neat series of steps described. Some of the steps may be omitted or bypassed, or may occur in a different order. Informal oral, e-mail, or other forms of communication may occur unpredictably, for example.

3 Search Strategy

One of the questions most frequently asked by chemists is how to conduct an effective information search—one that will yield optimum results quickly. Developing an answer to this question usually requires formulating answers to other, more specific questions. This process is helpful in developing an overall approach and philosophy for the use of chemical literature.

The first step is to formulate the goal or objective—the information being sought—as precisely as possible to save time and avoid wasted effort. The search should be delimited within those parameters that reflect the chemist's precise interest. Section 10.10 contains suggestions on ways to properly limit a search.

Some questions the chemist should ask in delimiting and conducting a search include the following:

1. Are my goals and objectives clearly defined? Do I know what I want, why I want it, and what I will do with it when I get it?
2. What information do I already have on hand? Have I looked thoroughly at this information to see what leads this might provide and also to avoid going over the same ground?
3. How soon do I need this information? How important is it to me or to my project? Answers to these questions will help determine how much time and effort to put into the search and will reflect priorities.
4. Before looking at the literature, have I talked with colleagues in my own organization who might have some information (or leads to information) at their fingertips? See also point 22 on outside contacts.
5. What time period do I need to cover? Must I go all the way back to the beginning of modern chemical literature, or would a search of the last few years suffice? Can I limit the search to the current year, at least at the outset? See also point 14 on time period as a factor.
6. Do I need to make a search international in scope? Or can I limit it to a specific country, language, organization, or geographic area where I believe most of the important work has been or is being done?
7. Can I limit the search to certain kinds of documents? For example, am I interested in patents only or in nonpatents only?
8. What specific aspect of the field, the chemical, or the chemistry am I interested in? (If I am interested in all aspects of a large field, I could be taking on more than is ordinarily possible.) An examination of the latest Collective Index to *Chemical Abstracts* (*CA*) will give some ideas of what and how much has been published on the topic.

9. What sources appear to be most fruitful in attacking the problem and obtaining the needed information? Before beginning a search it is important to list sources likely to be most productive. If completeness is an objective, all pertinent sources should be consulted; this also helps avoid bias that can come from consulting just one source. Use of multiple sources has many advantages.
10. How readily available are those sources to me, and do I know how to use them properly? How specific are they to the subject at hand? (I might prefer a smaller, highly specialized source or service in an area of specific interest to a larger, more generalized source or service.)
11. In my overall search plan, have I worked out a strategy wherein those sources that I know will take a long time to respond (such as persons I need to correspond with) will have been contacted at the outset so that I will not be delayed?
12. Before beginning a detailed search of such sources as all of *CA,* will I check selectively such potential sources of quick answers as, for example,
    a. *Kirk–Othmer Encyclopedia of Chemical Technology* (see Section 12.2.A),
    b. desk handbooks (see Section 15.11),
    c. *Beilstein* (see Section 12.4.A),
    d. *Gmelin* (see Section 12.5.A),
    e. review resources (e.g., review articles) (see Chapter 11),
    f. *CA* online for recent years, and
    g. monographs and treatises on the subject (see Chapter 12)?
13. With regard to point 12, are there recently published reviews, encyclopedias, or monographs that contain much or all of the data I need and that may shorten (or even eliminate) need for additional searching at least for the immediate present?
14. Will I start with the latest available source? (This is usually the best procedure rather than use an archaic source). If I use a source such as *CA,* will I work *backward* in time, that is, start with the latest years and work backward, if needed, to the earlier years? This is the preferred method. After all, why go all the way back to 1907 (the beginning of *CA*) if I have good reason to believe that everything I need was published during the last five years, for example.
15. Have I developed a subject search plan and a flexible iterative subject search policy? This means, first, systematically developing an array of subjects or key words that I believe are most likely to cover my areas of interest. It then means—based on experience as the search progresses—modifying, as necessary, my original array by adding or deleting search terms and tools. Some of these changes may be a result of a need for greater breadth; others may reflect a need for being more specific. Such tools as the *Index Guide* to *Chemical Abstracts* (see Section 6.5) help in construction of an array.
16. Do I consider all significant sides of the question under investigation?
17. Should I consider attacking the problem through approaches other than subject such as author, originating organization ("corporate source"), journal reference or patent number, or molecular formula?
18. Have I asked a surrogate, such as a chemical information scientist or similar person, to review my research strategy and plans before starting in order to get the benefit of that person's expert advice?

19. Do I keep a systematic record of my progress? (Sources looked at, information found, and index entries used should be meticulously recorded in a standard laboratory notebook rather than on loose scraps of paper.)
20. Am I alert to the possibility of serendipity—of unexpectedly finding information pertinent to another of my interests or otherwise important and stimulating but not directly related to my immediate question? Am I prepared to react to and use such information if I find it? See also point **23**.
21. In implementing my strategy, do I take into account the personal qualities needed to achieve best results? Examples of such qualities include patience, orderliness, attention to detail, reasonable persistence (see point **24**), and imagination. Ability to cope with some initial frustration is also helpful.
22. Are there reliable sources outside my organization and my immediate circle of professional colleagues that I could appropriately contact to get the data needed? Such contacts may include manufacturers of chemicals, equipment, and instruments. They may also include carefully selected scientists in other facilities such as in certain university or government laboratories. Of course, confidentiality or proprietary concerns may not make such contacts advisable or feasible. See also section 5.9 on personal contacts.
23. Have I taken full advantage of the useful clues that can sometimes be obtained from studies that report negative results or results that are opposite from those expected by the researcher? For example, when looking for applications for new chemicals, products that fail as candidates for one end use may be ideal for the reverse or opposite of that use or for a related use under different conditions.
24. Do I know when to stop? Do I know when I have all that I need? Can I recognize the point of diminishing return?

Finally, it is worth stating the obvious point that, in any review of the literature and patents, the full original patents or journal papers (not just an abstract or a citation) that are pertinent should be consulted, especially if decisions are to be made or actions are to be taken on the basis of the literature. The importance of consulting the full original documents cannot be overemphasized. In cases that are especially crucial and critical, the investigator would do well to consider resolving any questions by personally communicating with the author of the original source, if this is feasible, appropriate, and within the bounds of confidentiality.

Chapter 10, especially Section 10.10, contains additional hints on search strategy. Although Chapter 10 relates primarily to online and Internet searching, some of the same principles apply in other situations.

4 Keeping Up To Date: Current Awareness Programs

4.1. INTRODUCTION

With more than 700,000 articles, patents, books, reports, and other documents of chemical interest published annually, how can chemists and engineers keep up with new information in their fields? The task is difficult even for those who specialize in narrow fields. Fortunately, experience has shown that many persons can develop and implement manageable programs for keeping up to date.

At the outset, the chemist or engineer must accept the likelihood that 100% coverage, even in a relatively limited field of interest, probably is not achievable in any current awareness effort. There will be pertinent material that will be identified too late or not at all.

With this important limitation in mind, the first step is to clearly define current professional interests—areas in which to develop or maintain knowledge. It is better to state these interests in positive terms (to state what you are interested in) than in negative terms (to state what you are not interested in).

This statement is usually called a *profile.* It may include not only subjects but also authors and organizations whose activities the chemist wishes to follow because they are known to be active in specific fields, and citations to key references previously located.

The chemist or engineer planning a current awareness program needs to consider carefully the finite resources available. Other demands on resources as well as priorities also need to be taken into account. One of the most valuable assets any person has is time. Budgeting time spent in keeping up to date is essential to achieving career objectives. About 5 hours per week is a good rule of thumb for most chemists who have full-time jobs. That figure can easily be doubled or tripled for full-time students.

Chemists embarking on new projects about which they may know little or nothing will require considerably more time than the rule of thumb, especially in the initial phases. A full-fledged literature and patent search should precede all new laboratory projects to avoid duplicating previous efforts by others. Such a search could easily take several days or even weeks, depending on the project. Laboratory chemists and engineers might want to have this kind of study done by a chemical information specialists (or other qualified surrogate.) This should save time and would be an appropriate allocation of skills. There are, however, important advantages to doing at least some of one's own searches (see Chapter 1).

In planning current awareness efforts, the chemist needs to consider available funds. Current awareness programs, particularly those based on computers, can become costly. Several hundred dollars per year is common for a basic computer-based awareness program for a chemist in industry.

The key, then, is to take stock of available resources, both personal and those of one's organization, and to develop a current awareness effort consistent with those resources and appropriate to needs. With a time-and-dollar budget clearly defined (but flexible enough to meet the unexpected), the chemist can plan with confidence, and will be better able to cope with continuing proliferation of chemistry literature and chemical information services, many of which are highly specialized and expensive. The widespread availablility of chemical information on the Internet adds another major factor to the equation that needs to be considered. Each chemist needs to make careful decisions as to which publications and other sources are to be scanned or to be read cover to cover regularly, and for how many hours a week. The balance of the literature (except for special items identified through *Selective Dissemination of Information* programs as described later in this chapter) will simply need to be largely ignored, difficult though that may be from the perspective of personal discipline. Unless this kind of decision is made, the chemist can spend a lifetime doing nothing else but reading the literature, both printed and electronic.

Flexibility is, however, important too. Journals and services improve, merge, become more expensive, deteriorate, or collapse, and new ones are introduced. Personal and organization interests, budgets, and policies change. Accordingly, reevaluation every 6–12 months is imperative.

4.2. JOURNAL CIRCULATION AND READING CLUBS

Some organizations manually circulate journals received to persons in their organizations who are interested in these journals. This is a function typically managed by an organization's librarian. It is a simply way to achieve current awareness, and it has worked. The limitations include the fact that only those journals subscribed to are circulated; many other publications may potentially be of interest. Another problem is that some persons on the routing list may fail to read and circulate the journals promptly, thereby inconveniencing others on the circulation list; constant and systematic policing, backed up by a management directive, is required to keep the journals moving promptly. Manual journal routing will probably be largely supplanted by electronic alerting as to tables of contents as described in this chapter.

In a reading club, each member of a team working on the same project accepts responsibility for "following" certain journals and/or electronic sources and sharing key findings with the others. This takes discipline, but it works.

4.3. PERSONAL PLAN OF ACTION

In developing a plan of action for keeping up to date, the chemist will find it helpful to visualize a triangle as shown in Figure 4.1.

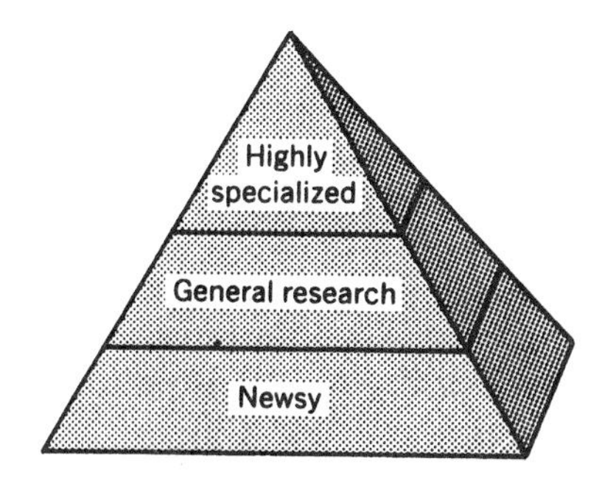

Figure 4.1. The current awareness triangle.

The base of the triangle consists of general *newsy* (but important) material with which most chemists and engineers should maintain some awareness. The middle part of the triangle consists of relatively broad areas of chemistry within which there may be some specialization. The apex represents specific materials closely allied to unique individual interests and on which most current awareness time will be spent.

A. Base of the Triangle: Chemical News Magazines

The general material of most widespread interest is reported in chemical *news* magazines. The chemist who wants to be well informed should systematically and promptly scan news magazines such as:

1. *Chemical & Engineering News* (weekly publication of the American Chemical Society).
2. *Chemical Week* (weekly publication of Chemical Week Associates).
3. *Chemical Market Reporter* (weekly publication of Schnell Publishing Co.).
4. *Chemical Engineering* (monthly publication of McGraw-Hill, Inc.).
5. *Chemistry and Industry* (semimonthly publication of the British Society of Chemistry and Industry which contains news, research, and other features).
6. *European Chemical News* (weekly publication of Reed Business Publishing).

It is wise for the chemist who wants to keep as current as possible to include more than one weekly chemical news magazine as part of a carefully thought-out reading program.

A basic and virtually indispensable weekly chemical news magazine of choice for any American chemist or chemical engineer (or, for that matter, any English-reading chemist or engineer in any country) is *Chemical & Engineering News* (*C&EN*). The coverage of this publication is more balanced and broader in scope than any other weekly chemical news magazine. Under the editorship of Madeleine Jacobs, *C&EN* has substantially improved and revitalized its coverage. *C&EN* shines in its timely coverage of important new scientific and technical developments and of professional and career matters. Industry news developments are well covered. In addition, it includes statistical compilations on chemical industry performance over time, with insightful commentary and perspective. It also offers extensive and candid coverage of the U.S. legislative and regulatory scene. Included are listings of new publications, and there are occasional book reviews as well. *C&EN*'s listing of new software products of interest to chemists is one way to keep up with this software on a regular basis.

The letters to the editor often contain the first announcements of new safety hazards. A valuable adjunct feature is that the full text of *C&EN* is online through STN International (see Chapters 6 and 7) and has been searchable since 1991.

However, *C&EN* alone is not enough. For American chemists and chemical engineers working in industry, either *Chemical Market Reporter* (*CMR*) or *Chemical Week* or both should be added to get other perspectives, especially from the point of view of the chemical industry. Each of these three leading American weekly chemical news magazines frequently has different details and slants on the same developments, and each has valuable special feature stories in depth. The latter two also offer valuable printed buyer's guides (see Chapter 16) as part of the subscription; these guides are also on the Internet sites of the respective publications.

As much as any other weekly chemical news magazine, *Chemical Market Reporter,* founded in 1871, gives the reader the feel of being in close touch with important current chemical industry marketing and other business developments and thinking, especially in the United States. Its weekly coverage of chemical pricing data is unique, and the chemical import data presented are also valuable. Its *Chemical Profile* series offers convenient snapshots of important industrial chemicals. Large segments of the publication are covered in several different online sources, and there is an Internet presence (www.chemexpo.com). The editor is Michael McCoy, and the publisher is the Schnell Publishing Company, New York, NY.

Chemical Week is another widely read weekly news magazine. Auxiliary products provide especially fast delivery options and include *Chemical Week Newswire,* which is faxed or mailed to subscribers on a daily basis; it contains the most significant business news in two to three pages. In addition, there is a biweekly *Chemical Week Asia* newsletter and a weekly *Chlor-Alkali Marketwire.* The full text of *CW* is available online through LEXIS-NEXIS; a CD-ROM version is available, and *CW* is on the Internet (www.chemweek.com).

Chemical Week offers an Executive Edition on the Internet, which includes news of interest to chemical industry executives. *Chemical Week* also offers a special twice-a-month edition entitled *Electronic Chemicals News.* As indicated by its title, the focus is on news about chemicals utilized in electronics applications such as in the manufacture of semiconductors and related materials. Content includes news about companies, products, meetings, regulatory developments, and financial developments.

Another U.S. magazine that provides excellent coverage of the chemical industry is *Chemical Engineering,* but it now appears only monthly. Regularly appearing news briefs in the front of the magazine do a unique job of highlighting some key developments, and the technical articles zero in on key topics of interest to chemical engineers everywhere.

The best news magazines for coverage of European chemical industry developments are the British magazines *European Chemical News* (*ECN*) and *Chemistry and*

Industry. ECN includes such features as spot prices for bulk chemicals, and new plant construction projects, worldwide, are tabulated. Several times a year, a special supplement (*Chemscope*) explores in detail a different geographical region or industry sector. For Asian developments, the best choices include *Japan Chemical Week* and *Asian Chemical News.*

See also Chapter 12, Section II, which discusses journals.

Except for *C&EN,* the chemical news publications are oriented primarily toward the industrial or business aspects of chemistry. But the boundary between this kind of chemistry and the more scientific aspects is blurred. Each kind spills over into and has implications for the other. Chemists who want to further their careers will develop a balanced current awareness program and will keep informed on both the scientific and business aspects, although they will usually concentrate on one or the other.

B. Middle Part of the Triangle: General Research Publications, Title Announcements, and Abstracts Groupings

A balanced reading program should contain two or three general research publications within which there may be some specialization and that can be skimmed selectively. Examples include

1. *Angewandte Chemie—International Edition* (in English)—a top-quality publication (issued 22 times per year) that includes important, fastbreaking developments, abstracts of important work published elsewhere, and some original research papers.
2. *Science*—almost all branches of science are covered in this weekly, although emphasis is on the biological aspects. Important breakthroughs are frequently announced here first.
3. One or more of the American Chemical Society's primary journals (see Chapter 12, Section II) and/or selected publications of the American Institute of Chemical Engineers such as *Chemical Engineering Progress* or *AIChE Journal. Chemical Engineering Progress,* a monthly publication, includes publishes some 60 pages of technical papers each month. In addition, there are number of special features that are offered. The December issue each year offers an extensive directory of software. (See Section 10.5.B). Other special features are Regulatory Update; Washington Update (under consideration); Patent Update (intellectual property issues): Web Sights (Internet sites of interest to chemical engineers); and professional development and career advancement.
4. Other publications, including commercial and trade magazines.

In cases of uncertainty, the chemist or engineer should call on the local chemistry librarian or chemical information specialists for advice in making a selection. Educators and more experienced colleagues can also help "fine-tune" choices to achieve a manageable portfolio.

The following examples also belong in the middle part of the triangle. Reading time and costs usually permit a selection of only one of these examples:

1. *Current Contents, Physical, Chemical, and Earth Sciences Edition,* published by the Institute for Scientific Information. The *Physical, Chemical & Earth Sciences* edition of this weekly service reproduces tables of contents of approximately 900 of the most important scientific and technical journals (as well as many books) in these sciences. The service permits the chemist to quickly scan contents of many journals of potential interest. When a pertinent article is noted, the chemist can consult the full original document in a local chemistry library, can request reprints from the authors, or can order copies conveniently and quickly using *ISI's The Genuine Article* document delivery service, which is described in further detail in Section 5.2; this service is to eventually become part of *ISI Document Solution.* The cost of this *Current Contents* edition in printed form for 1998 is $580. In addition to the contents pages of journals, there are four indexes. These include a list of all journals in each issue, significant words and phrases from titles, author index and address directory, and publisher address directory. All *Current Contents* editions—totaling seven in various scientific and other disciplines—are available in print, diskette, and magnetic tape formats, and also online. The *Physical, Chemical & Earth Sciences* edition is also available in CD-ROM format. Some of the electronic formats include additional features such as the option of English-language author abstracts for every article that includes one, additional keywords as based on article bibliographies ("KeyWords Plus") and author and publisher names and addresses. *Current Contents on Diskette* is available in a Microsoft Windows version as an optional alternative to the previously available versions. Enhancements include a multiple issue search feature, capable of handling 16 issues at a time; a new quick-search option for novice users; and an optional cumulation on CD-ROM called *Current Contents Reference Edition.* Another related publication, *Current Contents, Engineering, Computing and Technology* (*ECT*), is concentrated in areas that are particularly relevant to chemical engineering, materials engineering, mechanical engineering, and electrical and electronic engineering. ISI has recently expanded journal coverage of this product with the addition of 200 additional titles, bringing the number of technical and applied research journals covered to more than 1000. To reflect the new editorial focus, the name of the product has been changed. The former name was *Current Contents Engineering, Technology & Applied Sciences. Current Contents Search* is online in the DIALOG system covering from 1990 to date. It includes all of the printed editions for this time period and thus covers not only the chemical and other physical sciences, including the earth sciences, but also engineering and applied sciences; the clinical medicine and life sciences; agricultural, biological, and environmental sciences; and social and behavioral sciences. Arts and humanities are also included. Other online systems that carry this product are DataStar (March 1992 to date) and Ovid Technologies (2-year rolling file). The file includes English-language author abstracts when available, as well as the "KeyWords Plus" feature. The product is also available from Silver Platter Information, Cambridge, MA, either on magnetic tape or via a remote Internet server, and from Information Access Company, Foster City, CA, through Internet access, FTP delivery. Appropriate client/server software is also available through Lotus Development Corporation, (Lotus Notes), which is located in Cambridge, MA. The most recent development regarding *Current Contents* is that ISI is currently (October 1977) introducing Internet and intranet access under the product name *Current Contents Connect* for all seven editions.

Chemical Titles is a similar publication issued every 2 weeks by Chemical Abstracts Service (CAS). Nearly 800 of the world's most important chemically oriented journals are covered. Each article is indexed by author and keyword with a reference to the journal in which it appears. Every article selected for coverage in *Chemical Titles* is subsequently reported in *Chemical Abstracts* (*CA*). Orientation is more strongly chemical than in any other comparable U.S. publication. Cost to American Chemical Society (ACS) members was $225 per year (1997). The disadvantage of most services that are based primarily on titles is that titles are often unreliable indicators as to content of the full articles. Title services do, however, offer such features as speed of production, ease of scanning, and relatively low cost, in most cases.

2. *Advance ACS Abstracts,* published semimonthly, is an alerting publication that contains early drafts of abstracts of papers accepted for publication in 24 ACS journals 2–12 weeks in advance of publication. Tentative dates of publication are shown with each abstract. The cost to ACS members is only $29 (1997). At this writing, this product is not online, but it could be in the future. Authors for ACS journals who plan to file for patents or take other intellectual property actions need to be aware of this publication.*

3. One of the Section Groupings (see also Section 6.6) of *Chemical Abstracts* provides an excellent mechanism for keeping informed in broad or specialized areas by appropriate scanning. Available groupings include (a) applied chemistry and chemical engineering; (b) biochemistry; (c) macromolecular chemistry; (d) organic chemistry; and (e) physical, inorganic, and analytical chemistry. Because many sections are large, it is more feasible to skim quickly (to identify material of special interest) than to read exhaustively. The Section Groupings have unique advantages of offering broader coverage than any other comparable source, full abstracts, and keyword indexes—all at a reasonable cost. Cost to ACS members was $490 per year (1997).

4. Scanning of one or more sections of the Derwent *Chemical Patents Index* is often appropriate (see Section 13.14). This highly regarded tool offers comprehensive, rapid coverage of patents for most important industrial nations. Chemists and engineers working in industry find this tool of special value, since many important technological developments are first reported in patents. Cost is variable, depending on section and other factors, but is much more expensive than the other sources mentioned unless an organization already has a basic subscription and simply desires to order multiple copies.

C. Apex of the Triangle: Specialized Journals, Selective Dissemination, and Other Approaches

The most vital part of any current awareness program is the apex of the triangle as depicted in Figure 4.1. This part represents the specific areas in which the chemist wants to keep as up to date as possible and that most closely correspond with the statement of interests or profile.

*This publication was discontinued as of 1998 because the full texts of all "publishable" papers are now on the Internet site of ACS.

If there is a specialized journal or other publication that zeros in on this profile, this is one approach. Unfortunately, in many fields of chemistry, there are several journals in fields of specific interest, and some of these are often too thick, no matter how specialized. Special consideration should be given to so-called letters journals, such as *Tetrahedron Letters,* in which material is more likely to appear quickly than in the large, omnibus journals. A totally unique journal—the only one of its kind in the field—would be the prime candidate, although such journals are relatively rare. Chapter 12, Section II discusses journals in more detail.

In addition to journals, there are many specialized newsletters. Many of these cover not only the technology but the business aspects as well. An example is *Stereochemical Technology News,* which is published monthly by Business Communications Co., Norwalk, CT. A selection of these newsletters can be made by consulting an information professional, contacting the trade associations in the field, or by consulting directories such as *Ulrich's.* See Section 12.19.

Also falling within the apex are computer-based tools targeted specifically at current awareness. These tools are based on input of a statement of interests (profile) matched by computer versus the total volume of input of chemical interest covered in a specific database for the most recent update. When a "match" or "hit" occurs, this is automatically recognized by the computer, printed, and mailed to the persons whose profiles are represented. This service is usually referred to as *selective dissemination of information* (SDI).

The first known computer-based SDI effort in chemistry is the work of Maizell and colleagues reported in 1964 (1). These investigators employed *Chemical Titles* as the basis for automatically keeping chemists informed in crucial areas of research.

One SDI approach uses so-called predefined *standard interest profiles* or *macroprofiles.* These are profiles selected by an outside organization to meet the common interests of a broad group of chemists.

A good example is *Research Alert* Topics, which is published and sold by the *Institute for Scientific Information,* Philadelphia (see also Section 8.3). Over 200 topics in the sciences and social sciences are available. Topics of special interest to chemists are shown in Figure 4.2. The list, which is revised and updated annually, is representative of what was available in 1996.

Entries in the report look like the example shown in Figure 4.3. Results are mailed to subscribers at a cost of $315 per year (1996 price for users in United States, Canada, and Mexico; cost is higher elsewhere). Users receive weekly computer-based reports containing complete bibliographic information for new items published on topics they select.

Principal advantages include speed of issue and availability of the document delivery service of ISI for convenience in getting copies of full documents (see Section 5.2).

Research Alert Topics are, however, limited to publications covered by the *Institute for Scientific Information.* Patents, which are especially important to industrial research workers, are not included. The chemist is limited to information supplied in the computer printout; there are no abstracts in these reports.

An extremely important, valuable, and popular current awareness tool is CA *SELECTS,* which is published by CAS. This is a further development and enhancement of the now discontinued macroprofile service initiated by the United Kingdom Chemi-

Bioconjugates
Carcinogenesis–Chemical
Catalysis
Chemical Hazards–Health and Safety
Chemical Reaction Mechanisms and Kinetics
Chromatography
Chromatography–High Speed Liquid
Colloid and Surface Chemistry
Computers in Chemistry
Detection and Identification of Narcotic Drugs and Poisons
Electrochemistry
Food Additives
Free Radicals
Fullerenes
Liquid and Molecular Crystals
Mass Spectrometry
Molecular Asymmetry
Molecular Orbital Calculations
Paints, Enamels, and Related Products
Polymers–Fibers and Films
Polymers–Plastics and Elastomers
Polymers–Preparation of New Monomers and Polymers
Polymers–Science and Technology
Spectroscopic Techniques–Atomic Absorption, Mossbauer, Gamma-ray, X-ray
Spectroscopic Techniques–IR, UV, Laser, Raman
Spectroscopic Techniques–EPR, ESR, NMR, PMR
Thin Film Research and Technology

Figure 4.2. Research ALERT chemistry topics available from the Institute for Scientific Information as of 1996. (Additional topics of interest to chemists appear in the biotechnology and pharmaceutical and medicinal chemistry categories, which should also be considered.) Reprinted with permission of the Institute for Scientific Information.

cal Information Service in 1971. *CA SELECTS* provides complete *CA* abstracts and bibliographic citations. Well over 200 topics are now covered in this service, which was introduced in 1976. Recent (1997) prices are only $70 per topic (26 issues a year) for ACS members. Advantages include excellent abstracts, comprehensiveness, and specific focus. The list of topics covered in 1996 is shown in Table 4.1. It is antici-

RESEARCH ALERT (S0038) 9114 ACCT NO
MASS SPECTROMETRY

SAMPLE REPORT 2 S0038

REPORT FOR 17 MAR 95 COL. 1

26,766 PUBLISHED ITEMS SEARCHED THIS WEEK

MASS SPE **SELECTIVE DETERMINATION OF PHOSPHOPEPTIDE BETA-CN(1-25) IN A BETA-CASEIN DIGEST BY ADDING IRON - CHARACTERIZATION BY LIQUID-CHROMATOGRAPHY WITH ONLINE ELECTROSPRAY-IONIZATION MASS-SPECTROMETRIC DETECTION**
GAUCHERO.F MOLLE D LEONIL J MAUBOIS JL
32 REFS
J CHROMAT B 664(1): 193-200,FEB 3 1995
-----> CHECK TO ORDER THE ARTICLE ----->( ) #QG484
F GAUCHERON, INRA,RECH TECHNOL LAITIERE LAB,65 RUE ST BRIEUC, F-35042 RENNES, FRANCE

SIMULTANEOUS DETERMINATION OF ALCOHOLS AND ETHYLENE-GLYCOL IN SERUM BY PACKED-COLUMN OR CAPILLARY-COLUMN GAS-CHROMATOGRAPHY
LIVESEY JF PERKINS SL TOKESSY NE MADDOCK MJ
29 REFS
CLIN CHEM 41(2): 300-305,FEB 1995
THESE ITEMS IN THIS PROFILE WERE CITED:
SCHUBERTH J BIOL MASS SPECTROM 20 699 91
-----> CHECK TO ORDER THE ARTICLE ----->( ) #QG652
JF LIVESEY, OTTAWA CIVIC HOSP,DEPT LAB MED,1053 CARLING AVE, OTTAWA, ON K1Y 4E9, CANADA

MODIFICATION OF A CONVENTIONAL HIGH-PERFORMANCE LIQUID-CHROMATOGRAPHY AUTOINJECTOR FOR USE WITH CAPILLARY LIQUID-CHROMATOGRAPHY (TECHNICAL NOTE)
SIMPSON RC 5 REFS
J CHROMAT A 691(1-2): 163-170,FEB 3 1995
THESE ITEMS IN THIS PROFILE WERE CITED:
EVANS JE BIOL MASS SPECTROM 22 331 93
-----> CHECK TO ORDER THE ARTICLE ----->( ) #QG491
RC SIMPSON, SMITHKLINE BEECHAM PHARMACEUT,DEPT DRUG METAB & PHARMACOKINET,POB 1539, KING OF PRUSSIA, PA 19406

SIMULTANEOUS SORPTION AND ANALYTICAL DERIVATIZATION ON A POLYSTYRENE DIVINYLBENZENE POLYMER - PREPARATION OF CHROMOPHORIC AND FLUOROPHORIC DERIVATIVES OF THE PROSTAGLANDINS
ROSENFEL.JM FANG XC 11 REFS
J CHROMAT A 691(1-2): 231-237,FEB 3 1995
THESE ITEMS IN THIS PROFILE WERE CITED:
WADDELL KA BIOMED MASS SPECTROM 10 83 83
SHINDO N BIOMED ENVIRON MASS 15 25 88
-----> CHECK TO ORDER THE ARTICLE ----->( ) #QG491
JM ROSENFELD, MCMASTER UNIV,DEPT PATHOL,1200 MAIN ST W, HAMILTON, ON L8N 3Z5, CANADA

MASS SPE **COMPOSITIONAL ANALYSIS OF THE PHENYLTHIOCARBAMYL AMINO-ACIDS BY LIQUID-CHROMATOGRAPHY ATMOSPHERIC-PRESSURE IONIZATION MASS-SPECTROMETRY WITH PARTICULAR ATTENTION TO THE CYST(E)INE DERIVATIVES**
SCHMEER K KHALIFA M CSASZAR J FARKAS G
BAYER E MOLNARPE.I 7 REFS
J CHROMAT A 691(1-2): 285-299,FEB 3 1995
-----> CHECK TO ORDER THE ARTICLE ----->( ) #QG491
I MOLNARPERL, LORAND EOTVOS UNIV,INST INORGAN & ANALYT CHEM,POB 123, H-1518 BUDAPEST 112, HUNGARY

MASS SPE **SEPARATION AND ANALYSIS OF PROTEINS BY PERFUSION LIQUID-CHROMATOGRAPHY AND ELECTROSPRAY-IONIZATION MASS-SPECTROMETRY**
BANKS JF 14 REFS
J CHROMAT A 691(1-2): 325-330,FEB 3 1995
THESE ITEMS IN THIS PROFILE WERE CITED:
LEE ED BIOMED ENVIRON MASS 18 844 89
-----> CHECK TO ORDER THE ARTICLE ----->( ) #QG491
JF BANKS, ANALYT BRANFORD INC,29 BUSINESS PK DR, BRANFORD, CT 06405

SUBSTRATE UTILIZATION DURING THE FIRST WEEKS OF LIFE
SAUER PJJ CARNIELL.VP SULKERS EJ VANGOUDO.JB
18 REFS

RESEARCH ALERT (S0038) 9114 ACCT NO
REPORT FOR 17 MAR 95 COL. 2

ACT PAEDIAT 83(S405): 49-53,DEC 1994
THESE ITEMS IN THIS PROFILE WERE CITED:
JONES DM BIOL MASS SPECTROM 20 641 91
-----> CHECK TO ORDER THE ARTICLE ----->( ) #QG455
PJJ SAUER, ERASMUS UNIV ROTTERDAM,UNIV ROTTERDAM HOSP,SOPHIA CHILDRENS HOSP,DEPT PEDIAT, 3015 GJ ROTTERDAM, NETHERLANDS

METABOLIC PROCESSING OF PAF (REVIEW)
SNYDER F 170 REFS
CLIN R ALL 12(4): 309-327,WIN 1994
THESE ITEMS IN THIS PROFILE WERE CITED:
YASUDA K BIOMED ENVIRON MASS 16 137 88
-----> CHECK TO ORDER THE ARTICLE ----->( ) #QG635
F SNYDER, OAK RIDGE ASSOCIATED UNIV,OAK RIDGE INST SCI & EDUC,DIV MED SCI,POB 117, OAK RIDGE, TN 37831

STRUCTURAL AND (PATHO)PHYSIOLOGICAL DIVERSITY OF PAF (REVIEW)
PINCKARD RN WOODARD DS SHOWELL HJ CONKLYN MJ
NOVAK MJ MCMANUS LM 94 REFS
CLIN R ALL 12(4): 329-359,WIN 1994
THESE ITEMS IN THIS PROFILE WERE CITED:
YASUDA K BIOMED ENVIRON MASS 16 137 88
-----> CHECK TO ORDER THE ARTICLE ----->( ) #QG635
RN PINCKARD, UNIV TEXAS,HLTH SCI CTR,DEPT PATHOL, SAN ANTONIO, TX 78284

MASS SPE **MATRIX-ASSISTED LASER-DESORPTION IONIZATION TIME-OF-FLIGHT MASS-SPECTROMETRY CHARACTERIZATION OF POLY(BUTYL METHACRYLATE) SYNTHESIZED BY GROUP-TRANSFER POLYMERIZATION**
DANIS PO KARR DE SIMONSIC.WJ WU DT
34 REFS
MACROMOLEC 28(4):1229-1232,FEB 13 1995
THESE ITEMS IN THIS PROFILE WERE CITED:
KARAS M MASS SPECTROM REV 10 335 92
DANIS PO ORG MASS SPECTROM 28 923 93
HILLENKAMP F BIOL MASS SPECTROM 40 90
-----> CHECK TO ORDER THE ARTICLE ----->( ) #QG795
PO DANIS, ROHM & HAAS CO,RES LABS,727 NORRISTOWN RD, SPRING HOUSE, PA 19477

MASS SPE **INDUCTIVELY-COUPLED PLASMA-MASS SPECTROMETRY FOR SEQUENTIAL DETERMINATION OF TRACE-METALS IN RAIN AND RIVER WATERS USING ELECTROTHERMAL VAPORIZATION**
SANTOSA SJ TANAKA S YAMANAKA K 20 REFS
ANAL LETTER 28(3): 509-534,1995
-----> CHECK TO ORDER THE ARTICLE ----->( ) #QH121
S TANAKA, KEIO UNIV,FAC SCI & TECHNOL,DEPT APPL CHEM,KOUHOKU KU,3-14-11 HIYOSHI, YOKOHAMA, KANAGAWA 223, JAPAN

NEW DELTA(5,9)-FATTY-ACIDS IN THE PHOSPHOLIPIDS OF THE SEA-ANEMONE STOICNACTIS-HELIANTHUS
CARBALLE.NM MEDINA JR 18 REFS
J NAT PROD 57(12):1688-1695,DEC 1994
THESE ITEMS IN THIS PROFILE WERE CITED:
DJERASSI C ACCOUNTS CHEM RES 24 69 91
-----> CHECK TO ORDER THE ARTICLE ----->( ) #QG536
NM CARBALLEIRA, UNIV PUERTO RICO,DEPT CHEM,POB 23346, SAN JUAN, PR 00931

7'-HYDROXYSEIRIDIN AND 7'-HYDROXYISOSEIRIDIN, 2 NEW PHYTOTOXIC DELTA(ALPHA,BETA)-BUTENOLIDES FROM 3 SPECIES OF SEIRIDIUM PATHOGENIC TO CYPRESSES (TECHNICAL NOTE)
EVIDENTE A SPARAPAN.L 18 REFS
J NAT PROD 57(12):1720-1725,DEC 1994
THESE ITEMS IN THIS PROFILE WERE CITED:
PORTER QN MASS SPECTROMETRY HE 278 85
-----> CHECK TO ORDER THE ARTICLE ----->( ) #QG536
A EVIDENTE, UNIV NAPLES FEDERICO II,DIPARTIMENTO SCI CHIM AGR,VIA UNIV 100, I-80055 PORTICI, ITALY

MASS SPE **CHARACTERIZATION OF OLIGONUCLEOTIDES AND NUCLEIC-ACIDS BY MASS-SPECTROMETRY**
LIMBACH PA CRAIN PF MCCLOSKE.JA 67 REFS
CURR OPIN B 6(1): 96-102,FEB 1995
THESE ITEMS IN THIS PROFILE WERE CITED:
PHILLIPS DR INT J MASS SPECTROM 128 61 93
CRAIN PF MASS SPECTROM REV 9 505 90
EHRING H ORG MASS SPECTROM 27 472 92
SMITH RD BIOL MASS SPECTROM 22 493 93
-----> CHECK TO ORDER THE ARTICLE ----->( ) #QH391
PA LIMBACH, UNIV UTAH,DEPT MED CHEM,311A SKAGGS HALL, SALT LAKE CITY, UT 84112

You can receive the full text of articles listed in this report simply by checking off the articles you want and returning the report to ISI. *The Genuine Article®*, ISI's fast, inexpensive document delivery service, will provide the actual pages from the original journals or guaranteed, high-quality photocopies. Orders are processed within 24 hours of their receipt...or sooner through our FAX service. A wide range of ordering and delivery options are available. For complete information including pricing, call toll-free 1-800-336-4474 and ask to speak with *The Genuine Article* Account Representative, or write to the address below.

The Institute for Scientific Information® makes a reasonable effort to supply complete and accurate information in its information services, but does not assume any liability for errors or omissions.

A Service of **The Institute for Scientific Information®**, 3501 Market Street, Philadelphia, PA 19104 U.S.A. Tel: (800) 523-1850, (215) 386-0100 Fax: (215) 386-6362

Figure 4.3. Typical research ALERT output. (Copyright owned by the Institute for Scientific Information, Philadelphia, PA.)

TABLE 4.1. ***CA SELECTS*[a] Topics**

Agrochemicals
See also Biochemistry
Novel Pesticides & Herbicides
Pesticide Analysis

Analytical Chemistry
See also Chromatography, Spectroscopy
Analytical Electrochemistry
Automated Chemical Analysis
Catalytic & Kinetic Analysis
Chelating Agents
Chemical Instrumentation
Drug Analysis in Biological Fluids & Tissues
Electrophoresis
Enzyme Assays
Food and Feed Analysis
Inorganic Analytical Chemistry
Ion Exchange
Membrane Separation
Organic Analytical Chemistry
Surface Analysis
Thermal Analysis
Trace Element Analysis
X-Ray Analysis & Spectroscopy

Applied Chemstry
See also Materials, Polymers
Chemical Engineering Operations
Chemical Processing Apparatus
Chemical Vapor Deposition
Coal Science & Process Chemistry
Corrosion
Distillation Technology
Drilling Muds
Electrodeposition
Fluidized Solids Technology
Gaseous Waste Treatment
Laser Applications
Paper Chemistry
Plasma & Reactive Ion Etching
Solar Energy
Solvent Extraction
Stress Corrosion—Metals

Biochemistry
See also Agrochemicals, Environmental Chemistry, Health Hazards, Pathology
Animal Longevity & Aging
Artificial Sweetners
Biogenic Amines & the Nervous System
Fermentation Chemicals
Free Radicals (Biochemical Aspects)
Metallo Enzymes & Metallo Coenzymes
Molecular Modeling (Biochemical Aspects)
Nitrogen Fixation
Novel Natural Products
Photobiochemistry
Porphyrins
Prostaglandins
Psychobiochemistry
Steroids (Biochemical Aspects)
Steroids (Chemical Aspects)

Chemical Reactions
Alkylation & Catalysts
Catalyst Regeneration
Crosslinking Reactions
Electrochemical Organic Synthesis
Electrochemical Reactions
Emulsion Polymerization
Inorganic Chemicals & Reactions
Inorganic & Organometallic
Reaction Mechanisms
Isomerization & Catalysts
Laser-Induced Chemical Reactions
Natural Product Synthesis
Optimization of Organic Reactions
Organic Reaction Mechanisms
Organometallics in Organic Synthesis
Oxidation Catalysis
Phase Transfer Catalysts
Photocatalysts
Photochemical Organic Synthesis
Polymer Degradation
Polymerization Kinetics & Process Control
Radiation Chemistry
Radiation Curing

Chromatography
See also Analystic Chemistry
Ion Chromatography
Paper & Thin-Layer Chromatography

(continued)

TABLE 4.1. *(Continued)*

Energy
Energy Reviews & Books
Enhanced Petroleum Recovery
Fuel & Lubricant Additives
Synfuels

Environmental Chemistry
See also Agrochemicals, Biochemistry, Health Hazards
Acid Rain & Acid Air
Air Pollution (Books & Reviews)
Environmental Pollution
Indoor Air Pollution

General Chemistry
Computers in Chemistry
New Books in Chemistry

Health Hazards, Safety, & Toxicity
See also Environmental Chemistry
Drug & Cosmetic Toxicity
Flammability
Food, Drugs, & Cosmetics—Legislative & Regulatory Aspects
Food Toxicity
Occupational Exposure & Hazards

Inorganic & Physical Chemistry
Adsorption
Bismuth Chemistry
Chemiluminescence
Chemistry of Ir, Os, Rh, & Ru
Colloids (Applied Aspects)
Colloids (Macromolecular Aspects)
Colloids (Physicochemical Aspects)
Color Science
Crystal Growth
Electron Spin Resonance (Chemical Aspects)
Fullerenes & Clusters
Geochemistry
Infrared Spectroscopy (Physicochemical Aspects)
Liquid Crystals
Organo-Transition Metal Complexes
Platinum & Palladium Chemistry
Proton Magnetic Resonance
Quaternary Ammonium Compounds
Selenium & Tellurium Chemistry
Silicas & Silicates
Silver Chemistry
Solid State NMR
Surface Chemistry (Physicochemical Aspects)
Thermochemistry

Materials
See also Applied Chemistry, Polymers
Activated Carbon
Alkoxylated Oleochemicals
Aluminum–Lithium & Aluminum–Cerium Alloys
Antioxidants
Carbon Fiber Composites
Carbon & Graphite Fibers
Ceramic Materials (Journals)
Ceramic Materials (Patents)
Coatings, Inks, & Related Products
Colorants & Dyes
Composite Materials (Ceramic)
Composite Materials (Metallic)
Corrosion-Inhibiting Coatings
Cosmetic Chemicals
Electrically Conductive Organics
Electronic Chemicals & Materials
Emulsifiers & Demulsifiers
Fats & Oils
Fiber Optics & Optical Communication
Formulation Chemistry
Hot-Melt Adhesives
Lubricants, Greases, & Lubrication
Memory & Recording Devices & Materials
Metallic Glasses
Nonlinear Optical Materials
Oleochemicals Containing Nitrogen
Omega-3 Fatty Acids & Fish Oil
Optical & Photosensitive Materials
Organic Optical Materials
Oxide Superconductors
Paint Additives
Paper Additives
Photoresists
Shape Memory Alloys
Technical Ceramics
Water-Based Coatings

TABLE 4.1. ***(Continued)***

Organic Chemistry
Free Radicals (Organic Aspects)
Infrared Spectroscopy (Organic Aspects)
Novel Sulfur Heterocycles
Organic Stereochemistry
Organofluorine Chemistry
Organophosphorus Chemistry
Organosulfur Chemistry (Journals)
Organotin Chemistry
Phospholipids (Chemical Aspects)
Synthetic Macrocyclic Compounds

Pathology & Drugs
See also Biochemistry, Environmental Chemistry, Health Hazards
AIDS & Related Immunodeficiencies
Allergy & Antiallergic Agents
Alzheimer's Disease & Related Memory Dysfunctions
Antiarrhythmics
Antibacterial Agents
Anticonvulsants & Antiepileptics
Antifungal & Antimycotic Agents
Anti-Inflammatory Agents & Arthritis
Atherosclerosis & Heart Disease
Beta-Lactam Antibiotics
Blood Coagulation
Calcium Channel Blockers
Drug Interactions
Hypertension & Antihypertensives
Leukotrienes
Monoclonal Antibodies
New Antibiotics
Nutritional Aspects of Cancer
Osteoporosis & Related Bone Loss
Pharmaceutical Chemistry (Journals)
Pharmaceutical Chemistry (Patents)
Structure–Activity Relationships
Ulcer Inhibitors
Virucides & Virustats

Polymers
See also Applied Chemistry, Materials
Block and Graft Polymers
Composite Materials (Polymeric)
Conductive Polymers
Elastomers
Epoxy Resins
Fiber-Reinforced Plastics
Fluoropolymers
Heat-Resistant & Ablative Polymers
Ion-Containing Polymers
New Plastics
Novel Polymers from Patents
Photosensitive Polymers
Plastic Films
Plastic Additives
Plastics Fabrication & Uses
Plastics Manufacture & Processing
Polyacrylates (Journals)
Polyacrylates (Patents)
Polyesters
Polyimides
Polymer Blends
Polymer Morphology
Siloxanes & Silicones
Synthetic High Polymers

Spectroscopy
See also Analytical Chemistry
Atomic Spectroscopy
Electron & Auger Spectroscopy
Raman Spectroscopy
Spectrochemical Analysis
Ultraviolet & Visible Spectroscopy

[a]*CA SELECTS* is a series of current awareness bulletins published every 2 weeks, 26 times per year. Each *CA SELECTS* topic is a separate printed publication, designed to bring researchers only the scientific and technical information they need. Each topic includes *CA* abstracts and associated bibliographic information, selected by computer for the CAS database according to a precise, special-interest profile.

TABLE 4.2. ***CA SELECTS PLUS*** **Topics**

Adhesives
Amino Acids, Peptides & Proteins
Antitumor Agents
Asymmetric Synthesis & Induction
Batteries & Fuel Cells
Carbohydrates (Chemical Aspects)
Carbon & Heteroatom NMR
Carcinogens, Mutagens & Teratogens
Catalysis (Applied & Physical Aspects)
Chemical Hazards, Health & Safety
Controlled Release Technology
Detergents, Soaps & Surfactants
Drug Delivery Systems & Dosage Forms
Electrophoresis
Environmental Pollution
Enzyme Applications
Flavors & Fragrances
Forensic Chemistry
Fungicides
Gas Chromatography
Gel Permeation Chromatography
Herbicides
High Performance Liquid Chromatography
Insecticides
Liquid Waste Treatment
Mass Spectrometry
Organosilicon Chemistry
Pharmaceutical Analysis
Photochemistry
Pollution Monitoring
Polyurethanes
Recovery & Recycling of Wastes
Solid & Radioactive Waste Treatment
Ultafiltration
Water Treatment
Zeolites

pated that new topics will continue to be added and others dropped, depending on developments in chemistry and changing interests of chemists.

In October 1995, CAS introduced the *CA SelectsPlus* series. Content includes not only references from *CA* but, in addition, references from 1300 "core journals" (see Chapters 6 and 7) before they appear in *CA*. Topics covered are shown in Table 4.2. Subscription cost for ACS members is only $75 per year. Those topics that are in the

TABLE 4.3. ***CAS BioTech Updates***

Agriculture
Antibody Conjugates
Biochemical Immobilization & Biocatalytic Reactors
Biosensors
Cell & Tissue Culture
Commercial Fermentation
DNA & RNA Probes
DNA Formation & Repair
Environmental Biotechnology
Enzymes in Biotechnology
Genetic Engineering
Nucleic Acid & Protein Sequences
Pharmaceutical Applications
Product Purification & Separation
Slow-Release Pharmaceuticals

Reprinted with the permission of Chemical Abstracts Service.

CA SelectsPlus series are no longer available in the *CA Selects* series; not all *CA Selects* topics will necessarily be converted to *Plus* topics.

CAS also offers a related series *CAS BioTech Updates,* which was first introduced in 1986. Topics covered are shown in Table 4.3. The content includes input from both *CA* and *Chemical Industry Notes.* In addition to technical developments, coverage includes developments in production, processes, government activities, and people in the industry. Keyword and author indexes are provided. Subscription cost is $230 per year.

Other organizations, too numerous to mention here, also offer standard interest profile service. The principal advantages of the macroprofile concept are usually given as low cost, speed, and sometimes, breadth of coverage. The principal disadvantage is that the coverage may not focus sufficiently on material of specific interest to an individual chemist or engineer. This can result in too much nonpertinent material. Macroprofiles may also lack the flexibility needed by some chemists who are interested in more than one field, or whose interests vary sharply from time to time.

Customized SDI, based on a profile specifically designed to meet the needs of an individual chemist or a group of chemists, is usually considered more desirable than the macroprofile approach. Customizing costs more, but it is more likely to yield pertinent material, and it can be varied to meet changing interests. This service is available from CAS, the *Institute for Scientific Information,* and others. If the chemist's organization has a large computer center, SDI service may be available locally. The chemist can elect to have a profile run against a variety of databases. For many chemists and chemical engineers, *CA* is a very attractive choice if broad coverage of chemistry is desired.

CAS approaches to customized SDI (selective dissemination of information) include *Corporate Updates,* bulletins for which search profiles are prepared by CAS staff experts to suit the interests clients describe; and *Individual Search Service* (ISS), for which clients write their own profiles and may also call on CAS staff.

Corporate Updates include *CA* abstracts and, where applicable, structure diagrams. Abstracts may be grouped under subject or by CAS Registry Number, whichever is preferred. Printouts from the ISS service include *CA* abstracts and keyword index entries. CAS Registry Numbers or Volume Index entries may be optionally included.

The *Institute for Scientific Information* (*ISI*) provides another form of current alerting tailored to meet specific needs of individual chemists or groups of chemists. This important service is designated as *Customized Research Alerts* or *Research Alert Direct.* For the regular customized service, results are delivered in paper or magnetic tape form. For the direct customized service, results are delivered daily or weekly via Internet e-mail and include English-language abstracts when available. Levels of service include table of contents alerting for individuals and companies and customized profile-based alerting for individuals or groups. Patents are not included. A number of approaches can be utilized to search for and retrieve pertinent information. These include: title words and word stems, names of source journals or cited authors, source journal or cited journal titles, frequently cited references or books, and author's organization. Citation searching is a special feature that often locates items that might otherwise be missed (see also Section 8.3). Output is based on approximately 16,000 journals and other sources and includes authors, address of first author or reprint author, article title, complete journal citation, language codes, number of references, and the order number for *ISI*'s document delivery service (see Section 5.2). Items retrieved may include articles, book reviews, letters, editorials, meeting abstracts, and technical notes. Costs vary depending on scope as specified by the user. *Research Alert* is particularly distinguished by its speed of coverage and citation access.

Another alerting service is offered by Derwent Information, London, UK, and McLean, VA, through what is called their *Standard Interest Profiles.* Some 57 chemical topics are covered in this excellent service, which is, however, expensive and available only to subscribers (full or partial) to the printed Derwent abstract products. Abstracts are given for the worldwide patents that are included, and the frequency of issue is every 1–4 weeks, depending on the title.

The Royal Society of Chemistry, Cambridge, UK, offers a number of current awareness products. These include, for example, its *FOCUS SERIES.* These are monthly chemical business newsletters that combine brief comment and analysis with abstracts and citations to the literature, including such sources as trade magazines, company press releases, and annual reports. There are currently seven available topics: *Biopesticides Plus, Catalysts, Organic Dyes and Colours, Pigments, Solvents, Surfactants,* and *Powder Coatings.* A typical newsletter is eight pages in length and covers such topics as markets and business, company news, new materials and components, new technology, environment, bookshelf, and upcoming meetings. Subscription costs are reasonable enough ($450/year, 1997) to make these newsletters affordable for smaller organizations but beyond the budgets of most individual chemists. Unfortunately, the number of topics available is only a handful as noted above; a number of topics previously offered have been dropped.

Another RSC (The Royal Society of Chemistry) product is *Methods in Organic Synthesis.* Intended for the synthetic organic chemist, this is a key monthly current awareness publication that provides bibliographic citations and diagrams of reaction schemes with conditions and brief descriptive notes on conditions and yields as needed. Each issue index covers authors, products, reactions, reactants, and reagents. Approximately 100 worldwide journals are covered. The annual subscription cost of $472 (1997) is quite reasonable in comparison to the competition.

Natural Products Update, another monthly current awareness publication of the Royal Society of Chemistry, cites, diagrams, and provides other data on natural product isolation relating to new products and known products from new sources. Approximately 100 journals are covered. There are indexes by author, taxonomic names, nonplant sources, biological activity, classes of compounds, and common names.

In addition to the types of programs and services just described, chemists can initiate their own automatic SDI programs through most of the major online databases described in Chapter 10. In this approach, search strategies (profiles) are first created and then, after testing, can be saved. Whenever predesignated files are updated, these are automatically run against the saved profiles, and results are mailed or electronically transmitted to the user-defined address. Success of this approach depends on the skill with which databases are selected and with which strategies are created, implemented, and updated. A less automatic option is to go online at regular intervals (as selected by the user) and then scan across any updated files with the strategy selected. This affords a greater degree of control and flexibility than the fully automatic approach in that changes in profile scope may usually be made easily.

In addition to all the above, advances in computer technology have facilitated widespread availability of customized tables of contents products for thousands of journals. This service can be expected to be especially fast and relatively inexpensive. A potential shortcoming is that titles alone (without abstracts) are not reliable guides to the full contents of a journal paper or other publication. Some services may include abstracts.

For this type of product, the user simply designates which journals are to be included in the service. Delivery of results may be weekly or monthly, or, in some cases, may be daily. Copies of full documents may be ordered electronically in some cases, and in some cases, the user may enter an author or subject search as well.

The chemist or engineer now has a choice of several organizations that offer automatic tables of contents delivery. Among the leading suppliers of these for the chemist and engineer are Chemical Abstracts Service (see Chapters 6 and 7), Institute for Scientific Information (see Chapter 8), and the UnCover Company (see Chapter 5). Delivery of the tables of contents may be through one of several means depending on the vendor, but typically includes the user's e-mail address as one delivery option.

The CAS table of contents service is known as *CAplus TOC.* This provides fast access to over 1350 of the most important chemical journals. Results are sent via regular mail, Internet, *STNmail* (part of the STN system described later), or fax (facsimile). The ISI table of contents services are known as *Journal Tracker* and *Corporate Alert.* Abstracts are included when available. *Reveal* is the name of the product offered by UnCover.

A number of major journal publishers have table of contents information available on the Internet. In some cases, this information is searchable. Some of the information may be available to subscribers only, except for current issues. In some cases, full text is also available.

The American Chemical Society's Internet addresses for their journals (as well as for other valuable materials) include http://pubs.acs.org and www.ChemCenter.org. Elsevier's table of contents services, *ESTOC,* is also available on the Internet. It covers some 900 Elsevier journals. Abstracts and titles are searched to yield results, but abstracts are not shown on the screen. The Internet address is www.elsevier.nl/locate/estoc. The time period covered is from January 1995. See also Chapter 12, Section II.

4.4. MEETINGS

As part of any current awareness program, the chemist needs to allocate time for attending professional society and trade association meetings. These meetings offer the opportunity to listen to and meet persons at the leading edge of research and development and obtain information long before it is published. Further, the opportunity to meet informally with colleagues in the same field of interest is invaluable. Many chemists and engineers prefer to attend more specialized smaller meetings such as Gordon Research Conferences or special symposia and seminars devoted to relatively narrow fields.

In contrast, huge *omnibus* national meetings, which attempt to cover all aspects of chemistry or chemical engineering, can be too broad in scope, the quantity of papers can be overwhelming, and the number of persons attending can dilute opportunities for personal contacts as well as for good question-and-answer sessions. Attending large national meetings can be useful only if there is careful preplanning, for example, by making plans to attend several symposia on specialized topics of clear interest to the individual.

How many meetings per year can or should a chemist or engineer attend? Like reading journals, chemists could easily spend a lifetime attending meetings. Many organizations will send their key people to an average of one meeting each year, but this can vary widely, depending on meeting location, duration, and fees. The chemist should select the meetings to be attended as far in advance as feasible—at least several months.

A. Getting Copies of Meeting Papers

Because no one can attend all meetings, what can chemists and engineers do when they see announcements of papers of interest, but when no one from their organization can attend?

Some actions to get copies of meeting papers can be attempted, but results are often uncertain. Some presenters at chemical meetings have only an oral presentation and never put their papers in writing. In a few cases, an abstract appears, but the author withdraws the paper before it is presented at the meeting. Reliance on ultimately seeing a published version of a meeting paper is, therefore, uncertain. Only about half of all meeting papers are published according to some estimates. Those published often undergo extensive revision after oral presentation and may therefore vary significantly in content from the original version. The time lag between oral presentation and formal publication can be substantial—from several months to well over a year.

American Chemical Society (ACS) meeting papers are best accessed initially through the collection of abstracts of meeting papers that are published prior to national and some regional meetings. An index to the abstracts, based on words in the titles, speeds up identification of pertinent material. Beginning with the fall 1995 national meeting, ACS meeting abstracts are covered in *CAplus* (see Chapters 6 and 7), an online database. This fills an important void that had existed previously and provides a useful and enduring electronic record. In the case of AIChE (American Institute of Chemical Engineers), abstracts of meeting papers can be found on the World Wide Web (www.aiche.org).

Because meeting papers are not usually subject to rigorous review, and for other reasons, abstracts of them leave much to be desired. Chemists interested in learning more about the contents of a meeting paper can write the author and request a preprint. These are readily available, however, only for an estimated 50% or less of all meeting papers. Some suggestions on contacting authors for copies of their papers can be found in Section 5.3.

Obtaining meeting papers for several divisions of ACS is simplified because they publish preprints at the time of national meetings. At this writing, these divisions include

1. Environmental Chemistry
2. Fuel Chemistry
3. Petroleum Chemistry Inc.
4. Polymer Chemistry Inc.
5. Polymeric Materials: Science & Engineering Inc.

In addition to making available copies of papers presented at all national meetings and at some local meetings, the Rubber Division of the American Chemical Society has offered its own extensive library service for some years. This is under the direction of Mrs. Joan C. Long at the University of Akron, Akron, OH.

For AIChE papers, copies are often available directly from AIChE for several weeks after the meeting. Thereafter, photocopies are usually available from the Linda Hall Library, Kansas City, MO.

As an aid to deciding which meetings to monitor and possibly attend, the chemical news magazines publish lists on a regular basis. For example, in its July 1, 1996 issue, *Chemical and Engineering News* published an extensive list of domestic and international meetings of chemical interest for the next 12 months. This printed list is updated periodically and is also on the Internet. There are also publications that list future meetings of technical, scientific, medical, and management organizations and universities, for example, the quarterly *World Meetings* series compiled by Macmillan (now part of Simon and Schuster, New York, NY) and *Scientific Meetings,* published quarterly by Scientific Meetings Publications, San Diego, CA.

Conference Papers Index is an index of papers presented at scientific and technical meetings worldwide from 1973 to the present. It is available in both printed form and online (Internet). The publisher is Cambridge Scientific Abstracts, Bethesda, MD.

Meeting paper preprints that are not published elsewhere are among the many kinds of documents covered by *Chemical Abstracts.*

Index to Scientific and Technical Proceedings, published by the *Institute for Scientific Information,* can help chemists and others locate papers published in conference proceedings. The series covers approximately 4200 published conference proceedings each year—over 170,000 individual papers in all. It is published monthly with an annual cumulation, and in addition to the print version (back years available through 1978), ISTP is available on CD-ROM (rolling 5-year file), in magnetic tape form, and online (DIMDI).

The British Library Document Supply Center publishes *Index of Conference Proceedings* (*Conference Proceedings Index*) each month. It is also available online through BLAISE-LINE.

EventLine, a product of Elsevier Science, Amsterdam, the Netherlands, is available as CD-ROM and online (STN, DIALOG, and other hosts, see Chapter 10). It is a multidisciplinary listing of past, present, and future conferences, symposia, trade shows, and exhibits.

During recent years the ACS conducted experiments on alternative ways of providing chemists with access to meeting papers. The principal approach considered was known as *teleconferencing* or *audiographics.* In the ACS approach, chemists at sites distant from national meetings were provided with copies of slides in advance. At the time of delivery of the paper, chemists at these remote sites heard the speakers and could ask questions, employing modified telephone lines. Video access was not provided, except for the previously mentioned slides, which were displayed by local projectionists at the remote sites. The results were not encouraging, and ACS decided to discontinue this approach. However, remote access is still offered for some ACS short courses; material is presented via satellite television in a program known as *satellite seminars.* Attendees may ask questions during the course of the presentations. Videotapes are available for purchase following the seminars.

A few "electronic conferences or meetings" on chemical topics have begun to appear on the Internet. For example, these have been held on such topics as applications of technology in teaching chemistry (1993), chemometrics (1994), electronic computational chemistry (1994 and 1995), and trends in organic chemistry (1995). CD-ROM copies are available in at least one case. These meetings offer important time and cost-saving advantages because no travel is required. However, they lack the benefits afforded only by face-to-face contacts. There are other pros and cons as well. Descriptions of electronic conferences and discussion of the advantages will be found in S. M. Bachrach, Editor, *The Internet—a Guide for Chemists,* American Chemical Society, Washington, DC, 1996. See especially Section 10.15.I.

The tools just mentioned, and others that either exist now or will almost inevitably be developed in the future, are an encouraging indication that chemists should be able to get a much better "handle" on meeting and conference papers than ever before.

4.5. RESEARCH IN PROGRESS AND DISSERTATIONS

Identification of research in progress can be used by chemists and engineers to

- Avoid unwarranted duplication of research effort and expenditure.
- Locate possible sources of support for research on a specific topic.
- Identify leads to the published literature or participants for symposia.
- Obtain information to support grant or contract proposals.
- Stimulate new ideas for research planning or innovations in experimental techniques.
- Acquire source data for technological forecasting and development.
- Survey broad areas of research to identify trends and patterns or reveal gaps in overall efforts.
- Learn about current work of a specific research investigator, organization, or organizational unit.

Information about research underway is frequently difficult to obtain, since most investigators are understandably reluctant to allow premature disclosure of results. There are, however, sources that can help identify the scope of some of the work in progress.

Until the early 1980s, the leading source of this kind in the United States had been the Smithsonian Science Information Exchange, Inc. (SSIE), which covered primarily federally sponsored research in progress. However, after years of highly successful operation under the direction of David Hersey, SSIE was discontinued in 1981 as part of budget-cutting efforts at that time.

Fortunately, part of SSIE's functions were assumed by, and are still available through, the National Technical Information Service (NTIS) in the form of an online file known as *Federal Research in Progress* (*FEDRIP*), which may be accessed through such online systems as DIALOG and Knowledge Express Data Systems, and through a CD-ROM version being offered by Cartermill Publishing, London (see Chapter 10). In addition, the previously developed SSIE file is still online, at least for the present, and historical information therein complements *FEDRIP.*

Use of these two files is a convenient way to determine what research some federal agencies have funded in areas of interest. *FEDRIP* is, however, far from complete. U.S. government agencies participating that are of potential interest to chemists include Department of Agriculture, Department of Energy, Environmental Protection Agency, NASA, National Institutes of Health, NIST, National Science Foundation, and Geological Survey. Addition of all other pertinent federal departments is needed to make this database more fully useful.

Information included in *FEDRIP* consists of the following elements (if the submitting agency is able to provide all the data):

Sponsoring agency
Starting date
Completion date
Performing organization (doing the work)
Abstract
Progress
Subject index terms
References
Investigators
Title
Funding

A particularly notable subfile within *FEDRIP* is the "CRISP" (Computerized Retrieval of Information on Scientific Projects) set of the many research projects funded by the National Institutes of Health. See also Chapter 10.

Among many other features, *FEDRIP* also makes it possible to identify Small Business Innovation Research projects (SBIRs). These projects are initiated by small businesses and often contain among the most innovative of research ideas in chemistry and other technologies.

Unfortunately, abstracts in *FEDRIP* lack the important editorial review for uniformity and completeness of detail that had been provided by SSIE staff, many of whom had graduate degrees in chemistry and other sciences. In addition, deep indexing and highly skilled retrieval assistance previously supplied by SSIE is missing. The demise of SSIE left a crucial void that desperately needs filling. See also Chapter 9.

Current Research in Britain, published by Cartermill Publishing (London, www.cartermill.com) in conjunction with the British Library, is a national register of research being conducted in universities and other higher educational institutions in the United Kingdom. It is available in both printed and CD form. Cartermill also publishes *Current Research Worldwide CD-ROM, Current Research in Europe CD-ROM,* and *Current Research in Medicine Worldwide CD-ROM,* as well as a number of similar publications in agriculture, engineering, and materials, as well as in other fields. The firm also provides databases of inventions available for license and of faculty expertise.

Doctoral dissertations offer the chemist another way of obtaining information as soon as possible and can contain some information that may never be published in any other form. Although chemists can sometimes borrow dissertations through local librarians if the university where the work was done is known, this procedure is often time consuming, and no loan copies are available in many cases.

The preferred approach in looking for dissertations is to start with *Dissertation Abstracts International,* a product of the UMI operations of Bell and Howell. This source permits searching for pertinent dissertations by subject, author, title, and university. Online computer-based access is possible through the DIALOG and DataStar systems, as well as through STN, OVID, and OCLC's Epic. The online file covers the time period since 1861, and abstracts are included from 1980. Print and CD-ROM editions are also available. UMI can provide full copies of the dissertations listed.

The program is based on agreements between individual degree-granting institutions and UMI. Most American universities now participate in this file. In addition to U.S. dissertations, many from Canada are included. UMI has initiated an effort to expand its coverage of European dissertations, especially those in English. Some master's theses are included, but this coverage is very limited. Section 15.7 describes another source of information about master's theses.

CA coverage of dissertations is discussed in Section 7.17.

The American Chemical Society's *Directory of Graduate Research* (2) is another valuable source that can provide clues to ongoing research. Research interests and recent publications of faculty members in chemistry and chemical engineering are specified. All major colleges and universities in the United States and Canada are included. The compilation is available in both printed and electronic forms.

In addition to such directories of scientists such as *American Men and Women of Science,* there are more specialized sources that can give clues as to the research activities and interests of scientists.

For example, Synergistic Technologies, Inc. (PO Box 1487, Research Triangle Park, NC, 27709, 800-972-8501) publishes directories of university specialists in such areas of technology as semiconductors; biotechnology, advanced engineering materials; and air, waste, and environmental research. For each research specialist list, types of information given includes: location, phone, fax, research focus, research description, and technology developments (with near term commercialization potential), as well as other information. A new volume is a directory of research centers for the electronics industry that is published in cooperation with the Semiconductor Research Corporation. These volumes are currently available in both printed and diskette forms. Consideration is being given to CD-ROM and online versions in the future.

An important source for identifying university faculty member expertise and inventions is the Community of Science system, which is headquartered at 1615 Thames Street, Baltimore MD 21231 (phone 410-563-5382). The service was founded in 1988. Access to this system is available through Internet's World Wide Web (http://cos.gdb.org/). The product is also available in a fully networkable CD-ROM form that is updated quarterly.

Input is based on cooperative arrangements with some 125 universities in the United States and Canada, with input also from a few government and other nonuniversity facilities. For these cooperating universities and others participating, data given includes the following: descriptions of the expertise of individual faculty members; a list of university technologies that are available for license, as based on inventions;

and lists of university facilities. Also available are lists of current and past NIH, NSF, SBIR, ATP, and USDA grants. Also included are sources of future funding or upcoming funding opportunities. Future plans call for coverage of European universities and a growing European constituency.

Knowledge Express Data Systems (KEDS), Berwyn, PA also offers access to new technologies, both those offered by universities and by industry. The scope includes research in progress and scientific expertise. The Licensing Executives Society cooperates with KEDS in some activities. See Section 10.15.E.

A further source of expertise in universities is the *ProfNet* (*SM*) file, which is available by phone, the Internet, or fax. It is a cooperative of public information officers in universities and elsewhere linked via the Internet, originally intended to give journalists and authors the capability to search for faculty experts at over 700 universities and other research facilities, primarily in the United States. The network was originally coordinated by the Office of News Services at the State University of New York at Stony Brook, and is now a subsidiary of PR Newswire (phone is 1-800-PROFNET).

4.6. SUMMARY

This chapter makes the point that a workable personal plan of action for current awareness can be developed by any chemist or engineer who is eager to keep up to date. Persons who are able to develop and implement such plans can expect to find that timely information, of the right kind and in manageable amounts, can help provide an important competitive edge. Choice of which current awareness tools and services to use depends in large measure on career goals and on present or anticipated future employment. A chemist who does research and teaches in a university, and plans to stay there, would have a different personal plan of action for keeping up to date than a person working in industrial research. Another very important factor is, of course, the field of chemistry in which the chemist or engineer is working. Every person should, however, strive to have a well-rounded, balanced plan. This should typically include a combination of firsthand browsing and reading of selected sources, meaningful personal contacts at meetings and elsewhere, and perusal of automated (computer-based) output based on a carefully selected individual profile or an appropriate group or macroprofile. Writing the plan in a notebook, revising it as needed, and keeping of a log of current awareness strategies, efforts, and results can help provide structure, incentive, and a sense of accomplishment.

REFERENCES

1. R. R. Freeman, J. T. Godfrey, R. E. Maizell, C. N. Rice, and W. H. Shepard, "Automatic Preparation of Selected Title Lists for Current Awareness Services and as Annual Summaries," *J. Chem. Doc.*, **4,** 107–112 (1964).
2. American Chemical Society, *Directory of Graduate Research,* Washington, DC, Updated every few years.

List of Addresses

American Chemical Society
1155 16th St., NW
Washington, DC 20036
(800-227-5558)

American Chemical Society
Rubber Division
University of Akron
Akron, OH 44325
(330-972-7197)

BLAISE-LINE
Boston Spa
Wetherby, W. Yorkshire LS23 7BQ
England

Cambridge Scientific Abstracts
5161 River Road
Bethesda, MD 20816

Chemical Abstracts Service
PO Box 3012
Columbus, OH 43210
(800-848-6538)

Institute for Scientific Information
3501 Market Street
Philadelphia, PA 19104
(800-523-1850)

Macmillan Publishing Co., Inc.
866 Third Avenue
New York, NY 10022

Scientific Meetings
PO Box 81622
San Diego, CA 92138

UMI
300 North Zeeb Road
Ann Arbor, MI 48106
(800-521-0600, except Michigan)

5 How to Get Access to Articles, Patents, Translations, Specifications, and Other Documents Quickly and Efficiently

5.1. INTRODUCTION

Many chemists report that one of the most frustrating experiences they encounter in their professional work is gaining quick access to needed articles, patents, books, and other documents not available in a local chemistry library or bookstore. The problems of obtaining access to or copies of documents quickly and efficiently have been brought home to thousands of chemists and engineers by their use of new automatic and other alerting services (see Chapter 4). Some of these services are so fast that they list many materials before they are received by the local chemistry library. Some materials listed are so "exotic" that obtaining copies becomes a challenging, sometimes formidable task.

Difficulty in document access is most likely to be experienced by chemists who lack ready access to strong chemistry libraries or who are unfamiliar with the document delivery services described in this chapter. It is also more likely to occur when the document sought is from less frequently used sources, such as journals from remote countries, material written in unfamiliar languages, or from material that is older or out of print.

Another complication in locating papers is use of very brief abbreviations in some sources for identifying journals and other references. For example:

B.	*Berichte der Deutschen Chemischen Gesellschaft* or *Chemische Berichte*
C.r. or C.R.	*Comptes Rendus Hebdomadaires des Séances de l'Académie des Sciences*
C. or C.Z.	*Chemisches Zentralblatt*
JACS	*Journal of the American Chemical Society*

For U.S. patents, one frequently sees citations that utilize the letters "US" followed by the appropriate patent number (e.g., US 5,000,000). Patent documents of other countries may be found cited similarly as, for example, "EP" for European patent documents, "WO" for so-called World or Patent Cooperation Treaty patent documents, "GB" for British patent documents, or "JP" for Japanese patent documents, followed by the patent numbers. In the case of patent documents, it is especially helpful to others to give full citation information (patent document number, assignee, inventor, title, date of publication of the document, number and date of application) because of the potential confusion in the citing of patent applications and of issued patents. See also Chapter 13.

Citations in patent documents (references cited by patent examiners and by inventors) are notoriously "loose" and casual and may be especially difficult to decipher and locate. Information professionals can usually help solve these problems.

If a conventional citation is followed by a semicolon and the appropriate *Chemical Abstracts* or *Beilstein* reference (e.g., ———; *CA* 105:123456), this usually signifies that the person making the citation has not utilized or seen the original document, but rather has read an abstract or reference to it.

Fortunately, there are steps that the informed chemist or engineer can take to obtain copies of most of the publications that are needed for research.

5.2. DOCUMENT DELIVERY SERVICES

During recent years, there has been virtually a small explosion in the capabilities and variety of document delivery services and sources. These can provide high-speed copies of millions of documents from worldwide publications. The field has become so highly competitive that the chemists and engineers now have the luxury of selecting from a broad array of excellent choices as to document delivery. In addition, many documents may be viewed online through commercial systems or the Internet.

Factors involved in selection of a document delivery vendor include the following: payment of the required copyright fee, reasonable costs, speed, reliability, good quality of copies provided, and intelligence and know-how of staff. Some services provide an exceptional degree of know-how. For example, when a chemist requested of the British Library's Patent Express Service a copy of a specific British patent application, the staff was able to determine that this document was, in fact, at that time published only as a World Patent Cooperation Treaty (PCT) patent document and provided a copy of that World document (see Chapter 13 for a discussion of World and other patent documents). This degree of know-how is provided as a routine service. Some of the other competitive organizations provide only "mechanical" service. Willingness to search for a document that is hard to find or to provide assistance in cases in which the citation may not be fully complete is an important feature to consider in selecting a document delivery source.

Almost all major document delivery services will offer a variety of delivery options. These include conventional mail, fax, and e-mail. Another feature of a number of the services is high-speed turnaround. Same-day delivery is commonplace, and, in some cases, documents can be provided within the same hour requested. Delivery is, of course, subject to copyright requirements as noted below.

If an organization orders many documents over the course of a year, it is worth exploring whether a special contractual arrangement with one or more suppliers may offer cost or other advantages.

To help track documents costs and for overall efficiency, establishing of a single, central ordering point (say, a library staff member, an administrative assistant, or a technician) may help. However, this can introduce an element of red tape and bureaucracy, depending on the type of arrangement and the persons assigned to do this work.

The typical cost (in 1996–1997 dollars) for a document can range from as low as about $3 (for a patent copy) to about $20. These costs typically include a copy of the document, copyright charges (not required for patent copies), and delivery expenses. The user must usually pay a premium for same-day or other express service. If a document has more pages than is usually the case, this will also incur additional charges as a rule. If the requestor has considerable information about the document, such as *Chemical Abstracts* reference and the full citation, this will frequently help control the costs of copies.

Orders may typically be placed by toll-free telephone numbers, mail, or fax, or through the Internet and other online services.

Most of the services cannot translate documents, nor can they usually provide multiple copies because of copyright requirements. For the same reason, copies of monographs or published proceedings can seldom be provided, although in some cases, a loan of such books may be possible. Examples of some of the major document delivery services are given below.

One of the most important and reliable sources for any chemist seeking copies of more recent chemical documents not available in a local library or information center is the Document Detective Service (DDS) of Chemical Abstracts Service (CAS), which was initiated under the leadership of James L. Wood in 1980. The start-up of DDS marked the successful filling of a major void that had previously existed in obtaining of hard-to-get documents cited by CAS. DDS files are comprehensive, and the service is exceptionally efficient, knowledgeable, and courteous.

Documents covered include

Journal articles, conference proceedings, symposia, and edited collections published since 1975; documents from the former Soviet republics since 1970

U.S. patents since 1971 and patents from many other countries since 1978

DDS coverage is "limited" to the material covered in CAS's *CAplus* and *Chemical Industry Notes* products (see Chapters 6 and 7), but this is very extensive coverage, and includes not only chemistry but also the chemical aspects of a number of other fields such as biotechnology, engineering, health sciences, and the environment. Chemical Abstracts Service indexes and abstracts more journals in chemistry and related areas of science and technology than any other source, and its coverage of chemically related patents is also excellent.

The cost (1997) is $8 as the basic price for patents and $12 plus copyright fee for nonpatents. There are additional charges for special services such as express or fax delivery. As an alternative to ordering while on STN, orders may be placed using certain other online services, including the Internet, or by phone, mail, or fax. Hours of service are gratifyingly long: from 8 A.M. to 8 P.M., Eastern Time, 5 days a week. DDS is now part of the Client Services Department of CAS. As could be expected, DDS does not provide translations, multiple copies, or copies of complete monographs, proceedings, and the like. Documents for *CA* abstracts based on abstracts from another service are not provided.

SourceOne, part of the DIALOG family, offers exceptionally rapid fax or Internet delivery service of many patent documents and also quick delivery, as required, for many other types of documents through its *SourceOne* operations (800-238-3458). Orders may be placed online, or by phone, facsimile, or mail. Delivery modes include Internet, fax, courier, or mail. Hours are 8 A.M. to 8 P.M., Eastern Time. *SourceOne* says, remarkably, that it can provide copies of virtually any document, regardless of subject matter or when it was published, to the extent that copyright requirements permit. It says that it is a full-service, one-stop shopping document delivery organization, in contrast to its sister company UnCover (discussed later in this section), which is more of a self-service organization.

To receive documents through Internet e-mail from *SourceOne* requires an Internet e-mail address, MIME (multipurpose Internet mail extensions) e-mail system that allows inclusion of nontext data, and viewer software that supports TIFF (tagged image file format) images. Enough hard-drive space is also required to handle the transmission.

One of the more unique features of *SourceOne* is that it has an extensive collection of digitally stored U.S., European, and PCT (Patent Cooperation Treaty) patent documents. In addition, it has its own hard-copy collections of other documents, such as all the documents covered in the *Engineering Index* since 1986, and it has an extensive collection of medical, biomedical, and biochemical documents. Additional hardcopy collections, such as those of the American Petroleum Institute, are in the process of being added. Further, it is linked through cooperative agreements with a number of outstanding libraries both in the United States and Europe.

Delivery of patents by *SourceOne* can take just a few minutes after the user enters into the online system the patent numbers desired; this is because the transmission system is largely automated. The cost of digitally stored (laser disks) U.S. and certain international patents in its collections is just $5 each whether sent by regular mail or Internet; the cost is $12 for copies that are faxed. These digitally stored patents include all the graphics that appear in the original patent. The standard charge for a nonpatent document is $12.95 and, in addition, charges for copyright and any special delivery options, as appropriate. All prices are 1997.

Connected with *SourceOne* is another important relatively new source of document copies that also emphasizes fax delivery. This is the UnCover Company, Denver, Colorado (phone 800-787-7979), which is also now part of DIALOG. It may be accessed either by telephone, fax, or online such as via the Internet or through the DIALOG system. Service is excellent, and costs are highly competitive. Copies of documents from over 17,000 journals are available on the basis of cooperative agreements with over 20 libraries in several countries.

UnCover provides a means for both keeping up with the titles of new articles and for obtaining copies of most of these articles. The firm was originally established as a part of CARL, the Colorado Alliance of Research Libraries, and quickly established an excellent reputation for fast and low-cost document delivery. UnCover became part of DIALOG in 1995. The scope includes citations for millions of articles, most of which are in English and cover a variety of subjects ranging from science and technology to business to the humanities. In some cases, brief abstracts are also included. The file goes back to 1988 and is current, in some cases, to the current week's issue as it appears on library shelves.

It may be searched online by journal, subject, or author, and there is a very simple mechanism for placing online orders for fax delivery in the case of most of the articles.

Another excellent choice for the chemist for some years has been *The Genuine Article,* a product of the Institute for Scientific Information, Philadelphia, PA (phone 1-800-336-4474). This important service has covered essentially the ISI databases (see Chapter 8 and elsewhere in this book) for the most recent 5 years. These databases provide very good coverage of journals in chemistry, and they are strong as well in the biomedical, biochemical, and medical sciences. Some 7200 journals are included, but patents are not covered at this time. ISI says that it has catered to the hard-core researcher in its coverage. The service accepts orders by phone, mail, fax, or on-line. Some 40 per cent of requests have been filled through tear sheets direct from the original journal; this usually affords the best available visual quality, a detail that can be crucial with tables, figures, and other material that contains detail. ISI has been an early pioneer in ensuring honoring copyright requirements. Very fast response time has always been a hallmark of the service.

In 1997 ISI introduced a major, new full-service document delivery operation. This new service, *ISI Document Solution,* is based on a partnership between ISI and Infotrieve, Los Angeles, CA. The kinds of published document copies to be provided will be essentially unlimited as to field of science and technology or date of publication. It is significant that ISI has appointed a Vice President for Document Delivery (Lynn Sonk), which indicates the level of importance attached to this function by ISI. As to delivery of documents through the Internet, ISI is in the process of working with primary publishers for permission to distribute their materials electronically.

Another major document supply source, and, in fact, one of the largest in the world, is the British Library Document Supply Center. Along with document supply, a number of interesting current awareness options are offered by this organization. The Internet address is http://portico.bl.uk/dsc/.

5.3. REPRINTS

One traditional and inexpensive way of obtaining copies is writing authors for reprints. The first step is to be certain the reference and address are correct; this is simply a matter of proofreading in many cases. Beyond this, it is helpful to supply as much incentive as possible to encourage authors to fulfill requests for reprints quickly.

A politely worded, typed letter or e-mail communication is helpful, preferably with indication of why the reprint desired is of interest. Even better is reciprocity, as for example, enclosing a reprint of one's own work in the same or related fields. This can help establish an ongoing rapport that may result in access to information prior to pub-

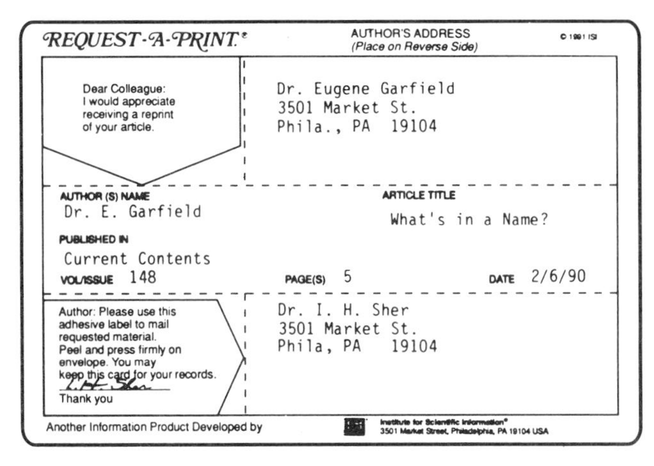

REQUEST-A-PRINT®

AUTHOR'S ADDRESS
(Place on Reverse Side)

© 1991 ISI

Dear Colleague:
I would appreciate receiving a reprint of your article.

Dr. Eugene Garfield
3501 Market St.
Phila., PA 19104

AUTHOR (S) NAME
Dr. E. Garfield

ARTICLE TITLE
What's in a Name?

PUBLISHED IN
Current Contents

VOL/ISSUE 148 PAGE(S) 5 DATE 2/6/90

Author: Please use this adhesive label to mail requested material. Peel and press firmly on envelope. You may keep this card for your records.
Thank you

Dr. I. H. Sher
3501 Market St.
Phila, PA 19104

Another Information Product Developed by Institute for Scientific Information®
3501 Market Street, Philadelphia, PA 19104 USA

Figure 5.1. Request-A-Print form. (Copyright owned by the Institute for Scientific Information, Philadelphia, PA.)

lication, depending on how close a professional bond can be established. It is also desirable to enclose a self-addressed stamped reply envelope and/or to use a tool such as *Request-A-Print* (see Figure 5.1).

Request-A-Print (RAP) cards are postcard-sized forms that eliminate the need to write tedious letters when requesting an article reprint from an author. RAP cards, available through the Institute for Scientific Information (ISI), Philadelphia, PA, are preprinted with the sender's name, address and the verbiage normally included in a reprint request. The name and address of the author and the bibliographic information of the article desired can be filled in, and the user can then peel off the label with the author's name and address and affix it to the address side of the card. The label with the bibliographic information can be kept for the user's records. (A duplicate impression remains on the card.) RAP cards can be ordered in bulk quantities.

Request-A-Print forms are also available for desktop computers for use with *Current Contents on Diskette, Current Contents on Diskette with Abstracts,* and *Focus On* products.

Some authors have an antipathy to answering correspondence, especially requests for reprints. In this event, a fax, or phone call may be necessary to get what is wanted.

Although the reprint approach is by far the most economical way to get copies of documents, the often lengthy time delays could seriously interfere with ambitious work plans and schedules. The uncertainty of whether a sought-after reprint will *ever* be received can be extremely frustrating.

Another alternative, especially for less recent material, is for the chemist to locate one or more abstracts of the document. This can be done by a search of the abstracting services. Abstracts can help confirm whether the document is pertinent and worth ordering. See Chapters 6–8.

Even if the document sought is not available in local chemistry libraries, *equivalents* may be (see also Section 5.7 on this point). It is common for an author to write about the same or similar work in several publications and languages.

Another option is to see if the author has written about the work as part of a series. In that case, an earlier or later paper may be readily available and may suffice.

Also, review articles or books on the subject may contain sufficient information or the work may be adequately described in publications of other authors writing on the same subject. See Chapter 11 on reviews and Section 13.20 on patent equivalents.

5.4. HELP AVAILABLE FROM CHEMISTRY LIBRARIANS

Chemistry librarians have a wide variety of tools for locating documents, for example, *CASSI* (1). This tool shows which major libraries in the United States and abroad receive publications cited by *CAS* since 1907. *CASSI* also includes references to journal literature cited by *Chemisches Zentralblatt* from 1830 to 1969 and several hundred titles covered in *Beilstein's Handbuch der Organischen Chemie** (see Section 12.4.A). Other information useful in locating and obtaining document copies is included. *CASSI* is online through the Questel•Orbit system. Another excellent source for locating information about journals is *Ulrich's International Periodicals Directory,* which is published annually by Reed Elsevier (Bowker), New Providence, NJ, and is also available online through DIALOG. *Ulrich's* gives considerable information about the periodicals included but does not give library holdings. Librarians can be of great help in identifying abbreviated or otherwise cryptic journal references using tools such as those just mentioned.

A common library practice has been to obtain a photocopy of a needed document from local or other libraries. Photocopying of most types of published documents (patent documents are one major exception) is, of course, subject to copyright provisions. In the United States, a revised copyright law went into effect in 1978, and major changes were effected by the Berne Convention Implementation Act in 1988, which became effective on March 1, 1989. The American Chemical Society has a Joint Board/Council Committee on Copyrights that holds open meetings at all ACS National Meetings, and symposia almost every year, mostly joint with the ACS Division of Chemical Information. The Committee publishes a free pamphlet entitled *Are You Up to Date on Copyright Issues.* A chapter on copyright appears in *The ACS Style Guide* which is published periodically by The American Chemical Society. The U.S. Copyright Office at the Library of Congress is, of course, an excellent source of copyright information.

Other actions that librarians can take include borrowing from another library (using the familiar procedure known as interlibrary loan), purchasing copies of documents, and helping chemists determine whether equivalents or translations are available (see Section 5.7).

5.5. ONLINE SERVICES AND RELATED ELECTRONIC SOURCES

A wide variety of services permit and encourage online ordering of documents (see also Chapter 10). This can be a convenient next step after completing a patent and literature search online. In addition, the full texts of a number of documents are avail-

*Now *Beilstein's Handbook of Organic Chemistry.*

able online and/or in other electronic formats as discussed in Chapter 10 and elsewhere in this book. These include many important journals, millions of U.S. and European patent documents, several major chemical encyclopedias, and the classic handbooks, *Beilstein* and *Gmelin* (see Chapter 12).

Thus, the full texts of a number of key journals are available online via STN (see Chapter 10) as follows:

Primary journals of ACS
Many journals of The Royal Society of Chemistry
Some of the journals of Elsevier Science Publishers
Some of the journals of VCH Publishers
The five polymer journals of John Wiley & Sons Publishers

A number of leading journals are also available from other electronic sources. Thus, primary ACS journals are available through the Chemical Abstracts Service *SciFinder* product. Some ACS journals are available in CD-ROM format. The full texts of a number of journals are now available to Internet subscribers, as is discussed further in this book. Individual organizations can subscribe to electronic editions of journals from such publishers as Elsevier Science. In addition, separately from journals, unrefereed "preprints" can be found posted by individual chemists on the Internet, usually to the dismay of journal editors.

5.6. PATENT DOCUMENTS

Patent copies are not subject to copyright laws, but they may sometimes be more difficult to obtain than copies of other types of documents, since many chemists, and the librarians who serve them, are less familiar with all the details of patent systems, especially non-U.S. patent systems. Patent application documents, for example, must be distinguished from examined patents.

One of the major improvements in locating document copies relates to copies of U.S. and European patent documents. The full texts of U.S. and (soon) European patent documents may be read (and searched) in online systems operated by DIALOG, STN International (see Chapters 6 and 7), and by Questel•Orbit. The STN file of U.S. patents (*USPATFULL*) includes graphics, so important in patents, from 1993, in addition to high quality *CA* indexing of chemical patents; Chemical Abstracts Service has the same file, renamed *Chemical Patents Plus,* on the World Wide Web. It is most easily accessible through a link on the CAS home page (http://www.cas.org). This CAS patents file includes all patents, not just chemical patents, for the time period covered (essentially 1974 to date, with partial coverage from 1971 to 1973). As discussed in Chapter 13, the full texts of U.S. patents from the 1970s are now available on the Internet in a free site operated by IBM.

Other sources of U.S. patents include selected federal, state, municipal, and university libraries that maintain extensive sets of U.S. patents. See Table 5.1 for a list of these libraries.

TABLE 5.1. Locations of U.S. Patent and Trademark Depository Libraries[a]

1. Science and Technology Division
   Akron-Summit County Public Library
   55 South Main Street
   Akron, OH 44326-0001
   (330-643-9075)

2. Reference Services
   New York State Library
   Cultural Education Center
   Albany, NY 12230
   (518-474-5355)

3. Centennial Science and Engineering Library
   The University of New Mexico
   Albuquerque, NM 87131-1466
   (505-277-4412)

4. Physical Sciences and Engineering Library
   Lederle Graduate Research Center Lowrise
   University of Massachusetts
   Amherst, MA 01003
   (413-545-1370)

5. Anchorage Municipal Libraries
   Z. J. Loussac Public Library
   3600 Denali Street
   Anchorage, AK 99503-6093
   (907-562-7323)

6. Media Union Library
   2281 Bonisteel Blvd.
   The University of Michigan
   Ann Arbor, MI 48109-2094
   (313-647-5735)

7. Technical Resources
   Library and Information Center
   Georgia Institute of Technology
   Atlanta, GA 30332-0900
   (404-894-4508)

8. Science and Technology Department
   Ralph Brown Draughon Library
   Auburn University
   Auburn University, AL 36849-5606
   (334-844-1747)

9. McKinney Engineering Library
   ECJ 1.300
   The University of Texas at Austin
   Austin, TX 78713
   (512-495-4500)

10. Reference Services
    Troy H. Middleton Library
    Louisiana State University
    141 Middleton Library
    Baton Rouge, LA 70803
    (504-388-5652)

11. Ferris State University
    Abigail S. Timme Library
    1201 S. State Street
    Big Rapids, MI 49307-2747
    (616-592-3620)

12. Government Documents Department
    Birmingham Public Library
    2100 Park Place
    Birmingham, AL 35203
    (205-226-3602)

13. Boston Public Library
    PO Box 286
    Boston, MA 02117
    (617-536-5400, Ext. 265)

14. Science and Technology Department
    Buffalo and Erie County Public Library
    Layfayette Square
    Buffalo, NY 14203
    (716-858-7101)

15. Patent Center
    Montana Tech of the University of Montana Library
    1300 West Park Street
    Butte, MT 59701
    (406-496-4281)

16. Natrona County Public Library
    307 East Second Street
    Casper, WY 82601-2598
    (307-237-4935)

TABLE 5.1. ***(Continued)***

17. Science and Technology Information Center
Business/Science/Technology Division
Chicago Public Library
400 South State Street, 3S-12
Chicago, IL 60605
(312-747-4450)
18. Science and Technology Department
The Public Library of Cincinnati and Hamilton County
800 Vine Street
Cincinnati, OH 45202-2071
(513-369-6936)
19. Reference Unit
R. M. Cooper Library
Clemson University
Box 343001
Clemson, SC 29634-3001
(864-656-3024)
20. Government Documents Department
Cleveland Public Library
325 Superior Avenue
Cleveland, OH 44114-1271
(216-623-2870)
21. Reference Services and Acting Head
Engineering and Physical Sciences Library
University of Maryland
College Park, MD 20742
(301-405-9157)
22. Documents/Maps/Microtext Department
Room 200-D
Sterling C. Evans Library
Texas A&M University
College Station, TX 77843-5000
(409-845-3826)
23. Information Services Department
Ohio State University Libraries
1858 Neil Avenue Mall
Columbus, OH 43210
(614-292-6175)
24. New Hampshire State Library
20 Park Street
Concord, NH 03301-6303
(603-271-2239)
25. Dallas Public Library
1515 Young Street
Dallas, TX 75201
(214-670-1468)
26. Business and Government Publications
Denver Public Library
10 West 14th Avenue Parkway
Denver, CO 80204-2731
(303-640-6220)
27. Information Services and Patent Depository
State Library of Iowa
East 12th & Grand
Des Moines, IA 50319
(515-281-4118)
28. Great Lakes Patent and Trademark Center
Detroit Public Library
5201 Woodward Avenue
Detroit, MI 48202
(313-833-3379)
29. Government Documents Department
Broward County Main Library
100 South Andrews Avenue
Fort Lauderdale, FL 33301
(954-357-7444)
30. Chester Fritz Library
University of North Dakota
PO Box 9000
Grand Forks, ND 58202
(701-777-4888)
31. Federal Documents Section
Hawaii State Library
478 South King Street
Honolulu, HI 96813
(808-586-3477)
32. Division of Government Publications and Special Resources
The Fondren Library—MS44
Rice University
6100 South Main Street
Houston, TX 77005-1892
(713-527-8101, Ext. 2587)

(continued)

TABLE 5.1. ***(Continued)***

33. Business, Science and Technology Division
    Indianapolis-Marion County Public Library
    PO Box 211
    Indianapolis, IN 46206
    (317-269-1741)
34. Mississippi Library Commission
    PO Box 10700
    Jackson, MS 39289-0700
    (601-359-1036)
35. Linda Hall Library
    5109 Cherry Street
    Kansas City, MO 64110
    (816-363-4600)
36. Engineering Library
    Nebraska Hall, 2nd floor west
    University of Nebraska—Lincoln
    Lincoln, NE 68588-0516
    (402-474-3411)
37. Arkansas State Library
    One Capital Mall
    Little Rock, AR 72201-1081
    (501-682-2053)
38. Science, Technology and Patents Department
    Los Angeles Public Library
    630 West Fifth Street
    Los Angeles, CA 90071-2097
    (213-228-7220)
39. Reference and Adult Services
    Louisville Free Public Library
    301 York Street
    Louisville, KY 40203-2257
    (502-574-1611)
40. Government Documents Department
    Texas Tech University
    Box 40002
    Lubbock, TX 79409-0002
    (806-742-2282)
41. Kurt F. Wendt Library
    University of Wisconsin-Madison
    215 North Randall Avenue
    Madison, WI 53706
    (608-262-6845)
42. Mayagüez Campus
    General Library
    University of Puerto Rico
    PO Box 5000
    Mayagüez, PR 00681-5000
    (787-832-4040, Ext. 3459)
43. Business/Science Department
    Memphis & Shelby County Public Library and Information Center
    1850 Peabody Avenue
    Memphis, TN 38104-4025
    (901-725-8877)
44. Business and Science Department
    Miami-Dade Public Library
    101 West Flagler Street
    Miami, FL 33130-2585
    (305-375-2665)
45. Milwaukee Public Library
    814 West Wisconsin Avenue
    Milwaukee, WI 53233
    (414-286-3051)
46. Technology and Science Department
    Minneapolis Public Library and Information Center
    300 Nicollet Mall
    Minneapolis, MN 55401
    (612-372-6750)
47. Evansdale Library
    West Virginia University
    PO Box 6105
    Morgantown, WV 26506-6105
    (304-293-2510, Ext. 113)
48. Reference Department
    University of Idaho Library
    Moscow, ID 83844-2361
    (208-885-6235)
49. Stevenson Science and Engineering Library
    Vanderbilt University
    419-21st Avenue South
    Nashville, TN 37240-0007
    (615-322-2717)
50. New Haven, CT
    Tentative plans call for a facility at the New Haven, CT Public Library to open in Feb. 1998

TABLE 5.1. ***(Continued)***

51. Science, Industry and Business Library
188 Madison Avenue
New York, NY 10016
(212-592-7000)
52. Reference Department
University of Delaware Library
Newark, DE 19717-5267
(302-831-2965)
53. Social Sciences, Sciences, and U.S. Government Publications Division
Newark Public Library
PO BOx 630, 5 Washington Street
Newark, NJ 07101
(201-733-7782)
54. Reference Department Library
PO Box 162666
University of Central Florida
Orlando, FL 32816-2666
(407-823-2562)
55. Science and Engineering Center
Raymond H. Fogler Library
University of Maine
Orono, ME 04469
(207-581-1678)
56. Government Publications Department
The Free Library of Philadelphia
1901 Vine Street
Philadelphia, PA 19103-1189
(215-686-5331)
57. Government Documents Department
Library of Science and Medicine
Rutgers University
PO Box 1029
Piscataway, NJ 08855-1029
(908-445-2895)
58. Science & Technology Department
The Carnegie Library of Pittsburgh
4400 Forbes Avenue
Pittsburgh, PA 15213-4080
(412-622-3138)
59. Paul L. Boley Law Library
Northwestern School of Law
Lewis & Clark College
10015 SW Terwilliger Boulevard
Portland, OR 97219
(503-768-6786)
60. Providence Public Library
225 Washington Street
Providence, RI 02903
(401-455-8027)
61. Research and Information Services
D. H. Hill Library
North Carolina State University
Box 7111
Raleigh, NC 27695-7111
(919-515-3280)
62. Devereaux Library
South Dakota School of Mines and Technology
501 East Saint Joseph Street
Rapid City, SD 57701-3995
(605-394-6822)
63. Government Publications Department
University Library
University of Nevada, Reno
Reno, NV 89557-0044
(702-784-6500, Ext. 257)
64. Virginia Commonwealth University
Box 842033
Richmond,, VA 23284-2033
(804-828-1104)
65. Government Publications Section
California State Library
Library-Courts Building
PO Box 942837
Sacramento, CA 94237-0001
(916-654-0069)
66. Business, Science & Technology Dept.
St. Louis Public Library
1301 Olive Street
St. Louis, MO 63103
(314-241-2288, Ext. 390)
67. Marriott Library
University of Utah
Salt Lake City, UT 84112
(801-581-8394)

(continued)

TABLE 5.1. ***(Continued)***

68. Science Section
San Diego Public Library
820 E Street
San Diego, CA 92101
(619-236-5813)

69. San Francisco Public Library
Civic Center
San Francisco, CA 94102
(415-557-4500)

70. Engineering Library
University of Washington
Box 352170
Seattle, WA 98195-2170
(206-543-0740)

71. Patent and Trademark Depository Library
Illinois State Library
300 South 2nd Street
Springfield, IL 62701-1796
(217-782-5659)

72. Government Documents Department
Oklahoma State University
501 Library
Oklahoma State University
Stillwater, OK 74078-0390
(405-744-7086)

73. Sunnyvale Center for Innovation, Invention and Ideas
465 South Mathilda Avenue, 3rd Floor
Sunnyvale, CA 94086
(408-730-7290)

74. Patent Library
Library LIB122
University of South Florida
4202 East Fowler Avenue
Tampa, FL 33620-5400
(813-974-2726)

75. Daniel E. Noble Science and Engineering Library/Science/Reference
Arizona State University
Tempe, AZ 85287-1006
(602-965-7010)

76. Science-Technology Department
Toledo/Lucas County Public Library
325 Michigan Street
Toledo, OH 43624
(419-259-5212)

77. Documents Section
C207 Pattee Library
Pennsylvania State University Libraries
University Park, PA 16802
(814-865-4861)

78. Founders Library
Howard University
500 Howard Place NW
Washington DC 20059
(202-806-7252)

79. Siegesmund Engineering Library
Potter Center
Purdue University
West Lafayette, IN 47907-1250
(317-494-2872)

80. Ablah Library
Campus Box 68
Wichita State University
Wichita, KS 67260-0068
(316-978-3155)

---

[a]Each of these libraries (listed in order by city) has extensive collections of U.S. patents, electronic and manual search tools, and a staff of specially trained librarians to help. This list is updated in the weekly *Official Gazette* of the U.S. Patent and Trademark Office. For additional updates on this list, see the U.S. Patent Office Internet site, www.uspto.gov. The list above is supplied by Martha Crockett Sneed who manages the PTDL program.

PTDLs have recently been opened at the State University of New York at Stony Brook (Stony Brook, New York) and at the University of Vermont, Burlington, Vermont. Another PTDL is currently being planned for Hartford, CT at the public library there.

The libraries listed in Table 5.1, designated as *Patent and Trademark Depository Libraries* (PTDLs), receive current issues of U.S. patents on microfilm and maintain collections of earlier issued patents. The collections vary in strength from coverage of only recent years to all or most of the patents issued since 1790. These facilities are open to the public without charge. Each PTDL includes not only copies of the patents as mentioned, but also extensive electronic and manual search tools, such as *CASSIS* (Classification and Search Support Information System), the *Manual of Classification, Index to the U.S. Patent Classification,* and *Classification Definitions.* Specially trained librarians are available to help the user. Facilities for making copies from either microfilm or bound books are generally provided for a fee. Some of the larger PTDLs also have copies of non-U.S. patent documents.

Owing to variations in the scope of patent collections among the Patent Depository Libraries and in their hours of service to the public, anyone contemplating use of the patents at a particular library is advised to contact that library, in advance, about its collection and hours, so as to avert possible inconvenience.

The most complete patent library in the United States is at the U.S. Patent Office, Arlington, VA (see Chapter 13). Many libraries in industrial and other chemical research laboratories have extensive files of chemical patents in microfilm or printed form. The chemist can also obtain copies of patents from government patent offices worldwide or from private organizations that specialize in procurement of patents. United States patents can be obtained for $3.00 each (1997 price) from the Commissioner of Patents and Trademarks.

Addresses of major patent offices around the world, along with prices for copies of patents, are listed in the Introduction to the Semiannual Volumes of *Chemical Abstracts* (found in the first issue of each volume). Most government patent offices cannot match the delivery speed of commercial procurement organizations. Examples of such organizations include

Derwent Information Limited, 14 Great Queen Street, London, WC2B 5DF; in North America at 1725 Duke Street, Alexandria, VA 22314 (phone 800-336-5010).

Document Detective Service of Chemical Abstracts Service, Columbus, OH (phone 800-631-1884). Previously mentioned also in Chapter 4.

Patent Express Service of the British Library, 25 Southhampton Buildings, London WC2A 1AW (phone 44 71 323 7926). This service is represented in the United States by The Library Connection, Saluda, VA (phone 804-758-3311).

5.7. TRANSLATIONS

One of the most vexing information problems facing chemists who are natives of the United States, Canada, the United Kingdom, Australia, and other English-speaking nations is that of coping with interesting material published in a foreign language. It is a particularly troublesome problem in the United States, where relatively few chemists are multilingual.

TABLE 5.2. Language of Publication of Journal Literature Abstracted in Chemical Abstracts (as Percentage of Total Journal Literature Abstracted)

Language	1961	1966	1972	1978	1984	1993	1994	1995
English	43.3	54.9	58.0	62.8	69.2	80.3	81.9	81.3
Russian	18.4	21.0	22.4	20.4	15.7	6.4	5.2	4.7
Chinese	—[a]	0.5	—[a]	0.3	2.2	2.9	4.6	4.4
Japanese	6.3	3.1	3.9	4.7	4.0	4.6	4.2	4.2
German	12.3	7.1	5.5	5.0	3.4	2.2	1.5	2.1
French	5.2	5.2	3.9	2.4	1.3	0.9	0.6	0.7
Korean	—[a]	—[a]	0.2	0.2	0.2	0.4	0.5	0.5
Polish	1.9	1.8	1.2	1.1	0.7	0.5	0.3	0.5
Spanish	0.6	0.5	0.6	0.7	0.6	0.4	0.3	0.3
Others	12.0	5.9	4.3	2.4	2.7	1.4	0.9	1.3

[a]Included in "others" for year.
Reprinted with the permission of Chemical Abstracts Service.

This is still important, despite the fact that about 80% of the world's chemical literature (papers) is in English. Over the past 50 years or so, the use of English in this form of chemical literature has made steady and continuous progress. Statistics accumulated by Chemical Abstracts Service (CAS) with reference to abstracted papers (patents and books are not included) are shown in Table 5.2.

Some publications appear totally or partially in English, even though this is not the native language of that country. Journals published in the Czech Republic, Japan, Sweden, and Germany are among countries where this is sometimes the case. Among the most outstanding examples are *Chemische Berichte* and *Liebigs Annalen.** Other examples include:

Acta Chemica Scandinavica
Angewandte Chemie, International Edition
Bulletin of the Chemical Society of Japan
Collection of Czechoslovak Chemical Communications
Japan Chemical Week
Kunststoffe–Plast Europe

Although English is by far the dominant language in journal papers, the situation is different with regard to patents. As Table 6.8 shows, Japanese patents constitute (1996) 57.3% of the total abstracted by CAS. With this type of document (patents), the need for translations is much more acute. Availability of English-language equivalents (see Chapter 13) in some cases can help considerably.

*These two historic journals are now part of the *European Journal of Organic Chemistry* and the *European Journal of Inorganic Chemistry,* both published by Wiley-VCH. These have essentially replaced the major national chemistry journals in several European countries.

What are some options open to the chemist or engineer who learns about intriguing research published in an unfamiliar foreign language?

There are several alternative courses of action that can often help resolve the apparent dilemma relatively quickly and at minimal cost. Some of these involve working closely with a chemical information specialist or research librarian. The following paragraphs list these actions.

1. *Locate and Use a Suitable Abstract.* This may be found in an abstracting service or sometimes in the original journal. (Many foreign-language journals publish abstracts in several languages, one of which is usually English.) Sometimes the abstract indicates that the material is not worth pursuing further—that it is not as pertinent as the title or other preliminary information indicated. Conversely, the abstract may suggest that the publication *is* of interest and worth delving into further. The abstract may contain sufficient data to satisfy immediate interest and need, but it is unwise to rely on the abstract alone. The preferred route, if the abstract indicates that the publication is of interest, is to try to look at the full text in a language the chemist can understand. How to achieve this is described in some of the following paragraphs.
2. *Locate and Use an Already Available Translation*
   a. Many Russian language journals in chemistry are translated cover to cover. One of the largest commercial publishers of such translations is Plenum Publishing Corporation's Consultants Bureau operation. Plenum is located at 233 Spring Street, New York, NY 10013. The English-language translations appear a few months after the originals. Examples include:

      Biochemistry
      Chemical and Petroleum Engineering
      Chemistry and Technology of Fuels and Oils
      Chemistry of Heterocyclic Compounds
      Chemistry of Natural Compounds
      Fibre Chemistry
      Journal of Applied Spectroscopy
      Journal of Structural Chemistry
      Pharmaceutical Chemistry Journal
      Russian Chemical Bulletin
      Theoretical and Experimental Chemistry

   In addition, Plenum distributes exclusively worldwide additional Russian-to-English translated journals from the Russian Academy of Sciences that are published by MAIK Nauka/Interperiodica Publishing, Moscow, concurrently with the Russian-language originals. Some of these publications of potential interest to chemists and chemical engineers include

      Applied Chemistry and Microbiology
      Colloid Journal

Doklady Biochemistry
Doklady Chemical Technology
Doklady Chemistry
Doklady Physical Chemistry
Glass Physics and Chemistry
High Energy Chemistry
Inorganic Materials
Journal of Analytical Chemistry
Journal of Evolutionary Biochemistry and Physiology
Kinetics and Catalysis
Protection of Metals
Radiochemistry
Russian Journal of Applied Chemistry
Russian Journal of Bioorganic Chemistry
Russian Journal of Coordination Chemistry
Russian Journal of Electrochemistry
Russian Journal of General Chemistry
Russian Journal of Organic Chemistry
Theoretical Foundations of Chemical Engineering

CASSI (1) is a convenient source for identifying and locating journals that are translated cover-to-cover.

b. Ask a chemistry librarian to determine if a translation may already exist. This may be the case with material of widespread or commercial interest or importance. There are several electronic and printed translations lists. For example, the *World Translations Index* provides extensive coverage of translations in all fields of science and technology. It is a product of the International Translations Center, Delft, the Netherlands and CNRS/INIST, Vandoeuvre-les-Nancy, France, and it can be searched online through DIALOG.

The U.S. National Translations Center, previously located at the John Crerar Library of the University of Chicago, was transferred for a short time to the U.S. Library of Congress, but unfortunately ceased operations in 1993. Its files for 1989–1993 were transferred to the Canada Institute for Scientific and Technical Information (CISTI), Ottawa, Canada, which continues to maintain an active translation file. Most translations prior to 1993 were transferred to the British Library Document Supply Center, Boston Spa, Wetherby, West Yorkshire, UK.

3. *Identify Equivalent Journal Articles.* As previously noted, the same author may publish on the same research in journals issued in several different countries and languages. Even authors whose native tongue is English sometimes publish first in foreign-language publications and subsequently publish the same or similar material in English. The best professional society and other publications will not accept outright double publication, but it occurs in journals with less rigid standards and in some trade publications on the applied aspects of chemistry. Identification

of such duplicates is straightforward; the chemist merely needs to check the author indexes to *Chemical Abstracts* and related publications.

4. *Locate Review Articles or Books That Adequately Describe the Work of Interest.* See Chapter 11 on reviews.
5. *Locate Equivalent or Corresponding Patent Documents Published in English.* Identification of corresponding patents can be done with concordances or patent indexes such as those of the International Patent Documentation Center, Derwent, *Chemical Abstracts,* and IFI (see Chapter 13 on patents). These concordances provide information about "families" of closely related patents applied for (or issued in) different countries.
6. *Forego Translations of the Full Document.* Instead work with those parts of the publication that can be handled with a limited amount of dictionary lookup. Examples of well-regarded dictionaries are cited in References 3–5. This approach involves scanning for chemical equations (which require no translation) and for such parts as tables, graphs, and abstracts that may require use of a dictionary for only a few keywords. (In some cases, journals in other languages will have already translated synopses or abstracts and captions into English.) It is surprising how often this simple approach will provide results at minimum cost and with acceptable time delay. However, this is not a substitute for the full text and can be risky.
7. *Identify People.* Identify a person within the chemist's own organization who can provide an on-the-spot oral translation. The advantage of this approach is that questions and answers between chemist and translator can readily be exchanged on points that require amplification. In addition, an oral translation session can emphasize parts of the publication most pertinent—key sections within the text that the chemist can identify to the translator in the course of their session. Such in-house translations are sometimes the only way to handle confidential or proprietary materials.
8. *Obtain a Custom-Made Translation.* These can be done by commercial translation houses or by freelance translators. Quality and accuracy of such translations vary widely. The work should be done by a translator who has a proven reputation for top-quality work. The translator must have knowledge of the languages to be translated "to" and "from," and must also have adequate training as a chemist or engineer. (Those organizations that translate cover-to-cover versions of chemical journals are good sources for custom-made translations or can make recommendations.) Translation rates vary, depending on such factors as language, type of document, and speed of delivery required. In 1997, commercial rates for a top-quality translation from languages such as German or French were approximately $120 and up per 1000 words translated into English. These rates include perfect typed copy. The comparable rate for Japanese was approximately $170 and up. One way to control the cost of translations is to request a *rough draft* rather than a finished product. By saving polishing and editing, this should help minimize costs. It is particularly appropriate for a field with which the chemist is already familiar. One way of identifying translators is to contact the American Translator's Association, which publishes a translation services directory that may be helpful.

This directory is published every 2 years, and includes some 1300 translation firms or individuals indexed by language and subject matter. Persons listed have the option of accreditation, but the Association recommends that references be checked in any event. Its booklet *A Consumer's Guide to Good Translation* (1994) can be very helpful in deciding on and evaluating the work of a translator.

Machine (computer-based) and machine-assisted translation has been available for some decades now and is still developing. It is believed by some that machine-assisted translation may provide a viable option in that the output from such processing can provide the human translator with copy that can be subsequently polished into a very useful end product. One of the most exciting developments is that there is now commercially available software that permits the user to scan or otherwise enter foreign-language material into a personal computer and then to obtain a preliminary translation, sufficient, it is claimed, to get a fairly good idea of the content of the document. Some of the companies that are active in machine translation efforts include, for example, Globalink (Fairfax, VA), Intergraph (Huntsville, AL), Language Engineering Corporation, Logos Corp. (Waltham, MA), and Systran (San Diego, CA). Globalink even has a product, Web Translator, that permits on-the-fly translation of material (German, French, and Spanish) as it is received on the World Wide Web. Multilingual Computing Solutions (Washington, DC; contact Chris Miller, 202-483-2229) consults with firms interested in utilizing machine translation on a large scale.

5.8. SPECIFICATIONS AND STANDARDS

The American Society for Testing and Materials, now known simply as ASTM, is the principal developer and publisher of U.S. standards, including analytical and other methods.

The American National Standards Institute (ANSI), is another good source, and can help identify both U.S. and foreign standards.

A National Standards System Network (NSSN) is being developed jointly by the American National Standards Institute (ANSI) and the National Institute of Standards and Technology. The Internet address can be accessed through http://www.ansi.org. ASTM is another important cooperating organization. The project is intended to help users find thousands of standards from private and government sources through a common index, and then to point to locations where these standards can be obtained. At the present time, this location contains links to home pages for a number of home pages for other standards organizations but does not yet contain the common index.

A major supplier of industry and government specifications, standards, and associated documents, including codes and manuals, is Information Handling Services (IHS).

IHS offers indexes to, and the full texts of, standards from a large number of leading standards organizations on CD-ROM. Coverage includes the full texts of thousands of U.S., European, and international standards. The customer can choose as to which collections of standards are desired. A related product covers U.S. military specifications. A daughter company, Global Engineering, can provide hard copies of individual standards.

5.9. PERSONAL CONTACTS

Chemists and engineers should always consider personal contacts with authors of papers and inventors of patents as a major source of important information. Requesting reprints (Section 5.3) is just a first step. A logical second or concurrent step is to talk with (or write to) authors or inventors of papers and patents of interest for specific detail about, for example, what was actually done, what worked best or worst and why, and what future plans are.

The reasons why this type of personal contact is so important are that: publications often contain only relatively limited detail on what is of interest; they are edited to conform to space limitations, legal, and other requirements; and they are often out of date. In addition, chemists are often willing to say over the phone or in other informal contacts what they may not be willing to put in a formal publication or any other written form. Face-to-face contact is always best, but is not always feasible.

The recommendation of personal contact is made regardless of whether authors or inventors being contacted are in colleges and universities or in industry, government, or other chemical facilities. Chances for success with this type of contact are best when authors are in colleges and universities. Nevertheless, it is an approach always worth trying. The worst that can usually happen is a polite "no," but the best can mean an important insight into a research effort or other project. A prerequisite is discretion on the part of the chemist making the inquiry and clear indication of some knowledge of, and interest in, the field of chemistry being discussed.

5.10. OTHER OPTIONS

It may be necessary for the chemist to personally visit a chemistry library in another city to obtain needed information as completely and quickly as possible. Prior arrangements should be made through the local chemistry librarian to increase chances of success on such visits.

A few examples of some of the stronger chemistry libraries in North America include the following:

1. Canada Institute for Scientific and Technical Information (CISTI), National Research Council, Ottawa, Canada K1A 0S2 (phone 800-688-1222). Margot Montgomery, Director General. CISTI is reportedly the largest "freestanding" scientific library in North America. Not only does this facility include an excellent scientific and technical library, but there are a number of outreach services. There is an online catalog that includes 50,000 serials and over 500,000 monographs. This free file can be located on the World Wide Web: http://www.cisti.nrc.ca/cisti/eps.cat.html. CISTI is the agent in North America (until 1999) for *SwetScan,* an online service of Swets and Zeitlinger, Inc. of the Netherlands that contains the tables of contents for over 13,000 journals. Based on its own holdings and those of UMI/InfoStore (Bell and Howell), CISTI offers an impressive document delivery service utilizing electronic scanning technology for both conventional and electronic delivery of document copies. In addition, current alerting services are available. For Canadians only, an extensive search

service is provided on a fee basis. CISTI does not have any patents in its collections, but says it can obtain these.

2. Linda Hall Library, 5109 Cherry Street, Kansas City, MO 64110 (phone 800-662-1545). Probably the premier library in the United States that is dedicated to coverage of engineering and other technical literature. Excellent coverage of the literature of chemistry. A broad range of services is available to the public.

3. University libraries in the United States. Almost all of the large, state-supported "flagship" universities in the midwestern United States such as the Universities of Illinois, Indiana, and Wisconsin, have outstanding chemistry libraries, and in addition, many have access to the unique and powerful online *CrossFire* system that includes *Beilstein* and *Gmelin* (see Chapter 12). Other fine chemistry libraries can be found at major universities throughout the United States.*

There are, of course, important chemical libraries in other countries. One of the most outstanding is The Royal Society of Chemistry's Library and Information Center, which is located at Burlington House, Picadilly, London (phone 44-171-437-8656). Services include document delivery and a technocommercial information service. A corporate membership plan is available. There is an extensive collection of journals, books, CD-ROMS, online databases, and other materials. The librarian is Peter Hoey.

Libraries of comparable strength are located in major European nations and Japan. In the United Kingdom, an outstanding source of scientific documents is the British Library Document Supply Center. Services are provided on a worldwide basis.

No discussion of document sources would be complete without mention of the U.S. National Technical Information Service (NTIS), which is discussed further in Chapter 9. Unclassified federally funded research reports and other documents are a prime emphasis of this large and important facility, which is part of the U.S. Department of Commerce. For hard-to-get articles and other documents published in other countries, the U.S. State Department in Washington may be able to help.

Many chemists belong to organizations that have laboratories, plants, or other representatives and contacts abroad. Persons at these overseas locations are resources who can be called on for aid in obtaining copies of documents.

REFERENCES

1. *CASSI,* Chemical Abstracts Service, Columbus, OH.
2. J. S. Dodd, Ed., *ACS Style Guide,* American Chemical Society, Washington, DC, 1986. Revised second edition published in 1997. See Chapter 11, *Copyright and Permissions.*
3. A. M. Patterson (revised by J. C. Cox and G. E. Condoyannis, Eds.), *Patterson's German-English Dictionary for Chemists,* 4th ed., Wiley, New York, 1992.
4. A. M. Patterson, *French–English Dictionary for Chemists,* 2nd ed., Wiley, New York, 1950.
5. L. I. Callaham, P. E. Newman, and J. R. Callaham, *Callaham's Russian-English Dictionary of Science and Technology,* 4th ed., Wiley, New York, 1995.

*The Library of Congress is another excellent resource.

List of Addresses

American Chemical Society
1155 Sixteen St. NW
Washington, DC 20036
{http://www.acs.org)

American Translators Association
1800 Diagonal Road
Alexandria, VA 22314

ANSI
1430 Broadway
New York, NY 10018
(http://www.ansi.org)

ASTM
100 Bar Harbor Drive
W. Conshohocken, PA 19428
(610-832-9500)
(http://www.astm.org)

British Library
Document Supply Center
Boston Spa, Wetherby
West Yorkshire LS23 7BQ
United Kingdom
(http://portico.bl.uk/dsc)

Chemical Abstracts Service
Document Detective Service
2540 Olentangy River Road
Columbus, Ohio 43210
(800-631-1884)
(http://www.info.cas.org/Support/DDS/front.html)

Commissioner of Patents and Trademarks
Washington, DC 20231
(http://www.uspto.gov)

Global Engineering Documents
15 Inverness Way East
Englewood, CO 80112
(800-854-7179)
(http://www.ihs.com)

Information Handling Services
15 Inverness Way East
Englewood, CO 80112
(http://www.ihs.com)

Institute for Scientific Information
3501 Market Street
Philadelphia, PA 19104
(http://www.isinet.com)

ISI The Genuine Article
PO Box 7649
Philadelphia, PA 19104
(215-386-4399)
(http://www.isinet.com/prodserv/tga/tgadoc.html)

Library of The Royal Society of Chemistry
Burlington House
London W1V 0BN
United Kingdom

SourceOne
75 Varick Street, 9th floor
New York, NY 10013
(800-238-3458)
(http://www.krsourceone.com)

UnCover
4801 E. Florida Avenue
Denver, CO 80210
(303-758-3030; 800-787-7979)
(http://www.uncweb.carl.org)

U.S. National Technical Information Service
4285 Port Royal Road
Springfield, VA 22161

6 Chemical Abstracts Service: History and Development

6.1. INTRODUCTION

The history of Chemical Abstracts Service (CAS) is the story of the development of the world's most outstanding tool of scientific and technical information. CAS development, from its founding in 1907 to the present, reflects accurately the history of chemistry as a science, and of other related sciences and technologies, extending into the biological, engineering, and computer sciences. Historically, the editors and directors of CAS (see list in Table 6.1), perhaps because of their unique positions, or more likely because of their unique abilities, have been outstanding individuals in the history of scientific and technical information and in the history of chemistry.

Since July 1992, the director of CAS has been Robert J. Massie, who came to the organization as a highly successful executive in the information industry and an attorney. He has largely reorganized the top management, with a number of the key people recruited from industry and elsewhere, and he has brought a new and innovative focus. A number of important new products have been introduced recently, and still others are planned. Significantly, quality is now defined as customer satisfaction, and a special group has been established to oversee this function.

Careful study of a recent historical perspective such as this should help chemists use CAS tools more effectively because it summarizes development of these tools over time and indicates when the most important policy changes took place. This perspective also provides an appreciation for the complexity and magnitude of the changes and improvements introduced, and the amount of research and other investigations involved. For example, the efforts that went into the hallmark transformation of CAS into its present computer-based mode rank near the top of management and research achievements in recent American chemical history. Figures 6.1 and 6.2 will help provide some perspective on related abstracting and indexing efforts worldwide. Chapter 7 details some specifics on the current use of *CA*.

6.2. PHYSICAL PLANT AND STAFF

Much of the growth of chemical information can be traced both in the number of pages and abstracts in *Chemical Abstracts* (*CA*) (Table 6.2A), and in the buildings occupied by CAS staff.

TABLE 6.1. Editors and Directors of CAS

Name	Years in Office
Directors	
E. J. Crane	1956–1958
Dale B. Baker	1958–1986
Ronald L. Wigington	1986–1991
Clayton F. Callis[a]	1991–1992
Robert J. Massie	1992–
Editors	
William A. Noyes, Sr.	1907–1908
Austin M. Patterson	1909–1913
John J. MIller	1914 (3 months)
E. J. Crane[b]	1914 (9 months)
E. J. Crane	1915–1958
Charles L. Bernier	1958–1961
Fred A. Tate[b]	1961–1967
Russell J. Rowlett Jr.	1967–1982
David W. Weisgerber	1982–

[a]Ad interim.
[b]Acting editor.

During its first year of operation (1907), *CA* was edited at the National Bureau of Standards in Washington, DC. In 1908, offices were moved to the University of Illinois at Urbana. Beginning in 1909, and for many years thereafter, quarters for CAS staff was a crowded (desk-to-desk) floor in the chemistry building of the Ohio State University (OSU) in Columbus. A new building on the campus was occupied in 1955, but that soon became inadequate despite addition of a floor in 1961. Finally, in 1965, CAS moved to its present American Chemical Society–owned Olentangy River Road

1830 Chemisches Zentralblatt and predecessors (discontinued 1969)

1907 Chemical Abstracts

1926 British Abstracts (discontinued 1953)

1927 Nippon Kagaku Soran (Second Series) (changed 1974)

1940 Bulletin Signalétique (discontinued 1983)

1953 Referativnyi Zhurnal, Khimiya

Figure 6.1. First abstract journals in chemistry and dates started. (Reprinted with the permission of Chemical Abstracts Service.)

Chemical Abstracts (1907)

Derwent Central Patent Index and predecessors (1951)

Referativnyi Zhurnal, Khimiya (1953)

Current Abstracts of Chemistry and Index Chemicus® (1960)

Chemischer Informationsdienst (1970)

Kagaku Gijutsu Bunken Sokuho, Kagaku, Kagakukogyo Hen (Current Bibliography on Science and Technology, Chemistry and Chemical Engineering) (1974)

Figure 6.2. Major abstract journals in chemistry today and dates started. The Derwent Central Patent Index is now better known as *Derwent World Patent Index*. (Reprinted with the permission of Chemical Abstracts Service.)

site, comprising 55 acres adjacent to the OSU campus, and in 1973 a second CAS building was constructed on that site.

These two buildings now house a total staff of some 1250 persons, as compared to employment levels of just 3 part-time staff in 1907, and 58 full-time and 32 part-time staff as late as 1950 (see Table 6.2B). This employment increase is easily explained by the explosive growth of chemical literature, the many functions and services that CAS now offers, and the fact that almost all abstracting operations are now done in-house.

The change to in-house abstracting took place gradually in the 1960s and 1970s and has largely replaced a system of using primarily volunteer abstractors and section editors from around the world, individuals who were paid a token fee for their efforts. This change was brought about by successful development of computer-aided production of *CA* that facilitated abstracting and indexing of documents in a single unified and highly efficient step. In addition, increasing use is being made of author abstracts.

The volunteer abstractor system, now almost entirely outmoded, served an important function in its day. There are no accurate statistics prior to 1963. In 1963, 77% of all abstracts were prepared by volunteer abstractors. From then on, the number of abstracts prepared by volunteers has steadily declined. The peak in the number of volunteer abstractors was reached in 1966 with 3292. As of 1997, a team of only 100 volunteers remained in Japan under the direction of JAICI (Japan Association for International Chemical Information). No attempt should be made to correlate the number of volunteer abstractors with the percentage of abstracts prepared by them. Other factors enter into play: growth of literature, productivity of individual abstractors, total journal coverage versus individual document assignment, and so on. Table 6.3 shows how the system has evolved over the years.

This arrangement (volunteer abstractors) made possible economical (although slow) abstracting of documents in many chemical specialties and in a variety of languages. To the volunteers, their function was a badge of honor and a matter of genuine pride. It was a way to perform a needed service for the chemical profession, and it was an excellent way to keep up with new developments in fields of interest.

TABLE 6.2A. Chemical Abstracts Publication Record 1907–1996

Year	Vol.	Number of Abstracts			Total Abstracts	Total Abstracts to Date	Patent Equiva-lents	Total Documents Cited	Total Documents Cited to Date	Pages of Abstracts	Issue Index Pages	Vol. Index Pages	Total Pages Published[a]
		Papers	Patents	Books									
1907	1	7,994	3,853	—	11,847	11,847	—	—	—	3,074	—	363	3,437
1908	2	11,414	3,658	97	15,169	27,016	—	—	—	3,416	—	473	3,889
1909	3	11,455	3,806	198	15,459	42,475	—	—	—	3,020	—	341	3,361
1910	4	13,006	3,754	785	17,545	60,020	—	—	—	3,314	—	727	4,041
1911	5	15,892	5,014	776	21,682	81,702	—	—	—	3,926	—	845	4,771
1912	6	15,740	6,919	535	23,194	104,896	—	—	—	3,544	—	799	4,343
1913	7	19,025	6,946	659	26,630	131,526	—	—	—	4,096	—	834	4,930
1914	8	16,468	7,920	727	25,115	156,641	—	—	—	3,872	—	725	4,597
1915	9	12,200	6,159	622	18,981	175,622	—	—	—	3,379	—	580	3,959
1916	10	10,519	5,265	324	16,108	191,730	—	—	—	3,180	—	492	3,672
1917	11	10,921	4,680	344	15,945	207,675	—	—	—	3,470	—	524	3,994
1918	12	9,283	4,074	524	13,881	221,556	—	—	—	2,712	—	503	3,215
1919	13	10,957	3,741	542	15,240	236,796	—	—	—	3,338	—	589	3,927
1920	14	13,619	4,432	1,275	19,326	256,122	—	—	—	3,826	—	846	4,672
1921	15	15,211	4,265	975	20,451	276,573	—	—	—	4,059	—	783	4,842
1922	16	18,070	5,142	866	24,098	300,671	—	—	—	4,365	—	1,156	5,521
1923[b]	17	19,507	4,749	1,059	25,315	325,986	—	—	—	3,924	—	1,008	4,932
1924	18	20,523	5,084	1,036	26,643	352,629	—	—	—	3,740	—	1,135	4,875
1925	19	20,951	5,475	671	27,097	379,726	—	—	—	3,618	—	1,155	4,773
1926	20	23,103	6,099	1,036	30,238	409,964	—	—	—	3,842	—	1,406	5,248
1927	21	25,037	7,872	582	33,491	443,455	—	—	—	4,098	—	1,413	5,511
1928	22	28,153	9,936	1,046	39,135	482,590	—	—	—	4,878	—	1,727	6,605
1929	23	29,082	17,867	1,344	48,293	530,883	—	—	—	5,614	—	1,821	7,435

(continued)

TABLE 6.2A. *(Continued)*

1930	24	32,731	21,246	1,169	55,146	586,029	—	—	—	6,066	—	2,142	8,208
1931	25	32,278	18,904	1,546	52,728	638,757	—	—	—	6,161	—	2,282	8,443
1932	26	37,403	20,678	1,380	59,461	698,218	—	—	—	6,184	—	2,270	8,454
1933	27	36,139	28,051	1,963	66,153	764,371	—	—	—	6,024	—	2,172	8,196
1934[c,d]	28	38,371	21,824	1,375	61,570	825,941	—	—	—	3,798	—	1,157	4,955
1935	29	42,593	19,241	1,579	63,413	889,354	—	—	—	4,204	—	1,330	5,534
1936	30	41,927	20,836	1,809	64,572	953,926	—	—	—	4,346	—	1,528	5,874
1937	31	44,032	19,006	1,697	64,735	1,018,661	—	—	—	4,498	—	1,485	5,983
1938	32	45,917	19,515	1,496	66,928	1,085,589	—	—	—	4,782	—	1,542	6,324
1939	33	45,414	19,893	1,801	67,108	1,152,697	—	—	—	4,860	—	1,607	6,267
1940	34	40,624	11,635	1,421	53,680	1,206,377	—	—	—	4,170	—	1,384	5,554
1941	35	35,588	17,176	1,330	54,094	1,260,471	—	—	—	4,184	—	1,464	5,648
1942	36	30,479	14,334	833	45,646	1,306,117	—	—	—	3,684	—	1,232	4,916
1943	37	30,523	11,473	1,673	43,669	1,349,786	—	—	—	3,470	—	1,200	4,670
1944	38	30,440	11,494	1,766	43,700	1,393,486	—	—	—	3,306	—	1,175	4,481
1945	39	22,824	9,357	1,491	33,672	1,427,158	—	—	—	2,782	—	1,137	3,919
1946	40	29,943	8,810	825	39,578	1,466,746	—	—	—	3,853	144	1,591	5,588
1947	41	30,461	7,925	902	39,288	1,506,024	—	—	—	3,909	142	1,496	5,547
1948	42	35,867	7,002	1,127	43,996	1,550,020	—	—	—	4,623	168	1,740	6,531
1949	43	40,612	11,390	1,439	53,441	1,603,461	—	—	—	4,769	183	1,991	6,943
1950	44	47,496	10,063	1,539	59,098	1,662,559	—	—	—	5,592	210	2,350	8,152
1951[c]	45	50,657	10,417	1,959	63,033	1,725,592	—	—	—	5,340	242	2,534	8,116
1952	46	56,419	12,185	1,543	70,147	1,795,739	—	—	—	5,890	265	2,574	8,729
1953	47	61,273	11,906	1,912	75,091	1,870,830	—	—	—	6,444	294	2,821	9,559
1954	48	67,606	11,083	1,926	80,615	1,951,445	—	—	—	7,151	297	3,237	10,865
1955	49	74,664	9,926	1,732	86,322	2,037,767	—	—	—	8,264	324	3,721	12,309
1956	50	78,009	12,350	2,037	92,396	2,130,163	—	—	—	8,768	355	4,522	13,645

1957	51	84,205	16,822	1,498	102,525	2,232,688	—	—	—	9,353	392	4,397	14,142
1958	52	95,736	21,920	1,274	118,930	2,351,618	—	—	—	10,628	486	4,873	15,987
1959	53	98,680	26,760	1,756	127,196	2,478,814	—	—	—	11,557	525	5,471	17,553
1960	54	104,484	27,675	2,096	134,255	2,613,069	—	—	—	13,014	552	6,686	20,252
1961	55	118,337	26,249	2,307	146,893	2,759,962	7,609	154,502	2,767,571	13,999	658	8,322	22,979
1962	56,57	140,168	26,467	2,716	169,351	2,929,313	5,787	175,138	2,942,709	16,725	758	9,373	26,856
1963[c]	58,59	141,016	26,240	4,148	171,404	3,100,717	8,400	179,804	3,122,513	15,298	1,245	9,136	25,679
1964	60,61	161,489	26,422	2,082	189,993	3,290,710	13,375	203,368	3,325,881	16,608	1,498	10,176	28,282
1965	62,63	165,770	29,225	2,088	197,083	3,487,793	19,312	216,395	3,542,276	17,963	1,820	11,074	30,857
1966	64,65	181,715	35,031	3,557	220,303	3,708,096	28,940	249,243	3,791,519	20,700	2,104	12,660	35,464
1967	66,67	202,684	36,797	3,046	242,527	3,950,623	26,766	269,293	4,060,812	22,815	2,683	14,023	39,521
1968	68,69	198,035	31,720	2,753	232,508	4,183,131	19,180	251,688	4,312,500	22,103	2,621	15,703	40,427
1969	70,71	210,344	39,424	2,552	252,320	4,435,451	33,026	285,346	4,597,846	23,533	3,220	18,071	44,824
1970	72,73	230,902	43,044	2,728	276,674	4,712,125	33,068	309,742	4,907,588	23,792	3,777	21,144	48,713
1971	74,75	262,127	43,405	3,444	308,976	5,021,101	41,129	350,105	5,257,693	24,690	5,744	21,151	51,585
1972	76,77	280,143	51,179	3,104	334,426	5,355,527	44,622	379,048	5,636,741	27,386	5,877	20,791	54,054
1973	78,79	269,711	48,683	2,611	321,005	5,676,532	35,544	356,549	5,993,290	25,865	5,708	20,123	51,696
1974	80,81	272,235	58,436	2,953	333,624	6,010,156	42,039	375,663	6,368,953	26,282	5,522	21,180	52,984
1975	82,83	317,472	68,471	6,291	392,234	6,402,390	62,011	454,245	6,823,198	31,100	6,711	25,158	62,969
1976	84,85	317,985	67,176	5,744	390,905	6,793,295	67,603	458,508	7,281,706	31,390	6,891	27,310	65,591
1977	86,87	348,059	55,441	6,637	410,137	7,203,432	68,088	478,225	7,759,931	32,253	7,263	28,980	68,496
1978	88,89	363,195	57,343	7,804	428,342	7,631,774	70,217	498,559	8,258,490	33,368	7,593	27,677	68,638
1979	90,91	370,771	58,738	7,378	436,887	8,068,661	78,854	515,741	8,774,231	34,878	8,010	30,002	72,890
1980	92,93	407,342	61,998	6,399	475,739[e]	8,544,400	72,937	548,676	9,322,907	38,188	8,697	32,197	79,064
1981	94,95	373,973	71,180	5,434	450,587	8,994,987	98,739	549,326	9,872,233	37,074	10,098	35,650	82,822
1982	96,97	381,257	70,774	5,758	457,789	9,452,776	99,658	557,447	10,429,680	37,970	10,595	36,072	84,637
1983[b]	98,99	371,389	74,948	5,416	451,753	9,904,529	95,811	547,564	10,977,244	34,394	10,498	35,368	80,260

(continued)

TABLE 6.2A. *(Continued)*

1984	100,101	380,692	73,907	5,970	460,569	10,365,098	111,239	571,808	11,549,052	34,876	11,216	39,135	85,227
1985	102,103	380,091	73,073	4,767	457,931	10,823,029	98,781	556,712	12,105,764	36,284	10,794	39,631	86,709
1986	104,105	384,141	85,767	4,521	474,429	11,297,458	104,453	578,882	12,684,646	37,826	11,380	42,913	92,119
1987	106,107	386,466	85,219	4,493	476,178	11,773,636	102,419	578,597	13,263,243	38,268	11,650	45,269	95,187
1988	108,109	389,685	80,795	4,065	474,545	12,248,181	103,583	578,128	13,841,371	38,968	12,048	45,711	96,727
1989	110,111	397,158	88,099	3,934	489,191	12,737,372	112,437	601,628	14,442,999	40,424	12,736	51,143	104,303
1990	112,113	394,945	91,082	3,490	489,517	13,226,889	118,113	607,630	15,050,629	41,097	13,005	54,059	108,161
1991	114,115	453,640	95,536	3,885	553,051	13,779,940	133,666	686,717	15,737,346	46,608	14,596	57,890	119,094
1992	116,117	430,247	98,505	3,685	532,437	14,312,377	135,831	668,268	16,405,614	45,394	14,716	54,394	114,504
1993	118,119	448,733	99,411	3,261	551,405	14,863,782	135,837	687,242	17,092,856	48,461	15,472	53,910	117,843
1994	120,121	542,511	107,226	3,318	653,055	15,516,837	122,752	775,807	17,868,657	58,331	17,472	66,364	142,167
1995	122,123	562,955	121,214	3,620	687,789	16,204,626	126,459	814,248	18,682,905	61,707	18,777	75,426	155,910
1996	124,125	579,251	121,682	5,336	706,269	16,910,895	141,374	847,643	19,530,548	65,481	18,635	80,114	164,230

[a]Total includes "List of Periodicals" pages, which are not shown separately, for those years in which that list was published in *CA*.

[b]Type size decreased.

[c]Page size increased.

[d]Two-column format adopted.

[e]About 17,000 of the abstracts published in 1980 resulted form a 14-day shortening of processing time.

Reprinted with permission of Chemical Abstracts Service.

TABLE 6.2B. Number of CAS Employees

Year	Part-Time	Full-Time	Year	Part-Time	Full-Time
1907	3	0	1947	33	63
1908	5	0	1948	40	59
1909	1	4	1949	34	48
1910	0	4	1950	32	58
1911	2	4	1951	40	63
1912	13	5	1952	59	75
1913	10	5	1953	55	86
1914	13	5	1954	29	78
1915	3	6	1955	49	95
1916	13	14	1956	36	97
1917	15	6	1957	46	126
1918	11	6	1958	42	126
1919	8	5	1959	44	204
1920	8	5	1960	51	240
1921	12	6	1961	67	297
1922	7	8	1962	60	375
1923	8	7	1963	36	431
1924	11	8	1964	34	544
1925	8	8	1965	29	691
1926	11	11	1966	37	927
1927	6	11	1967	—	945
1928	15	13	1968	—	935
1929	10	16	1969	—	954
1930	16	20	1970	—	978
1931	9	19	1971	—	984
1932	10	17	1972	—	1006
1933	9	18	1973	—	1080
1934	13	18	1974	—	1162
1935	7	21	1975	—	1148
1936	9	34	1976	—	1130
1937	13	35	1977	—	1142
1938	13	29	1978	—	1157
1939	11	31	1979	—	1159
1940	6	25	1980	—	1181
1941	14	24	1981	—	1189
1942	23	24	1982	—	1200
1943	23	25	1983	—	1223
1944	23	25	1984	—	1271
1945	31	31	1996	—	1250
1946	29	42			

TABLE 6.3. Transition to In-House Abstracting: Number of Volunteer Abstractors

Year	Total	Year	Total	Year	Total
1907	129	1934	442	1961	2250
1908	171	1935	339	1962	2870
1909	205	1936	428	1963	3137
1910	252	1937	449	1964	3030
1911	219	1938	438	1965	3232
1912	237	1939	448	1966[a]	3292
1913	219	1940	449	1967	3245
1914	248	1941	474	1968	3107
1915	203	1942	474	1969	2921
1916	187	1943	456	1970	2819
1917	172	1944	457	1971	2745
1918	159	1945	444	1972	2549
1919	155	1946	496	1973	2236
1920	221	1947	486	1974	2084
1921	230	1948	590	1975	1831
1922	253	1949	641	1976	1542
1923	276	1950	624	1977	1406
1924	260	1951	771	1978	1255
1925	231	1952	835	1979	1029
1926	255	1953	943	1980	949
1927	261	1954	1029	1981	830
1928	272	1955	1200	1982[b]	699
1929	338	1956	1406	1983	666
1930	390	1957	1283	1984	648
1931	415	1958	1396	1994	96
1932	438	1959	1536	1997	100
1933	338	1960	1668		

[a] In 1966 there were 1811 domestic and 1481 foreign volunteer abstractors.
[b] In 1982 there were 102 domestic and 597 foreign volunteer abstractors.
Reprinted with permission of Chemical Abstracts Service.

In addition to producing *CA,* the offices at Columbus now handle keyboarding for composition of all CAS publications and American Chemical Society (ACS) primary journals. This work was previously done at Mack Printing Company in Easton, PA. The change was accomplished in a gradual transition from 1965 to 1980. Furthermore, starting in 1974, CAS took over computer handling of all ACS membership records, subscription fulfillment, accounting, financial, personnel, and other business records for ACS headquarters in Washington, DC, as well as in Columbus. Other functions and services of CAS are described throughout this chapter. In 1984, the budget for CAS was approximately $70 million. In 1985, CAS's budget grew to $75 million. The budget for 1986 was $85 million, and for 1997 was over $125 million.

The staff in Columbus has remained at about the same size (about 1250 persons) since 1984, a remarkable achievement considering the significant increase in volume of literature covered and the additional products offered.

CAS has an agreement with Fachinformationszentrum Chemie GmbH (FIZ Chemie) in Berlin whereby a small staff at the CAS Input Center there abstracts and indexes some German, Austrian, Swiss, and Russian journal literature. Another agreement is with the Japanese Association for International Chemical Information (JAICI), which maintains a small CAS Input Center that abstracts about 40% of the Japanese patent documents covered in *CA*.

6.3. *CA* ABSTRACTS

For many years, *CA* had provided exceptionally full and complete abstracts, often running a full column or even more. Emphasis on briefer, more *findings-oriented* abstracts began to evolve beginning in about 1950. Unlike earlier abstracts, no attempt was made to record in the abstract all new data reported in original documents. In a few areas of synthetic organic chemistry and analytical chemistry, however, abstracts continued to provide lengthy, detailed procedures and extensive lists of numerical data. Finally, in 1970, instructions (reflected in the 1971 edition of the CAS *Directions to Abstractors*) were issued to have essentially all abstracts prepared in a concise, *findings-oriented* form. This was gradually implemented over the 1970–1971 period.

Abstracts are now concise and compact, although, fortunately, indexing remains as full as ever. The *CA* abstract today is intended to provide access to the original literature, but not to serve as a substitute for it, a very important distinction. Use of Registry Numbers and of newer type fonts and page arrangements helped achieve some of the compaction, although the primary means is simply use of fewer words. The average abstract text continues to be about 100 words long according to CAS estimates.

Few chemists would quarrel with this policy of briefer abstracts, particularly because the efficient CAS Document (Delivery) Detective Service is available (see Section 5.2). Problems could, however, potentially arise with (1) documents in less familiar foreign languages when there are no English language "equivalents," or (2) possibly in cases where large numbers of pages of the full text of documents may need to be obtained. In these cases, the older forms of more lengthy abstracts might still be useful.

CA staff make extensive use of English-language author abstracts as found in journal papers, perhaps with some minor modifications, if the staff finds that these are appropriate. The staff does write the abstracts for all non-English journal paper abstracts. Abstracts found in patent documents, regardless of language, are usually extensively or totally rewritten.

Since 1967, abstracts have been numbered sequentially in each volume. Before that time, the number linked to index entries represented a column, and the small letter, or still earlier a numerical superscript, described the fraction of the column in which the entry of interest could be found.

CA issues have been published weekly since 1967, with about half of the sections appearing one week and the other half the next week. Before that time, all sections were published together on a biweekly basis, and prior to 1961, on a twice-a-month basis. Beginning in 1997, all 80 sections will be published weekly.

6.4. *CHEMICAL ABSTRACTS* INDEXES

Chemists have always depended heavily on *CA*'s comprehensive, reliable, in-depth indexing based on the original documents. The expertise with which CAS indexes are constructed is the single most important reason why *CA* is such an indispensable tool for chemists and for scientists in other disciplines around the world. Use of these indexes in an online mode (with computers) further enhances their utility. See Figure 6.3 on the development of indexes to *CA*.

CA indexes had been published annually until 1962, when they began to be published semiannually (every 6 months). *CA* Collective Indexes combine into single, organized listings the content of 10 individual Volume Indexes. Ten-year (decennial) Collective Indexes were published for *CA* from 1907 to 1956; 5-year (quinquennial) Indexes have been produced since 1956 (see Tables 6.4 and 6.5). These also show the date of the start of each of the major types of *CA* Volume Indexes and of the associated *Index Guides*.

Use of Collective Indexes reduces the repetitive work that would otherwise be needed to search individual Volume Indexes. Thus, a search of *CA* index content from 1907 through 1981 requires separate searches in 10 Collective Indexes rather than in 95 Volume Indexes. Table 6.4 indicates the individual indexes that comprise each of the Collective Indexes.

Volume Indexes to *CA* are based on a controlled vocabulary. To provide chemists with more rapid indexing of the contents of individual *CA* issues, a form of quick indexing (designated as the Keyword Index) has appeared with each issue beginning in 1963. Because the Keyword Index is based on uncontrolled vocabulary, it is relatively superficial. It is not intended to provide the concise and reliable access of the Volume Indexes. However, this distinction is now less important for chemists who can search *CA* online since rapid access to both types of indexes is possible online.

6.5. *CHEMICAL ABSTRACTS* NOMENCLATURE

For years (since 1945), CAS has published guidelines to, and summaries of, its approach to naming of chemical compounds. But it was not until 1972 that CAS extended systematic nomenclature in its indexes to most substances and decided to use *trivial* or nonsystematic names only rarely. In almost all cases, nomenclature currently

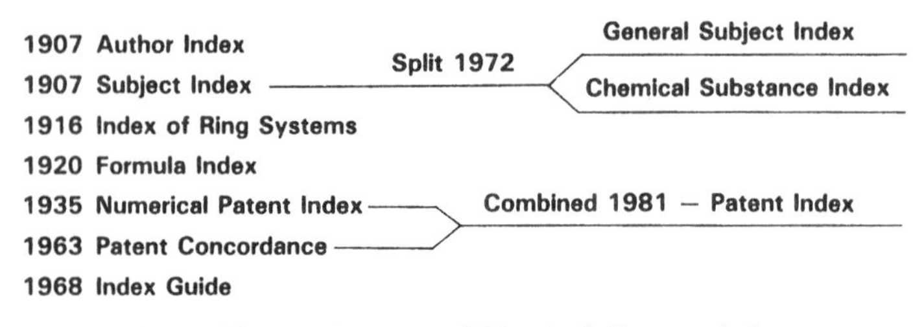

Figure 6.3. Development of *Chemical Abstracts* indexes.

TABLE 6.4. *CA* Collective Indexes: Historical Development

	Collective Index											
	1st	2nd	3rd	4th	5th	6th	7th	8th	9th	10th	11th	12th
Years	1907–1916	1917–1926	1927–1936	1937–1946	1947–1956	1957–1961	1962–1966	1967–1971	1972–1976	1977–1981	1982–1986	1987–1991
CA Volumes	1–10	11–20	21–30	31–40	41–50	51–55	56–65	66–75	76–85	86–95	96–105	106–115
Author Index	×	×	×	×	×	×	×	×	×	×	×	×
Subject Index	×	×	×	×	×	×	×	×				
Index Guide								×	×	×	×	×
Chemical Substance Index									×	×	×	×
General Subject Index									×	×	×	×
Formula Index		—[b]	—[b]	—[b]	×	×	×	×	×	×	×	×
Index of Ring Systems	—[a]	—[a]	—[a]	—[a]	—[a]	—[a]	×	×	×	×	×	×
Numerical Patent Index				—[b]	×	×	×	×	×			
Patent Concordance							×	×	×			
Patent Index										×	×	×

[a]For the 1st through the 6th Collective Indexes, Ring System Information was included in the Introduction to the *Subject Index*; for the 7th through the 12th Collective Indexes, the *Index of Ring Systems* was bound with the *Formula Index*.

[b]Two special Collective Indexes are available for searches of patents and formulas: The *10-Year Numerical Patent Index to Chemical Abstracts* (1937–1946) and the *27-Year Collective Formula Index to Chemical Abstracts* (1920–1946). The Special Libraries Association published a patent index covering 1907–1936.

TABLE 6.5. Collective Indexes to Chemical Abstracts

	1st 1907–1916	2nd 1917–1926	3rd 1927–1936	4th 1937–1946	5th 1947–1956	6th 1957–1961	7th 1962–1966	8th 1967–1971	9th 1972–1976	10th 1977–1981	11th 1982–1986	12th 1987–1991
Documents Cited												
Nonpatent abstracts	138,436	170,493	357,507	370,117	559,180	510,373	804,749	1,118,615	1,478,249	1,896,992	1,924,002	2,041,761
Patent abstracts	53,294	47,741	186,455	142,693	104,247	119,426	143,385	194,390	293,945	304,700	378,469	440,721
Total abstracts	191,730	218,234	543,962	512,810	663,427	629,799	948,134	1,313,005	1,772,194	2,201,692	2,302,471	2,482,482
Patent equivalents	—	—	—	—	—	7,609	75,814	153,169	251,819	400,081	509,942	570,218
Total documents cited	191,730	218,234	543,962	512,810	663,427	637,408	1,023,948	1,466,174	2,024,013	2,601,773	2,812,413	3,052,700
Author Index												
No. of author entries	230,000	284,000	653,000	615,000	795,000	820,000	1,800,00	3,020,000	4,668,000	5,847,000	6,399,039	7,966,801
Pages	1,980	2,452	3,095	3,541	5,074	6,164	9,607	13,845	19,128	24,684	26,756	32,659
Books	2	2	2	2	4	4	6	8	11	14	15	17
Patent Index												
No. of abstracted and equivalent patents	—	—	—	142,693	104,247	127,035	219,199	347,559	545,764	704,534	888,411	1,010,939
Pages	—	—	—	182	144	172	352	761	1,374	3,643	4,961	5,720
Books	—	—	—	1	1	1	1	—[a]	1	2	3	3
Subject Index												
Index Entries	307,000	620,000	1,378,000	1,609,000	3,241,000	3,289,000	4,735,000	6,876,000	—	—	—	—
Pages	2,843	4,139	4,885	6,386	13,740	12,659	24,632	33,548	—	—	—	—
Books	2	3	3	4	11	7	13	18	—	—	—	—

General Subject Index												
Index entries	—	—	—	—	—	—	—	—	3,746,000	5,345,000	6,334,017	7,520,022
Pages	—	—	—	—	—	—	—	—	16,797	25,508	33,424	42,893
Books	—	—	—	—	—	—	—	—	10	15	19	23
Chemical Substance Index												
Index entries	—	—	—	—	—	—	—	—	7,351,000	9,264,000	10,791,421	13,208,976
Pages	—	—	—	—	—	—	—	—	40,891	56,112	71,720	97,323
Books	—	—	—	—	—	—	—	—	25	32	40	51
Formula Index												
Index entries	—	—	—	560,000[b]	714,000	867,000	1,869,000	2,566,000	3,189,000	3,493,000	4,754,683	5,187,482[c]
Pages	—	—	—	2,077[b]	2,968	3,869	7,007	11,769	16,251	19,904	24,016	34,950
Books	—	—	—	2[b]	3	3	4	7	9	11	14	19
Index Guide												
Index entries	—	—	—	—	—	—	—	154,000	149,000	191,318	238,523	243,406
Pages	—	—	—	—	—	—	—	2,280	1,440	1,605	2,115	2,248
Books	—	—	—	—	—	—	—	2	1	1	2	2
Summary												
Total books	4	5	5	9	19	15	24	35	57	75	93	115
Total pages	4,823	6,591	7,980	12,186	21,926	22,864	41,598	62,203	95,881	131,456	162,992	215,793
Distribution time following completion of collective period					65 mo.	40 mo.	44 mo.	24 mo.	20 mo.	14 mo.	12 mo.	12 mo.

[a]Bound with Author Index.

[b]This Formula Collective covered the 27-year period from 1920 to 1946.

[c]Maximum number of references to be printed lowered from 50 to 25.

Reprinted with the permission of Chemical Abstracts Service.

used by CAS for chemical substance index headings is that of IUPAC (International Union of Pure and Applied Chemistry).

CAS staff believe that this has resulted in substance index names generally easier to interpret and easier to derive, and in improved groupings of structural families together in the alphabetically ordered index. Some chemists believe this policy change makes the use of *CA* unnecessarily complex, although public debates that followed original announcement of the change have subsided. Fortunately, however, the *CA Index Guide,* first introduced in 1968, provides a guide to trivial names and cross-references these to systematic names. Some cross-references were reintroduced into the Volume Indexes during the ninth collective period (1972–1976) for several hundred substances, but CAS could not afford to do this for all substances covered.

In addition to the *Index Guide,* other major aids to identification of correct nomenclature are the *Registry Handbook—Number Section,* first published in 1974, and the *Registry Handbook—Common Names* first published in 1978. Use of Registry Numbers ensures unambiguous identification regardless of the potential vagaries of nomenclature.

6.6. *CHEMICAL ABSTRACTS* SECTIONS

The evolution of *CA* sections is a particularly helpful way to trace the growth of *CA* and of chemistry as a science. The existence of sections, initiated with *CA* at the outset in 1907, is intended to make it possible for chemists and other uses of *CA* to limit scanning, reading, or searching of *CA* to specific sections, with a reasonable degree of assurance that most of what is pertinent will probably be found, providing the suggested cross-references are also pursued. However, no studies on the effectiveness of sections in regard to their intended function have been published to date.

Existence of sections also makes it possible to publish the Section Groupings. These provide a relatively affordable tool for chemists to keep up to date in broad, related areas of chemistry without the necessity of reading, or paying for, the full *CA*. The first Section Grouping (Biochemistry) was introduced in 1962, and the four additional Groupings (Organic Chemistry; Macromolecular; Applied Chemistry and Chemical Engineering; and Physical, Inorganic, and Analytical Chemistry) were introduced in 1963.

In 1945, the number of sections in *CA* was increased from the original 30 to 31 with the addition of the section *Synthetic Resins and Plastics.* In 1962 abstracts were arranged by subsection, and permanent section cross-references were provided. The number of sections was increased to 73 in 1962, to 74 in 1963, and to the present total of 80 in 1967.

The first published guide on the scope of each of the sections was made available to the chemical community in 1970. This guide became increasingly important as the development of online searching made it possible to more easily and efficiently limit searches to specific sections. Assignment of abstracts to sections by CAS staff is necessarily somewhat arbitrary. Accordingly, if a *complete* search is desired, limiting by section is not wise. Note that cross-references are provided at the end of each section and these should be pursued if the user is interested in a more complete search.

National concern with oil supply and prices and with other energy sources was reflected in reorganization of the *CA* sections on energy (51 and 52) in 1974. A separate section on safety, in response to a resurgence of interest in the same year, was, however, not provided. Instead, 1974 saw the emergence of emphasis on indexing of the safety aspects. At about the same time, CAS increased coverage of toxicology, although pinpointing of a specific year is not possible because this change took place gradually.

In 1982, *CA* sections were reorganized to more accurately represent additional changes in the development of chemistry. For example, the new sections *Biochemical Genetics* and *Biomolecules and Their Synthetic Analogs* were introduced. The complete list of *CA* sections at present is shown in Table 6.6.

6.7. PATENT DOCUMENTS

Improved and expanded coverage of patents by CAS reflects the recognition of chemists of the importance of patent documents, including issued patents and published patent applications, as unique and timely sources of chemical information (see also Chapter 13). This improvement may also reflect the stimulus provided by the competition of Derwent Information Ltd., generally acknowledged as worldwide leader in abstracting and indexing of patent document information for *all* technologies.

CAS now covers all patents of chemical or chemical engineering interest from 18 countries and two international organizations—an especially sharp increase over earlier years. In addition, for patents issued by another 9 countries, all patents of chemical or chemical engineering interest are covered for those patents issued to individuals or organizations residing in the granted country or resident in countries not listed (see Table 6.7).

Table 6.7 shows the development of CAS patent document coverage since 1907.

Patent documents are recognized by chemists as a source of much data that never appears elsewhere, although how much has never been definitively demonstrated. Patent documents present new information quickly, often for the first time anywhere, especially in the form of published unexamined patent applications. The popularity of the unexamined patent application has grown, and almost all patent offices now employ this approach. CAS recognized this trend by beginning to cover Netherlands applications in 1964, West German applications in 1968, and Japanese applications in 1972, to name the most outstanding examples.

In the 1970s, two new multinational patent systems (the European Patent Office and the Patent Cooperation Treaty) were organized to reduce the cost and effort involved in filing in individual countries. CAS began to cover these as soon as the first patent applications appeared in 1979.

Multiple filing (in many countries) shows the commercial importance of patent documents, and knowledge of the existence of applications in English can point the way to obtaining copies of English-language equivalents. To reflect this, CAS has provided a printed (not online) patent concordance since 1963. Prior to that, CAS provided *title-only* abstracts to equivalents.

TABLE 6.6. List of *CA* Sections

Abstract Sections[a]

Biochemistry Sections

1. Pharmacology
2. Mammalian Hormones
3. Biochemical Genetics
4. Toxicology
5. Agrochemical Bioregulators
6. General Biochemistry
7. Enzymes
8. Radiation Biochemistry
9. Biochemical Methods
10. Micobial, Algal, and Fungal Biochemistry
11. Plant Biochemistry
12. Nonmammalian Biochemistry
13. Mammalian Biochemistry
14. Mammalian Pathological Biochemistry
15. Immunochemistry
16. Fermentation and Bioindustrial Chemistry
17. Food and Feed Chemistry
18. Animal Nutrition
19. Fertilizers, Soils, and Plant Nutrition
20. History, Education, and Documentation

Organic Chemistry Sections

21. General Organic Chemistry
22. Physical Organic Chemistry
23. Aliphatic Compounds
24. Alicyclic Compounds
25. Benzene, Its Derivatives, and Condensed Benzenoid Compounds
26. Biomolecules and Their Synthetic Analogs
27. Heterocyclic Compounds (One Hetero Atom)
28. Heterocyclic Compounds (More Than One Hetero Atom)
29. Organometallic and Organometalloidal Compounds
30. Terpenes and Terpenoids
31. Alkaloids
32. Steroids
33. Carbohydrates
34. Amino Acids, Peptides, and Proteins

Macromolecular Chemisty Sections

35. Chemistry of Synthetic High Polymers
36. Physical Properties of Synthetic High Polymers
37. Plastics Manufacture and Processing
38. Plastics Fabrication and Uses
39. Synthetic Elastomers and Natural Rubber
40. Textiles and Fibers
41. Dyes, Organic Pigments, Fluorescent Brighteners, and Photographic Sensitizers
42. Coatings, Inks, and Related Products
43. Cellulose, Lignin, Paper and Other Wood Products
44. Industrial Carbohydrates
45. Industrial Organic Chemicals, Leather, Fats, and Waxes
46. Surface-Active Agents and Detergents

Applied Chemisty and Chemical Engineering Sections

47. Apparatus and Plant Equipment
48. Unit Operations and Processes
49. Industrial Inorganic Chemicals
50. Propellants and Explosives
51. Fossil Fuels, Derivatives, and Related Products
52. Electrochemical, Radiational, and Thermal Energy Technology
53. Mineralogical and Geological Chemistry
54. Extractive Metallurgy
55. Ferrous Metals and Alloys
56. Nonferrous Metals and Alloys
57. Ceramics
58. Cement, Concrete, and Related Building Materials
59. Air Pollution and Industrial Hygiene
60. Waste Treatment and Disposal
61. Water
62. Essential Oils and Cosmetics
63. Pharmaceuticals
64. Pharmaceutical Analysis

TABLE 6.6. ***(Continued)***

Abstract Sections[a]	
Physical, Inorganic, and Analytical Chemisty Sections	72. Electrochemistry
65. General Physical Chemistry	73. Optical, Electron, and Mass Spectroscopy and Other Related Properties
66. Surface Chemistry and Colloids	74. Radiation Chemistry, Photochemistry, and Photographic and Other Reprographic Processes
67. Catalysis, Reaction Kinetics, and Inorganic Reaction Mechanisms	75. Crystallography and Liquid Crystals
68. Phase Equilibriums, Chemical Equilibriums, and Solutions	76. Electric Phenomena
69. Thermodynamics, Thermochemistry, and Thermal Properties	77. Magnetic Phenomena
70. Nuclear Phenomena	78. Inorganic Chemicals and Reactions
71. Nuclear Technology	79. Inorganic Analytical Chemistry
	80. Organic Analytical Chemistry

[a]Guidelines used to assign abstracts to *CA* Sections according to their subject content are summarized in the publication, "Subject Coverage and Arrangement of Abstracts by Sections in Chemical Abstracts" (Subject Coverage Manual), 1992.

Derwent is now the only English-language service that provides the convenience of English-language abstracts for equivalent patents previously abstracted on a worldwide basis.

In 1981, CAS introduced a new *Patent Index,* which combined the content of the *Numerical Patent Index* and *Patent Concordance* and considerably expanded the identifying information presented for patents. Competition in the concordance approach to chemical patents is provided by Derwent and by INPADOC (International Patent Documentation Center).

Another evidence of the growing importance attached to patents is the 1979 decision by CAS to additionally index chemical substances from patent claims, even though these may be unsupported by examples, data, and so on. Prior to that, CAS had not provided indexing for this type of "paper chemistry." Even today, CAS is more selective in its coverage of patent information than is Derwent.

Another advantage of patents is that they are regarded by many as a good indication of the innovativeness of not only individuals and organizations, but also of nations. Tables 6.8 and 6.9 show trends in patent document coverage by CAS.

6.8. CHEMICAL ABSTRACTS SERVICE IN TRANSITION

The Baker–Tate–Rowlett era, which spanned more than two decades beginning in 1958, was among the most meaningful and successful in the history of CAS. This era saw transition to

a. Start of CAS's first computer-based publication, *Chemical Titles* (1961)
b. Start of computer-based current awareness services (1962–1963)

TABLE 6.7. *CA* Patent Document Coverage by Country and Type of Patent Document[a]

	Year																																	
	1963	1964	1965	1966	1967	1968	1969	1970	1971	1972	1973	1974	1975	1976	1977	1978	1979	1980	1981	1982	1983	1984	1985	1986	1987	1988	1989	1990	1991	1992	1993	1994	1995	1996
Australian	R	R	R	R	R	R	R	R	R	R	R	R	R	R	R	R	R	R	R	A	A	A	A	A	A	A	A	A	A	A	A	A	A	A
Australian (Petty)																				A	A	A	A	A	A	A	A	A	A	A	A	A	A	A
Austrian	R	R	R	R	R	R	R	R	R	R	R	R	R	R	R	R	A	A	A	A	A	A	A	A	A	A	A	A	A	A	A	A	A	A
Belgian	A	A	R	R	R	R	R	R	R	A	A	A	A	A	A	A	A	A	A	A	A	A	A	A	A	A	A	A	A	A	A	A	A	A
Brazilian Pedido														A	A	A	A	A	A	A	A	A	A	A	A	A	A	A	A	A	A	A	A	A
British	A	A	A	A	A	A	A	A	A	A	A	A	A	A	A	A	A	A	A	A	A	A	A	A	A	A	A	A	A	A	A	A	A	A
British Amended				A	A	A	A	A	A	A	A	A	A	A	A	A	A	A	A	A	A	A	A	A	A	A	A	A	A	A	A	A	A	A
British U.K. Patent Application																	A	A	A	A	A	A	A	A	A	A	A	A	A	A	A	A	A	A
Canadian	R	R	R	R	R	R	R	R	R	R	R	R	A	A	A	A	A	A	A	A	A	A	A	A	A	A	A	A	A	A	A	A	A	A
Canadian Application																											A	A	A	A	A	A	A	A
China																								A	A	A	A	A	A	A	A	A	A	A
Czechoslovakian	R	R	R	R	R	R	R	R	R	R	R	R	R	R	R	R	R	R	R	R	R	R	R	R	R	R	R	R	R	R				
Czech Republic																															R	R	R	R
Danish	R	R	R	R	R	R	R	R	R	R	R	R	R	R	R	R	R	R	R	R	R	R	R	R	R	R	R	R	R	R	R	R	R	R
Finnish	R	R	R	R	R	R	R	R	R	R	R	R	R	R	R	R	R	R	R	R	R	R	R	R	R	R	R	R	R	R	R	R	R	R
French	A	A	A	A	A	A	A	A	A	A	A	A	A	A	A	A	A	A	A	A	A	A	A	A	A	A	A	A	A	A	A	A	A	A
French Addition	A	A	A	A	A	A	A	A																										
French Demande							A	A	A	A	A	A	A	A	A	A	A	A	A	A	A	A	A	A	A	A	A	A	A	A	A	A	A	A
French Medicinal	A	A	A	A	A	A	A	A	A	A																								
French Addition to Medicinal	A	A	A	A	A	A	A	A	A	A	A																							
German (East)	R	R	R	R	R	R	R	R	R	R	R	R	R	R	R	R	R	R	R	A	A	A	A	A	A	A	A	A	A	A	A	A		
German Patenschrift	A	A	A	A	A	A	A	A	A	A	A	A	A	A	A	A	A	A	A	A	A	A	A	A	A	A	A	A	A	A	A	A	A	A
German Auslegeschrift	A	A	A	A	A	A	A	A	A	A	A	A	A	A	A	A	A	A	A	A	A	A	A	A	A	A	A	A						
German Offenlegungschrift								A	A	A	A	A	A	A	A	A	A	A	A	A	A	A	A	A	A	A	A	A	A	A	A	A	A	A
Hungarian		R	R	R	R	R	R	R	R	R	R																							
Indian	R	R	R	R	R	R	R	R	R	R	R	R	R	R	R	R	R	R	R	A	A	A	A	A	A	A	A	A	A	A	A	A	A	A
Israeli	R	R	R	R	R	R	R	R	R	R	R	R	R	R	R	A	A	A	A	A	A	A	A	A	A	A	A	A	A	A	A	A	A	A
Italian	R	R	R	R	R	R	R	R	R	R	R	R	R																					

Japanese Tokkyo Koho	R	R	R	R	R	R	R	R	R	R	R	R	R	A	A	A	A	A	A	A	A	A	A	A	A	A	A	A	A	A	A	A	A	A
Japanese Kokai Tokkyo Koho											R	R	R	A	A	A	A	A	A	A	A	A	A	A	A	A	A	A	A	A	A	A	A	A
Netherlands	R	R	A	A	A	R	R	R	R	R	R	R	R	A	A	A	A	A	A															
Netherlands Application		R	A	A	A	R	R	R	R	R	R	R	R	A	A	A	A	A	A	A	A	A	A	A	A	A	A	A	A	A	A	A	A	A
Norwegian	R	R	R	R	R	R	R	R	R	R	R	R	R	R	R	R	R	R	R	R	R	R	R	R	R	R	R	R	R	R	R	R	R	R
Polish	R	R	R	R	R	R	R	R	R	R	R	R	R	R	R	R	R	R	R	R	R	R	R	R	R	R	R	R	R	R	R	R	R	R
Romanian					R	R	R	R	R	R	R	R	R	R	R	A	A	A	A	A	A	A	A	A	A	A	A	A	A	A	A	A	A	A
South Africa						A	A	A	A	A	A	A	A	A	A	A	A	A	A	A	A	A	A	A	A	A	A	A	A	A	A	A	A	A
Spanish	R	R	R	R	R	R	R	R	R	R	R	R	R	R	R	R	R	R	R	R	R	R	R	R	R	R	R	R	R	R	R	R	R	R
Swedish	R	R	R	R	R	R	R	R	R	R	R	R	R	R	R	R	R	R	R	R	R	R	R	R	R	R	R	R	R	R	R	R	R	R
Swiss Patentschrift	R	R	R	R	R	R	R	R	R	R	R	R	R	R	R	A	A	A	A	A	A	A	A	A	A	A	A	A	A	A	A	A	A	A
Swiss Auslegeschrift																				A	A	A	A	A	A	A	A	A	A	A	A	A	A	A
United States	A	A	A	A	A	A	A	A	A	A	A	A	A	A	A	A	A	A	A	A	A	A	A	A	A	A	A	A	A	A	A	A	A	A
U.S. Patent Application												A	A	A	A	A	A	A	A	A	A	A	A	A	A	A	A	A	A	A	A	A	A	A
U.S. Published Patent Appl. (NTIS)												A	A	A	A	A	A	A	A	A	A	A	A	A	A	A	A	A	A	A	A	A	A	A
U.S. Reissue	A	A	A	A	A	A	A	A	A	A	A	A	A	A	A	A	A	A	A	A	A	A	A	A	A	A	A	A	A	A	A	A	A	A
U.S. Defensive Publication							A	A	A	A	A	A	A	A	A	A	A	A	A	A	A	A	A	A	A									
U.S. Statutory Invention Registration																								A	A	A	A	A	A	A	A	A	A	A
U.S.S.R.	R	R	R	R	R	R	R	R	R	R	R	R	R	R	R	A	A	A	A	A	A	A	A	A	A	A	A	A	A	A	A			
European Patent Application																	A	A	A	A	A	A	A	A	A	A	A	A	A	A	A	A	A	A
PCT International Application																	A	A	A	A	A	A	A	A	A	A	A	A	A	A	A	A	A	A

Key: A = all patents of chemical or chemical engineering interest (i.e., patents issued to nationals or nonnationals); R = chemical and chemical engineering patents issued to individuals or organizations resident in the granting country (i.e., nationals) or resident in countries not listed above.

Plans have been announced to cover patent documents from Latvia, Lithuania, Slovakia, and Moldova as soon as these become available. In addition to its coverage of conventional patent documents, CAS also covers publications such as *Research Disclosure* which is further discussed in Chapter 13.

[a]For coverage of patent documents in selective years from 1907–1963, see the Second Edition of *How to Find Chemical Information*, pages 70 and 71.

TABLE 6.8. Country of Issue of Patent Documents Abstracted[a] in *Chemical Abstracts* (as Percentage of Total Patents Abstracted)

	1960	1965	1970	1975	1980	1985	1990	1993	1994	1995	1996
Japan	5.6	6.0	11.8	40.4	43.4	54.9	56.7	60.3	61.0	57.6	57.3
United States	30.2	24.9	23.7	15.1	11.3	8.8	7.8	7.1	8.5	7.4	8.0
European Patent Organization	—	—	—	—	4.0	8.7	12.9	9.8	8.3	7.6	7.4
World Intellectual Property Organization	—	—	—	—	0.4	1.4	3.4	7.2	8.0	9.0	11.2
Federal Republic of Germany	23.6	8.3	23.5	20.3	12.1	7.7	4.8	3.9	4.6	4.1	4.8
USSR	4.8	6.9	7.7	8.8	9.7	5.3	4.7	4.9	2.8	6.7	4.2
People's Republic of China	—	—	—	—	—	—	1.6	1.6	1.8	3.2	1.7
France	3.3	16.0	15.3	4.0	1.7	2.1	0.7	0.7	0.7	0.6	0.5
Canada							0.2	0.9	1.1	0.8	0.7
United Kingdom	13.8	11.4	8.9	3.3	4.9	1.0	0.8	0.5	0.6	0.6	0.6
Poland	0.7	0.8	0.6	0.3	1.7	0.8	1.5	0.8	0.4	0.8	1.0
All others	18.0	25.7	8.5	7.8	10.8	9.3	4.9	2.3	2.2	1.6	2.6

[a]CAS abstracts the first disclosure it receives on a particular invention. Subsequently issued members of the family of patent documents on the invention are cited in the *CA* Patent Index (Patent Concordance prior to 1981).

Reprinted with the permission of Chemical Abstracts Service.

TABLE 6.9. Country of Issue of Equivalent Patents[a] Cited in *Chemical Abstracts* (as Percentage of Total Equivalent Patents Cited)

	1965	1970	1975	1980	1985	1990	1993	1994	1995	1996
Japan	1.3	0.9	19.8	22.4	30.4	28.9	32.7	30.6	29.9	26.3
European Patent Organization	—	—	—	3.3	15.0	20.6	17.6	20.1	20.5	18.6
United States	16.7	18.1	15.8	12.4	11.6	13.7	13.1	13.5	12.7	12.7
Canada	0.1	0.1	7.7	10.7	6.3	4.3	9.9	8.4	7.4	7.4
South Africa	—	4.5	4.0	2.5	3.0	2.2	1.5	0.1	1.9	5.4
Spain	—	—	—	—	—	0.5	2.3	5.9	4.0	5.0
Austria	—	—	3.2	1.4	3.0	3.3	4.1	3.0	1.5	4.1
People's Republic of China	—	—	—	—	—	4.3	1.9	2.8	3.2	3.7
World Intellectual Property Organization	—	—	0.4	0.4	1.2	1.2	1.8	2.8	2.7	2.7
Australia	—	—	0.1	0.2	2.4	4.3	3.7	3.5	3.0	2.5
Hungary	—	0.1	0.1	0.3	2.3	2.6	1.8	1.4	1.8	1.6
Federal Republic of Germany	14.2	32.0	2.2	3.2	1.3	1.5	1.4	1.3	1.5	1.5
Brazil	—	—	—	2.6	1.8	1.9	1.4	1.3	1.1	1.4
France	16.7	16.0	14.1	7.6	2.1	1.4	1.2	1.2	1.4	1.4
United Kingdom	27.3	23.6	16.6	13.7	6.6	2.4	1.0	1.2	1.3	1.1
Israel	—	—	—	—	—	0.7	1.0	0.3	1.0	0.7
Norway	—	—	—	0.4	0.9	1.1	0.8	0.3	1.3	0.6
All others	23.7	4.7	16.4	16.9	10.1	5.1	2.8	2.3	3.8	3.3

[a]CAS abstracts the first disclosure it receives on a particular invention. Subsequently issued members of the family of patent documents on the invention are cited in the *CA* Patent Index (Patent Concordance prior to 1981).

Reprinted with the permission of Chemical Abstracts Service.

c. Initiation of the CAS Registry System (1965)
d. Beginning of computerized production (1967)
e. Largely in-house abstracting (1970s)
f. Start of *CA SELECTS* (1976)
g. Beginning of *CAS ONLINE* and Document Delivery Service (1980)
h. Beginning of STN International (1983–1984)

With (1) the inclusion of abstracts in *CAS ONLINE* beginning in April, 1983; (2) the start of full-scale subject searching of *CA* in December 1983 through *CAS ONLINE*; and (3) the start of the STN international network for scientific and technical information with West Germany in September 1983, CAS was well along in its transition from a relatively passive organization to one of great vigor. CAS is now a full service organization. It offers a more complete range of chemical information services than any other organization.

An extensive chronology of key events in the history of CAS, including major policy changes, appears in Table 6.10.

6.9. COMPUTERIZATION

The most important change in the history of CAS is its adoption of computer-based production and, associated with that, development of computer-based tools for searching and current alerting. This trend began in the early 1960s, aided initially by a series of extensive grants from the National Science Foundation for CAS to act as a prototype for automation of secondary information services in the United States. Many of the major objectives were achieved in the 1970s, and a landmark was attained with initiation of *CAS ONLINE* in 1980.

When CAS officials first started to share their computerization goals with other members of the ACS (primarily through papers and open forums at ACS National Meetings, as well as in the pages of *Chemical & Engineering News*), there were many who listened with considerable skepticism, perhaps based on concern that these efforts represented an undue departure from established goals, or perhaps based on a lack of understanding. Opinions pro and con ran high, and there was spirited discussion at almost every ACS National Meeting.

In any event, the magnitude of the success ultimately achieved in this complex and difficult task cannot be overestimated. It required outstanding management, first-class research, and development of computer technology that has made CAS a leader far beyond the chemical community.

As evidence of this, in 1977 the EPA (Environmental Protection Agency) selected CAS as the organization best qualified to compile EPA's inventory of toxic substances. In 1983, the U.S. Patent Office stated its interest in employing the CAS approach for the automatic composition (input) and retrieval of information found in all U.S. patents, and CAS was subsequently chosen as part of the team to develop and install an automated system for processing patent applications.

TABLE 6.10. Milestones in the History of Chemical Abstracts Service (CAS)[a]

1907

Chemical Abstracts, Volume 1, Number 1, January 1, 1907

W. A. Noyes, Sr., Editor

Office in National Bureau of Standards, Washington, DC; later (in September) at University of Illinois, Urbana

Covered the world's publicly available chemical and chemical engineering literature and grouped the abstracts into 30 sections by subject matter

Covered all granted patents from Great Britain, France, Germany, and the United States [after 1949, the British documents covered were the examined applications; after 1979, unexamined applications were also covered. French unexamined applications began to be covered in 1969. German (West) unexamined applications began to be covered in 1970. United States unexamined applications available through NTIS began to be covered in 1974]

Volume *Author* and *Subject Indexes* introduced (Vol. 1) with uninverted chemical compound names as index headings

1908

List of Journals Abstracted published in *CA* Volume 2, Issue 1

1909

Austin M. Patterson, *CA* Editor (2nd)

CA office moved to The Ohio State University (OSU), Columbus, Ohio

1910

Patent coverage extended

Canada, granted

1912

Volume *Numerical Patent Index* introduced (Vol. 6) (discontinued in 1915)

1913

Patent coverage extended

Austria, granted

Denmark, Norway, Sweden, granted (after 1968, examined applications)

1914

John J. Miller, *CA* Editor (3rd) (3 months); E. J. Crane, Acting Editor (9 months)

1915

E. J. Crane, *CA* Editor (4th)

Year added as part of journal reference in abstract heading

1916

Volume 10 *Subject Index* Introduction provided with list of organic radicals and Index of Ring Systems

Volume 10 *Subject Index*—format changed so that each entry began on a separate line; systematic index nomenclature introduced for chemical compounds (inverted names); page fractions added to *CA* reference

1917

First Decennial Collective Index (Vols. 1–10) (1907–1916) included *Subject* and *Author Indexes* with Ring Index in *Subject Index* Introduction

1918

Patent coverage extended

Japan, examined applications (beginning in 1972, unexamined applications were also covered)

(Continued)

TABLE 6.10. ***(Continued)***

1920

Used *title only* abstracts for patent documents that were equivalent ro previously abstracted documents; cross-references to the abstracts were given (discontinued in 1961)

Volume *Formula Index* introduced (Vol. 14)

1921

Patent coverage extended

Belgium, granted

Netherlands, granted (in 1964 unexamined applications began to be covered)

Switzerland, granted (in 1982 examined applications were also covered)

1922

List of Periodical Abstracted by Chemical Abstracts included list of library holdings

1928

CA office moved to expanded facilities at new McPherson Chemistry Building on OSU campus

1929

Chemical Abstracts, Volume 23, Number 1 given a new cover with *CA* logo

Patent coverage extended

Australia, granted (after 1949 the examined applications were covered)

Hungary, granted (in 1971 examined applications were added and in 1972 unexamined applications)

Italy, granted (discontinued 1976)

Union of Soviet Socialist Republics, granted

1930

The Little CA house organ first issue (discontinued after issue 110 in 1966)

1934

CA Issue page in two columns; abstracts indexed by column number and numerical page fraction

1935

Volume *Numerical Patent Index* reintroduced (Vol. 29)

1937

1,000,000th abstract published

1940

The Ring Index published

1942

CA established an office in the Library of Congress to facilitate the acquisition of hard-to-get publications

1944

30-Year Collective Numberical Patent Index (1907–1936) published by Special Libraries Association

1945

Volume 39 *Subject Index* Introduction included Discussion of the Naming of Chemical Compounds for Indexing; published separately as *The Naming and Indexing of Chemical Coumpounds by Chemical Abstracts*

CA Sections expanded from 30 to 31

ACS Photocopy Service introduced via U.S. Department of Agriculture (discontinued in 1956)

TABLE 6.10. ***(Continued)***

1947

CA Issues—letters substituted for column fraction numbers

1948

Patent coverage expanded
  India, examined applications

1949

10-Year Collective Numerical Patent Index to CA (Vol. 31–40) (1937–1946) published

1951

27-Year Formula Index (Vols. 14–40) (1920–1946) published

1953

Patents coverage expanded
  Spain, granted

1955

CA located in building on OSU campus exclusively for *CA*
R&D Department established
2,000,000th abstract published

1956

Chemical Abstracts Service (CAS) established
ACS Member personal use pledge on *CA* subscription started (Member rate dropped in 1965)
E. J. Crane. first director of CAS
CAS on self-supporting basis
Patent coverage extended
Czechoslovakia, granted

1957

Fifth Decennial Index (Vols. 41–50) (1947–1956) contained for the first time *Numerical Patent Index* and *Formula Index*

1958

Dale B. Baker, CAS Director (2nd)
Charles L. Bernier, *CA* Editor (5th)
Patent coverage extended
  Israel, unexamined applications
CA Issues Patent Index introduced (Vol. 52)
Bibliography of Chemical Reviews (name changed to *Bibliography of Reviews in Chemistry* in 1962 and discontinued that year)

1959

Patent coverage extended
  East Germany, granted
First Index produced by Varitype-Fotolist
  (Vol. 53 *Formula Index*)

1960

Fouth floor added to CA building on OSU campus
Patent coverage extended
  Finland, granted (in 1968 the examined applications were covered)
  Poland, granted
The Ring Index, 2nd ed. published

(Continued)

TABLE 6.10. ***(Continued)***

1961

Fred A. Tate, Assistant Director and Acting Editor (6th)

CA Issues published every 2 weeks

See references for patent duplicates included in Numerical Patent Index (discontinued in 1963)

Diacritical marks dropped and special characters converted into English equivalents

Chemical Titles introduced; computer produced with Keyword-out-of-Context (KWOC) Index

CAS Russian Photocopy Service introduced (since 1980, under umbrella of CAS Document Delivery Service now Document Detective Service^{SM})

NewsCASter issued for informed staff

ACS purchase of 60-acre property along Olentangy River Road

1962

CA Sections expanded from 31 to 73

CA Issues have abstracts arranged by subsection; permanent section cross-references included

Biochemistry Sections published as separate *CA* Section Grouping

Volume *Subject Index*

assumption notes introduced; nomenclature and indexing formalized for coordination, inorganic, and organometallic compounds, addition compounds, charge-transfer complexes, salt of organic bases, ferrocene and its analogs; multiple entries (e.g., esters) provided

Formula Index included separate entries for each positional isomer

Sixth Collective Index (Vols. 51–55) (1957–1961); first five-year Collective Index

1963

CA Sections expanded from 73 to 74

Four additional *CA* Section Groupings offered

Organic Chemistry

Macromolecular

Applied Chemistry and Chemical Engineering

Physical and Analytical Chemistry [in 1982 changed to Physical, Inorganic, and Analytical Chemistry]

CA Issue Keyword Index introduced (Vol. 58)

Patent Concordance introduced (Vol. 58); elimination of *see* references in *Numerical Patent Index*

Issue number added as part of journal reference in abstract heading

CAS Open Forum presented at ACS National Meeting

As they became available, began including national or international patent classifications in the bibliographic information in patent document abstracts with the inclusion of U.S. Patent Classification Numbers in U.S. patent abstracts

3,000,000th abstract published

1964

CAS Advisory Board held first meeting (discontinued in 1982)

1965

CAS building located on Olentangy River Road

First Marketing Plan developed

Language designation added to abstract heading for journal articles

Chemical Titles available in computer readable form

Chemical Abstracts available on microfilm

CAS Registry System installed

TABLE 6.10. ***(Continued)***

Chemical-Biological Activities (*CBAC*) printed and computer-readable file introduced; this was the first service to include Registry Numbers (printed service discontinued in 1971)

1966

Desk-Top Analysis Tools produced

Board of Directors Committee on CAS started

Steroid Conjugates Bibliography published

1967

Russell J. Rowlett, Jr., CA Editor (7th)

CA Issues published weekly

CA Sections expanded from 74 to 80

CA Abstracts numbered sequentially with check digits

Patent coverage extended

Romania, granted

Polymer indexing begun for specific homopolymers and copolymers; Polymer Structural Repeating Unit (SRU) indexing

Nomenclature included (*E*)–(*Z*) and (*R*)–(*S*) systems

Subject Index included subdivision of large headings

B and R used in Volume *Subject Index* for references to books and reviews

First index to be computer produced (Vol. 66 *Formula Index*)

Hetero-Atom-in-Context (*HAIC*) *Index* introduced (Vol. 66) (discontinued in 1971)

Polymer Science and Technology (*POST*) printed and computer-readable file introduced (printed service discontinued in 1971)

CAStings first issue (discontinued in 1972)

Selective Dissemination of Information began experimentally; (name changed to *Basic Journal Abstracts* in 1968; discontinued at the end of 1972)

Polymer and Plastic Business Abstracts introduced (name changed to *Plastics Industry Notes* in 1968)

Seventh Collective Index (Vols. 56–65) (1962–1966) contained for the first time *Patent Concordance*

1968

CA Issue Indexes formatted in upper- and lowercase type font (Vol. 69)

Abstract heading changes

Author's name inverted (last name first)

Year of publication follows journal title

Patent coverage extended

South Africa, unexamined applications

Index Guide introduced (Vol. 69)

Registry II version of CAS Registry System installed

CA Condensates introduced (incorporated into *CA SEARCH* in 1978)

4,000,000th abstract published

1969

B used as prefix to references for book titles in Volume *Author Index*

Volume 71 *Subject and Formula Indexes* include Registry Numbers for the first time

Registry Number Index introduced (Vol. 71) (discontinued in 1971)

ACCESS introduced [name changes to *Chemical Abstracts Service Source Index* (*CASSI*) in 1970]

(Continued)

TABLE 6.10. *(Continued)*

Naming and Indexing of Chemical Compounds (8CI *Blue Book*) published
Provided illustrative Introduction to *CA* Issues
Patent Concordance available in computer-readable form
1,000,000th Registry Number assigned

1970

Findings-oriented abstracts introduced
Unified Document Analysis begun
CA Collective Indexes (1–7 Collective) available on microfilm
Chemical Abstracts Service Source Index (*CASSI*) available in computer-readable form
Subject Coverage and Arrangement of Abstracts by Section in CA available

1971

Expanded coverage of *Plastics Industry Notes* and changed name to *Chemical Industry Notes*
5,000,000th abstract published

1972

CA Sections 1–5, 35–46 were the first computer-produced abstracts with Registry Numbers and highlighting (italics) of chemical substance names
Indexing of chemical reactants in synthetic and preparative studies extended
Systematic nomenclature introduced for most substances
Structural diagrams included in *CA* Issues and Volume Indexes according to new guidelines
Volume *Subject Index* divided into *Chemical Substance Index* and *General Subject Index*
Eighth Collective Index (Vols. 66–76) (1967–1971) computer produced
CA on microfiche available experimentally (offered as a service in 1977)
Naming and Indexing of Chemical Substances for *Chemical Abstracts* during the Ninth Collective Period (1972–1976) (Section IV, Vol. 76 Index Guide)
CAS Report introduced
CA Integrated Subject File (*CAISF*) produced (discontinued at the end of 1978)
2,000,000th Registry Number assigned

1973

Second CAS building on Olentangy River Road site constructed
First Workbook *Using Chemical Abstract Service Printed Information Services* available
First workshop offered on how to use *CA*
CA Subject Index Alert (*CASIA*) produced (incorporated into *CA SEARCH* in 1978)
Chemical Industry Notes (*CIN*) available in computer-readable form

1974

CA Sections on energy (Sections 51 and 52) reorganized
Safety indexing emphasized
Bibliographic Guide for Editors and Authors (*BIOSIS/CAS/Ei*) published
Registry III version of CAS Registry System installed
Registry Handbook—Number Section published
First foreign CAS workshop presented in West Germany
6,000,000th abstract published

1975

Special Bibliographies in *Chemical Titles* published (precursors of *CA SELECTS*)
CA Issues completely computer produced
Registry Numbers in abstracts extended to 34 Sections
CAS Editorial Advisory Board established (discontinued in 1982)

TABLE 6.10. ***(Continued)***

Computer-readable Abstract Text Services offered (discontinued in 1981)
  Ecology and Environment
  Food and Agricultural Chemistry
  Energy
  Materials
CBAC expanded with three additional *CA* Sections
Volume Index entries added to *CBAC* and *POST*
CAS assumed responsibility for ASTM CODEN assignment
CA Collective Indexes (1–8 Collectives) available on microfiche
First stand-alone Workbook *CAS Printed Access Tools* published
3,00,000th Registry Number assigned

1976

Volume Number included as prefix to *CA* abstract number in Issues
CA SELECTS introduced
Hierarchies of *General Subject Index* headings included in *Index Guide*
Patent coverage extended
  Brazil, unexamined applications
Registry Numbers in abstracts extended to 40 *CA* Sections

1977

Volume Formula Index coverage extended to unspecific derivatives
International CODEN Directory published
Parent Compound Handbook published (discontinued in 1983)
7,000,000th abstract published
4,000,000th Registry Number assigned

1978

CA Issue Keyword Index format includes indentures
REG/CAN FILE available (discontinued in 1981)
CA BIBLIO FILE available (discontinued in 1981)
CA SEARCH introduced
"P" for patent in Issue Keyword and Author Indexes
Registry Handbook—Common Names available in microfilm and microfiche
CAS Document Copy Service available (July–December); now part of Document Detective Service

1979

Indexing of chemical substances from patent claims (unsupported by examples, data, etc.) initiated
Patent coverage extended
  European Patent Applications
  Patent Cooperation Treaty International Applications
Individual Search Services (ISS) bulletins introduced
CAS Updates introduced
8,000,000th abstract published
CAS Library Document Copy Service (noncopyrighted documents) available; now part of Document Delivery Service

(Continued)

TABLE 6.10. ***(Continued)***

1980
CAS ONLINE, a chemical structure search and display system, introduced
Search Assistance Desk established
5,000,000th Registry Number assigned
CAS Document Delivery Service introduced

1981
Volume and Issue *Patent Index* introduced (combined *Numerical Patent Index* and *Patent Concordance*)
BIOSIS/CAS SELECTS published

1982
David W. Weisgerber, *CA* Editor (8th)
CA Sections reorganized
Tenth Collective Index published (Vols. 86–95) (1977–1982); includes *Patent Index* for the first time
IUPAC names given for new and hypothetical elements
9,000,000th abstract published

1983
Abstracts available on *CAS ONLINE*
CAS User Councils formed
6,000,000th Registry Number assigned
CA File component (bibliographic and indexing information) of *CAS ONLINE* introduced; STN International, the Scientific and Technical Information Network, established (STN became operational in May 1984)

1984
10,000,000th abstract published by *CA*
CAOLD File component (pre-1967 references to chemical substances) of *CAS ONLINE* introduced
Ring Systems Handbook published

1985
Keyword-Out-of-Context (KWOC) Index to *Chemical Abstract Service Source Index* (*CASSI*) published, available on microfiche
7,000,000th Registry Number assigned
Patent coverage extended
People's Republic of China (first abstracts appear in 1986)
Abstract texts searchable experimentally via *CAS ONLINE*
Markush searching for specific chemicals made possible in *CAS ONLINE*

1986
Ronald L. Wigington, CAS Director (3rd)
CAS BioTech Updates introduced
STN Service Center established in Tokyo
11,000,000th abstract published
8,000,000th Registry Number assigned

1987
U.S. Patent 4,642,762 issued to ACS for Markush storage/retrieval method
STN Express introduced
STN Mentor introduced (dormant as of 1996)

TABLE 6.10. ***(Continued)***

Polymer registration expanded to include alternating, block, graft distinction
Superconductor Update introduced (discontinued in 1990)

1988

AgPat and *PharmPat* introduced on STN (coverage began in 1987; discontinued in 1989)
CASREACT File introduced on STN (coverage began in 1985)
Capreviews File introduced on STN
Sequence seaching available
12,000,000th abstract published
9,000,000th Registry Number assigned
Began coverage of errata and retractions from primary journals

1989

STN Help Desk established
BIOSIS/CAS Selects discontinued; 15 of 35 topics converted to *CA Selects*

1990

MARPAT File introduced on STN (coverage began in 1988)
Protein and Peptide Sequence Search introduced on STN
13,000,000th abstract published
Completed the backfile registration project which added almost 700,000 substances indexed during the sixth and seventh collective indexing periods (1957–1966) to the Registry database
Alloy search and display enhancements introduced in Registry File on STN
10,000,000th Registry Number assigned

1991

Clayton F. Callis, CAS ad interim Director (4th)
CAS Governing Board established
11,000,000th Registry Number assigned

1992

Robert J. Massie, CAS Director (5th)
Nucleic Acid Sequence introduced on STN
Sterochemical parity data began to be introduced into the Registry database structural records
Stereo Display introduced in Registry File on STN
CAST-3D File of three-dimensional structures
14,000,000th abstract published
CAS assumed from API ownership and responsibility for producing the *CHEMLIST* database of regulated chemicals (available on STN)

1993

Twelfth Collective Index (Vols. 106–115) (1987–1991) available on CD-ROM
Twelfth Collective Index (Vols. 106–115) (1987–1991) abstracts available on CD-ROM
12,000,000th Registry Number assigned
Polymer classs terms introduced in Registry File on STN
Advance ACS Abstracts introduced (joint ACS Pub/CAS publication); discontinued end of 1997
CASurveyor introduced on CD-ROM with five topics
Sterosearch introduced in Registry File on STN
Functional group searching capability added to CASREACT
MARPAT Previews File introduced on STN
CASLINK introduced on STN
First computer-generated *CA* chemical substance index names added to the Registry database

(Continued)

TABLE 6.10. ***(Continued)***

1994

Four Online Teaching Partnerships dedicated (Northwestern, Rutgers, SUNY Buffalo, University of California at San Diego)

15,000,000th abstract published

13,000,000th Registry Number assigned

CASurveyor: Regulatory Compliance introduced on CD-ROM

CA index text modifications changed from an inverted, highly structured form to a natural-language phrase

Preparation "P" suffix to CAS Registry Number in *CA* File converted from algorithmic assignment to intellectual assignment; also added as new data element to *CA* Search File

Extended registration to many siloxane polymers previously indexed only as General Subjects

1995

CAplus introduced on STN

SciFinder, a client/server application for accessing CAS databases and ACS primary journals introduced

Introduced preparation flags "*pr*" in printed Chemical Substance Index (V.121)

Began coverage of electronic-only source documents (i.e., electronic journals, electronic conferences)

CASSI on CD-ROM introduced

Role Indicators introduced for chemical substances in CAS Files on STN International

Index of Ring Systems (part of *CA* Formula Index) discontinued

CA Selects Plus introduced

100,000th ring system registered

14,000,000th Registry Number assigned

16,000,000th abstract published

1996

CA available on CD-ROM

15,000,000th Registry Number assigned

STN Easy available on the Internet

Chemical Patents Plus available on Internet

1997

New version of *SciFinder* introduced with substructure search module

16,000,000th CAS Registry Number assigned (February)

17,000,000th CAS Registy Number assigned (October)

17,000,000th abstract published

SciFinder Scholar available to academic community

Thirteenth Collective Index (1992–1996) available in both printed and CD-ROM forms

CA Student Edition available to academic community online through OCLC

ChemPort plans announced; will make full texts of ACS and other chemical journal available on Internet with link to CAS online files

---

[a]Some of the coverage changes (changes in policy), such as extension of coverage to conference proceedings, government reports, and the like, cannot be pinpointed to date implemented because they occurred gradually, and without much fanfare, as did expansion of coverage of areas such as toxicology. Many indexing and nomenclature changes also took place gradually. In many cases, the type and tone of literature covered molded indexing policies rather than specific decisions on specific dates.

TABLE 6.11. CAS R&D Directors and When They Held Office

Year	Name
1955–1959	Karl F. Heumann
1959–1963	G. Malcolm Dyson
1963–1967	Kenneth H. Zabriske, Jr.
1967–1968	Robert W. White[a]
1968–1984	Ronald L. Wigington
1984–1994	Nick A. Farmer
1995–	Robert L. Swann[b]

[a]Acting director.
[b]Robert L. Swann's formal title is Director, Information Systems.

In retrospect, it appears that the CAS thrust toward computerization started with formation of a research and development office in 1955 under the leadership of Karl F. Heumann. Subsequent directors included the brilliant British chemist, G. Malcolm Dyson, Kenneth H. Zabriskie, Jr., Ronald L. Wigington, and Nick A. Farmer. The present R&D Director is Robert L. Swann, who was appointed to that position in 1995. Table 6.11 summarizes these changes.

G. Malcolm Dyson, along with Hans Peter Luhn of International Business Machines Corporation, conceived the idea of *Chemical Titles* (*CT*) in April 1959. This was to be the first computer-readable service produced by CAS. The initial sample copy of *CT* was produced in 1960, and full-scale production was started in 1961. A novel computer-produced keyword-in-context (KWIC) index appeared in each issue.

In 1965, the second computer-based CAS service began to appear. This was *Chemical–Biological Activities* (*CBAC*), which contained abstracts, index entries, and full bibliographic information.

In the drive toward CAS computerization, the appointment of Fred A. Tate as Assistance Director and Acting Editor in 1961 provided aggressive leadership and, in addition, much of the necessary technical interface with the outside world.

The first successful demonstration of use of CAS tapes outside CAS was achieved by Robert E. Maizell (Olin Corporation) and Charles N. Rice (Eli Lilly and Company) in 1962–1963 (1). They invented, developed, and implemented a method to use *CT* tapes for alerting of individual chemists or groups of chemists. This was the precursor for all computer-based alerting and searching in chemistry as practiced today. At about the same time, Purdue University ran sample *CT* tapes for students and faculty.

In 1965, CAS offered its first batch-search service based on the *CT* file. In 1967, the first experimental computer information center using CAS files was established at the University of Nottingham by The Chemical Society (London). In 1973, the first public online services based on CAS files were offered by System Development Corporation, Informatics, Inc., and Science Information Associates. A few years later, in 1980, CAS started its own online service *CAS ONLINE,* and, in 1984, the STN information network become operational.

TABLE 6.12. CAS Computer File Availability—Batch and Online: Selected Key Dates

1962–1963	First experiments by Maizell and Rice using batch searches of *CT* for current awareness
1965	CAS offered first batch-search service based on *CT* file
1967	First experimental computer information center using CAS files established at the University of Nottingham by the Chemical Society (London)
1973	First online services based on CAS files offered by System Development Corporation (*CA Condensates* file), Informatics Inc., (NLM *TOXLINE* file, which includes *Chemical-Biological Activities*) and Science Information Associates (both *CA Condensates* and *TOXLINE*); Lockheed *DIALOG*® was licenced to offer online searches in 1974, and *Bibliographic Retrieval Services* (*BRS*®) in 1976.
1980	Initiation of *CAS ONLINE*
1984	STN International operational
1997	*STN Easy* available on Internet

One of the many benefits of computerization to chemists and other users of *CA* has been significant improvement in the promptness and appearance of the indexes. For example, the Collective Index for 1962–1966 appeared 44 months after completion of the collective period. For 1977–1981, this was reduced to just 14 months, an improvement of an astonishing 30 months, even though the size of the Collective index had more than tripled. Because of availability of CAS computerized input to online search files, *CA* index entries are now available online simultaneously with, or earlier than, publication of the printed abstract issues. Evolution of CAS computer files is summarized in Table 6.12.

Online searching of *CA* files has almost totally transformed the information seeking patterns and habits of thousands of scientists and engineers throughout the world. This capability gives chemists the potential of outstanding searching power and speed unheard of just 10–15 years ago.

6.10. CHEMICAL ABSTRACTS SERVICE CHEMICAL REGISTRY SYSTEM

G. Malcolm Dyson, Director of Research at CAS during 1959–1963, originally conceived the idea of the Registry. He brought his structurally encoded files of boron and fluorine compounds to CAS, and this was eventually greatly modified and expanded into the present Registry. In 1965, Harry L. Morgan of the CAS staff, building on outstanding work supplied by D. J. Gluck of the du Pont Company (2), perfected an algorithm for generating a unique and unambiguous computer representation of a chemical substance, that is, a unique connection table (3). This algorithm is the foundation of the CAS Chemical Registry System.

Initiation of the CAS Chemical Registry System in 1965 marked a quantum leap forward not only in the development of CAS but also in the development of all chemical information and related tools. CAS Registry Numbers are now widely used in a number of other publications and are recognized as the most uniquely accurate and unambiguous manner of identifying a chemical structure, other than by drawing and communicating the full chemical structure.

The principal component of the Registry structure record is a topological description in the form of a connection table listing of atoms and bonds that make up a substance's two-dimensional structure. Other portions of the record define known stereochemical characteristics of the molecule, identify any labeled atoms or atoms with abnormal valences, and indicate presence of charges in salts and complexes. Table 6.13 shows the growth of the Registry System over the years.

Efforts to develop a substructure search system started in the late 1960s. The interest and imagination of chemists had been stimulated by presentations on substructure searching at the open forums sponsored by CAS at ACS National Meetings. Substructure searching via nomenclature was first demonstrated as soon as computer-readable tapes containing indexed and named chemical substances became available. A full-fledged structure and substructure search system based on the CAS Registry file, *CAS ONLINE* (now known as the *Registry File* on the STN International Network), was introduced in 1980.

CAS involvement with substructure search systems began in the late 1960s, with the development of an experimental system to search Registry II structure files. This system had screen and atom-by-atom search capabilities comparable to those of the present *CAS ONLINE* system but was restricted by the computer technology of the time; it was a tape-based batch mode system, with screen and atom-by-atom search manually encoded by the searcher and with answers limited to Registry Numbers. While the system did show the feasibility of substructure search procedures built around the CAS Registry structure files, it was hampered by lack of sufficiently powerful computer hardware and by the inability to provide structure diagrams for answers. As a consequence, the system was never promoted as a CAS service, although it was released on an experimental basis to several outside organizations. It was also used by the National Cancer Institute in a chemical information system to support their biological screening programs.

A few years later, several Swiss chemical firms—Ciba Ltd., J. R. Geigy Ltd., F. Hoffmann–LaRoche & Co., Ltd., and Sandoz Ltd.—decided to jointly develop a computer system for chemical information based on the CAS Registry System and formed the Basel Information Center for Chemistry (BASIC). Their system used the experimental CAS substructure search system to search a CAS Registry structure file licensed from CAS as well as private structure files built and maintained by a system based on CAS Registry II programs. The BASIC group made a number of improvements to the original CAS search system, developing several new screen types and revising the screen dictionary; these improvements reduced the costs of search screen generation and (by providing better screen-out) atom-by-atom searching, the two time-consuming aspects of substructure search. At the same time, BASIC upgraded the private structure file aspects of their system with the CAS Registry III programs.

In the late 1970s, CAS reexamined the possibilities of substructure search as a CAS service. While the earlier work discussed above had demonstrated the feasibility of using an initial screen search followed by atom-by-atom search, it was clear that the large size of the file (then already over 5 million substances) would present problems. The inverted file organization usually used for online searching could be expected to lead to unacceptably slow response times, due to the very large file size. Some of the screens which would normally be generated from a query would be assigned to very large numbers of substances, leading to the necessity to intersect very large lists of potential retrievals.

The result of the investigation was a new approach to large-file searching. [Reprinted with permission from the American Chemical Society (P. G. Dittmar et al., "The CAS ONLINE Search System") published in the *Journal of Chemical Information and Computer Sciences* **23,** 93–102 (1983).]

Content of records in the online *Registry* file includes Registry Numbers (including current, alternate, and deleted numbers); *CA* index names (present and previous); synonyms (including many trade names); molecular formulas; other files in STN that contain the Registry Numbers in question; structure diagrams; the number of references in *CA* and *CAOLD*; and the references, abstracts, and index entries for the 10 most recent *CA* references. One of the most powerful features of this STN file is that substances within the file may be searched for by structure or substructure. The searcher can, of course, decide in each case whether all of the file record, or just selected parts, will be searched and displayed.

The current scope of the file is from 1957 to the present for chemical substances of all types in the literature (journal articles, patents, conference proceedings, and substances on regulatory lists). The substances covered include: alloys, biosequences, coordination compounds, minerals, mixtures, polymers, and salts. A more detailed description of what CAS considers to be chemical substances, and that it indexes and registers, is as follows:

> Chemical elements and their isotopes, including hypothetical elements; chemical compounds and their stereoisomers, including fully defined and incompletely defined derivatives; ions; free radicals; specific antibiotics, enzymes, hormones, polypeptides, polysaccharides, and polynucleotides; polymers of specific compounds; mixtures resulting from an intentional admixing (prior to use) of individual components; elementary particles, including certain class designations; alloys of specific metals; specific minerals (as distinct from rocks); and substances possessing alphanumeric and trade-name designations (except a very few that have come to be regarded as generic).

Note the provision for inclusion of hypothetical elements, ions, and incompletely defined derivatives, all of which can be the subject of research reported in the litera-

TABLE 6.13. Growth of the CAS Chemical Registry System

Year	Substances Registered	Substances on File at Year End
1965	211,934	211,934
1966	313,763	525,697
1967	270,782	769,479
1968	230,321	1,026,800
1969	287,048	1,313,848
1970	288,085	1,601,933
1971	351,514	1,953,447
1972	277,563	2,231,010
1973	437,202[a]	2,668,212
1974	319,808	2,988,020
1975	372,492	3,360,512
1976	347,515	3,708,027
1977	369,676	4,077,703
1978	364,226	4,441,929
1979	346,062	4,787,991
1980	353,881	5,141,872
1981	424,230	5,566,102
1982	361,706	5,927,808
1983	418,905	6,346,713
1994	563,390[b]	6,910,103
1985	544,618[b]	7,454,721
1986	628,966[b]	8,083,687
1987	610,480[b]	8,694,167
1988	602,465[b]	9,296,632
1989	615,987[b]	9,912,619
1990	663,342[b]	10,575,961
1991	684,252	11,260,213
1992	690,313	11,950,526
1993	680,230	12,630,756
1994	777,212	13,407,968
1995	1,186,334	14,594,302
1996	1,269,246	15,863,548[c]

[a]Registry III implementation added 77,650 substances.

[b]Input of substances indexed prior to 1965 added 140,760 in 1984, 151,431 substances in 1985, 165,705 substances in 1986, 119,888 substances in 1987, 51,253 substances in 1988, 44,112 substances in 1989, and 14,550 substances in 1990.

[c]In 1997, the number of substances totalled over 17 million.

Reprinted with the permission of Chemical Abstracts Service.

ture. This is a composite statement based on paragraphs in the *CA Chemical Substance Index* Introduction, the *CA Index Guide,* and several published papers on the CAS Registry System. The CAS Registry System is discussed further in Chapters 7 and 10.

6.11. *CA SELECTS*

In 1976 CAS began publication of what has proved to be the extremely popular *CA Selects* series. This grew to a total of 231 different topics available in 1996 (see Table 4.1). The total for 1996 is up from 164 topics in 1986.

In 1995, CAS introduced *CA Selects Plus,* which provides more current references than *CA Selects.* This new product is based on the *CAPlus* online database. Thus, the new series includes not only fully indexed references from CA, but also essentially cover-to-cover citations from more than 1350 chemistry and chemistry-related "core" journals within one week of receipt in Columbus and before these appear in the printed *CA.* These citations include all research and other papers, as well as letters to the editor, product reviews, and other items not ordinarily covered in *Chemical Abstracts.* A total of 17 topics is currently available in this important new series, with further topics planned. In addition, CAS offers 15 different *CAS BioTech Updates* topics.

Anyone walking through the facilities of almost any modern chemical R&D organization will almost surely see copies of *CA Selects* in both laboratories and office areas. Not even in its most recent heydays did the full *CA* enjoy this kind of personal use and identification with individual chemists.

Reasons for this popularity are many. The major reason is probably the handy grouping in individual, rather modest-sized, easy-to-read booklets of recent abstracts pertaining to specific, limited areas of chemistry that are the subjects of the most intense current research of commercial interest. CAS staff have worked hard at keeping the list of *CA Selects* topics flexible and consistent with the changing interests of chemists, and they both add and delete topics accordingly.

The novelty of *CA Selects* is that it transcends *CA* section boundaries; an abstract may be included in more than one topic, whether the emphasis of the document is primary or secondary to the topic. *CA Selects* focuses on very specialized areas. On the other hand, *CA* sections reflect subdisciplines of chemistry, and an abstract is placed in only one section, based on primary emphasis of the original document, with cross-reference to related sections. For a further discussion of *CA Selects,* see Chapter 4.

6.12. COOPERATIVE ACTIVITIES

CAS cooperation with other abstracting and indexing services is important because this reflects the increasingly interdisciplinary trends in science and technology and because of the potential of improvements in coverage of some selected areas. Cooperation frequently takes place within the framework of activities of such organizations as the International Council for Scientific and Technical Information (ICSTI) and the National Federation of Abstracting and Information Services (NFAIS). Details on cooperative activities are given in the sections that follow:

A. National Library of Medicine

In 1966, CAS developed a special registry of chemicals associated with foods, drugs, pesticides, and cosmetics under the joint sponsorship of the National Library of Medicine (NLM) and the National Science Foundation. This project established an experimental link between the CAS Chemical Registry and NLM's online *MEDLARS* service. Subsequently, CAS began providing information for certain NLM online files and continues to do so.

B. The Chemical Society; The Royal Society of Chemistry

An agreement was signed between ACS and The Chemical Society (now The Royal Society of Chemistry) in 1969 under which the British society provided CAS with abstracts and index entries for some British scientific papers and patents and marketed CAS publications and services in the United Kingdom and Ireland. Except for the marketing efforts, this agreement has been discontinued.

Earlier, The Chemical Society's research unit had been one of the first organizations to work with CAS's experimental computer-readable files and developed some of the first search techniques and programs for these files.

C. Japan Association for International Chemical Information

In 1977, an agreement was signed with the Japan Association for International Chemical Information (JAICI) under which the Japanese organization took over the marketing of CAS publications and services in Japan and was permitted to develop publications and services of its own from the CAS database for use in Japan.

JAICI provides CAS with abstracts and indexing for many Japanese patents and for about one-third of Japanese journals.

D. FIZ Karlsruhe

In 1984 the ACS initiated cooperative efforts with the West German organization Fachinformationszentrum Energie, Physik, Mathematik GmbH (FIZ Karlsruhe). The cooperative online system through which files are available in North America, Europe, and Japan is STN (Scientific and Technical Information Network).

Cooperative arrangements, such as those just described, permit CAS to more effectively keep up with literature from other countries and to add new databases or files not previously available. See also Chapter 8 for further descriptions of some of the abstracting services mentioned in this Chapter.

E. IUPAC

CAS staff members are active in IUPAC. One of the IUPAC groups in which CAS staff is active is the Committee on Print and Electronic Publications. Members of this committee are concerned with standards and guidelines, with a special focus on IUPAC data. One of the standards being considered relates to a format (called CXF) for chemical structures so as to facilitate interchange of chemical information between software packages, such as between ISIS and STN Express. Another topic of interest is a standard (called *Chemical MIME*) for incorporating additional information, such as chemical structures, into electronic mail messages. As might be expected, CAS staff are also very active in IUPAC nomenclature activities.

6.13. PRICING AND MARKETING

The changes in the pricing structure of *CA* (see Table 6.14) are fascinating. Impinging on it have been many complex factors, including increases in the cost of doing business, and in the number of documents covered and number of pages printed, as well as increasingly heavy reliance on computers as the principal tools used by the staff for producing abstracts and indexes. Another factor is the widespread use of *CA* online.

When *CA* was first published in 1907 publication costs were subsidized from dues paid by ACS members, and members who wished to, received it free of charge. The basic subscription cost to nonmembers was $6.00. In 1956, it was decided that CAS was to be self-supporting, and in 1966 CAS dropped the individual member rate entirely. The subscription cost for 1997 was $18,900, over double the 1986 price of $9,200 per year. There are reduced prices for universities, small colleges, and least-developed countries. A decision was made by the ACS Board of Directors in 1978 that more than half of CAS future revenues must come from sources other than the printed *CA* by 1983. This was achieved. Thus, the full printed *CA* has evolved from what was once a very affordable tool for individual chemists to one that is now quite frankly and intentionally priced at the institutional level.

The dichotomy is that, at the same time, CAS does offer a number of newer, intensely personal tools, most notably *CA SELECTS,* which is available to individual ACS members at only $70/year.

Without the changes in pricing structure just described, CAS would have been unable to continue its outstanding coverage of worldwide chemical literature, and indeed might well have been forced to sharply curtail or even to go out of business, a fate that was met by the U.K. counterpart of *CA, British Abstracts* in 1953, by its German counterpart, *Chemisches Zentralblatt* in 1969, and by the French *Bulletin Signalétique* at the end of 1983.

In any event, time has clearly demonstrated that high-quality chemical information, delivered in large volume and on a prompt basis, is expensive. Most chemists seem to have fully accepted this, but it remains to be demonstrated conclusively how the Internet will affect user expectations. The Internet is discussed more fully in Chapter 10.

No discussion of *CA* subscription costs would be complete without mention of the CA Section Groupings that include keyword indexes. In 1997, these were priced at $470 per year each to individual ACS members (limit of two different Groupings per member) and $2340 to nonmembers who were not subscribers to *CA*.

Related to changes in pricing policies at CAS is the establishment of a strong Marketing Division (the current Director is Suzan Brown) in 1978. It has become clear

TABLE 6.14. Subscription Price[a] and Number of Subscriptions to Chemical Abstracts

	Annual Subscription Price[a]			Number of Subscriptions	
Year	Base ($)	College or University ($)	ACS Member ($)	Total	ACS Member
1934	6	—	6	11,883	9,768
1935	6	—	6	12,020	9,766
1936	6	—	6	12,463	10,007
1937	6	—	6	13,098	14,480
1938	6	—	6	13,408	10,595
1939	6	—	6	13,810	10,703
1940	12	—	6	13,934	10,780
1941	12	—	6	14,249	11,006
1942	12	—	6	13,605	11,228
1943	12	—	6	14,651	12,067
1944	12	—	6	15,428	12,463
1945	12	—	6	16,729	13,138
1946	12	—	6	19,428	14,622
1947	12	—	6	21,524	15,758
1948	15	—	7	21,705	15,586
1949	15	—	7	22,675	15,919
1950[b]	20	—	10	21,627	14,631
1951[b]	60	—	15	19,232	12,701
1952[b]	60	—	15	19,174	11,639
1953[b]	60	—	15	19,361	11,487
1954[b]	60	—	15	19,510	11,283
1955[b]	60	—	15	19,475	11,023
1956	350	80	20	16,907	11,688
1957	350	80	20	17,216	11,172
1958	350	80	20	17,301	11,036
1959	350	80	20	17,351	11,133
1960	570	150	32	16,107	10,234
1961	925	200	40	15,183	9,747
1962	925	200	40	14,271	8,901
1963	1,000	500	500	6,758	1,458
1964	1,000	500	500	6,802	1,426
1965	1,200	700	700	6,808	1,378
1966	1,200	700	—[c]	6,672	—[c]
1967	1,200	700	—	6,653	—
1968	1,550	1,050	—	6,451	—
1969	1,550	1,050	—	6,401	—
1970	1,950	1,450	—	NA[e]	—
1971	1,950	1,450	—	NA	—
1972	2,400	1,900	—	NA	—
1973	2,400	1,900	—	NA	—
1974	2,900	2,400	—	NA	—

(Continued)

TABLE 6.14. *(Continued)*

1975	2,900	2,400	—	NA	—
1976	3,500	3,000	—	NA	—
1977	3,500	3,000	—	NA	—
1978	4,200	3,700	—	NA	—
1979	4,200	3,700	—	NA	—
1980	5,000	4,500[d]	—	NA	—
1981	5,500	5,000[d]	—	NA	—
1982	6,200	5,200[d]	—	NA	—
1983	6,800	5,800[d]	—	NA	—
1984	7,500	6,400[d]	—	NA	—
1985	8,500	7,200[d]	—	NA	—
1986	9,200	7,800[d]	—	NA	—
1996	19,100	16,920[d]	—	NA	—
1997	18,900	17,010[d]	—	NA	—

[a]The nonmember subscription price of *CA* was $6 per year form 1907 through 1939. Prior to 1934, all ACS members who wanted *CA* received it free of charge. *CA* publication expense was subsidized from ACS member dues from 1907 through 1953. Subscription totals prior to 1934 are not available and since 1970 have not been made public.
[b]*CA* publication expense was subsidized in part from ACS member dues through 1955.
[c]Special subscription price for ACS members was eliminated in 1966.
[d]Since 1980 smaller U.S. colleges have been eligible for substantial grants toward *CA* subscriptions.
[e]NA = not available from CAS (proprietary).

that establishment of this division, although originally criticized by some as too commercial, was necessary as CAS products became more technologically sophisticated, as they continued to grow in size, and as competition and economic pressures became more severe.

Achievements of the marketing staff include development of a complete battery of training programs and manuals. This division played an important role in the development of a revised fee structure for CAS products. Most importantly, marketing has determined the need for new and improved products, especially those that are computer-based, as mandated by the ACS policy decision of 1978 that the primary sources of future revenues are to be publications and services other than the printed *Chemical Abstracts.*

6.14. OTHER CHEMICAL INFORMATION SERVICES

Some aspects of CAS can be better appreciated when considered in the perspective of other major chemical information services. CAS is the only remaining major international literature and patent service of its type, with the exception of *Referativnyi Zhurnal, Khimiya,* published in the Soviet Union and now Russia since 1953. The French *Bulletin Signalétique,* founded in 1940, was discontinued at the end of 1983. *British Abstracts* (*BA*), originally founded in 1871, ceased publication as such in

1953, and the venerable *Chemisches Zentralblatt* (*CZ*), founded in 1830, was discontinued as such in 1969. Both *BA* and *CZ* underwent a number of name changes over the years, and both, as well as the French publication, have been succeeded by publications with new names in greatly watered down, modified form. See also Chapter 8.

Discontinuance of the original German and British services has avoided much unnecessary duplication of *CA*. Both are, however, still valuable today for years not covered by *CA*, or for those subject areas or types of documents where they may complement *CA*. For example, *CZ* is noted for its coverage of some earlier patents, especially German patents, and for the detail of its abstracts compared to those of *CA*.

The only other major chemical abstracting and indexing service published in the United States is *Index Chemicus*, published in the United States since 1960 by the *Institute for Scientific Information* (*ISI*) in Philadelphia. This publication provides only partial coverage of journals and does not cover patents, but it does have a number of attractive and unique features. *ISI* also offers a broad range of other important chemical information services. See also Chapter 8.

A leading British chemical information service is the battery of products offered by Derwent Information, Ltd. This worldwide patent information service has an outstanding worldwide reputation, especially among industrial chemists and engineers. See also Chapter 13. The Royal Society of Chemistry is another leading British chemical information source. It publishes a series of specialized abstracting and indexing services primarily in such fields as analytical chemistry and safety. See also Chapters 8 and 18.

6.15. OTHER REMARKS

CAS has emerged surely the most dominant factor in the chemical information industry worldwide. As with any other situation of comparable reputation, power, and influence, there are those who are somewhat uneasy about the outlook for the future. For example, will there continue to be room for small innovators in this field—can they hope to get into the arena now? Can (or should) additional healthy competition be nurtured to provide novel and diverse approaches? Would additional chemical information sources, if not really needed, simply dilute talent and financial resources and confuse unnecessarily? In 1997, the answers that emerged to these questions seem to point to increased innovation and a plethora of new chemical information products from both small and large sources as spurred by advances in personal computer, CD-ROM, and modern technology, as well as by the growth of the Internet as a vehicle. The net result, however, could be some confusion for many chemists as to which are the best or most cost effective sources to use in specific situations.

Other questions include the following. What kind of influence can users expect to have in the future on shaping of CAS programs and policies? Now that chemical information has become so complex and sophisticated, how important are grass roots opinions—perhaps what count most are opinions of a relatively few who make the effort to become highly informed. How can average chemists most effectively express their opinions and recommendations on handling of literature and patents that they and their colleagues generate and use? How can CAS best be persuaded to do something promptly if a significant segment of the chemical community thinks it should be done? Have all of the major advances at CAS made chemical information access significantly easier, or "merely" helped chemists keep their heads above the waters of the flood of chemical publications?

These are complex and controversial issues. ACS leadership and CAS staff welcome chemical community input on all of these questions. With the discontinuance of the CAS user councils that had existed for a number of years, the ACS Committee on Chemical Abstracts Service has been acting as a liaison between users of CAS products and ACS/CAS management. In the spring of 1996, the formation of a new Governing Board for Publishing was announced by ACS; direction of CAS will be among its functions. In addition, online users often leave helpful comments with the STN Help Desk, and numerous formal and informal contacts with users and CAS staff occur at ACS national meetings and other meetings.

REFERENCES

1. R. R. Freeman, J. T. Godfrey, R. E. Maizell, C. N. Rice, and W. H. Shepherd, "Automatic Preparation of Selected Title Lists for Current Awareness Services and or Annual Summaries," *J. Chem. Doc.*, **4,** 107–112 (1964).
2. D. J. Gluck, "A Chemical Structure Storage and Search System Developed at du Pont," *J. Chem. Doc,* **5,** 43–51 (1965).
3. H. L. Morgan, "The Generation of a Unique Machine Description for Chemical Structures—A Technique Developed at Chemical Abstracts Service," *J. Chem. Doc.*, **5,** 107–113 (1965).

SELECTED PAPERS OF HISTORICAL INTEREST RELATING TO *CA* (in chronological order)*

1. C. L. Bernier and E. J. Crane, "Indexing Abstracts," *Ind. Eng. Chem.*, **40,** 725–730 (1948).
2. A. H. Emery, E. J. Crane, and G. G. Taylor, "The Future of *Chemical Abstracts,*" *Chem. Eng. News,* **28**(30), 2517–2520 (1950).
3. E. J. Crane, "Scientists Share and Serve (The Priestley Medal Address)," *Chem. Eng. News,* **29**(42), 4250–4253 (1951).
4. E. J. Crane, "Chemical Abstracts," in *A History of the American Chemical Society;* C. A. Browne and M. E. Weeks, Eds., American Chemical Society, Washington, DC, 1952, pp. 336–367.
5. E. H. Volwiler and A. C. Cope, "Chemical Abstracts—Millstone or Milestone," *Chem. Eng. News,* **33**(25), 2636–2639 (1955).
6. E. J. Crane, "The Chemical Abstracts Service—Good Buy or Good-by," *Chem. Eng. News,* **33**(26), 2752–2754 (1955).
7. "CA Today: The Production of Chemical Abstracts," American Chemical Society, Washington, DC, 1958.

8. G. M. Dyson, "Research Expansion at the Chemical Abstracts Service," *Chem. Eng. News,* **37**(36), 128–131 (1959).
9. G. M. Dyson, "Closing the Gap in Chemical Documentation," *Chem. Eng. News,* **38**(19), 70–72 (1960).
10. D. B. Baker, "Growth of Chemical Literature," *Chem. Eng. News,* **39**(29), 78–81 (1961).
11. E. J. Crane and C. C. Langham, "On-the-Job Training at Chemical Abstracts Service," *J. Chem. Doc.,* **2,** 199–204 (1962).
12. R. R. Freeman and G. M. Dyson, "Development and Production of *Chemical Titles,* a Current Awareness Index Publication Prepared with the Aid of a Computer," *J. Chem. Doc.,* **3,** 16–20 (1963).
13. G. M. Dyson and M. F. Lynch, "Chemical-Biological Activities—A Computer-Produced Express Digest," *J. Chem. Doc.,* **3,** 81–85 (1963).
14. R. R. Freeman, J. T. Godfrey, R. E. Maizell, C. N. Rice, and W. H. Shepherd, "Automatic Preparation of Selected Title Lists for Current Awareness Services and as Annual Summaries," *J. Chem. Doc.,* **4,** 107–112 (1964).
15. W. E. Cossum, M. L. Krakiwsky, and M. F. Lynch, "Advances in Automatic Chemical Substructure Searching Techniques," *J. Chem. Doc.,* **5,** 33–35 (1965).
16. H. L. Morgan, "The Generation of a Unique Machine Description for Chemical Structures—A Technique Developed at Chemical Abstracts Service," *J. Chem. Doc.,* **5,** 107–113 (1965).
17. D. P. Leiter, Jr., H. L. Morgan, and R. E. Stobaugh, "Installation and Operation of a Registry for Chemical Compounds," *J. Chem. Doc.* **5,** 238–242 (1965).
18. D. B. Baker, "Chemical Literature Expands," *Chem. Eng. News,* **44**(23), 84–86 (1966).
19. F. A. Tate, "Progress Toward a Computer-Based Chemical Information System," *Chem. Eng. News,* **45**(4), 78–90 (1967).
20. H. Siegel, D. C. Veal, and D. A. McMullen, "Polymer Science & Technology (POST): A Computer-Based Information Service," *J. Polym. Sci., Part C,* (25), 191–196 (1968).
21. D. B. Baker, "Chemical Abstracts Service," in *Encyclopedia of Library and Information Service,* A. Kent and H. Lancour, Eds.; Marcel Dekker, New York, 1970, Vol. 4, pp. 479–499.
22. D. B. Baker, "World's Chemical Literature Continues to Expand," *Chem. Eng. News,* **49**(28), 37–40 (1971).
23. N. Donaldson, W. H. Powell, R. J. Rowlett, Jr., R. W. White, and K. V. Yorka, "Chemical Abstracts Index Names for Chemical Substances in the Ninth Collective Period (1972–1976)," *J. Chem. Doc.,* **14,** 3–14 (1974).
24. *Proceedings of the Symposium on Chemical Abstracts in Transition,* Chicago, IL, August 28, 1973, Chemical Abstracts Service, Columbus, OH, 1974.
25. D. B. Baker, "Recent Trends in Growth of Chemical Literature," *Chem. Eng. News,* **54**(20), 23–27 (1976).
26. Chemical Abstracts Service, in *A Century of Chemistry: The Role of Chemists and the American Chemical Society,* H. Skolnik and K. M. Reese, Eds.: American Chemical Society, Washington, DC, 1976, pp. 126–143.
27. D. B. Baker, J. W. Horiszny, and W. V. Metanomski, "History of Abstracting at *Chemical Abstracts* Service," *J. Chem. Inf. Comput. Sci.,* 20, 193–201 (1980).
28. D. B. Baker, "Recent Trends in Chemical Literature Growth," *Chem. Eng. News,* **59**(22), 29–34 (1981).

29. D. B. Baker, "Chemical Abstracts Service," in *Encyclopedia of Library and Information Science,* A. Kent, Ed.; Marcel Dekker, New York, 1983, Vol. 36, Suppl. 1, pp. 167–182.
30. R. J. Rowlett, Jr., "An Interpretation of Chemical Abstracts Service Indexing Policies," *J. Chem. Inf. Comput. Sci.,* **24,** 152–154 (1984).
31. D. W. Weisgerber, "Applications of Technology to CAS Data-Base Production," *Inf. Serv. Use,* **4,** 317–325 (1984).
32. R. J. Rowlett, Jr., "Perspectives on Editorial Operations of Chemical Abstracts Service," *J. Chem. Inf. Comput. Sci.,* **25,** 61–64 (1985).
33. R. J. Rowlett, Jr., "Abstracts and Other Information Filters," *J. Chem. Inf. Comput. Sci.,* **25,** 159–163 (1985).
34. D. B. Baker, "Chemical Abstracts Service's Secondary Chemical Information Services," *J. Chem. Inf. Comput. Sci.,* **25,** 186–193 (1985).
35. D. F. Zaye, W. V. Metanomski, and A. J. Beach, "A History of General Subject Indexing at Chemical Abstracts Service," *J. Chem. Inf. Comput. Sci.,* **25,** 392–399 (1985).
36. D. W. Weisgerber, "Improvements in Access to Information in the Chemical Abstracts Service Database," *Tagungsber. Vortragstag. Fachgruppe Chem.-Inf.* [*Ges. Dtsch. Chem.*], 2nd Meeting, 119–131 (1985).
37. R. L. Wigington, "Applications of Computer Technology to Science Information Services," in *The Role of Data in Scientific Progress,* P. S. Glaeser, Ed.; Elsevier, Amsterdam, 1985, pp. 505–509.
38. D. B. Baker, "Chemical Information Flow across International Borders: Problems and Solutions," *J. Chem. Inf. Comput. Sci.,* **27,** 55–59 (1987).
39. R. E. Stobaugh, "Chemical Abstracts Service Chemical Registry System. 11. Substance-Related Statistics: Update and Additions," *J. Chem. Inf. Comput. Sci.,* **28,** 180–187 (1988).
40. W. V. Metanomski, "Ten Million—and Growing," *Chem. Int.,* **13,** 7 (1991).
41. W. V. Metanomski, "Twelve Million—and Growing," *Chem. Int.,* **15,** 205 (1993).
42. P. M. Giles, Jr. and W. V. Metanomski, "The History of Chemical Substance Nomenclature at Chemical Abstracts Service (CAS)" in *Organic Chemistry: Its Language and Its State of the Art,* M. V. Kisakürek, Ed.; Verlag Helvetica Chimica Acta, Basel/VCH, Weinheim, 1993, pp. 173–196.
43. W. V. Metanomski, "Thirteen Million—and Still Growing," *Chem. Int.,* **18,** 8–9 (1996).

*List compiled by W. Val Metanomski, Chemical Abstracts Service.

7 Essentials of *Chemical Abstracts* Use

7.1. INTRODUCTION

As indicated in Chapter 6, by far the most heavily used and valuable information tool in chemistry, and the one best known to chemists and chemical engineers, is *Chemical Abstracts* (*CA*), the chief product of CAS (Chemical Abstracts Service). CAS is a division of the American Chemical Society (ACS).

Even though in the introductions to the printed *CA,* its indexes, and the *Index Guide* (see Section 7.7.A), there are many pages explaining in detail policies and proper use, especially for the indexes, the key to successful use of *CA*—as for any tool—is *hands-on* use on a regular basis. It requires both study and experience to use *CA* to full advantage.

To help its customers learn about and utilize its major products, CAS offers various types of assistance. One of these is an ongoing series of workshops at national ACS meetings and elsewhere on the use of the online *CA* files on STN International (see Chapter 10 for further discussion of the online files). Some of these workshops are offered at no charge. The content of certain workshops is more geared to laboratory (practicing) chemists and engineers just beginning to utilize STN, whereas other workshops are more appropriate for information professionals or for laboratory chemists and engineers who wish to go into considerable detail.

In addition, there is extensive written and other documentation, much of which is available at no charge, on the use of *CA* and other files on STN. CAS sells *STN Beginning Searching Video,* which covers the basics, and *Using CAS Databases on STN.* The latter is a printed publication that includes an instructor package and student manuals. Further, STN offers online training (or "learning") files, such as *LCA, LCAS-REACT, LMARPAT,* and *LREGISTRY,* which permit the user to practice with a small subset at very low cost; use of these files is highly recommended as a training–learning method. Customers of STN receive free periodic newsletters (*STNews*) that describe special features, new databases, and system improvements and changes.

In addition to the above, information on the use of certain CAS products and services is available on the CAS WWW (World Wide Web) server (http://www.cas.org). See Chapter 10 for further discussion about the Internet. This information is contained in four sections: About CAS, Products, Support, What's New, and Press Releases. The CAS web site includes CAS and other STN database summary sheets, CAS and other STN database descriptions, copies of *STNews,* and pointers to STN training workshops worldwide. In addition, the CAS site provides links to other CAS and ACS Internet product sites, e.g., *STN Easy* (access to selected STN databases), *Chemical Patents Plus* (chemical and nonchemical U.S. patents from the early 1970s), *ChemPort* (a site with access to full-text of ACS and certain other journals), and *ChemCenter.* The last-named, in turn, provide additional links to ACS Journals (tables of contents and full-text) and *STN International* (Telnet access), in addition to some of the sources previously mentioned. All of these are more fully described elsewhere in this book.

Finally, CAS maintains an exceptionally talented "Help Desk," which is discussed in further detail in Chapter 10.

Paradoxically, the significant increases in the size and particularly the capabilities of the CAS files also lead to increases in complexity of *CA* use, especially of the online product. This makes the new *SciFinder* product of CAS a notably welcome addition. *SciFinder* is discussed further in Section 7.3.

A lengthy book could be written about the use of *CA* alone, much less the balance of the important tools on chemical information. Accordingly, this chapter focuses on essentials, with emphasis on points less well understood by many chemists, on newer developments, and on important recent changes. In addition, understanding of the historical aspects, as described in the previous chapter, is essential to use of *CA.*

7.2. CAS PRODUCTS

CAS products are offered in a number of ways:

1. Print: *CA, CA Section Groupings, Chemical Titles, CA Index Guide, CA SELECTS, CA SELECTS PLUS, CAS BioTech Updates, Chemical Industry Notes, CAS Source Index, Registry Handbook-Number Section, Ring Systems Handbook,* Guides and Support Materials.
2. CD-ROM: *CA, CASurveyor,* Collective Index and Abstracts, *CAS Source Index.*
3. Online through STN: *CA, CAplus, Registry, CAOLD, CHEMCATS, CASREACT, CHEMLIST, CIN, MARPAT, MARPATprev.* Also Internet access to STN, *STN Easy, Chemical Patents Plus.*
4. Microfilm: *CA,* Collective Indexes, *CAS Source Index* (*Keyword-Out-of-Context Index* only), *Registry Handbook-Common Names, Registry Handbook-Number Section,* International CODEN Directory.
5. Software: *STN Express, Messenger.*
6. License: *CA* Search (a version of the *CA* file without abstracts) is licensed to DataStar, DIALOG, Ovid, and Questel•Orbit. *Registry* data and *Chemical Industry Notes* are licensed to DataStar, DIALOG, and Questel•Orbit. (Questel offers algorithmically derived structure diagrams from connection tables.) *CASSI* is licensed to Questel•Orbit. *CA Index Guide* is licensed to DIALOG. *CAST-3D* [a file of two- or three-dimensional (2D or 3D) chemical structures] is a licensable file, but there are no public vendors at this time.
7. SciFinder: client/server technology.

8. Services: CAS Document Detective Service, Corporate Updates, Individual Search Service, CAS Search Service, CAS Registry Services.

Most of these products are discussed in this chapter or elsewhere in this book. There are also specially priced products for colleges and universities only.

7.3. CAS ONLINE FILES AND *SCIFINDER*

In addition to the printed CAS products, the chemist has a choice of several important online or Internet files that are offered by CAS, as noted above. Each is briefly described below:

1. *Registry.* This key file contains records on all unique chemical substances that are included in CA since 1957. There are no current plans to go back to earlier years. A complete record includes the present Registry Number and any older discontinued Registry Numbers, the "official" *CA* name that is used for indexing, alternative names, molecular formula, component class identifier, other STN files that include the structure, structure diagram, and number of references in *CA* and *CAOLD.*

> The most powerful feature of this file is that users can search it by structure or substructure. If one is searching for any aspect of a chemical (synthesis, manufacture, properties, reactions, uses, etc.), it is always recommended practice to start a search with this file and then to automatically utilize pertinent results in subsequent searching of other files such as *CAplus* and any other appropriate STN file that contains Registry Numbers.

The approximate breakdown of substances by type in *Registry* is shown in Figure 7.1.

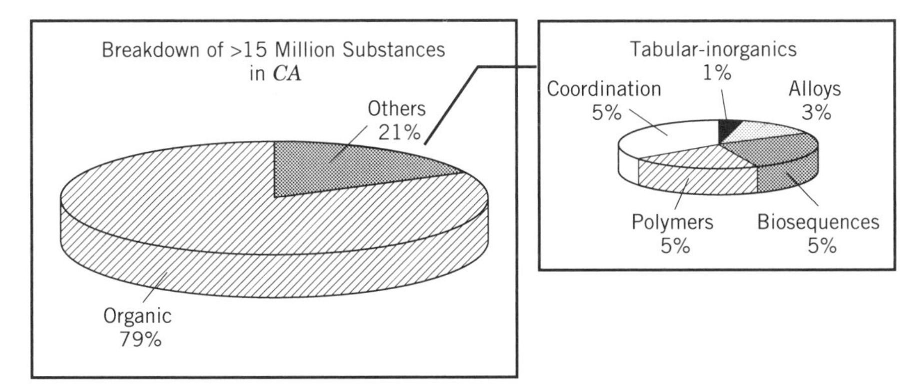

Figure 7.1. Breakdown of over 15 million substances in *Chemical Abstracts.* Tabular inorganics are those entered by CAS in the form of a table in such cases where, for example, bonding is unknown or in the case of non-stoichiometric compounds (an example is Registry Number [178927-84-9]). This process makes these compounds searchable. The most recent total is 2%. The percentage of biosequences has jumped to 13% because of the increased emphasis by CAS and occurrence in the literature. The total number of substances registered by CAS is now over 17 million. (Reprinted with the permission of Chemical Abstracts Service.)

2. *CAplus.* This is the most current and comprehensive of the online CAS bibliographic files. Its most significant feature is daily updating based on over 1350 core journals that are referenced cover-to-cover, in addition to all the other literature that is covered by CA since 1967. *CAplus,* introduced in 1995, provides references to every significant item in the table of contents of an issue of a core journal: scientific articles, biographies, book and product reviews, editorials, errata, letters to the editor, news announcements, and miscellaneous items. The core journal material will likely include some material outside the normal scope of *CA* (e.g., letters to the editor, product reviews, and clinical studies). Author abstracts are included for all core journal items when these abstracts exist. *CAplus* includes abstracts for many items that fall outside of *CA* coverage, in addition to abstracts for items that fall within *CA* coverage. A complete record includes accession number, full bibliographic citation, abstract, and all indexing, including any Registry Numbers. Since *CAplus* includes material not in *CA,* it cannot be broken down by Section Grouping. This file is exclusive to STN.

3. *CA.* Same as *CAplus,* except that electronic tape updating is weekly and coverage is based on *CA*'s customary editorial rules of coverage, abstracting and indexing. Thus, *CAplus* is both more current and more comprehensive. *CA* is searchable through other hosts, but only *CA* and *CAplus* on STN provide full CAS abstracts for all covered materials that fall within editorial policies.

4. *CAOLD.* Covers literature and patents from 1957 to 1966 as based on Registry Number only. Results given are citations to *CA* for this period (e.g., *CA*64:8033H), the type of document (for patents), and the Registry Numbers. There are no abstracts or bibliographic citations. Some consideration is being given to adding this information, but this may or may not happen. The current situation requires the user to consult the printed *CA* volume to get the citation and abstract. This file is exclusive to STN.

5. *MARPAT.* Contains Markush structure records for patents published since 1988 and covered by *CA*. A complete record includes accession number, title, inventor(s), patent assignee, patent number, date, patent application and date, language, abstract and diagrams of all Markush structures and related text (see Chapter 13). This file is exclusive to STN.

> *MARPATprev* contains the most recent Markush structures, bibliographic information, and in some cases, the abstract and Registry Numbers. When the complete record is added to *MARPAT,* the *MARPATprev* record is no longer available.

Note that the *CASLINK* cluster permits the user to concurrently search *REGISTRY, MARPAT, MARPATprev, CA,* and *CAplus.* It creates an answer set with duplicates removed.

6. *CASREACT.* This file provides details of reactions included in the Organic Sections of *CA*; journals from 1985, and patents from 1991 to date. This file is exclusive to STN.

7. *CHEMLIST.* This file provides information on regulatory status of chemical substances that are subject to regulation and is exclusive to STN. (See Chapter 14.)

8. *CIN* (*Chemical Industry Notes*). This file includes abstracts of worldwide chemical business news since 1974. (See Chapter 16.)

9. *CHEMCATS.* This is essentially a fine chemicals buyers' guide that began in October 1995. This file is exclusive to STN. (See Chapter 16.)

In addition to being on STN, some of these files are also available on other online systems, but for the most part, without some of the key features.

The online CAS files provide chemists and others with an outstanding solution to the issue of working with the sheer bulk of the printed *CA* volumes. In addition to providing for much more rapid searching than is possible manually, the online files permit types of searching that are not possible in printed files.

Thus, one can search for Registry Numbers, substructures, Markush structures, and logical combinations, such as preparation of a compound by a certain investigator, for a specific application, as reported in a European patent document that was published in a specific year.

Optimum use of the online files, except for *SciFinder* (see below), requires extensive training and experience, but thousands of chemists have mastered the techniques. Chapter 10 discusses some of the essentials of online use strategies.

However, outstanding though they are, the online *CA* files do not contain all of the information in the printed *CA*. The online files include abstracts only since 1967. Older literature and patents as covered in the printed *CA* prior to 1967 still contains information of great value and should always be considered, as appropriate. The printed books, especially the indexes, are well suited for unlimited browsing, a technique sometimes neglected today, but still worthwhile.

In addition, of the various *CA* products, only the printed volume indexes contain information about patent equivalents. This information is, unfortunately, not given in the online *CA* files because of an agreement with the European Patent Office, which is the source of the patent equivalent information.

One of the newest CAS products, a powerful research tool designated as *SciFinder,* originated in the New Product Development Department at CAS, and was introduced commercially in 1995 after extensive worldwide testing. It is intended and designed primarily for direct personal use by the laboratory scientist or engineer. Use of *SciFinder* requires no training, according to CAS, and is intended to be largely intuitive in its use. *SciFinder* is designed to operate on a client/server computer system in organizations employing a number of chemists. Initial reception is said by CAS to be enthusiastic. The client is the computer [Macintosh or IBM (or compatible) PC with Microsoft Windows] on the scientist's desk, and the server is a computer at CAS. For use of *SciFinder,* CAS recommends a direct connection to the Internet, or a leased line to CAS. Alternatively, dialup connection through Point-to-Point Protocol (PPP) or through an Internet provider is acceptable.

Special features of *SciFinder* include very simple ways to ask search questions that essentially automatically rephrase or expand search questions as required; a capability to browse tables of contents for some 1350 key journals; and "Keep Me Posted," a user-friendly automatic current awareness feature. A straightforward chemical drawing package is included. The system automatically searches for variations in chemical names, and will offer a list of options if the exact name cannot be found. Other user-friendly features include handling of molecular formulas regardless of the order of the elements and automatic insertion of hyphens into Registry Numbers. Further, *SciFinder* automatically suggests options for spelling of author names when the author name is not clear to the searcher.

Content includes the following major files: *CAplus* from 1967 to date, *REGISTRY*, and the full text of 23 ACS journals from 1992 to date. A new addition is data from Derwent Information's *World Drug Alerts and Patents Preview* (world patents, scientific literature, and conferences relating to drug discovery with abstracts and structures of pharmacologically active compounds, their synthesis, and intermediates). However, these files are essentially transparent to the user who does not need to specify a file name. A new feature, added in 1997, is substructure search capability.

SciFinder is not priced for the individual user, but rather at the level of the organization. Pricing may be based on the number of scientists served, or, alternatively, through a "task package" plan which is intended for smaller organizations or less frequent users. Under the latter plan, a "task" is defined as searching, analysis, and output for one question completed in *SciFinder*.

For additional comments about *SciFinder*, see the list of references at the end of this Chapter.

7.4. COVERAGE

The official statement of what *CA* covers is important and worth quoting in its entirety; the following statement in the table of contents of each *Chemical Abstracts* issue:

> It is the careful endeavor of *Chemical Abstracts* to publish adequate and accurate abstracts of all scientific and technical papers containing new information of chemical or chemical engineering interest and to report new chemical information revealed in the patent literature, but the American Chemical Society is not responsible for omissions or for such mistakes as may be made in abstracts and index entries.

Although the implication is that *CA* covers only new information, its editors cannot possibly check all material included to see if it is new. They can only make the *careful endeavor* noted in their policy statement. *CA* editors necessarily rely to a large extent on original statements as to novelty of information in the literature.

Subject areas covered by *CA* are shown in Figure 7.2. Figure 7.3 shows the types of source documents included in *CA*.

7.5. ABSTRACT AND INDEX CONTENT

CA contains informative abstracts of original documents. As mentioned in Chapter 6, these abstracts are not intended to replace the original, nor are they critical or evaluative. Rather, the abstracts are filters, with the intent of providing the user with enough

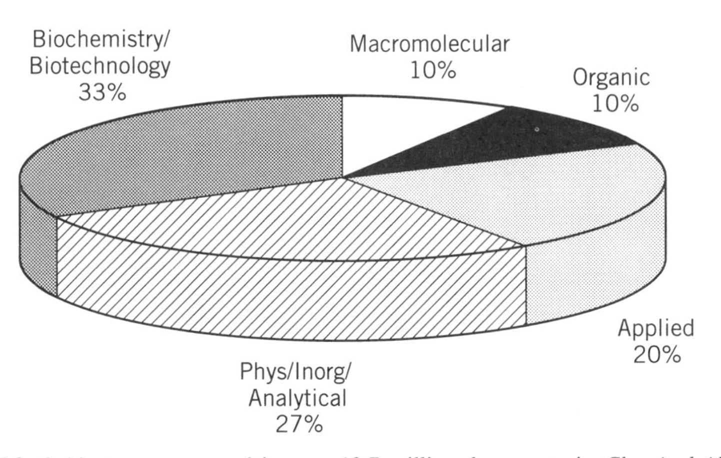

Figure 7.2. Subject areas covered in over 13.7 million documents in *Chemical Abstracts*. (Reprinted with the permission of Chemical Abstracts Service.)

information on content to permit one to decide whether to consult the originals. Any scientist who repeats a procedure based purely on the *CA* abstract is making a mistake. Moreover, not all new substances and subjects reported in the original appear in the abstract; all are, however, covered by the Volume and Collective Indexes, according to *CA* policy.

"Index density" (number of index entries per abstract) is a direct reflection of the depth and thoroughness of indexing. *CA* staff began significantly increasing indexing density in 1972; earlier years were not as deeply indexed. Beginning with the ninth collective period (1972–1976), indexing of reactants and intermediates was extended to all subject areas covered by *CA*, thereby further strengthening the indexes. This

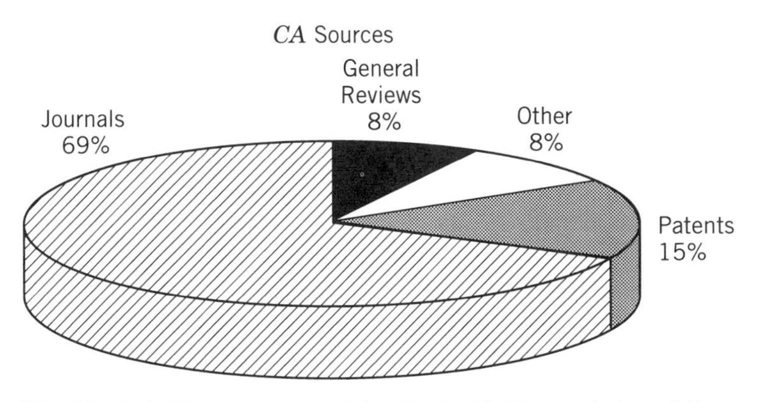

Figure 7.3. *Chemical Abstracts* sources. (Reprinted with the permission of Chemical Abstracts Service.)

was, of course, in addition to the products of reactions that are usual focal points of interest.

The most recent figures available from CAS (based on 1994 data) show that overall subject and substance index density averages 7.86 for all 80 *CA* sections. (The overall average has remained approximately the same for over 10 years; the average index density in 1984 was 7.4). In 1994, there are, on the average, 4.69 chemical substance entries and 3.17 general subject entries.

Index density varies depending on the *CA* section, as might be expected. For example, for organic sections, the 1994 average is 19.86 chemical substance entries, whereas in biochemistry, the average is 3.14 chemical substance entries.

7.6. SPEED OF COVERAGE AND INDEXING

The median of complete indexing and abstracting of all documents included continues to be approximately 3 months as based on the cover-page date of the publication and the cover date of the printed *CA*. More specific figures are as follows: 100 days for all documents, 85 for all journals, and 60 for core journals (defined as those nearly 800 journals covered in *Chemical Titles*). It is clear that patent documents are covered more slowly than are the journals.

CA editors continue to make efforts to further improve the timeliness of the printed CA. For example, the complete set of the semiannual (6-month) volume indexes is now mailed by the printer within about 4 months after completion of the semiannual volumes. This compares with about 6 months in the 1970s. The speed of appearance of the printed abstracts and indexes may be approaching the optimum that can be achieved with present technology, human capabilities, and financial resources.

As might be expected, currency for the online files is considerably better than for the printed *CA*. Thus, for the 1350 core journals that are covered in *CAplus,* CAS says that the user can expect to see online entries for those items selected within one week after journal receipt at CAS. Initially, these entries consist of the bibliographic citations and abstracts in most cases. Index entries are added later. (*CAplus* journals are very high yield as compared with the total of some 8000 journals that are included in *CA* coverage.) The *CAplus* files are updated daily.

The weekly issues of *File CA* are loaded online a day or two before the printer mails the corresponding printed issues. Of course, the *File CA* loading includes all the volume subject index entries. (At present, the printer of the CA issues and indexes is Metromail, Lincoln, NE, a division of R. R. Donnelley and Sons Company. The transfer from long-time printer Mack Printing Company was made in the first quarter of 1995.)

7.7. *CHEMICAL ABSTRACTS* INDEXES AND SOME ASPECTS OF SEARCHING

The following is a broad summary of the contents of the Volume, weekly, and Collective Indexes to the printed *CA* (see also Chapter 6):

A. Index Guide

This index (since 1992, new edition issued after 1st, 5th, and 10th volumes of a 5-year collective index period) details the major points of *CA* indexing policy for the appropriate collective period and provides cross-references from chemical substance names and general subject terms used in the literature to the terminology used. The *CA Index Guide* contains the following parts:

Introduction describes cross-references, homograph definitions, and indexing policy notes listed in the main portion of the *Index Guide.*

Alphabetical sequence of cross-references and indexing policy notes is the main portion of the *Index Guide.* Beginning with the 1985 edition this section includes all the valid general subject index headings (excluding latinized genus and species names) and diagrams for stereoparents.

Volume Indexes to Chemical Abstracts: Organization and Use discusses the relationship between and use of the chemical substance, general subject, and formula indexes and index of ring systems.

Selection of General Subject Headings discusses the content of the general subject index.

Chemical Substance Index Names summarizes the rules used in deriving the *CA* index names for chemical substances, and includes a section on converting *CA* index names to structural diagrams.

Hierarchies of General Subject Headings aids in generic searching.

This publication is invaluable in developing search strategies for both the online and printed *CA* files.*

B. Chemical Substance Index

This semiannual index relates the *CA* index names of chemical substances and their CAS Registry Numbers to *CA* abstract numbers for documents in which the substances are mentioned. Included with the *CA* index name is a brief description of the document's context. This index is initiated with the ninth collective period (1972–1976), prior to which chemical substances and general subjects were in a single index.

C. General Subject Index

This semiannual index relates index entries that do not refer to specific chemical substances to the corresponding *CA* abstracts. These entries include concepts, general classes of chemical substances, applications, uses, properties, reactions, apparatus, processes, and biochemical and biological subjects.

*CAS on the Internet now contains substantial and recent index term assistance, e.g., casweb.cas.org.vocabulary/index.html. The material on the Web pertains to the *General Subject Index.*

D. Formula Index

This semiannual index relates the molecular formulas for chemical substances with the *CA* chemical substance index names, CAS Registry Numbers, and corresponding *CA* abstract numbers.

E. Index of Ring Systems

(This index has been discontinued; the last such index appears in Volume 121, July–December 1994.) Ring composition, ring size, and number of rings are listed providing a means for determining the systematic *CA* index names for specific ring systems as well as the nonsystematic *CA* index names for cyclic natural products containing these ring systems.

F. Author Index

Authors, patentees, and patent assignees are listed in alphabetical order in this semiannual index with titles of their articles or patent specifications and *CA* abstract numbers.

G. Patent Index

This semiannual index includes (1) entries for all newly abstracted patent documents, (2) cross-references to the first-abstracted patent on an invention when more than one patent describes that invention, and (3) a listing at the first-abstracted patent of all the equivalent patents. The first *CA* numerical patent index was introduced in 1912, but was discontinued in 1915. It was reintroduced in 1935. A *Patent Concordance* was started in 1963, and the present combined numerical and concordance index begins coverage with the tenth collective period (1977–1981). In addition, the Special Libraries Association has published a collective numerical patent index to *CA* covering 1907–1936.

H. Weekly Issue Indexes

Each weekly issue of *CA* contains several indexes. The issue indexes include (1) keyword index, (2) patent index, and (3) author index.

Chemists can use the keyword index to search specific issues of *CA* or CA *Section Grouping* issues for subject content. As mentioned in Chapter 6, keyword index vocabulary is uncontrolled, that is, nonsystematic and not standardized. It reflects author terminology and the content of the abstract. This permits the index to be produced quickly and published as a part of each issue, but it makes for substantially less detail and much more superficial index coverage than is found in the printed semiannual volume indexes. Both the keyword index terms and the systematic, standardized volume index entries are in the online files of *CA*.

The patent index, which includes both abstracted and equivalent patents, is cumulated semiannually in the Volume Indexes and every 5 years in the Collective Indexes. The issue author index, consisting of author name and initials and the *CA* abstract number, is less detailed than the semiannual volume author index and is designed to serve the issue in which it appears. Beginning with Volume 124, the Issue Author Index includes the *full* author name.

TABLE 7.1. Development of *Chemical Abstracts* Collective Indexes

1st–4th Decennial Indexes each contain author and subject indexes
1st Collective Formula Index covers 1920–1946
Ten Year Numerical Patent Index covers 1937–1946
5th Decennial Index contains author, subject, numerical patent, and formula indexes
6th Collective Index contains author, subject, numerical patent, and formula indexes and index of ring systems
7th Collective Index contains author, subject, numerical patent, and formula indexes; index of ring systems; and patent concordance
8th Collective Index contains author, subject,[a] numerical patent, and formula indexes; index of ring systems; patent concordance; and index guide
9th Collective Index contains author, chemical substance, general subject,[b] numerical patent, and formula indexes; index of ring systems; patents concordance; and index guide
10th, 11th, and 12th Collective Indexes contain author, chemical substance, general subject,[b] patent, and formula indexes; index of ring systems; and index guide; the 13th Collective is the same except there is no longer an index of ring systems

[a]The *Subject Index* includes *Registry Handbook—Number Section* issued as part of the Eighth Collective Index.

[b]The *Subject Index* was subdivided into *General Subject* and *Chemical Substance Indexes* beginning with the Ninth Collective Index period.

I. *CA* Collective Indexes

Tables 6.4, 6.5, and 7.1 depict the development of *CA* Collective Indexes.

J. Searching *CA:* General Remarks

A complete search of *CA* will utilize both the printed volumes, for older information going back to 1907, and the online files, for the more recent information, sometimes within a few days of actual publication date of the original documents. The online files are usually the preferred sources to start, since they can be searched relatively quickly and easily as compared to the printed volumes. If the focus of the search is a chemical substance, *File Registry* is the place to start in order to obtain the proper CAS Registry Numbers, both current and discontinued. The results from this file can then be automatically "crossed over" and utilized for searching in *File CA* or *File CAplus.* The printed *CA* tools, especially the *Index Guide,* can still be very useful in structuring a complete search of the online files; this point is discussed further in this chapter. If for any reason the chemist does not have ready access to the online files, the printed indexes and abstracts can still yield excellent results for both recent and older information. The printed volumes are also always useful for browsing before a formal search is initiated.

Figure 7.4 depicts the field codes for search and display in the *CA* and *Registry* online files. Figure 7.5 shows the basic commands of the STN system.

To determine whether a search strategy for use with *CA* or *CAplus* on STN is likely to yield the types of results desired, a recommended procedure is to first display and examine a sample (say, 10–20) of the document titles found. This can be economically done by a command such as "D SCAN TI HITIND," where "D" means dis-

CA

Search and Display Field Codes

Fields that allow left truncation are indicated by an asterisk (*)

General Search Fields

Search Field Name	Search Code	Search Examples	Display Codes
Basic Index * (contains single words from title (TI), supplementary term (ST), index term (IT), and abstract (AB), fields, as well as CAS Registry Numbers)	None (or /BI, or /IA)	S 50-21-5 S TRANSGENIC COTTON S ?FLUOROCARBON? S (WATER (S) OIL)/BI	AB, IT, ST, TI
Abstract	/AB	S (WATER(1W)OIL)/AB S LD50/AB S HIGH TEMP?/AB S (HIV(S)TREAT?)/AB	AB
Accession Number	/AN	S 109:141853/AN	AN
Author (inventor)	/AU	S BARTON J?/AU S (DUCHEYNE P? (S) EDITOR#)/AU S ANON/AU	AU, IN
CA Section Cross Reference **(1)** (number and title)	/SX	S 1/SX S ANALYTICAL/SX S RADIATION CHEMISTRY/SX	CC
Classification Code **(1)** (contains CA section-subsection number, section title, and section group codes)	/CC	S 1/CC S 80-6/CC S TOXICOLOGY/CC S RADIATION CHEMISTRY/CC S L1 AND BIO/CC	CC
Controlled Term	/CT	S NEOPLASM INHIB?/CT	CT, IT
Controlled Word	/CW	S OPTIC?/CW	CT, IT
Corporate Source **(1)** (organization name and location)	/CS	S MERRELL/CS S MERRELL DOW/CS S USA DOW/CS S DOW CHEM MIDLAND/CS S "DOW CORNING"?/CS	CS, PA
Country of Author	/CYA	S USA/CYA	CS, CYA, PA
Crossover Key (CODEN, volume, issue, first page)	/CK	S JACSAT-104-1-318/CK	CK
Document Type (code and text)	/DT	S P/DT S PATENT/DT S REVIEW/DT S GR/DT	DT
Entry Date **(2)**	/ED	S ED>970211 S ED>FEB 11, 1997	Not displayed
Field Availability	/FA	S L1 AND ABS/FA	Not displayed
File Segment	/FS	S BIO/FS	FS
Index Term **(3)**	/IT	S 75-28-5(2W)CRACKING OF/IT S DETN OF/IT	IT
International Standard Document Number (contains CODEN and ISSN)	/ISN	S JOCRAM/ISN S 0021-9673/ISN	ISN, SO
Issue Number of Publication **(2)**	/IS	S 1-3/IS AND 32/VL	SO

March 1997 (Revised)

Figure 7.4. *Chemical Abstracts* search and display field codes; figure continued on pages 119–123. (Reprinted with the permission of Chemical Abstracts Service.) For use with STN.

CA

Search and Display Field Codes (cont'd)

Search Field Name	Search Code	Search Examples	Display Codes
Journal Title	/JT	S J CHROMATOGR/JT	JT, SO
Language (code and text)	/LA	S L1 AND EN/LA S L1 AND ENGLISH/LA S L1 NOT DE/LA	LA
Other Source	/OS	S CJACS/OS	OS
Publication Date **(2)**	/PD	S PD>960000 S JUNE 1992-SEPT 1993/PD	PI
Publication Year **(2)**	/PY	S 1991-1992/PY	PI, PY, SO
Publisher	/PB	S ACADEMIC/PB	PB
Publisher Item Identifier	/PUI	S "S 0014-5793(96)01227-6"PUI	PUI
Role	/RL	S 99685-96-8 (L) SPN/RL S 99685-96-8/SPN S FULLERENES (L) SPN/RL S FULLERENES/SPN	IT, RL
Source (contains publication title, date, publisher, conference title, meeting date, volume, issue, pagination, CODEN, ISSN and URL)	/SO	S INORG CHEM/SO S JOCRAM/SO S 0021-9673/SO S AM CERAM SOC/SO S 1992/SO	SO
Supplementary Term	/ST	S LIVER METAB?/ST	ST
Title	/TI	S LIVER/TI S SPIN SPIN/TI S (METABOLISME(S)VEGETAUX)/TI	TI
Uniform Resource Locator	/URL	S "HTTP://WWW.BIOSCIENCE.ORG/ BIOSCIENCE/1996/V1/D/CHINTALL/ HTMLS/324-339.HTM"/URL	SO, URL
Update Date **(2)**	/UP	S UP>961220 S UP>DEC 20, 1996	Not displayed
Volume and Issue of CA	/VI	S 107-4/VI	AN
Volume Number of Publication **(2)**	/VL	S 32-33/VL AND INOCAJ/SO	SO

(1) Search with implied (S) proximity is available in this field.
(2) Numeric search field that may be searched with numeric operators or ranges.
(3) There are no stopwords in this field.

Patent Search Fields

Search Field Name	Search Code	Search Examples	Display Codes
Designated States	/DS	S FR/DS S R DE/DS	DS
International Patent Classification (includes Main and Secondary IPCs)	/IC	S C07C/IC S C07C015/IC S C07C015-04/IC S CYANOGEN/IC	IC
International Patent Classification, Additional or Supplementary	/ICA	S B01J/ICA S B01J027/ICA S CYANOGEN/ICA	ICA
International Patent Classification, Index or Complementary	/ICI	S A61K/ICI S A61K031/ICI S AMMONIA/ICI	ICI
International Patent Classification, Main	/ICM	S A01N/ICM S A01N025/ICM S AMMONIA/ICM	ICM
International Patent Classification, Main Group, Range Searchable **(1)**	/MGR	S 10-20/MGR (S) C07C/IC	IC

March 1997 (Revised)

Figure 7.4. *(Continued)*

CA

Patent Search Fields (cont'd)

Search Field Name	Search Code	Search Examples	Display Codes
International Patent Classification, Secondary	/ICS	S C02F/ICS S C02F001/ICS S AMMONIA/ICS	ICS
International Patent Classification, Subgroup, Range Searchable **(1)**	/SGR	S SGR=>30000 (S) C01B031/IC	IC
Inventor Name	/IN	S PATTON JERRY R/IN	IN
National Patent Classification	/NCL	S 106035000/NCL	NCL
National Patent Classification, Range Searchable **(1)**	/NCLR	S 106020000-106040000/NCLR	NCL
Patent Application Country	/AC	S DE/AC	AI
Patent Application Date **(1)**	/AD	S AD>920100 S AD>JANUARY 20, 1993	AI
Patent Application Number **(2)**	/AP	S EP83-304630/AP S 83EP-0304630/AP S JP87-10001/AP S 87JP-0010001/AP	AI
Patent Application Year **(1)**	/AY	S 1990-1992/AY	AI
Patent Assignee **(3)**	/PA	S PFIZER/PA S PFIZER CORP/PA S "PFIZER CHAS"?/PA	PA
Patent Country	/PC	S WO/PC	PI
Patent Kind Code	/PK	S DEA1/PK	PI
Patent Number **(2)**	/PN	S EP536930/PN S EP-536930/PN S WO8402426/PN S JP04000104/PN S JP62000031/PN	PI
Priority Application Country	/PRC	S US/PRC	PRAI
Priority Application Date **(1)**	/PRD	S PRD>910600 S June 20 1991/PRD	PRAI
Priority Application Number **(2)**	/PRN	S US91-635890/PRN S 91US-0635890/PRN	PRAI
Priority Application Year **(1)**	/PRY	S 1990-1992/PRY	PRAI
Publication Date **(1)**	/PD	S PD>890000 S JUNE 1992-SEPT 1993/PD	PI

(1) Numeric search field that may be searched with numeric operators or ranges.
(2) Either STN format or Derwent format may be used.
(3) Search with implied (S) proximity is available in this field.

Super Search Fields (1)

Search Field Name	Super Search Code	Fields Searched	Search Examples	Display Codes
International Patent Classifications	/IPC	/IC ,/ICA ,/ICI	S A01B/IPC S A01B001/IPC	IPC
Patent Application and Priority Number **(2)**	/APPS	/AP, /PRN	S DE84-3400052/APPS S 84DE-3400052/APPS	AI, PRAI
Patent Countries	/PCS	/PC, /DS	S DE/PCS	DS, PI
Patent Numbers **(2)**	/PATS	/PN	S EP536930/PATS S EP-536930/PATS S WO8402426/PATS S JP04000104/PATS S JP62000031/PATS	PI, SO

(1) Enter a super search code to execute a search in one or more fields that may contain the desired information. Super search fields facilitate crossfile and multifile searching. EXPAND may not be used with super search fields. Use EXPAND with the individual field codes instead.
(2) Either STN format or Derwent format may be used.
March 1997 (Revised)

Figure 7.4. *(Continued)*

REGISTRY
(Dictionary Searching)

Search and Display Field Codes

Fields that allow left truncation are marked with an asterisk (*).

Search Field Name	Search Code	Search Examples	Display Codes
Basic Index (contains name fragments, molecular formula fragments, and Collective Index codes) **(1)**	None (or /BI)	S TOSYL S DIMETHYL ADIPATE S 6CI S 1,1(W)DICHLORO S C5H10BR2O2	AF, CN, IN, MF
CAS Registry Number	/RN	S 97-77-8/RN S 97-77-9	RN, AR, DR, PR
Class Identifier (codes or terms as a bound phrase)	/CI	S MXS/CI S ALLOY/CI	CI
Component Registry Number	/CRN	S 79-10-7/CRN	CRN
Definition	/DEF	S HYDROCARBONS/DEF	DEF
Entry Date **(2)**	/ED	S 970210/ED	Not displayed
Field Availability (codes or terms as a bound phrase)	/FA	S RSD/FA AND L5 S MATERIAL COMPOSITION/FA	Not displayed
File Segment (acronyms or single words)	/FS	S 3D/FS S PROTEIN/FS	FS
Number of References in the CA File **(2)**	/REF.CA	S L1 AND REF.CA<=10	REF
Number of References in the CA File for Non-Specific Derivatives **(2)**	/REF.CAD	S L3 AND 1/REF.CAD	REF
Number of References in the CAOLD File **(2)**	/REF.CAOLD	S L2 AND 1-5/REF.CAOLD	REF
Number of References in the CAplus File **(2)**	/REF.CAPLUS	S L2 NOT REF.CAPLUS>10	REF
Polymer Class Term (code or text)	/PCT	S POLYAMINE/PCT;S PM/PCT	PCT
Registry Number Locator	/LC	S TSCA/LC	LC
Update Date **(2)**	/UP	S UP>=970212	Not displayed

(1) Formula fragments searched in the Basic Index must be entered without spaces.
(2) Numeric search field that may be searched using numeric operators or ranges.

Nomenclature Fields

Search Field Name	Search Code	Search Examples	Display Codes
Chemical Name	/CN	S 1-CHLORO-1,3-BUTADIENE/CN	CN, IN
Chemical Name Segment* **(1)**	/CNS	S IMINO/CNS S ?QUAT?/CNS NOT AQUA	CN, IN
Heading Parent	/HP	S BENZOIC ACID/HP	CN, IN
Index Name Segment - Heading Parent	/INS.HP	S METHYLETHYL/INS.HP	CN, IN
Index Name Segment - Non-Heading Parent	/INS.NHP	S ACRYLO/INS.NHP	CN, IN
Other Name Segment	/ONS	S ANILINE/ONS	CN

(1) With left truncation, the input term must contain at least 4 characters.

March 1997 (Revised)

Figure 7.4. *(Continued)*

REGISTRY
(Dictionary Searching)

Search and Display Field Codes (cont'd)

Molecular Formula Fields

Search Field Name	Search Code	Search Examples	Display Codes
Atom Count **(1)**	/ATC	S 5/ATC	Not displayed
Element Count **(1)**	/ELC	S 7-9/ELC	Not displayed
Element Count for Substance **(1)**	/ELC.SUB	S ELC.SUB>=8	Not displayed
Element Formula **(2)**	/ELF	S AL CO LA O/ELF	AF, MF
Element Ratio, xx **(1)** (where xx = CH, CN, CO, HC, HN, HO, NC, NH, NO, OC, OH, or ON)	/ELR.xx	S 3.1666667/ELR.CH S 1-2/ELR.CN S ELR.CO<=1	Not displayed
Element Symbol	/ELS	S B/ELS AND H/ELS	Not displayed
Element Symbol for Multi-Component Formula	/ELS.MCF	S (N (XA) P)/ELS.MCF	Not displayed
Formula Weight **(1)**	/FW	S 420-460/FW	Not displayed
Material Composition **(3)**	/MAC	S 1-5 ND/MAC	STR
Molecular Formula **(4)**	/MF	S C7H3BR2FO2/MF S C4H4O4.2NA/MF S C24 H37 OS P3/MF	AF, MF
Number of Components **(1)**	/NC	S F/ELS NOT NC>=2	Not displayed
Periodic Group	/PG	S B6/PG S LNTH/PG	Not displayed
Relative Composition	/RC	S FE.CR.NI/RC	Not displayed
Specific Element Count **(1)**	/Element Symbol	S 7/SI	Not displayed

(1) Numeric search field that may be searched using numeric operators or ranges.
(2) Formulas must be entered with spaces between the elements.
(3) Combined numeric and text field. Composition terms are numeric and may be searched using numeric operators or ranges. Component terms are text terms.
(4) Formulas may be entered with or without spaces.

Figure 7.4. *(Continued)*

play, "SCAN" signifies a display in random order, "TI" signifies document titles and "HITIND" yields any index entries that contain the search terms.

If this step shows that the types of results desired have been produced, then more complete and useful data can be obtained by a command such as "D L1 1-5 CBIB ABS," where "L1" indicates search statement 1, "1–5" indicates the number of documents to be displayed, "CBIB" will yield a compact bibliographic citation, and "ABS" denotes the full abstract. It is frequently desirable to display all index entries as well; inclusion of "IND" in a display command will yield this result. Index entries have a special utility when the abstract does not contain any of the terms employed in the search; search terms are often found only in the index entries.

Search methods are discussed later in this chapter and in Chapter 10. See also the excellent books by Ridley (1) and Schulz (2).

K. Searching *CA:* Preparative Methods

Many chemists have special interest in preparative (synthetic) methods, ranging from simple formation, to synthesis, to commercial-scale routes. *CA* editors have recognized this for years. Beginning in 1983, preparation has been denoted in the online *CA File* by the symbol "P" immediately following the Registry Number in the *CA* index entry, as based on a computer algorithm that analyzed index entry content. Un-

REGISTRY
(Dictionary Searching)

Search and Display Field Codes (cont'd)

Ring Analysis Data

Search Field Name	Search Code	Search Examples	Display Codes
Elemental Analysis for Ring System **(1)** (and number of occurrences of EA in a component structure)	/EA	S C4N-C5N/EA S 2 C3NO-C6/EA	RSD
Elemental Analysis for Smallest Ring **(1)** (and number of occurrences of EAS in a ring system)	/EAS	S C5NO4/EAS S >9 C6/EAS	Not displayed
Elemental Sequence for Ring System **(1)** (and number of occurrences of ES in a component structure)	/ES	S NCOC2-C6/ES S 1-3 O2C4/ES	RSD, SRSD
Elemental Sequence for Smallest Ring **(1)** (and number of occurrences of ESS in a ring system)	/ESS	S FE3/ESS S >=2 SC2SC2/ESS	Not displayed
Number of Ring Systems **(2)**	/NRS	S 7/NRS	Not displayed
Number of Ring Systems in a Component **(2)**	/CNRS	S 4-5/CNRS	Not displayed
Number of Rings **(2)** (number of smallest rings)	/NR	S 10/NR	Not displayed
Number of Rings in a Component **(2)** (number of smallest rings)	/CNR	S CNR>=12	Not displayed
Number of Rings in Ring System **(2)**	/NRRS	S 5-6/NRRS	Not displayed
Ring Atom Count **(2)**	/RATC	S 4/RATC	Not displayed
Ring Element **(1)** (and number of occurrences of REL in a ring system)	/REL	S SE/REL S 5 P/REL	Not displayed
Ring Element Count **(2)**	/RELC	S 6/RELC	Not displayed
Ring Elemental Formula **(1,3)** (and number of occurrences of RELF in a component structure)	/RELF	S C N O P/RELF S >3 C N O/RELF	Not displayed
Ring Identifier **(1)** (and number of occurrences of RID in a component structure)	/RID	S 31779.1.2/RID S 1938/RID S >=2 1949.52/RID	RSD, SRSD
Ring Size of Smallest Ring **(1,2)** (and number of occurrences of SZS in a ring system)	/SZS	S 8/SZS S 5 4/SZS	Not displayed
Ring System Formula **(1)** (and number of occurrences of RF in a component structure)	/RF	S C20AGN4/RF S 5 C10/RF	RSD
Size for the Ring System **(1)** (and number of occurrences of SZ in a component structure)	/SZ	S 3-4-5/SZ S 3 5-5-6/SZ	RSD

(1) The number of occurrences must be entered first in the search field. It is a numeric term and may be searched using numeric operators or ranges.
(2) Numeric search field that may be searched using numeric operators or ranges.
(3) Formulas must be entered with spaces between the elements.

March 1997 (Revised)

Figure 7.4. *(Continued)*

Commands

STN has NOVICE and EXPERT versions of the command language. The NOVICE version (at least the first four letters of the command) includes prompts for all information necessary to process the command. The EXPERT version (at most three letters and in some cases one letter of the command) does not prompt you for additional information. You should supply all information that is not a default option following the EXPERT command. Both NOVICE and EXPERT versions are shown in the COMMAND column. Not all commands are shown below. See the *STN Guide to Commands* for complete command information.

COMMAND	FUNCTION	EXAMPLE
ACTIVATE ACT	Recall saved queries (/Q), saved answer sets (/A), and saved L-number lists (/L) for use in the current sessions.	**=>ACT CONTRACT/Q =>ACT NMR/A**
DELETE DEL	Delete saved items or items in the current session. To delete all L-numbers in an online session, enter DELETE HISTORY.	**=>DEL ?COBALT?/A =>DEL HISTORY**
DISPLAY D	Display answers. Non-consecutive answer numbers must be separated by commas or spaces.	**=>D1-10,15,21 =>D L6 1 TI ID**
DISPLAY BROWSE D BRO	Browse through an answer set. You can view consecutive answers, nonconsecutive answers, change formats, and view additional answers.	**=>DISPLAY BROWSE L5 =>D BROWSE L5**
DISPLAY COST D COST	Display the estimated cost of your session. Options are ON, BRIEF, and FULL.	**=>D COST =>D COST FULL**
DISPLAY HISTORY D HISTORY	Display the session history. A summary of the files entered and commands that have been used and the information associated with them is displayed.	**=>D HISTORY =>D HIS L1-L10 =>D HIS L#**
DISPLAY SAVED D SAVED	Display a list of the names of all saved queries (/Q), answer sets (/A), L-number lists (/L), BATCH (/B), or SDI (/S) requests.	**=>D SAVED =>D SAVED/B**
EDIT EDI	Change field codes of terms in lists of E-numbers, usually resulting from the SELECT command.	**=>EDIT E1-E10 /RN /BI**
EXPAND E EXPAND BACK E BACK	Look at the index around a term. Twelve terms are shown by default. To continue down the same index, enter E <RETURN> at the next arrow prompt. To expand in the reverse direction, enter E BACK followed by a term.	**=>E YATES, C/AU =>E BACK METHYL**
FILE FIL	Enter a database or cluster to search or display records.	**=>FILE BEILSTEIN =>FIL TOXICOLOGY**
HELP ?	Request online help. Enter HELP and a command name for help on how to use a specific command. Enter HELP MESSAGES for a list of all online help messages available.	**=>HELP PRINT =>HELP MESSAGES**
INDEX IND	Access STNindex and identify the files to be searched.	**=>INDEX ALLBIB =>IND CA SCISEARCH**
LOGOFF LOG Y	End the online session.	**=>LOGOFF =>LOG Y**
LOGOFF HOLD LOG H	End the online session and hold your entire search session for 60 minutes at no charge.	**=>LOGOFF HOLD =>LOG H**
NEWS NEW	Display current news headlines on STN International. To see one of the news items enter NEWS followed by the number or name of the news item at the arrow prompt.	**=>NEWS =>NEWS 10 =>NEWS FILE**
ORDER	Order an original document or copy from the various STN Document Delivery Services. Extensive prompts guide you through this command.	**=>ORDER**
PRINT	Print answers offline or deliver them to an electronic mailbox. You are prompted for all information.	**=>PRINT**
SAVE SAV SAVE ALL	Save an L-numbered query, answer set, or L-number list in your long-term storage. A monthly fee is charged for saved items. You must enter an L-number and a name ending in /Q for a query, /A for an answer set, or /L for an L-number list.	**=>SAVE L15 COMPUTER /Q =>SAV L1 TOX/A =>SAVE ALL C13/L**
SAVE TEMP SAV TEMP	Save an L-number query, answer set, or L-number list on a temporary basis. You must assign a name that ends with /Q for query, /A for an answer set, or /L for a L-number list.	**=>SAV TEMP L5 NMR/A**
SCREEN SCR	Define a screen number for searching. This command is available only in some structure-searchable files.	**=>SCREEN 2043**
SDI	Request searches be run automatically by STN at file updates. In files that have SDI service, additional prompts appear.	**=>SDI**
SEARCH S	Perform a search. If AUTOSEARCH is set to ON, the SEARCH command is assumed.	**=>S MOTORIST =>ACID RAIN**
SELECT SEL	Extract terms from display fields that can then be used as search terms.	**=>SELECT L1 1-5 RN**
SET	Set various terminal parameters and options. Enter HELP SET for a list of all the SET options.	**=>SET STEPS ON =>SET COST OFF PERM**
STRUCTURE STR	Create a structure query for searching. This command is available only in structure-searchable files.	**=>STRUCTURE**

Figure 7.5. STN Commands. (Reprinted with the permission of Chemical Abstracts Service.)

fortunately, this sometimes resulted in a number of frustratingly incorrect retrievals (false "drops") that did not truly relate to preparation of the compound in question, but rather to some other aspect, such as formulations utilizing the compound. In many cases, the potential for a false drop could be readily determined and avoided by the user who realized the computer basis for the assignment and who examined the full index entry. In other cases, the full document might need to be examined before the misassignment became apparent. (In any event, *CA* staff believe that the assignment was at least 99% accurate and became even more accurate when *CA* document analysts were aware of the algorithm and the need for greater precision in framing of indexing text modifications.)

In recognition of this situation, CAS staff made an important decision that eliminated this problem, beginning with the subject index for July–December 1994 (Vol. 121). Indexing as to preparations is now done on the basis of an intellectual decision on the part of the document analyst, and the symbol *pr* is utilized in the printed index to denote preparation (and such closely related concepts as manufacture, purification, recovery, and synthesis, as well as extraction, generation, isolation, and secretion when there is preparative intent). This change has resulted in much more reliable and precise results. As to studies dealing with "formation," most such documents are nonpreparative and do not receive the intellectually assigned preparation flag. However, those occasional cases of formation where there is preparative intent or interest do receive this flag, such as nuclear formation of unstable isotopes. In the online *CA* and *CAplus* files, addition of the symbol "/p" to an "L" number (search statement number) "crossed-over" from the *Registry File* into one of the *CA* files will continue to retrieve documents that deal with preparation, with the entries later than September 1994 reflecting the change noted above. The first step is to search for the correct CAS number for a chemical in *File Registry* and then to cross into a *CA* file for execution of the search relating to preparation. Thus, "S L1/p," in which "S" signifies search, will retrieve documents pertaining to preparation ("p") of the chemical identified in "L1" as obtained from *File Registry.*

In *CA,* the most likely indicator that a document relates to a commercial route is to search for a combination of the *pr* indicator (printed files) or /p indicator (online files), utilized in combination with patent documents as designated by p/dt (document type = patents). Most patents have at least the potential goal of commercial manufacture; otherwise the inventor or the inventor's organization would not go to the expense of filing for a patent.*

The suffix "d" in *CA* online indicates a generic or unspecified derivative, and thus "dp" indicates the preparation of a generic or unspecified derivative.

Closely related to the preceding discussion is the recent introduction by CAS of role indicators, a topic that is discussed next.

L. Searching *CA:* Role Indicators

A role indicator, as employed by CAS, denotes the function or role of a chemical substance in a document. Role indicators were first used extensively in the indexing of chemical documents for commercial or public use by such organizations as IFI and

*In the printed *CA* indexes, entry numbers preceded by "P" denote patents.

the American Petroleum Institute (see Chapter 13) in the indexing of their databases. The advantage of role indicators is that their use by the searcher greatly enhances precision of retrieval. Thus, for example, if a searcher is interested in the use of a chemical only as a reactant or as a catalyst that can be specified, the number of false "drops" can thereby be minimized. There is, of course, always a danger of excessive limitation by the searcher so that answers could be lost.

Beginning in late 1995, *CA* editors introduced role indicators for all chemical substances. These indicators now go back to the limit of the online *CA* file, 1967. At this point, they are not available for the printed *CA*. Since July 1994 (Vol. 121), role indicators have been intellectually assigned by CAS document analysts. Assignment of role indicators for previous years was done by computer algorithm; this was the only practical way to proceed for the backfiles.

CA staff utilize a total of 45 role indicators: 7 broad roles ("superroles") and 38 specific roles, all of which are "upposted" to the respective superroles under which each is posted. In addition, there are 3 specific roles not upposted to any super role. Figure 7.6 depicts the role indicators utilized. The role indicators assigned by *CA* indexers for any document are shown at the end of the index entries for that document in the online *CA* files preceded by the designation "RL."

7.8. *CHEMICAL ABSTRACTS* NOMENCLATURE AND SEARCHING *CHEMICAL ABSTRACTS* FOR CHEMICAL SUBSTANCES

As previously mentioned, beginning with Volume 76 (1972), CAS made a significant change to systematic IUPAC nomenclature. Note that this change applies to the boldface index headings, not to index modifications in which original author nomenclature is used.

Examples of the old and new policies follow:

Old Policy	New Policy
Acetic acid	No change
Acetylene	Ethyne
Aniline	Benzenamine
Carbonic acid	No change
Ethylene	Ethene
Ethylenediamine	1,2-Ethanediamine
Ethylene glycol	1,2-Ethanediol
Ethylene oxide	Oxirane
Ethylenimine	Aziridine
Ethyl ether	Ethane, 1,1′-oxybis-
Ethyl methyl ketone	2-Butanone
Formic acid	No change
Phenol	No change

Some chemists believe that this policy change makes it more difficult for them to use *CA*. They contend that such heavily studied chemicals as aniline, for example, should

There are a total of 45 roles--7 broad roles (super roles) and 38 specific roles.

ANST Analytical Study

Code	Role
ANT	Analyte
AMX	Analytical Matrix
ARG	Analytical Reagent Use *
ARU	Analytical Role, Unclassified

BIOL Biological Study

Code	Role
ADV	Adverse Effect, Including Toxicity
AGR	Agricultural Use *
BMF	Bioindustrial Manufacture *
BAC	Biological Activity or Effector, Except Adverse
BOC	Biological Occurrence *
BPR	Biological Process *
BPN	Biosynthetic Preparation *
FFD	Food or Feed Use *
MFM	Metabolic Formation *
THU	Therapeutic Use *
BUU	Biological Use, Unclassified *
BSU	Biological Study, Unclassified

FORM Formation, Nonpreparative

Code	Role
GFM	Geological or Astronomical Formation
MFM	Metabolic Formation *
FMU	Formation, Unclassified

OCCU Occurrence

Code	Role
BOC	Biological Occurrence *
GOC	Geological or Astronomical Occurrence
POL	Pollutant
OCU	Occurrence, Unclassified

PREP Preparation

Code	Role
BMF	Bioindustrial Manufacture *
BPN	Biosynthetic Preparation *
BYP	Byproduct
IMF	Industrial Manufacture
PUR	Purification or Recovery
SPN	Synthetic Preparation
PNU	Preparation, Unclassified

PROC Process

Code	Role
BPR	Biological Process *
GPR	Geological or Astronomical Process
PEP	Physical, Engineering, or Chemical Process
REM	Removal or Disposal

USES Uses

Code	Role
AGR	Agricultural Use *
ARG	Analytical Reagent Use *
CAT	Catalyst Use
DEV	Device Component Use
FFD	Food or Feed Use *
MOA	Modifier or Additive Use
POF	Polymer in Formulation
TEM	Technical or Engineered Material Use
THU	Therapeutic Use *
BUU	Biological Use, Unclassified *
NUU	Nonbiological Use, Unclassified

Code	Role
PRP	Properties**
RCT	Reactant**
MSC	Miscellaneous**

*The same specific role can appear under one or more super role headings. For example, "Therapeutic Use" (THU) appears under USES and BIOL.

**PRP (Properties), RCT (Reactant), and MSC (Miscellaneous) are specific roles and are not upposted to any super roles.

Figure 7.6. Summary of CAS roles. (Reprinted with the permission of Chemical Abstracts Service.)

be indexed under that name and not under *benzenamine,* which few chemists use in conversation, normal reporting, or commerce. Although the fully systematic name for aniline is cross-referenced from aniline in the *CA Chemical Substance Index,* most trivial names are not similarly cross-referenced, except in the *Index Guide* and in the online *File Registry.* See also *CAS* sites on the Internet.

The primary intent of this effort was to group structural families together in the alphabetically ordered *CA Volume Chemical Substance Index.* Other reasons behind the use of more fully systematic index names include:

1. They are generated more consistently.
2. They are translated more readily to complete structures.

3. They are edited and verified more rapidly and consistently.
4. They reduce manual and computer search efforts.
5. They employ fewer and more systematic nomenclature principles.

A simple example of past confusion is 4-methoxytoluene, also known as 4-methylanisole. Anisole and toluene are both descriptive *trivial* names. Under the new fully systematic index policies, there is no confusion. The compound is now designated by *CA* as *1-methoxy-4-methylbenzene.*

CA editors have provided several powerful tools to help chemists adjust to the new nomenclature and make most effective use of the general subject and chemical substance indexes published under current policies. These include (a) the *Index Guide* (*IG*), (b) the *Formula Index,* and (c) the Chemical Registry System and *Registry Handbooks.*

As previously indicated, the *IG* is a collection of cross-references and explanations of headings, which should be consulted if the current *CA* index policy for a substance is not known. New for the 1985 edition was the addition to the alphabetical sequence of all the general subject index headings (except latinized genus and species names) and diagrams for acyclic and cyclic stereoparents. (Prior to the initiation of the *IG,* the chemist found such cross-references and notes scattered throughout the *Subject Index.*) The *IG* is key to efficient use of the *General Subject* and *Chemical Substance Indexes,* as well as the use of the online files, with special value in helping the chemist *translate* trivial names to the more fully systematic index names used by *CA.* Chemists using the printed *CA* should use the more recent *IG* first and then the *IG* for the 13th (1992–1996), 12th (1987–1991), 11th (1982–1986), 10th (1977–1981) or 9th (1972–1976) collective period when they are searching retrospectively in these periods. There is also an *IG* for the 8th collective period (1967–1971), but much of the latter *IG* does not apply to current practice.

In a search of the printed *CA,* the *Formula Index,* published semiannually and cumulated every five years, provides the chemist with a reliable route of access to names of individual substances as listed in the *Chemical Substance Index.* The *Formula Index* is easier and more straightforward to use than the *IG,* especially if an accurate molecular formula is already in hand. Under each molecular formula, the chemist will find the *CA* systematic name for a compound with that empirical formula, the CAS Registry Number (for positive identification), and an abstract number in which the compound is mentioned in *CA* issues. A physical property, such as boiling point, is given when an original document does not provide enough information for an unambiguous name. Chemists should, however, not let these comments discourage them from use of the *IG;* calculation of molecular formulas can also be error prone.

In use of the online files, the *Registry File* is the preferred source for identifying "correct" chemical substance names, and it is also the most current source. However, as mentioned earlier, the printed *IG* continues to be very useful in planning any online search of *CA,* as well as for planning any search of the printed volumes.

The next step after use of the *Formula Index* is to consult the *Chemical Substance Index* if textual information is desired to narrow a search for a given substance. The

Chemical Substance Index also provides references for derivatives of the substance in question.

It is helpful to understand the relationship between the *CA Formula Index* and the *Chemical Substance* and *Subject* Indexes.

Currently, as stated in the Introduction to the *CA Formula Index:*

> Cross-references to the *Chemical Substance Index* appear in the *Formula Index* for very common substances where long lists of *CA* references would otherwise be necessary. The user is best served in these cases by reference to the *Chemical Substance Index,* where study-related descriptions assist in narrowing the search to items of particular interest.

For practical purposes, CAS considers a *common substance* a substance that in a given 6-month volume has more than 20 abstract references. Thus, in Volume 99 *Formula Index* (1983), **Silver chloride** (**AgCl**) and **Silver iodide** (**AgI**) have cross-references, but under **Silver fluoride** (**AgF**), six *CA* abstract numbers are listed.

A similar policy applies to a 5-year collective period. The difference is that *a common substance* is one that would have more than 25 abstract references in the *Collective Formula Index.* Thus, in the *Tenth Collective Formula Index* (1977–1981), **Silver, isotope of mass 105,** has a cross-reference, but under **Silver, isotope of mass 106,** 50 *CA* abstract numbers are listed.

Historically, the same yardstick has not always been used. The introduction to the first *CA Formula Index* in Volume 14 (1920), stated:

> Cross-references to the *Subject Index* have been used for all simple inorganic compounds, for all minerals of definite composition, and for the organic compounds more commonly met with, in general whenever it seemed likely that users of *Chemical Abstracts* would predominatingly refer to the *Subject Index.* The *Subject Index,* because of the modifying phrases entered there, is better to use in such cases.

Thus, for certain compounds such as *simple inorganic compounds* [e.g., **AgF** (silver fluoride), $\mathbf{BaN_2O_6}$ (barium nitrate), and $\mathbf{O_4SZn}$ (zinc sulfate)] and *minerals of definite composition* [e.g., $\mathbf{Ag_3S_3Sb}$ (pyrargyrite), $\mathbf{CCa_5O_{11}Si_2}$ (spurrite), and $\mathbf{Pb_2S_5Sb_2}$ (jamesonite)], cross-references had been made a priori even when the number of entries for each such compound was small. It was believed that the average searcher would turn to the *Subject Index* for information on such compounds rather than to the *Formula Index.*

As far as *common organic compounds* were concerned, for the first *Collective Formula Index* (1920–1946), it was assumed that approximately 50 entries for a compound would justify a cross-reference to the *Subject Index,* but that arbitrary decision was applied with discretion. Entries were made in that *Formula Index* when it seemed likely that the average individual would turn to the *Formula Index* rather than to the *Subject Index* for information.

As far as the early annual *Formula Indexes* were concerned, a cross-reference to the *Subject Index* was usually substituted for entries in the *Formula Index* when the same formula and name appeared more than 10 times on cards that were edited for a given *Volume Formula Index.* Derivatives such as esters and oximes were exceptions to that rule. Their *Formula Index* entries often contained more than 10 *CA* references, and no cross-references were inserted.

Prior to Volume 71 (1969), there existed a list of permanent *Formula Index* cross-references that corresponded to the categories listed previously. The use of that list was discontinued in 1969. For the remainder of the eighth collective period (1967–1971) and for the collective periods that followed, the distinction between *simple inorganic compounds, minerals of definite composition,* and *common organic compounds* was no longer made.

Another aspect of the *Formula Index* is that, under the parent compound's formula, the reader is alerted to the existence of "derivatives" that are incompletely specified and therefore cannot be structured or systematically named, and hence cannot yet have their own separate entries in the *Chemical Substance Index* under that parent compound's name.

For example, in the *Formula Index,* there is the entry

C_2HCl
  Ethyne, chloro- *[593-63-5]*
    For general derivs. see
    Chemical Substance Index

which directs the user to entries in the corresponding *Chemical Substance Index,* such as

Ethyne
  —, chloro- *[593-63-5]*
    alkyl derivs., model compd., . . . 10668b

This type of "general-derivative cross-reference," as *CA* calls it, was first introduced in Volume 86 (Jan.–June 1977) at the beginning of the 10th collective period.

Since Volume 71 (1969), each specific chemical substance with its assigned Registry Number has been treated in the same way by subjecting it to a count of entries in a given Volume or Collective Index. Whenever more than 20 entries in a Volume 25 or in a Collective Index are encountered, a cross-reference to the *Chemical Substance Index* is inserted. The limit was reduced from 50 to 25 entries with the 12th Collective Index. While the changes may be confusing, fortunately it is unnecessary for the user to remember them since cross-references are always present in the *Formula Index* to guide the user when appropriate.

Thus, as a general rule of thumb, to ensure complete coverage of *CA,* the *Chemical Substance* or *Subject Index* needs to be consulted.

The CAS Registry System, developed by CAS beginning in 1965 as described in Chapter 6, reached a total of 17 million substances in 1997. In this system, each substance is assigned a unique, unambiguous number that is independent of the vagaries of nomenclature. For example, the CAS Registry Number for benzene is 71-43-2. In

1995 alone, over one million substances were added to the system, as compared to some 700,000 substances in 1994. (In 1995 growth was attributed by CAS, in a press release dated December 18, 1995, to the "general increase in literature and patents, but many new registrations resulted from the growth in biotechnology and genome research." More than one-third of the total resulted from human genome research, "while another third came from published work about stereochemistry and chiral drugs.")

With a Registry Number in hand, a chemist can consult the online *Registry* file to obtain such information as the official *CA* index name, synonyms, structure diagram, molecular formula, number of entries in *Files CA* and *CAOLD,* and names of other files on STN that contain this Registry Number. If the name of the chemical or its structure is known, but not the Registry Number, the *Registry* online file may be searched by name, structure, substructure, or molecular formula to determine the correct Registry Number. In addition, Registry Numbers can be found in *Registry Handbook-Number Section,* a printed publication of CAS, and in other widely available sources such as noted below.

CAS Registry Numbers have been very widely accepted by the chemical community. They can be found in such extensively used tools as the CRC *Handbook of Chemistry and Physics, Kirk-Othmer Encyclopedia of Chemical Technology, Dictionary of Organic Compounds, Merck Index,* and in many online and printed databases and files, especially those dealing with regulation by government agencies (regulatory lists). Some of these tools and databases are discussed elsewhere in this book. Registry Numbers are often found also on shipping labels, containers, company trade literature, material safety data sheets, and other items of commerce.

A few Registry Numbers may change, particularly when the original document (article, patent, etc.) supplied incomplete or partial information. In some cases, multiple Registry Numbers may be assigned to what later proves to be the same structure. As CAS learns more about such structures, old numbers are replaced by the preferred numbers. The *Registry* online file contains a complete record of current Registry Numbers, and also includes all discontinued numbers. Information about changes also appears in the cumulative printed *CAS Registry Handbook—Registry Number Update,* issued annually.

Note that it is not possible to find associated *CA* abstracts for all CAS Registry Numbers, because some of these numbers resulted from the registration of special data collections such as the EPA's *TSCA Inventory.*

7.9. *REGISTRY HANDBOOK—COMMON NAMES*

Beginning in 1977, CAS initiated the *Registry Handbook—Common Names* on microform. This tool provides access to CAS Registry Numbers through the variety of substance names that are commonly used in the chemical literature and in chemical and allied industries. Molecular formulas are also given when known.

There are two parts: the name section and the number section. Emphasis is on simple, less systematic names, such as so-called author names, literature names, common names, and trivial names. The chemist can use the number section to identify sets of synonymous common names for the same substance.

7.10. *RING SYSTEMS HANDBOOK*

Another tool available from CAS is the *Ring Systems Handbook.* This tool, which first appeared in 1984, updates and expands the coverage of the *Ring Index* (and its supplements) and the *Parent Compound Handbook,* which it replaces. Uses include the following:

- The *Ring Systems Handbook* contains information about all ring and cage systems presently in the CAS Chemical Registry System. The entries are in ring analysis order.
- Cumulative supplements to the *Ring Systems Handbook,* issued every 6 months, provide a current awareness service for alerting to new ring and cage systems entered into the CAS Chemical Registry System during processing of the primary literature for *CA*. A new, updated complete edition of the *Handbook* is issued every 3–4 years. The 1993 edition is the latest base book.
- It helps determine the *CA* index name of a ring system or a cage system; the *CA* index name can then be used to enter the *Chemical Substance Index* to *CA* to find references for documents reporting substances containing the ring or cage system.
- It helps determine whether a ring system exists and what similar rings there may be.
- When searching computer-readable files, the chemist can use information from the *Ring Systems Handbook* to formulate substructure search strategies—that is, strategies that allow one to search for a group of substances that have in common certain ring systems or products of interest.

The *Ring Systems Handbook* includes, for a ring system:

1. Ring analysis data: number of rings, size of rings, and elemental analysis of rings.
2. Chemical structural diagram illustrating the nomenclature locant numbering system.
3. CAS Registry Number.
4. Current *CA* index name used in *Chemical Substance Index* to *CA*.
5. Molecular formula.
6. *CA* reference [if the ring system was referenced in a *CA* abstract after Vol. 78 (1973)].

A sample entry is shown in Figure 7.7.

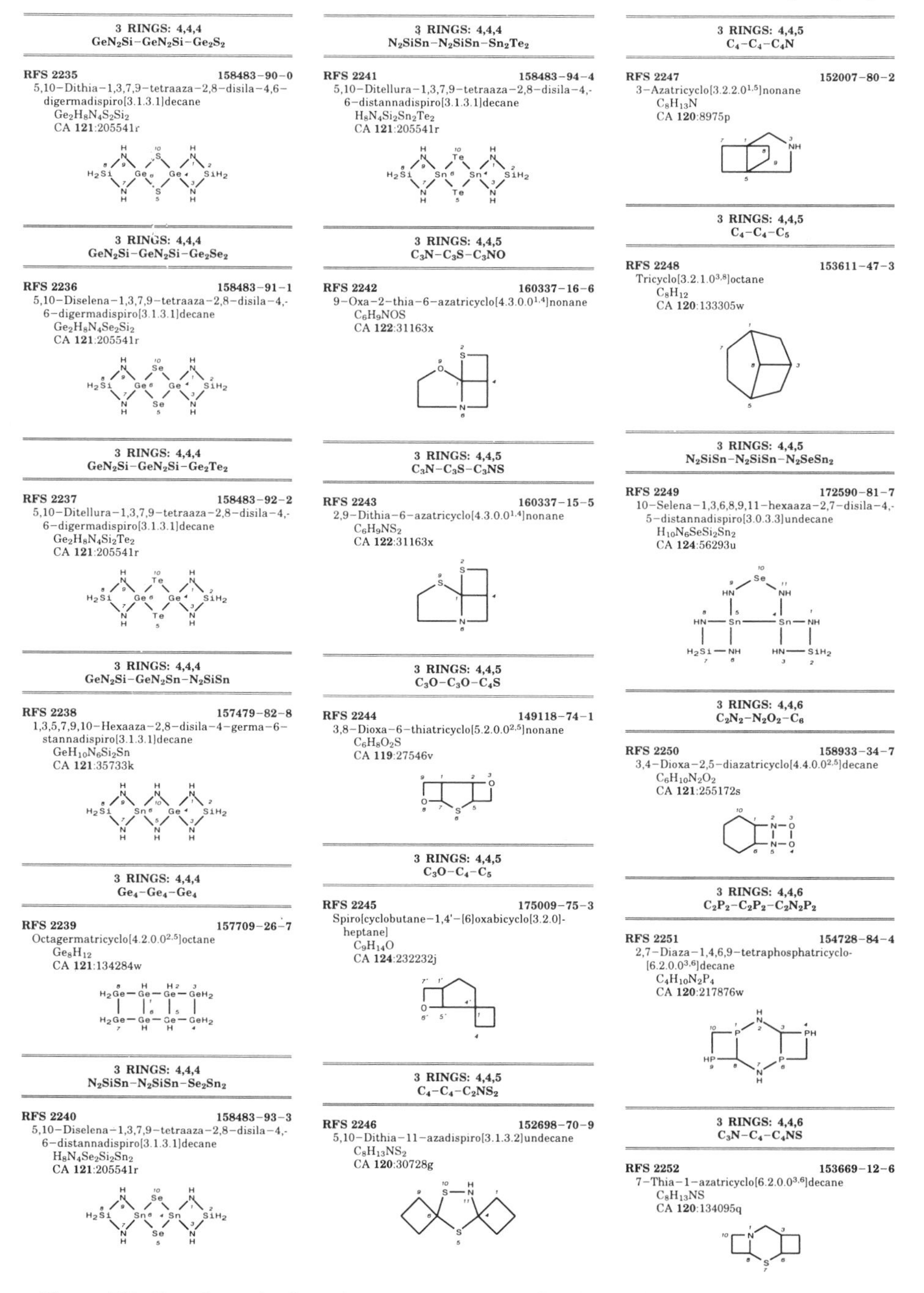

133 RSF

3 RINGS: 4,4,6
$C_3N-C_4-C_4NS$

3 RINGS: 4,4,4
$GeN_2Si-GeN_2Si-Ge_2S_2$

RFS 2235 **158483-90-0**
5,10-Dithia-1,3,7,9-tetraaza-2,8-disila-4,6-digermadispiro[3.1.3.1]decane
$Ge_2H_8N_4S_2Si_2$
CA **121**:205541r

3 RINGS: 4,4,4
$GeN_2Si-GeN_2Si-Ge_2Se_2$

RFS 2236 **158483-91-1**
5,10-Diselena-1,3,7,9-tetraaza-2,8-disila-4,-6-digermadispiro[3.1.3.1]decane
$Ge_2H_8N_4Se_2Si_2$
CA **121**:205541r

3 RINGS: 4,4,4
$GeN_2Si-GeN_2Si-Ge_2Te_2$

RFS 2237 **158483-92-2**
5,10-Ditellura-1,3,7,9-tetraaza-2,8-disila-4,-6-digermadispiro[3.1.3.1]decane
$Ge_2H_8N_4Si_2Te_2$
CA **121**:205541r

3 RINGS: 4,4,4
$GeN_2Si-GeN_2Sn-N_2SiSn$

RFS 2238 **157479-82-8**
1,3,5,7,9,10-Hexaaza-2,8-disila-4-germa-6-stannadispiro[3.1.3.1]decane
$GeH_{10}N_6Si_2Sn$
CA **121**:35733k

3 RINGS: 4,4,4
$Ge_4-Ge_4-Ge_4$

RFS 2239 **157709-26-7**
Octagermatricyclo[4.2.0.0^{2,5}]octane
Ge_8H_{12}
CA **121**:134284w

3 RINGS: 4,4,4
$N_2SiSn-N_2SiSn-Se_2Sn_2$

RFS 2240 **158483-93-3**
5,10-Diselena-1,3,7,9-tetraaza-2,8-disila-4,-6-distannadispiro[3.1.3.1]decane
$H_8N_4Se_2Si_2Sn_2$
CA **121**:205541r

3 RINGS: 4,4,4
$N_2SiSn-N_2SiSn-Sn_2Te_2$

RFS 2241 **158483-94-4**
5,10-Ditellura-1,3,7,9-tetraaza-2,8-disila-4,-6-distannadispiro[3.1.3.1]decane
$H_8N_4Si_2Sn_2Te_2$
CA **121**:205541r

3 RINGS: 4,4,5
$C_3N-C_3S-C_3NO$

RFS 2242 **160337-16-6**
9-Oxa-2-thia-6-azatricyclo[4.3.0.0^{1,4}]nonane
C_6H_9NOS
CA **122**:31163x

3 RINGS: 4,4,5
$C_3N-C_3S-C_3NS$

RFS 2243 **160337-15-5**
2,9-Dithia-6-azatricyclo[4.3.0.0^{1,4}]nonane
$C_6H_9NS_2$
CA **122**:31163x

3 RINGS: 4,4,5
$C_3O-C_3O-C_4S$

RFS 2244 **149118-74-1**
3,8-Dioxa-6-thiatricyclo[5.2.0.0^{2,5}]nonane
$C_6H_8O_2S$
CA **119**:27546v

3 RINGS: 4,4,5
$C_3O-C_4-C_5$

RFS 2245 **175009-75-3**
Spiro[cyclobutane-1,4'-[6]oxabicyclo[3.2.0]-heptane]
$C_9H_{14}O$
CA **124**:232232j

3 RINGS: 4,4,5
$C_4-C_4-C_2NS_2$

RFS 2246 **152698-70-9**
5,10-Dithia-11-azadispiro[3.1.3.2]undecane
$C_8H_{13}NS_2$
CA **120**:30728g

3 RINGS: 4,4,5
$C_4-C_4-C_4N$

RFS 2247 **152007-80-2**
3-Azatricyclo[3.2.2.0^{1,5}]nonane
$C_8H_{13}N$
CA **120**:8975p

3 RINGS: 4,4,5
$C_4-C_4-C_5$

RFS 2248 **153611-47-3**
Tricyclo[3.2.1.0^{3,8}]octane
C_8H_{12}
CA **120**:133305w

3 RINGS: 4,4,5
$N_2SiSn-N_2SiSn-N_2SeSn_2$

RFS 2249 **172590-81-7**
10-Selena-1,3,6,8,9,11-hexaaza-2,7-disila-4,-5-distannadispiro[3.0.3.3]undecane
$H_{10}N_6SeSi_2Sn_2$
CA **124**:56293u

3 RINGS: 4,4,6
$C_2N_2-N_2O_2-C_6$

RFS 2250 **158933-34-7**
3,4-Dioxa-2,5-diazatricyclo[4.4.0.0^{2,5}]decane
$C_6H_{10}N_2O_2$
CA **121**:255172s

3 RINGS: 4,4,6
$C_2P_2-C_2P_2-C_2N_2P_2$

RFS 2251 **154728-84-4**
2,7-Diaza-1,4,6,9-tetraphosphatricyclo-[6.2.0.0^{3,6}]decane
$C_4H_{10}N_2P_4$
CA **120**:217876w

3 RINGS: 4,4,6
$C_3N-C_4-C_4NS$

RFS 2252 **153669-12-6**
7-Thia-1-azatricyclo[6.2.0.0^{3,6}]decane
$C_8H_{13}NS$
CA **120**:134095q

Figure 7.7. Sample entries from the *Ring Systems Handbook*. (Reprinted with the permission of Chemical Abstracts Service.)

7.11. USE OF SUBDIVISIONS

For hundreds of commonly reported chemicals and some classes of chemicals—such as alkanes, benzene, boric acid, carbohydrates and sugars, propane, 1-propanol, pyridine, steel, sulfuric acid, and tellurium—for which there are many index references, the chemist will find the searching task simplified by subdivision of the index heading into these general categories (called *qualifiers*) in printed *CA* indexes:

Analysis
Biological studies
Formation (nonpreparative)—introduced in Volume 121 (July–Dec. 1994)
Miscellaneous—introduced in Volume 116 (Jan.–June, 1992; previously part of Uses and miscellaneous)
Occurrence
Preparation
Processes—introduced in Volume 121 (July–Dec. 1994)
Properties
Reaction
Uses—see note on Miscellaneous (above)

This practice was initiated with the eighth collective period (1967–1971).

For years prior to 1967, and for compounds for which the index is not subdivided as shown, searching the printed indexes can become both onerous and time-consuming. The chemist or engineer who desires to identify all pertinent references in *CA*, should read all index entries under headings that are not subdivided. This advice holds true for important searches, but does not necessarily apply to *everyday* searches.

Even subsequent to use of subdivisions of the heading, the chemist can never be completely sure that a search under any one of the subdivisions will yield every item of desired information. A document that emphasizes one aspect of a subject may also have valuable data on other aspects. For example, in the indexing of a document on *preparation* and *properties* of a substance, either of these subdivisions may be used. The *CA* document analyst uses judgment as to which subdivision best fits emphasis of the document. Accordingly, the user may need to consult both subdivisions to retrieve all of what is needed. The subdivision system just described can be a valuable time saver. Hopefully, the system can eventually be extended to include virtually all chemicals indexed by *CA*.

In the online file, roles serve this purpose.

In addition to the subdivisions described, index headings with large numbers of entries are also divided into chemical functional group subdivisions. These include the following:

Acetals
Anhydrides
Hydrazones
Lactones

Anhydrosulfides	Mercaptals
Compounds	Mercaptoles
Derivatives (general)	Oxides
Esters	Oximes
Ethers	Polymers
Hydrazides	

7.12. SIZE

The sheer physical bulk of the printed *CA* is imposing, almost forbidding to some. The indexes, in particular, are so large that considerable time can be required to locate what is needed in the printed volumes. For example, 75 books made up the *Tenth Collective Index* (1977–1981), the *Eleventh Collective Index* (1982–1986) comprised 93 books, and the *Twelfth Collective Index* (1987–1991) contains 115 books that occupy approximately 247 in. (over 20 ft) of horizontal shelf space.

Use of *CA* online, on CD-ROM, or in microform can greatly ameliorate the size–bulk situation and facilitate speed and ease of use. Familiarity with *CA,* and experience in its use, is also important.

Tables 7.2 and 7.3 summarize some trends in the growth of *CA*. During the early 1980s it appeared that the rate of growth of the number of chemical literature and patent documents had decreased significantly (it was essentially flat) as compared to earlier years. However, the data show that chemical information (patent and literature) sources have apparently resumed the upward trend that was the case for so many years.

TABLE 7.2. Rate of Growth of Abstracts in *CA*

Years	Average Annual Rate of Growth (Percentage)
1951–1960	8.6
1961–1970	7.6
1971–1975	7.5
1976–1980	4.0
1981–1985	−0.7
1986–1990	1.4
1991–1995	7.3

Note: The numbers of abstracts in 1995 was 687,789. The figures in the table reflect such factors as increases or decreases in the literature of chemistry and changes in national publication practices, as well as changes in CAS literature acquisition, editorial production processes, workflow, level of work-in-process, and printed publication schedules.

TABLE 7.3. *CA* Index Growth

Collective Index	Years Covered	CA Volumes	Number of Documents Referenced Including Equivalent Patents Cited
1st	1907–1916	1–10	192,000
2nd	1917–1926	11–20	218,000
3rd	1927–1936	21–30	544,000
4th	1937–1946	31–40	513,000
5th	1947–1956	41–50	663,000
6th	1957–1961	51–55	637,000
7th	1962–1966	56–65	1,024,000
8th	1967–1971	66–75	1,466,000
9th	1972–1976	76–85	2,024,000
10th	1977–1981	86–95	2,602,000
11th	1982–1986	96–105	2,812,413
12th	1987–1991	106–115	3,052,700

Reprinted with the permission of Chemical Abstracts Service.

7.13. COMPLEXITY

The development of the science of chemistry, coupled with the use of large numbers of automatic laboratory devices, have combined to produce substantially more data in recent years. This requires additional routes of access by CAS if the total literature is to be covered. *CA* today contains more data, and more indexing results, than ever before. The tools now offered by CAS are, of themselves, not necessarily more complex. It is simply a matter of more tools the user needs to master that, in the long run, should make the task easier.

7.14. PATENT DOCUMENT COVERAGE

CA utilizes INPADOC (International Patent Documentation Center of the European Patent Office) tapes as the basis for selecting patents and determining equivalents (family relationships). The agreement with INPADOC permits *CA* to provide information about family relationships in the printed *CA* only. INPADOC is discussed further in Chapter 13.

CA does not cover all chemical patent documents in its abstracts (see also Section 13.15). For example, if a patent is published first in a country other than the United States, and subsequently issued in the United States (or any other country), these equivalents (see Section 13.20) will not be reabstracted. Rather, they appear only in the *Patent Index* to *CA*. This policy resulted in the provision of cross-references in the numerical patent index for 1961–1962, and in issuing of a *Patent Concordance* beginning in 1963—an index that was eventually integrated into the *Patent Index* in the *Tenth Collective Index* (1977–1981). Prior to 1961, there was full bibliographic reference in *CA* issue text to the abstract of an equivalent patent. This policy (of not reabstracting equivalent patents) is intended to reduce costs. But it makes for considerably more effort on the part of the chemist trying to locate equivalents in printed issues of *CA,* and, unlike Derwent (see Section 13.14), it does not provide *automatic* notification when equivalents appear.

A total of 121,212 patent documents were covered in CA in 1995, representing 17.6% of the total number of documents. The corresponding figures for 1994 were 107,226 and 16.4%. The percentage of patents covered as compared to the total number of documents has remained about the same for recent years.

CA patent document coverage has improved significantly, especially since about 1960. Prior to that, patent coverage is spotty, and, at least for composition of matter patents, *Chemisches Zentralblatt* is a better source for those earlier years.

7.15. REVIEWS

CA provides excellent coverage of review papers. The *CA* definition of a review is based primarily on whether the author of the original document so designates it. If new experimental information is also provided, however, the document is not treated as a review, but as a regular research report and is indexed thoroughly.

7.16. CHEMICAL MARKETING AND BUSINESS INFORMATION

This is seldom included directly in *CA*. It appears instead in the subsidiary publication issued by CAS known as *Chemical Industry Notes* (*CIN*). *CIN* is excellent, but its abstracts (better termed "extracts") and indexes are not as good as the full *CA*. Material appears in *CIN* within a few weeks after the original. It is available both in printed form and online. A typical issue contains the following sections:

Production	Corporate activities
Pricing	Government activities
Sales	People
Facilities	Keyword index
Products and processes	Corporate name index

CIN is not as complete as some other marketing and business information sources (see Chapter 16), but it does include CAS Registry Numbers as appropriate, a valuable feature.

7.17. DISSERTATIONS

CA has provided coverage of American and Canadian dissertations since 1960. Rather than provide an abstract of the dissertation, the user is usually referred by *CA* to *Dissertation Abstracts* (see Chapter 4). This requires an often inconvenient second step, since *Dissertation Abstracts* is not readily available in many chemistry libraries and information centers, except in the online version. The content of *Dissertation Abstracts* is copyrighted, and for this reason, abstracts from this source are not given in *CA*.

7.18. COVERAGE OF DOCUMENTS FROM THE SOVIET UNION AND RUSSIA

Many Soviet documents were fully abstracted and indexed by *CA*. Some, however, were accessible only through the Russian abstracting and indexing service *Referativnyi Zhurnal* (*Ref. Zh.*). Because the abstracts in *Ref. Zh.* are copyrighted, only the titles and bibliographic information, including the *Ref. Zh.* data, were published in *CA*. The statement "Title only translated" was used instead of an abstract. Keywording and indexing were done from the title only. The use of *Ref. Zh.* has been discontinued by *CA* with the 1993 volumes of that service being the last to be utilized. Russian literature is now covered directly by *CA*.

7.19. MISTAKES

Mistakes, although infrequent, will be found in the abstracts, indexes, and other publications of CAS. Its staff members are human, and perfection is too much to ask for in such a large and complex work. Also, huge volumes of literature must be handled in relatively short periods of time. Correspondence with the editors of *CA* can almost always clear up any suspected mistake or other discrepancy.

Whenever possible, CAS staff handles mistakes in abstracts by republishing the corrected abstracts in later issues with reference to the earlier versions. Index references are then made to the correct abstracts only.

Errors in semiannual indexes are corrected in the printed Collective Indexes and online. This provides another excellent reason for discarding semiannual indexes when corresponding Collective Indexes appear. The only reason for retaining the semiannuals could be the convenience of using smaller volumes with fewer pages when date of coverage can be well identified, but this leaves open the increased possibility of being misled by any uncorrected errors.

REFERENCES

1. D. D. Ridley, *Online Searching: A Scientist's Perspective,* Wiley, New York, 1996.
2. H. Schulz and U. Georgy, *From CA to CAS Online-Databases in Chemistry,* 2nd ed. (in English), Springer-Verlag, Berlin, 1994.

Papers about *SciFinder*

1. J. Williams, "Information at the Desktop for Scientists," *ONLINE* **19,** 60{–}66 (July–Aug. 1995).

2. R. Cain and K. Schwall, "Guiding Your Literature Searching," *CHEMTECH* 25(8), 8–11 (Aug. 1995).
3. J. L. Macko and J. M. Steffy, "A New Pathway to Health and Safety Information," *Chem. Health Safety* **2**(5), 14–17 (Sept.–Oct. 1995).

8 Selected Other Abstracting and Indexing Services of Interest to Chemists

8.1. TRADITIONAL BRITISH, GERMAN, FRENCH, AND SOVIET OR RUSSIAN SOURCES

In addition to *Chemical Abstracts* (*CA*), a number of very important chemical abstracting and indexing services have been published for many years in countries other than the United States. These have been described in detail by Skolnik (1), and others and include, for example, *British Abstracts* (*BA*), *Chemisches Zentralblatt* (*CZ*), *Bulletin Signalétique,* and *Referativnyi Zhurnal, Khimiya* (*RZh* or *Ref. Zh., Khim.*) (see also Section 6.14).

Of these, only *CA* and *RZh* are still published as such. However, for complete coverage of early literature and patents, both *BA* and *CZ* need to be consulted, in addition to use of the *Beilstein* and *Gmelin Handbooks* (see Chapter 12), especially *CZ* for the literature published prior to the start of *CA* in 1907.

There is a question as to ready availability of these discontinued and older, but still important abstracting services. It is to be hoped that most better and larger libraries have space and funds needed to retain the volumes on their shelves. Even if a set of *CZ* is located, there can be an important, but unfortunate, language barrier for many English reading chemists; this barrier does not, of course, apply to use of *BA*.

Besides the advantage of earlier years of coverage, there are factors relating to content and scope of coverage. For example, as indicated in Section 6.14, *CZ* emphasizes earlier patents, especially from Germany, more than *CA* does, even for some years when the two services overlapped. In any event, because the abstracters and indexers of *BA* and *CZ* were usually different than those of *CA,* additional important information may be found in these abstracts and indexes. For example, *CZ* abstracts were noted for their length and detail as compared to other services. As could be expected, both *CZ* and *BA* were especially outstanding in coverage of publications from Germany and the United Kingdom, respectively, although their scope was international.

The following paragraphs from pages 133–134 of the book by E. J. Crane (editor of *CA* for many years), Austin M. Patterson, and Eleanor B. Marr (2) give a good comparison of *CA* and *CZ* and outline some of the history of the latter, but do not mention discontinuance of *CZ* in 1969 because it occurred after these words were written (in 1957):

This important German abstract journal [*CZ*] has value because of its early appearance, almost continuous publication, and good abstracts. There have been several changes of name, the original one being *Pharmaceutisches Centralblatt* (1830–1850). This was followed by *Chemisches–Pharmaceutisches Centralblatt* (1850–1856) and *Chemisches Zentralblatt* (1856–). *Zentralblatt* was spelled *Centralblatt* from 1856 to 1897. The *Deutsche Chemische Gesellschaft* published it from 1897 to 1945. Publication was suspended for a short time in 1945. Then two editions were issued in parallel from 1945 through 1950, one from the Eastern and the other from the Western Zone of Germany; the two editions were combined in 1951 to form a journal sponsored jointly by several scientific societies in each zone. It appears weekly. The usual form of reference to it is *Chem. Zentr.* or *Chem. Centr.*

Coverage. Up to 1919, *Chemisches Zentralblatt* was not a comprehensive abstract journal because it limited its abstracts to papers dealing directly or indirectly with pure chemistry, and it covered German chemical work more thoroughly than that of other countries. In 1919, the abstract section of *Angewandte Chemie* (then called *Zeitschrift für angewandte Chemie*) was made a part of *Chemisches Zentralblatt* and ever since it has endeavored to cover the world's periodical literature thoroughly for both pure and applied chemistry, though with some delays and omissions as an effect of World War II.

Before 1919 only German patents were abstracted by *Chemisches Zentralblatt,* and then only part of them, the leaning being toward patents on organic chemical substances. From 1919 to date *Chemisches Zentralblatt* has covered the world's chemical patents quite thoroughly. Throughout the years its patent coverage has not always been the same as that of *Chemical Abstracts* because the two journals have not made abstracts from the patents of all of the same countries. For example, *Chemisches Zentralblatt* does not make abstracts of Japanese patents whereas *Chemical Abstracts* has done so since 1917, and *Chemical Abstracts* has not always covered Russian patents, even in part. Because neither journal attempts to abstract all of the chemical patents of the world, the two sets of abstracts supplement each other in this respect.

Since 1926 new books have been announced by title in each of the divisions of the abstracts, usually at the beginning of the division or subdivision, but *Chemisches Zentralblatt* does not abstract or review books. The total number announced each year is smaller than reported in *Chemical Abstracts*.

Chemisches Zentralblatt's coverage of the world's chemical literature was adversely affected by World War II and its aftermath, from 1939 through 1951. By 1952 it was making abstracts from papers in 4,925 periodicals. It is now [1957] approaching its prewar standard and the abstracts are reasonably prompt. Abstracts of some papers originating in the USSR and its satellite countries appear sooner in *Chemisches Zentralblatt* than in *Chemical Abstracts*.

It is very unusual to find an abstract of a paper in *Chemisches Zentralblatt* that is not also in *Chemical Abstracts* from 1907 on because of the care that the editors of the latter use to prevent omissions. If it does occur, it is not likely to represent a paper of major chemical important but rather one of little chemical interest in a borderline field between chemistry and another science.

The same book gives a good description of *British Abstracts* (pp. 135–136):

Both the Chemical Society (London) and the Society of Chemical Industry (London) published abstracts in their representative journals, the *Journal of the Chemical Society* (London) and the *Journal of the Society of Chemical Industry,* until the end of 1925 (see below). In 1926 the two societies combined their abstracting activities and founded *British Chemical Abstracts,* which continued under this name through 1937. In 1938 *Physiological Abstracts* was merged with *British Chemical Abstracts* and the name was changed to *British Chemical and Physiological Abstracts.* In 1939 a section on anatomy was added. The name was changed again in 1946, *British Abstracts* becoming the new name.

Coverage. Up to 1938, as its history would lead one to expect, *British Abstracts* covered only the chemical field with considerable thoroughness, but after that date it became something more than a chemical abstract journal, with anatomy and nonchemical phases of physiology as the principal additional fields of coverage. The coverage of chemistry was extensive but not as comprehensive as that of *Chemical Abstracts* for the same period, 1926–1953. Applied as well as pure chemistry was abstracted, and many patents were covered. The excellent, promptly appearing abstracts, once informative, tended to become descriptive about the time of World War II.

Classification of Abstracts. From 1926 through 1944 there are two parts, A (Pure Chemistry) and B (Applied Chemistry). From 1944 through 1953 there are three parts: A, B, and C (Analysis and Apparatus). The parts have been subdivided in somewhat different ways at different times; they were issued and bound separately.

Indexes. Annual indexes cover subjects, authors, and patents. Collective indexes for parts A and B (1923–1932 and 1933–1937) cover authors and subjects. The 1923–1932 index includes abstracts that were published in the *Journal of the Chemical Society* (London) and in *The Journal of the Society of Chemical Industry* (1923–1925). The subject indexes are somewhat less extensive than those of *Chemisches Zentralblatt* and of *Chemical Abstracts.*

As previously mentioned, *CZ* ceased publication as such in 1969 and *British Abstracts* in 1953.

The Russian abstract journal *RZh, Khimiya* (Chemistry) started publication in 1953 and is one of a series of abstract journals published by VINITI (All-Russian Institute for Scientific and Technical Information). Its stated goal has been the coverage of all the world's chemical literature.

There are major impediments to the use of *RZh* in many Western countries, because of the lack of ready availability (except in major libraries) and the widespread lack of understanding of the Russian language. Comparison with *CA* shows that *RZh* coverage is not as extensive. *CA* does cover most of the available chemical literature from Russia and other countries of the former Soviet Union.

The balance of this chapter deals with relatively smaller abstracting and indexing services. Potential benefits of use of some of these services, as compared to *CA,* may include one or more of the following:

1. Broader depth of coverage in the area of concentration, for example, more specialized publications, very brief reports, abstracts of presentations, letters to the

editor, or industry news and developments that may not be covered by *CA* despite its broad scope. However, when taken as a whole, coverage of *all* publications of chemical interest by *CA* is superior to that of any other service, and *CAplus* does cover all of the materials on the title pages of the core journals.
2. More detailed abstracts that may be slanted to and written from the perspective of the particular need of a special interest group within chemistry or of another discipline outside chemistry. In some cases, however, abstracts may be identical with *CA* abstracts because of special arrangements that may have been made.
3. Indexes that are easier to use because they have more limited scope, are less complex, and are smaller than those of *CA*. Corresponding electronic databases may be quicker and easier to search for the same reason. Furthermore, indexing may be more specialized and in depth than that of *CA* in fields covered and may use jargon or technical terms more familiar to chemists and others working in the fields.
4. Individual issues that are easier to visually scan than those of *CA* because of brevity.
5. Unique presentation of abstracts and/or indexes that may be particularly well suited to more specialized purposes. A good example is in publications of the *Institute for Scientific Information* as described later in this chapter.
6. Staffs that may be able to provide specialized services on demand, including assisting in and performing of evaluations of state of the art, obtaining translations of documents that may not be readily available elsewhere, and referrals to outside practitioners and other experts who may be able to help solve laboratory and other problems.

For reasons such as these, chemists should determine whether abstracting and indexing services exist in their fields of specialization. This can usually be done with the assistance of the local chemistry librarian. Many such services are readily available online and, if so, an appraisal of any benefits is made easier. A full appraisal, however, can best be made by study of both printed and online versions. Parallel comparative evaluation with *CA* should help the chemist make a decision.

Criteria recommended for evaluating and selecting abstracting and indexing services include

1. Reputation in scientific community.
2. Quality of staff and sponsoring organization or publisher.
3. Quality and depth of abstracts and indexes.
4. Depth of coverage, including number and kinds of publications covered.
5. Ease of use.
6. Speed of coverage (abstracts and indexes).
7. Reasonableness of price.
8. Availability on online, Internet, or CD-ROM version.

9. Assistance to users, including knowledgeable *help* (toll-free) telephone lines, well-written user guides, copy service, and other user aids. (Abstracting and indexing services often offer auxiliary services on a fee basis. Thus, they will frequently provide copies of articles and patents included in their files. In addition, many will search their databases for customers on a "bureau" basis. Many offer customized alerting products and related assistance. Some also offer translation services/clearinghouses.)

8.2. INSTITUTE OF PAPER SCIENCE AND TECHNOLOGY, CAB INTERNATIONAL, AND OTHER "SMALLER" ABSTRACTING AND INDEXING SOURCES

One good example of a "smaller" abstracting and indexing service is found in the work of the Institute of Paper Science and Technology (IPST), which is located on the campus of the Georgia Institute of Technology, Atlanta, GA, where it moved in 1989 after many years in Appleton, WI. The operations of the Institute include research and development, an educational program with faculty and students, and information services. IPST offers a complete array of information services especially geared to chemists and engineers doing research and other work in pulp-and-paper chemistry and engineering. Funding is based primarily on income from member companies as based on tonnages of product produced by these companies.

The Director of the Information Services Division is Robert G. Patterson, and the Chief Editor, responsible for the abstracting and indexing products, is Rosanna M. Bechtel (404-853-9500, 800-558-6611). Some 25 persons are dedicated to abstracting and indexing, including 15 freelance abstracters. In the past, *Chemical Abstracts* was the source of about 2–3% of the abstracts, but this practice was discontinued in 1991.

The principal information service products include

1. *Abstract Bulletin of the Institute of Paper Science and Technology (ABIPST).* The monthly printed version started in 1931, and the online version, known as *PAPERCHEM,* begins coverage with 1967. The *Bulletin* is also available in microform and CD-ROM versions. About 20,000 abstracts are included each year, including articles, conference proceedings, and worldwide patents. There are monthly and annual subject indexes, and there is also an index of trade names, and names of companies, organizations, and products. Indexing is relatively deep: approximately 10–12 terms per document on the average. All abstracts are now written specifically for the *Bulletin,* and are, of course, written from the perspective of the pulp-and-paper industry. The product is online through both DIALOG and STN (please see Chapter 10). It is also on the Internet at www.paperchem.com. The CD-ROM version, called *PaperSearch,* covers a 5-year period and is updated quarterly; it includes approximately 75,000 abstracts. Subjects of particular interest to chemists and chemical engineers include papermaking chemicals, cellulose chemistry, pulping, bleaching, and effluent treatment. The 1997 subscription price was $1200/volume in North America (nonmember rate).

2. *Graphic Arts Bulletin (GABIPST).* This service, inherited from the Rochester Institute of Technology, was first published by IPST in 1993. It covers approximately 12,000 abstracts per year, although most of the topics covered relate to printing processes and the management of printshops, GABIPST also abstracts articles on the chemicals utilized in inks and printing and their recovery. Subscription cost is $400 per volume for nonmembers in North America.

3. *Paper Technology Updates.* These appear on a monthly basis and include abstracts relating to three topics: recycling, bleaching, and the environment. The cost for subscribing to these is much more economical than subscribing to the full bulletin; each of the update series costs $225/year in North America.

4. *PaperClip.* This product contains the tables of contents of the leading 100 journals in the field of paper science and technology. It appears twice a month. North American subscription cost is $200 per year.

5. *Forthcoming Meetings.* Appearing every other month, this lists and indexes upcoming international meetings in the field (1997 nonmember price $100).

The popular Bibliographic Series, previously offered for many years, was unfortunately discontinued in the late 1980s. This consisted of hundreds of annotated bibliographies on a series of key topics, most of which were of great interest in the paper industry. However, the Institute does offer a retrospective searching capability.

In addition to its various publications, the Information Services Division includes the American Museum of Papermaking and the William Haselton Library. The library contains a unique collection of documents on pulp and paper and provides photocopies to worldwide users. Translation services are also offered. More information about the Information Services Division is available on the Internet at http://www.ipst.edu/isd.

In the field of agricultural and related sciences, an outstanding abstracting and indexing service, in addition to *CA*, is *CAB Abstracts.* This is a product of CAB International, formerly known as the Commonwealth Agricultural Bureaux, now the Center for Agriculture and Biosciences International (CABI) located in Wallingford, UK. CABI says that it maintains the world's largest agricultural database and says that it is far broader in scope than the U.S. Department of Agriculture's *Agricola* product, which emphasizes U.S. agriculture. CAB abstracts cover virtually every branch of agricultural science and are available in both printed and online forms. An expert staff also provides literature searches, bibliographies, translations, and document copies. Printed abstract journals are published covering highly specialized areas, for example, rice or seeds or soybeans. Broader areas are also covered, as, for example, soils and fertilizers, or weeds or applied entomology. Many of the publications of CAB are succinct enough to be very conveniently scanned. In all, there are 24 main abstract journals, 17 specialist abstract journals, 6 primary journals, 7 serial publications, and 5 news and information periodicals. CABI also publishes books. Online access to the comprehensive *CAB* abstracting and indexing database is available from 1973 to the present through a number of online hosts: CAN/OLE, DIMDI, European Space Agency, DIALOG, DataStar, and STN. CAB has a staff of over 500 persons, and conducts field research and training in addition to its publishing activities. The Director General is James H. Gilmore. CAB is an intergovernmental nonprofit organization owned and governed by its member countries, which now total some 40 nations worldwide. The mission is "to help improve human welfare worldwide through the dissemination, application, and generation of scientific knowledge in support of sustainable development, with emphasis on agriculture, forestry, human health, and the management of natural resources, and with particular attention to the needs of developing countries." There is a North American office at 198 Madison Avenue, New York, NY 10016 (800-528-4841, attention Mrs. Pam Sherman).

An interesting and unique abstract service is *Maro Polymer Notes,* formerly *Drexel Polymer Notes,* which consists of highly concise abstracts of selected worldwide journal articles and patents as edited by a staff lead by Roger D. Corneliussen, formerly a professor of chemistry at Drexel University. Publication was initiated in 1985, and the publisher is Craig Technologies, PO Box 38, Folcroft, PA 19032 (610-461-8800). The product is also available in diskette form from 1989 to date.

The unique features are that each document included is not only read and abstracted concisely, but then also ranked as to depth (comprehensiveness) and as to probable interest to the reader. Approximately 12,000 documents are covered each year. Because of the time involved in the evaluation feature, material appears about 45 days after the original article. Availability on the World Wide Web portion of the Internet is planned for 1998 via Chapman and Hall Publishers.

Services include searches of the database and copies of articles and patents. Another service consists of weekly or monthly alerts on specific topics of interest to the user. There are also "prepackaged" Maro Special Reports available, including such topics as acrylics, adhesives, antioxidants, catalysts, and a number of others. Each Report pinpoints significant articles and patents over recent years.

The product is also available in a form suitable for use with a special computer drive, that is, a 3.5 in. external disk that operates at very high speed and high capacity. The drive can store 100 megabytes (Mbytes) or more. This hardware is made by such companies as Iomega and Syquest.

Another service of interest to polymer chemists and engineers, especially those in industry, is offered by Technomic Publishing, Lancaster, PA. Products include *Urethane Abstracts,* which is published monthly. Another is *Polymer Blends, Alloys, and Interpenetrating Networks-Abstracts,* also monthly. In both cases, journals, magazines, and patents are covered, and both publications are also searchable online through *PLASPEC* as noted in Section 15.16.

Technomic also publishes the newsletters *Plastics in Building Construction* and *Urethane Plastics and Products,* both monthly. The first of these is online through *PLASPEC.*

There are a number of other "smaller" services, especially in the more applied or technical aspects of chemistry. These are too numerous to mention here, but this section helps illustrate potential advantages. Several important relatively large services that deserve the attention of the chemist are noted in the pages that follow.

8.3. INSTITUTE FOR SCIENTIFIC INFORMATION

The Institute for Scientific Information (ISI), founded in 1960 by Dr. Eugene Garfield, is an outstanding for-profit organization that offers a wide variety of services to the chemist. ISI is located in Philadelphia, PA, with a European sales office located in Uxbridge, England, a Tokyo office, and data processing facilities in Cherry Hill, NJ and Limerick, Ireland. ISI has more than 700 employees worldwide. The Web site is www.isinet.com.

Effective April 3, 1992, The Thomson Corporation, Toronto, Canada, acquired a majority interest in ISI from JPT Publishing Group. Thomson's Publishing and Information Group comprises more than 120 companies located throughout the world. Thomson also owns Derwent Information (described in Section 13.14) and a number of other leading organizations in the information industry.

The success of ISI is due in large measure to the innovative talents of Garfield, now Chairman Emeritus, who continues to direct the Company's efforts. Michael Tansey is President and Chief Executive Officer. Shelly H. Rahman is Director of the Chemical Information Division.

About 50% of users of ISI are academic, and the balance are equally divided between industry and government and other nonprofit organizations (about 25% each).

Some ISI publications and other services are listed throughout this book. Several related to abstracting and indexing are described in the following paragraphs.

The cornerstone of ISI's coverage of chemistry is *Index Chemicus* (*IC*), formerly *Current Abstracts of Chemistry and Index Chemicus* (*CAC&IC*). This is a highly regarded weekly abstracting and indexing service, published since 1960, which tells what new organic compounds and syntheses are reported in over 100 of the most important journals in organic chemistry. About 200,000 new compounds (including intermediates) are reported annually; over 5.7 million compounds have been recorded since 1960. ISI says that this represents over 90% of all new organic compounds reported in journal literature. Coverage is prompt; ISI says its goal is to see that articles appear in *IC* within 60 days after the article is published. *IC* is also available as a database with monthly cumulative updates.

Graphic and narrative abstracts are prepared from the original article selected from the source journal. Flow diagrams are used extensively to facilitate rapid scanning and discovery of material relevant to the user's research. "Use profile" and data-alert symbols signal the user as to whether the author of the original mentions potential or tested biological activities of the compounds, and highlight the presence of labeled compounds, explosive reactions, and new synthetic methods. Further details on new methods covered in *Index Chemicus* appear in another ISI publication, *Current Chemical Reactions,* described later in this chapter.

Each *Index Chemicus* issue contains indexes by author, labeled compounds, biological activity, and keywords. There is a list of journal titles covered in each issue. Indexes are cumulated annually. The cumulations include an annual organizational source index.

The product is available on CD-ROM, print, and ISIS database formats. A 22-year microform cumulation of *IC* is also available, containing abstracts from 1960 to 1981;

included are cumulated indexes for the years 1962 to 1981 on subjects, authors, and journals.

The strong points of *IC* are speed of coverage (lag time of receipt to publication is said to be 60 days or less) and easy-to-read, highly graphic abstracts that include structural diagrams and reaction flows.

As noted, although many important journals are covered, many journals are not covered. However, studies at ISI show that most new compounds are reported in only a relative handful of journals. Patents are not included. *IC* was previously available online, but this is no longer the case.

The subscription cost (1998) for *IC* in print is $7200 for industrial users. In an educational setting it is $4325. Higher rates prevail in Japan.

IC is now available as a relational database using MDL Information Systems *ISIS* software in conjunction with Oracle (by Oracle Systems Corp.). This permits searching both structures and textual data. Further additional information is present, such as the reference for a preliminary communication, identification of a natural product source, bioassays performed, proposed biological activities, and a controlled vocabulary of index terms. This database is available from 1993 to date.

IC has also been released as a standalone CD-ROM running under Windows. The CD permits substructure and text searching, and has the graphical summaries on each article as found in the print edition. A typical abstract is shown in Figure 8.1.

Details on new and newly modified synthetic methods appear in the related monthly publication, *Current Chemical Reactions* (*CCR*), which began publication in 1979. This product contains data from over 100 leading organic chemistry journals. In *CCR,* reaction schemes, experimental data, and yields are included in addition to bibliographic information and author abstracts. An index section, containing journal, author, subject terms (reaction name and type, starting materials, products, reagents, catalysts, explosive reactions, biologically reactive products, or newly synthesized labeled compounds), and corporate address indexes, is incorporated into each monthly issue and is cumulated annually. More than 500 articles that relate to new or newly modified chemical reactions appear in a typical monthly issue (see Fig. 8.2).

The purposes of *IC* and *CCR* are different. *IC*'s purpose is to report new compounds, while *CCR* reports new synthetic methods. *IC* and *CCR* have the same core journals for the print products, although *CCR* print also highlights review articles from the core journals and additional review journals not covered in *IC.* For articles covered in both issues, *IC* may have a reaction flow, but there will be more detail about the reaction in *CCR,* such as time, temperature, pressure, yields, advantages, limitations, and key steps. Reaction descriptor headings are included for each abstract.

Also available is the *CCR Database,* a machine-readable database. Key features include all that is in the printed format, and, in addition: experimental data, including specific conditions and yields; manual atom-to-atom mapping and bond highlighting; the ability to access intermediate steps of multiple-step syntheses as individual reactions; and searchable English-language author abstracts (since 1991). Coverage is also broader than the print version; this database covers over 350 journals and U.S. patents since 1988. The core journals that are in the print edition are manually

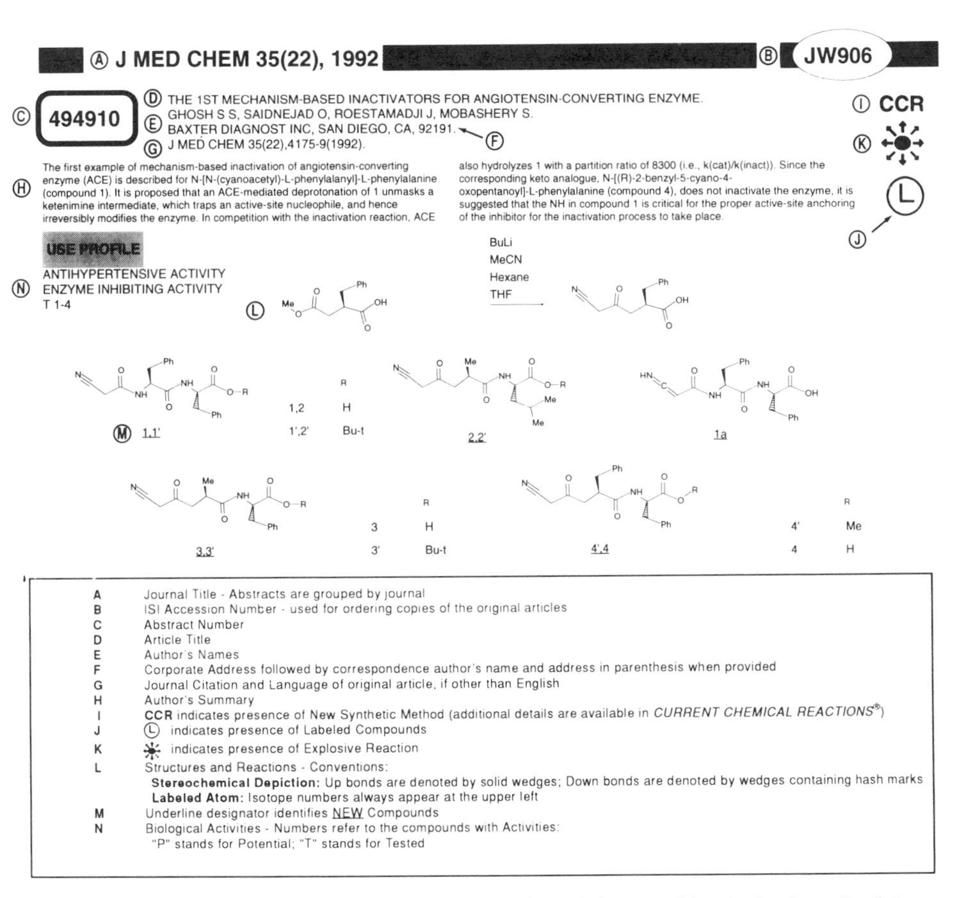

Ⓐ J MED CHEM 35(22), 1992 Ⓑ JW906

Ⓒ 494910

Ⓓ THE 1ST MECHANISM-BASED INACTIVATORS FOR ANGIOTENSIN-CONVERTING ENZYME.
Ⓔ GHOSH S S, SAIDNEJAD O, ROESTAMADJI J, MOBASHERY S.
BAXTER DIAGNOST INC, SAN DIEGO, CA, 92191. Ⓕ
Ⓖ J MED CHEM 35(22),4175-9(1992).

Ⓘ CCR

Ⓗ The first example of mechanism-based inactivation of angiotensin-converting enzyme (ACE) is described for N-[N-(cyanoacetyl)-L-phenylalanyl]-L-phenylalanine (compound 1). It is proposed that an ACE-mediated deprotonation of 1 unmasks a ketenimine intermediate, which traps an active-site nucleophile, and hence irreversibly modifies the enzyme. In competition with the inactivation reaction, ACE also hydrolyzes 1 with a partition ratio of 8300 (i.e., k(cat)/k(inact)). Since the corresponding keto analogue, N-[(R)-2-benzyl-5-cyano-4-oxopentanoyl]-L-phenylalanine (compound 4), does not inactivate the enzyme, it is suggested that the NH in compound 1 is critical for the proper active-site anchoring of the inhibitor for the inactivation process to take place.

USE PROFILE
ANTIHYPERTENSIVE ACTIVITY
Ⓝ ENZYME INHIBITING ACTIVITY
T 1-4

BuLi
MeCN
Hexane
THF

Ⓜ 1,1'

	R
1,2	H
1',2'	Bu-t

2,2'

1a

3,3'

	R
3	H
3'	Bu-t

4',4

	R
4'	Me
4	H

A	Journal Title - Abstracts are grouped by journal
B	ISI Accession Number - used for ordering copies of the original articles
C	Abstract Number
D	Article Title
E	Author's Names
F	Corporate Address followed by correspondence author's name and address in parenthesis when provided
G	Journal Citation and Language of original article, if other than English
H	Author's Summary
I	**CCR** indicates presence of New Synthetic Method (additional details are available in *CURRENT CHEMICAL REACTIONS®*)
J	Ⓛ indicates presence of Labeled Compounds
K	indicates presence of Explosive Reaction
L	Structures and Reactions - Conventions: **Stereochemical Depiction:** Up bonds are denoted by solid wedges; Down bonds are denoted by wedges containing hash marks **Labeled Atom:** Isotope numbers always appear at the upper left
M	Underline designator identifies NEW Compounds
N	Biological Activities - Numbers refer to the compounds with Activities: "P" stands for Potential; "T" stands for Tested

Figure 8.1. Model abstract from *Index Chemicus.* (Copyright owned by the Institute for Scientific Information, Philadelphia, PA.)

scanned, article by article, during selection. Several examples of a new synthetic method may be included for better retrieval. The supplemental journals are scanned by a broad alert profile that highlights articles for possible inclusion, and these are then scanned by the chemical indexers. It is updated annually, and back files since 1986 are available. The database currently includes over 300,000 reactions, and is updated by an average of about 30,000 new entries each year. There is a cumulative file from 1986 to 1994; the second release for 1995 is available as a full year or as part of a 1986–1995 cumulation.

The *CCR Database* is available for REACCS, ISIS, and Daylight reaction softwares as well as a generic file format. As compared to the print edition, the database product permits reaction substructure searches and other similarity searches.

Cost (1998) to subscribers in the United States is $1325 for a basic print subscription. The annual lease for one year of the *CCR* database varies according to the number of users.

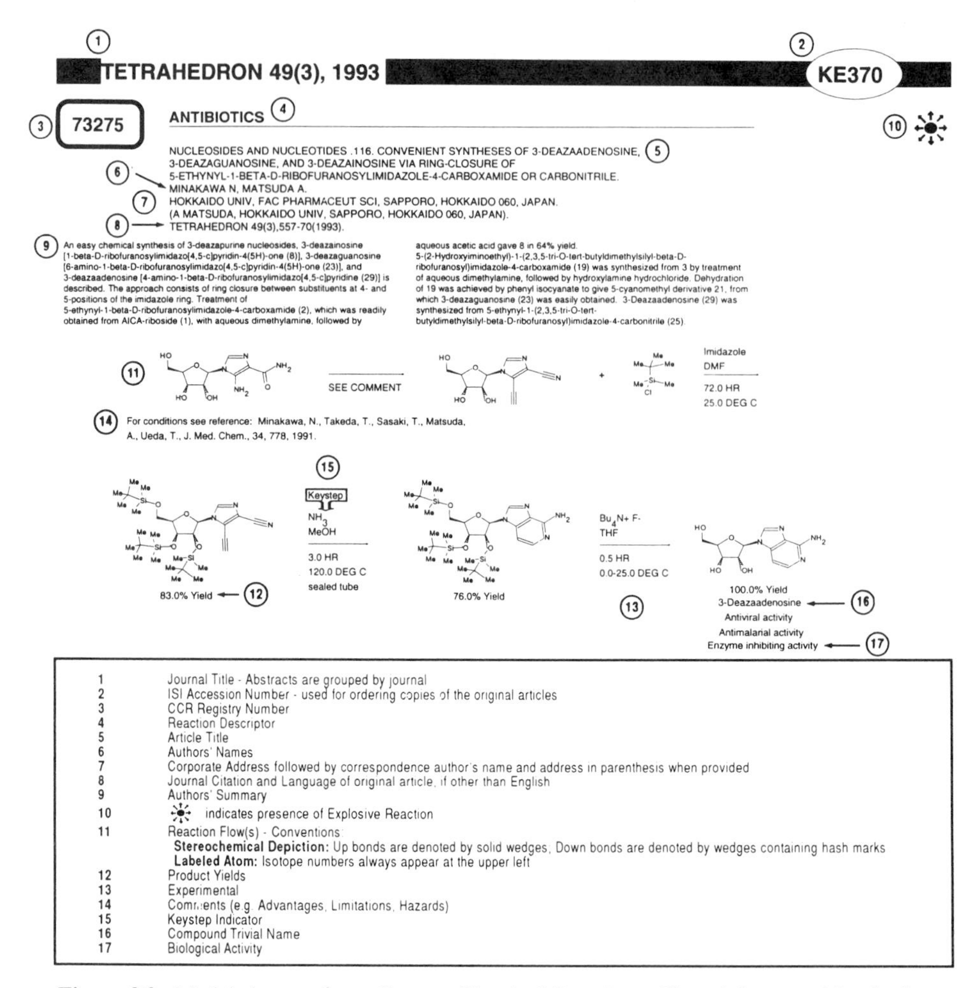

1	Journal Title - Abstracts are grouped by journal
2	ISI Accession Number - used for ordering copies of the original articles
3	CCR Registry Number
4	Reaction Descriptor
5	Article Title
6	Authors' Names
7	Corporate Address followed by correspondence author's name and address in parenthesis when provided
8	Journal Citation and Language of original article, if other than English
9	Authors' Summary
10	indicates presence of Explosive Reaction
11	Reaction Flow(s) - Conventions: **Stereochemical Depiction:** Up bonds are denoted by solid wedges, Down bonds are denoted by wedges containing hash marks **Labeled Atom:** Isotope numbers always appear at the upper left
12	Product Yields
13	Experimental
14	Comments (e.g. Advantages, Limitations, Hazards)
15	Keystep Indicator
16	Compound Trivial Name
17	Biological Activity

Figure 8.2. Model abstract from *Current Chemical Reactions.* (Copyright owned by the Institute for Scientific Information, Philadelphia, PA.)

ISI offers three Specialty Citation Index products that should be of interest to chemists: the *Chemistry Citation Index,* the *Biochemistry and Biophysics Citation Index,* and the *Materials Citation Index.* Briefly stated, a citation index captures cited references within the published literature. Thus, these products, like ISI's other citation products, allow the searcher to take a known, relevant paper and find other, more recent papers that cite it. Citation indexing brings an added dimension to searching the literature by allowing the user to search forward in time. A cited reference search can often lead the user to relevant, recently published research that cannot be readily found by other methods, such as a traditional subject search.

The *Chemistry Citation Index* provides information from virtually all areas of chemical research. Within each bimonthly issue of this database on CD-ROM, the searcher will discover information about major current developments in chemistry. It indexes items in the journals, books, and conference proceedings selected for coverage and that are considered key to chemistry. In addition, ISI's entire database of nearly 8000 journals—as well as thousands of books and proceedings—is scanned daily for relevant material outside the core literature. The (1998) price for new subscribers is $2320, which includes data back to 1991.

A new subscription to *Biochemistry and Biophysics Citation Index,* which includes data back to 1992, is $1170, while the *Materials Citation Index,* which includes back-year data to 1991, is $1170. There are networking options as well for all Specialty Citation Indexes.

In addition, ISI's Citation Indexes offer another search feature known as "Related Records." This is a unique mechanism that increases retrieval of relevant articles by extending the power of citation indexing. This extension links all the articles that have one or more references in common—references that indicate subject relationships not always evident by article titles. When one record of interest is displayed, with a single keystroke the searcher can receive additional articles that are related to the search.

ISI also offers multidisciplinary citation databases, such as the *Science Citation Index* (*SCI*), which is unique and valuable because of the inclusion of the cited reference feature in combination with the multidisciplinary coverage. It is available in print, CD-ROM, magnetic tape and online. The online version is now accessible through DIALOG and DataStar, DIMDI, and STN (*SCISEARCH* is the usual name for the online file). If using this file on STN, the user can take advantage of STN's command language, which adds special features for citation searching and crossfile searching to give the searcher further powerful techniques. Charles E. Huber (3), of the University of California, Santa Barbara, has presented a brilliant discussion of how this tool can best be utilized. He also points out some areas for further improvement.

For example, searchers who find one or more references of interest in *File CA* (or any of a number of specific other major files available through STN) have the option of crossing over to *File SCISEARCH* to identify any papers in *SCISEARCH* that cite the references in question. One can then return to *CA,* if desired, to get the abstracts of these newly found citing papers. It is reasonable to assume that these papers are related to the one originally found and that the findings of the search are thus extended in a way that would be difficult to do any other way.

In addition, when *SCISEARCH* is utilized on STN, one can perform a Related Records search. As explained above, given a reference of interest, this feature links other references that share one or more references in common—references that indicate subject relationships not always evident by article titles. It is assumed that the more references in common, the closer the relationship between the articles is likely to be. STN is the only online host in which this operation can be performed, although it is also available, in less powerful form, in the CD product. On STN, articles may be ranked based on the number of cited references hit or the number of references shared.

The *Science Citation Index* on CD-ROM is available with or without abstracts and provides instant, electronic access to current bibliographic data and cited references. User-friendly help functions enable the searcher to master searching techniques with ease. The chemist can create customized search profiles, which can then be saved and modified, enabling one to save hours of research time each week. Boolean logic and truncation features can be utilized to broaden or narrow the search.

Science Citation Index in print is $15,020 for 1998; CD-ROM is $16,190 for the 1998 series; CD with abstracts is $20,890; print and CD combinations are also available.

Arrangement of indexes is as follows:

1. *Citation Index.* This important tool lists documents referred to (cited) in the current literature alphabetically by name of author. It permits the user who already knows one or more authors who have written in a field of interest to identify other items citing the previously known pertinent work. This can lead to a network or chain of related references. For example, users may wish to know who has cited their own work or the work of colleagues.
2. *Source Index.* This index is alphabetically arranged by author, and there is also a separate index by organization. It is a way of keeping track of work by specific authors and organizations.
3. *Permuterm Subject Index.* This is an index based on the original words in the titles of items covered. All significant words in a title are coupled or paired together.

The print format of *SCI* is published six times a year and cumulated annually going back to 1945. There are also cumulations covering 1945–1954, 1955–1964, 1965–1969, 1970–1974, 1975–1979, 1980–1984, 1985–1989, and 1990–1994. This can be a unique and powerful tool when appropriately utilized. The most significant feature is, of course, the citation indexing employed.

In 1995, ISI released an exciting and powerful new tool, the *Reaction Citation Index.* This database combines the features and benefits of ISI's *Current Chemical Reactions* and *Science Citation Index* to provide a unique literature navigation tool not previously available to chemists. Reactions are covered back to 1986, and literature is covered back to 1981. There are no immediate plans to include patent citations. This magnetic tape database is updated twice a year for reactions and monthly for bibliographic information.

By incorporating cited references and Oracle's relational database management system (RDBMS), the *RCI* establishes conceptual links between chemistry. As a result, the chemist can perform current or retrospective searches of the information related to associated synthetic methods and their reaction structures.

The *Reaction Citation Index* is believed to be the only chemical database that can offer this capability. This is because the publisher captures cited references of the articles in the chemistry journals covered—the "footnotes" of every article in the database—and makes them available for searching related pieces of research.

The *Reaction Citation Index* provides the same benefits associated with *Current Chemical Reactions* and *Science Citation Index* (searchable structures, complete bibliographic data, full-length, English-language author abstracts). In addition, this database enables the user to "hypernavigate" from core journal articles to conceptually related articles so a researcher can discover parallel information, "hypernavigate" forward in time from one discovery to other related developments so as to see how a reaction has been improved, find a critical correction or locate relevant review articles, and "hypernavigate" backward in time to identify research that has contributed to present discoveries so as to trace the history of an idea.

In a recent development, it was announced that ISI's *Science Citation Index Expanded* (offers greater coverage than online or disk versions) can be searched through the proprietary software known as *Web of Science,* a Web-browser interface.

Other important ISI services include *Research Alerts* and *Current Contents,* which are discussed in Chapter 4, and document delivery services. ISI has also developed indexes to reviews, proceedings, and multiauthored scientific books.

Unlike Chemical Abstracts Service, ISI does not offer "bureau" service; that is, ISI does not conduct searches on behalf of the user or customer—it is to be hoped that this decision could be reconsidered by ISI.

8.4. THE ROYAL SOCIETY OF CHEMISTRY

The Royal Society of Chemistry offers significant abstracting and current awareness publications.

Chemical Engineering and Biotechnology Abstracts(*CEABA*) is an abstracting product of The Royal Society of Chemistry, the DECHEMA, the Institute of Chemical Engineers, and FIZ Chemie. The intended audience is chemical engineers and biotechnologists, and the content ranges from theory to practical applications and operating concerns, including corrosion, plant safety, and personnel. It is available in printed form, online (DataStar, DIALOG, Fiz Technik, Questel•Orbit, and STN), and as a CD-ROM. Subsets of the printed version include *Theoretical Chemical Engineering*; *Environmental Protection and Process Safety*; *Process and Chemical Engineering*; *Biotechnology: Apparatus, Plant, and Equipment*; and *Current Biotechnology*. *CEABA* covers chemical engineering and biotechnology from theoretical studies through plant operation including also safety engineering and related environmental areas.

The Royal Society of Chemistry also produces abstracting services of special interest to analytical chemists. These are discussed separately in Chapter 18. Royal Society of Chemistry current awareness products are described in Chapter 4, and those specific to environment and safety are noted in Chapter 14. In addition, RSC offers a broad variety of other publications, notably both journals and books, as described elsewhere in this volume. Further, its Library offers extensive chemical information services and document delivery.

8.5. *RAPRA*

Rapra Abstracts is an important abstracting and indexing service that is a product of the large research and consulting firm, Rapra Technology Ltd., formerly the Rubber and Plastics Research Association, which is located in Shawbury, Shropshire, UK. Rapra's agent in the United States for its information and software products is Plastics Design Library, 3 Eaton Avenue, Norwich, NY (607-337-5000).*

This abstracting service not only includes technology information, including the chemistry and equipment and machinery, but also covers the business and commercial aspects. Thus, in searches relating to plastics and rubbers, this is a vital source to search, in conjunction with a search of *Chemical Abstracts,* and even more so when trade and commercial data is sought or when the only clues available are linked to tradenames or company names.

In addition to being available online through such systems as STN International and DIALOG (see Chapter 10), *Rapra Abstracts* is also available in CD-ROM form. Coverage of literature is from 1972 to date, but coverage of patents did not begin until 1994 (European, U.S., and World patent documents only). There is a directory section that is an excellent source of tradenames and company names, addresses and telephones.

Subsets of the CD-ROM file are available covering, respectively, plastic materials, polyurethanes, packaging and film, rubber materials, and adhesives and sealants.

A printed version of *Rapra Abstracts* is still available; this is published on a monthly basis, with about 300 pages in each issue.

Related to the above is Rapra's *Review Report* series, each of which presents comprehensive state-of-art overviews of specific topics. Numbers and titles of some recent Rapra Review Reports include the following:

74 Specialty Rubbers
75 Plastics and the Environment
76 Polymeric Precursors for Ceramic Materials
77 Advances in Tire Mechanics
78 PVC—Compounds, Processing and Applications
79 Rubber Compounding Ingredients—Need, Theory, Innovation, Part I
80 Anti-Corrosion Polymers: PEEK, PEKK, and other Polyaryls
81 Thermoplastic Elastomers—Properties and Applications
82 Advances in Blow Molding Process Optimization
83 Molecular Weight Characterization of Synthetic Polymers
84 Rheology and its Role in Plastics Processing
85 Ring-Opening Polymerization
86 High Performance Engineering Plastics
87 Rubber to Metal Bonding
88 Plasticisers —Selection, Applications and Implications
89 Polymer Membranes —Materials, Structures and Separation Performance
90 Rubber Mixing
91 Recent Developments in Epoxy Resins

*Rapra now has a U.S. office in Charlotte, NC. See www.rapranet.com.

92 Continuous Vulcanisation of Elastomer Profiles
93 Advances in Thermoforming
94 Compressive Behaviour of Composites
95 Thermal Analysis of Polymers
96 Polymeric Seals and Sealing Technology

Each report is supplemented by references and abstracts to the published literature on the subject. The abstracts have been taken from the *Rapra Abstracts* database. The present cost for each report is $130.

In addition, Rapra produces software products, specifically, for example, flow analysis software for injection molding. Rapra also publishes a number of books including industry directories.

Rapra is developing what it calls a knowledge-based approach to more effectively utilize its information products. This system is to include three components: (a) selected technical references from the Rapra plastics and rubbers database; (b) a series of expert systems with interactive access that address particular issues or problems such as materials selection, fault and failure diagnosis, process optimization, and sensitivity analysis, and (c) a system that permits users to input their own data into the files. This system is now operational.

Rapra offers a document delivery service, Copyquest. Many of the documents are not available from other sources, as for example, a very large collection of company literature and data sheets.

Dr. David Wright is Rapra's Technical Director. Chapter 15 contains information on other Rapra products.

8.6. *CHEMINFORM* AND OTHER FIZ CHEMIE PRODUCTS

ChemInform, subtitled *Selected Abstracts in Chemistry,* and previously titled *Chemischer Informationsdienst,* is an abstracting and indexing service that is a product of FIZ Chemie, Berlin, Germany. Publication began in 1970 under the aegis of the German Chemical Society. This is a weekly service that covers some 200 journals; patents are not included. The focus is on new or improved organic reactions and syntheses, as well as applications of known reactions to synthesis of new compounds, although other topics such as some inorganic reactions and physical–inorganic chemistry are also included. Key results as to structural formulas and reaction schemes are presented graphically and in words. When given by the authors, information provided includes overall yield and/or yield of the different products or intermediates, enantiomeric excess, and reaction temperature, time, and pressure. One-step and multistep reactions are included, but polymers are not covered. As again noted in Chapter 11, a special feature is semiannual compilation of all review references that are included in each issue. Each issue contains author, substance, and subject (in classified form) indexes. All material is in English. Helga Lehmann-Seider is editor.

ChemInform is, in some respects, a partial successor to the highly regarded, but discontinued (end of 1969), *Chemisches Zentralblatt* (p. 140), although by no means a complete replacement since *CZ* was far broader in scope. Flambard and Weiske have written about the history of FIZ Chemie, *CZ,* and *ChemInform* (4).

Over 17,000 abstracts, 60,000 reactions and 100,000 compounds are covered annually. In addition to the printed version, there is a CD-ROM version, and there is a database version *ChemInform RX* (not as complete as the printed or CD versions) that is available online through the STN International network (see Chapter 10). There is also an inhouse electronic version that is available through MDL Information Systems (see Chapter 10). Coverage with the online version begins in 1991. Reactants, products, reagents, etc., are structure-searchable on STN.

FIZ Chemie is additionally engaged with several chemical engineering-oriented databases, including *CEABA* (p. 153), *KKF,* and *VtB,* searchable online through STN International. *KKF* (*Kunststoffe, Kautschuk, Fasern*), produced in cooperation with Deutsches Kunststoff-Institut (DKI), covers worldwide literature on plastics and polymers. *VtB* (*Verfahrenstechnische Berichte*), produced in cooperation with BASF and Bayer, covers worldwide literature in the field of chemical process engineering and related fields. Patents are not included in these databases.

Most FIZ Chemie products and databases are available in English, but the printed and database versions of *KKF* and *VtB* are in German. The titles are additionally in English, and the controlled terms in *VtB* are in both German and English. *KKF* offers an online thesarus which provides the German controlled terms and their English equivalents.

FIZ Chemie also offers thermophysical and thermodynamic substance data through *DETHERM,* which they say is the world's largest data collection for plant planning and construction and for process control, and is available for both inhouse use and online through STN; it is a joint product with DECHEMA (the German Society for Chemical Engineering, Chemical Apparatus, and Biotechnology, which is located in Frankfurt).

FIZ Chemie (Fachinformationszentrum Chemie) is a specialized information center whose focus is chemistry. It was founded as such in December 1981, having been derived from a former office of the German Chemical Society. There are approximately 65–70 employees. In addition to its products as mentioned above, a search service is offered, and there are training workshops on use of online databases. FIZ Chemie is a nonprofit organization that is state-funded. The Web address is www.chemistry.de/index.e.html.

8.7. NATIONAL LIBRARY OF MEDICINE DATABASES; *EMBASE*

In medicinal chemistry, major sources in addition to *CA* include the databanks and publications of the National Library of Medicine (NLM), Bethesda, MD. The two principal online files of NLM are *MEDLINE* (medical literature) and *TOXLINE* (published human and animal toxicity studies and some additional reports and submissions). The overall approach of NLM is, of course, from a medical perspective, but many chemists find its files, especially *TOXLINE,* helpful. One of the newer online files available from NLM is *DIRLINE,* which is a directory of information resources (organizations and experts) who can be contacted. This file was originally developed by the National Referral Center at the Library of Congress. Melvin Spann is the leader of the efforts of chemical interest at NLM. These and other NLM online files are discussed in more detail in Section 9.5.

In addition to *MEDLINE,* there are several other major related services from other sources that cover fields of interest to medicinal chemists and biochemists and that are important to be consulted to make searching more complete. One of the leading examples is *EMBASE,* which is an *Excerpta Medica* database, produced by Elsevier Science, Amsterdam, The Netherlands. The online service is about 50% more complete than the corresponding printed product, *Excerpta Medica,* which continues to be available in some 41 subsets. Approximately 3500 journals are covered in *EMBASE,* as are some proceedings that are printed, but patents and monographs are not covered at this time. Currency is said to be excellent; that is, abstracts are available within 12–15 working days after the journal is received. Coverage begins in 1974. There is considerable emphasis on coverage of drugs and other chemicals. Approximately 53% of the input is from European sources, and 30% is from North America. Coverage from the Pacific Rim and Japan is also said to be very good. *EMBASE* may be searched online through several systems: STN, DIALOG, DataStar, Ovid, LEXIS-NEXIS, and DIMDI. Several of the printed abstract subsets are available on the Internet.

There is some overlap between *MEDLINE* and *EMBASE,* especially with regard to North American sources, but there are also many references in each of these databases that do not appear in the other. As mentioned, neither covers patents, and this is a major gap. *MEDLINE,* as a publicly supported product, is quite inexpensive or free (depending on system utilized) to search, whereas *EMBASE,* which is privately published, is relatively expensive. *MEDLINE* has an especially knowledgeable help desk with reference to use (direct searching through NLM) of *MEDLINE* and the other databases in the NLM online family, which is known as *MEDLARS. MEDLINE* coverage began in 1966, with extension to earlier years; *EMBASE* coverage begins in 1974.

BIOSIS Previews, mentioned below, also contains important information on medical topics, and is said to be especially strong as to conferences and proceedings. (This service covers, as well, "traditional" areas of biology and interdisciplinary areas such as biochemistry.) Another example of an important medical (and other sciences) information source is *SCISEARCH* with its strong citation searching capability.

In addition, there are a number of other excellent abstracting sources, most of which deal with pharmaceuticals, especially from the perspective of patents and the business aspects.

8.8. BIOSIS

BIOSIS is a not-for-profit organization based in Philadelphia, PA. The staff of approximately 300 persons produces a number of information products of interest to biochemists, medicinal chemists, and other chemists interested in the interface of biology with chemistry. BIOSIS was founded in 1926. See www.biosis.org.

The principal online product is *BIOSIS Previews,* a well-known and highly regarded abstracting and indexing service that is searchable online through a number of systems: DIALOG and DataStar, STN International, Ovid, OCLC, and DIMDI. Almost 11 million records are contained in the file. This is the online equivalent of the printed *Biological Abstracts* and *Biological Abstracts/RRM* (*Reports, Reviews, Meetings*). Coverage begins in 1969. Some 6000 journals are covered, and about 2000 meetings are regularly monitored. Books and book chapters (if individually authored) are covered (coverage of individual book chapters is unusual). Patents are not included except for U.S. patents, 1986–1989 only.

A current awareness service is also offered. This is designated as *BITS* (*BIOSIS Information Transfer System*). BIOSIS staff provide a monthly diskette based on the user's delineation of fields of interest.

Another related product is an online file known as *BioBusiness,* which covers the business aspects of technical developments in the life sciences. References to scientific journals and to business, scientific, and trade publications, as well as to newspapers, meetings, and U.S. patents, are included. Coverage begins in 1985. Online availability is through DIALOG, DataStar, STN, and Ovid. Emphasis is on biotechnology, health care, and pharmaceuticals.

BIOSIS offers a document delivery service in partnership with another organization, but does not provide "bureau" searching—that is, the staff will not conduct searches for customers. The president is John E. Anderson.

8.9. *ENGINEERING INDEX*

Engineering Index, also known as *Ei,* was founded in 1884 and continues to be published by Engineering Information, Inc. Types of literature included are journal articles, technical reports, and conference papers and proceedings. Patents and books are not included. Coverage is multidisciplinary, with chemical engineering among the many broad engineering areas that are included. Besides applied engineering, coverage extends also to manufacturing, quality control, and engineering management issues.

The computer-readable form of *Engineering Index* is *Ei Compendex,* with coverage beginning in 1970. Online access is readily available either directly on a fee basis through the Internet (*Ei Compendex Web*) or through a number of major systems, including DIALOG and DataStar, STN International, and Questel•Orbit. A CD-ROM and site licensing are also available. On CD, KR OnDisc offers *EiChemDisc,* a subset of *Ei Compendex* concentrating on chemical engineering.

The World Wide Web site is http://www.ei.org. The firm also offers a Web-based subscription service known as *Engineering Information Village.* This service includes access to many other prescreened, organized, annotated, and monitored Internet sites of interest to engineers; a roster of senior engineers available as consultants; about 200 engineering topics culled from *Ei Compendex* and updated weekly; a compilation of "annotated and filtered" peer newsgroups and listservs on the Internet; technical newswire reports; online document delivery; "Ask A Librarian" (reference questions are answered); and transactional pricing access to some 150 other databases on the Internet, including *Ei Compendex,* plus the option of unlimited access to *Compendex.* Another special feature in Ei Village is "ILI StandardsWeb," which offers bibliographic industrial and military standards information leading to full-text fulfillment via document delivery. ILI is a United Kingdom-based firm. *EiVillage* is powerful but expensive; fees begin at $3,000 and depend on number of users and level of service. However, individual pricing is also available.

Engineering Information offers a document delivery service, and bureau searching (custom searching for a fee) is available.

The staff of only approximately 40 persons adds 220,000 new abstracts, with index entries, annually. Engineering Information, Inc. is located on the campus of the Stevens Institute of Technology, Hoboken, NJ. It is a for-profit organization that is employee-owned. The president and chief executive officer is John J. Regazzi.

The special utility of *Ei* to chemists and chemical engineers is when engineering topics other than those covered by *Chemical Abstracts* are encountered. There is some overlap between the two services, especially in the area of chemical processing.

8.10. *INFORMATION SERVICE FOR PHYSICS, ELECTRICAL ENGINEERING, AND COMPUTING (INSPEC)*

Information Service for Physics, Electrical Engineering, and Computing (*INSPEC*) is a product of the Institute of Electrical Engineers in the United Kingdom and is represented in the United States by the IEE/INSPEC Department at the IEEE Operations Center. INSPEC provides excellent coverage of the interface of chemistry with electrical and electronics engineering and computer technology, as well as physics. This source is especially recommended as a complement to the use of *Chemical Abstracts* for any chemist or engineer interested in semiconductors or other electronics material or chemicals. The major printed publications (also available online through a number of major systems such as those listed in Chapter 10) are *Electrical & Electronics Abstracts, Computer & Control Abstracts,* and *Physics Abstracts.*

IEE offers an array of journals and books. Among the book offerings is the *Electronic Materials Information Service* (*EMIS*) *Datareviews* series, which consists of data compilations such as properties of metal silicides, gallium arsenide, silicon, and indium phosphides.

Current awareness publications include *Current Papers in Physics, Current Papers in Electrical & Electronics Engineering,* and *Current Papers on Computing & Control.* A special low-cost SDI (Selective Dissemination of Information) service is also offered; materials science is one of the areas of coverage.

INSPEC abstracts are meaningful and complete, and indexing is excellent. One specific advantage of *INSPEC* is the indexing, which is from an electronics perspective; terms used are usually those familiar to workers in electronics. Unfortunately, however, patents are not included at this time, although they were covered in earlier years, until about 1976.

8.11. AMERICAN VERSUS BRITISH SPELLINGS

American chemists who use *INSPEC* (or any other British publication) need to constantly remind themselves of minute differences in spelling that could potentially result in missing significant information, especially when using printed or online indexes. For example, English (British) spelling of *aluminum* is *aluminium, analog* is *analogue, color* is *colour, fiber* is *fibre, ionization* is *ionisation, molding* is *moulding, sulfur* is *sulphur,* and *vapor* is *vapour.* Similarly, British chemists who use American publications need to remind themselves of these variations. Table 8.1 contains a more extensive list of spelling differences.

8.12. CAMBRIDGE SCIENTIFIC ABSTRACTS

Another major provider of abstracting and indexing services is Cambridge Scientific Abstracts (CSA), Bethesda, MD (800-843-7751). This firm publishes a number of abstracting and indexing services as well as other information products. The areas of emphasis of this firm are:

1. Environmental sciences and pollution management information (see Chapter 14)
2. Aquatic and marine sciences
3. Biology, biotechnology, and bioengineering
4. Materials information

A privately held firm that was originally founded in 1958, CSA has recently significantly expanded its offerings as evidenced, for example, by its recent entry into material information through acquisition of information products from ASM and the Institute of Materials. Almost all CSA offerings are available through the Internet, as well as through the DIALOG and STN International online systems. The CSA *Internet Database Service* includes titles produced both by Cambridge Scientific Abstracts and by others, including products of the U.S. National Library of Medicine (*MEDLINE* and *TOXLINE*), the American Society of Health System Pharmacists, and *Excerpta Medica Abstracts Journals* from Elsevier Science (15 titles). CSA's award-winning *Environmental RouteNet,* an Internet product, is discussed separately on page 379. The firm's chairman is Robert Snyder, and the president is James McGinty. The number of employees is now more than 100. The Web site is www.csa.com.

With regard to CSA's new thrust in materials information, the background is as follows. A few years ago ASM had formed a joint venture, Materials Information (MI), with the Institute of Materials (London, UK) to produce three abstracting and indexing services. The Institute of Materials incorporates The Institute of Metals, The Institute of Ceramics, The Plastics and Rubber Institute, and the British Composites Society. The current editorial director is Carol Houk. In 1996 this information operation was sold to Cambridge Scientific Abstracts. (Relationships with ASM and the

TABLE 8.1. American Variant Spellings for Terms in the *INSPEC Thesaurus*[a,b]

A-centers	use	A-centres
aluminum	use	aluminium
aluminum alloys	use	aluminium alloys
aluminum compounds	use	aluminium compounds
amorphization	use	amorphisation
analog computer applications	use	analogue simulation
analog computer circuits	use	analogue computer circuits
analog computer methods	use	analogue simulation
analog computer programming	use	analogue computer programming
analog computers	use	analogue computers
analog differential analyzers	use	differential analysers
analog digital computers	use	hybrid computers
analog–digital conversion	use	analogue–digital conversion
analog–digital convertors	use	analogue–digital conversion
analog memories	use	analogue storage
analog simulation	use	analogue simulation
analog storage	use	analogue storage
analog, direct	use	direct analogue
analyzer, differential	use	differential analysers
anodization	use	anodisation
anodized coatings	use	anodised layers
anodized layers	use	anodised layers
anodized thin films	use	anodised layers
apodization	use	acoustic imaging optic images
atomic polarizability	use	atomic polarisability
attenuation equalizers	use	attenuation equalisers
auroral ionization	use	auroral ionisation
autoionization	use	autoionisation
beta-ray polarization	use	beta-ray polarisation
biological effects of ionizing particles	use	biological effects of ionising particles
circular polarization	use	polarisation
carbon fiber reinforced composites	use	carbon fibre reinforced composites
carbon fibers	use	carbon fibres
channeling	use	channelling
chemical vapor deposition	use	chemical vapour deposition
code converters	use	code convertors
color	use	coiour
color blindness	use	colour vision
color cameras, television	use	colour television cameras
color centers	use	colour centres
color display tubes, television	use	colour television picture tubes
color filters	use	optical filters
color model	use	colour model
color perception	use	colour vision

(continued)

TABLE 8.1. ***(Continued)***

color photography	use	colour photography
color picture tubes, television	use	colour television picture tubes
color receivers, television	use	colour television receivers
color television	use	colour television
color television cameras	use	colour television cameras
color television picture tubes	use	colour television picture tubes
color television receivers	use	colour television receivers
color TV	use	colour television
color TV receivers	use	colour television receivers
color vision	use	colour vision
computerized air traffic control	use	air traffic computer control
computerized communications control	use	communications computer control
computerized control	use	computerised control
computerized industrial control	use	industrial computer control
computerized instrumentation	use	computerised instrumentation
computerized manufacturing control	use	manufacturing computer control
computerized materials handling	use	computerised materials handling
computerized monitoring	use	computerised monitoring
computerized navigation	use	computerised navigation
computerized numerical control	use	computerised numerical control
computerized pattern recognition	use	computerised pattern recognition
computerized picture processing	use	computerised picture processing
computerized power station control	use	power station computer control
computerized power system control	use	power system computer control
computerized process control	use	process computer control
computerized signal processing	use	computerised signal processing
computerized spectroscopy	use	computerised spectroscopy
computerized test equipment	use	automatic test equipment
computerized traffic control	use	traffic computer control
computerized transport control	use	transport computer control
converters	use	convertors
copolymerization	use	polymerisation
corporate modeling	use	corporate modelling
crystallization	use	crystallisation
demagnetization	use	demagnetisation
demagnetization, adiabatic	use	demagnetisation + magnetic cooling
deuteron polarization	use	deuteron polarisation
dialog programming	use	interactive programming
dielectric depolarization	use	dielectric depolarisation
dielectric polarization	use	dielectric polarisation
differential analyzers	use	differential analysers
digital-analog converters	use	digital-analogue conversion
digital-analog conversion	use	digital-analogue conversion
digital differential analyzers	use	digital differential analysers
digitizers	use	analogue-digital conversion
direct analogs	use	direct analogues
diverters	use	divertors

TABLE 8.1. ***(Continued)***

electron ionization	use	electron impact
electron microprobe analyzers	use	electron probe analysis
electron probe analyzers	use	electron probe analysis
electrosynchronization	use	synchronisation
elliptical polarization	use	polarisation
equalizers	use	equalisers
F-centers	use	F-centres
F_2-centers	use	M-centres
F_3-centers	use	R-centres
F_a-centers	use	A-centres
fiber optics	use	fibre optics
fibers	use	fibres
fiber reinforced composites	use	fibre reinforced composites
field ionization	use	field ionisation
file organization	use	file organisation
flow visualization	use	flow visualisation
fluidized beds	use	fluidised beds
fluidized powders	use	powders
frequency converters	use	frequency convertors
gamma-ray polarization	use	gamma-ray polarisation
glass fiber reinforced plastics	use	glass-fibre reinforced plastics
glass fibers	use	glass fibres
graphitization	use	graphitisation
graphitizing	use	graphitising
H-centers	use	H-centres
heat of crystallization	use	heat of crystallisation
heat of vaporization	use	heat of vaporisation
image converters	use	image convertors
impact ionization	use	impact ionisation
impedance converters	use	impedance convertors
information analysis centers	use	information centres
information centers	use	information centres
inverters	use	invertors
ion microprobe analyzers	use	ion microprovbe analysis
ionization	use	ionisation
ionization chambers	use	ionisation chambers
ionization gauges	use	ionisation gauges
ionization of atoms	use	ionisation of atoms
ionization of gases	use	ionisation of gases
ionization of liquids	use	ionisation of liquids
ionization of molecules	use	ionisation of molecules
ionization of solids	use	ionisation of solids
ionization potential	use	ionisation potential
ionization time	use	ionisation

(continued)

TABLE 8.1. ***(Continued)***

isobaric analog resonances	use	isobaric analogue resonances
isobaric analog states	use	isobaric analogue states
isomerization	use	isomerisation
latent heat of crystallization	use	heat of crystallisation
latent heat of vaporization	use	heat of vaporisation
lattice localized modes	use	lattice localised modes
light polarization	use	light polarisation
linearization techniques	use	linearisation techniques
liquid-vapor transformations	use	liquid-vapour transformations
localized modes in crystals	use	lattice localised modes
localized electron states	use	localised electron states
localized states, electron	use	localised electron states
M-centers	use	M-centres
magnetization	use	magnetisation
magnetization reversal	use	magnetisation reversal
magnetohydrodynamic converters	use	magnetohydrodynamic convertors
magnitude converters	use	magnitude convertors
mercury vapor lamps	use	mercury vapour lamps
mercury vapor rectifier	use	mercury vapour rectifier
metal vapor lamps	use	metal vapour lamps
metallization	use	metallisation
metallizing	use	metallising
MHD converters	use	magnetohydrodynamic convertors
minimization	use	minimisation
minimization of switching nets	use	minimisation of switching nets
modeling	use	modelling
modeling, computer	use	computer-aided analysis
molecular electron impact ionization	use	molecular electron impact ionisation
molecular polarizability	use	molecular polarisability
monitoring, computerized	use	computerised monitoring
multichannel analyzers	use	pulse rate analysers
negative impedance convertors	use	negative impedance convertors
network analyzers	use	network analysers
network equalizers	use	equalisers
neutron polarization	use	neutron polarisation
normalizing	use	normalising
nuclear isobaric analog resonances	use	isobaric analogue resonances
nuclear isobaric analog states	use	isobaric analogue states
nuclear particle track visualization	use	particle track visualisation
nuclear polarization	use	nuclear polarisation
nuclear polarization in solids	use	nuclear polarisation in solids
OH^--centers	use	OH^--centres
optical fibers	use	optical fibres
optimization	use	optimisation
orthogenelized plane wave calculation	use	OPW calculations

TABLE 8.1. ***(Continued)***

parametric up-converters	use	parametric devices
particle track visualization	use	particle track visualisation
Penning ionization	use	Penning ionisation
phase converters	use	phase convertors
photoionization	use	photoionisation
photoionization of gases	use	photoionisation
photon polarization	use	photon polarisation
polarizability	use	polarisability
polarization	use	polarisation
polarization in nuclear reactions and scattering	use	polarisation in nuclear reactions and scattering
polymerization	use	polymerisation
power converters	use	power convertors
power utilization	use	power utilisation
proton polarization	use	proton polarisation
pulse amplitude analyzers	use	pulse height analysers
pulse analyzers	use	pulse analysers
pulse height analyzers	use	pulse height analysers
pulverized coal	use	pulverised fuels
pulverized fuels	use	pulverised fuels
quantization	use	quantisation
R-centers	use	R-centres
recrystallization	use	recrystallisation
recrystallization annealing	use	recrystallisation annealing
recrystallization texture	use	recrystallisation texture
renormalization	use	renormalisation
rotary converters	use	rotary convertors
self-optimizing systems	use	self-adjusting systems
self-organizing storage	use	self-organising storage
signal processing, computerized	use	computerised signal processing
solid–vapor transformations	use	solid–vapour transformations
spectral analyzers	use	spectral analysers
spectrum analyzers	use	spectral analysers
spontaneous magnetization	use	spontaneous magnetisation
stabilization	use	stabilisation
stabilizers	use	controllers
standardization	use	standardisation
storage, analog	use	analogue storage
storage organization	use	file organisation
sulfur	use	sulphur
superconducting ionization	use	superconducting ionisation
surface ionization	use	surfrace ionisation
synchronization	use	synchronisation
synchronizing reactors	use	current limiting reactors

(continued)

TABLE 8.1. ***(Continued)***

textile fibers	use	fibres
thermally stimulated depolarization	use	thermally stimulated currents
thermally stimulated polarization	use	thermally stimulated currents
track visualization, particle	use	particle track visualisation
transient analyzers	use	transient analysers
trapped electron centers	use	F-centres
trapped hole centers	use	V-centres
tunneling	use	tunnelling
tunneling spectra	use	tunnelling spectra
U-centers	use	U-centres
V-centers	use	V-centres
V_h-centers	use	H-centres
vaporization	use	vaporisation
vaporizing	use	vaporisation
vapor density	use	density of gases
vapor deposited coatings	use	vapour deposited coatings
vapor deposition	use	vapour deposition
vapor-deposited thin film	use	vapour deposited coatings
vapor–liquid transformations	use	liquid–vapour transformations
vapor phase epitaxial growth	use	vapour phase epitaxial growth
vapor pressure	use	vapour pressure
vapor pressure measurements	use	vapour pressure measurements
vapor–solid transformations	use	solid–vapour transformation
visualization, particle track	use	particle track visualisation
volatilization	use	vaporisation
wave analyzers	use	wave analysers
Z-centers	use	Z-centres

[a]While this list is intended for use with a database covering electronics and electrical engineering, it should also be of value in the use of some databases in chemistry.

[b]For a number of the terms contained in the *INSPEC Thesaurus*, the normal American spelling differs from the English form as used by INSPEC. The differences are generally small and almost all fall into the following categories:

"Z" in place of "S"
  e.g., "ionization" and "ionisation"
"or" in place of "our"
  e.g., "color" and "colour"
"er" in place of "re"
  e.g., "center" and "centre"
"g" in place of "gue"
  e.g., "analog" and "analogue"

To assist users of the INSPEC database this table gives a list of all such terms contained in the *Thesaurus*, giving the American spelling and the equivalent English form. It is planned to include the American variants as lead-in terms in future editions of the *INSPEC Thesaurus*. (Note: Both forms are likely to be used in the free-indexing.)

Source: Reprinted by permission of *INSPEC*. The Institution of Electrical Engineers.

Institute of Materials continue on a "dotted-line" basis.) There are some 80 abstractors worldwide, and there are 15 indexers, the majority of whom are on site in the Cleveland, OH, area. Some of the products of this operation are the following:

Metals Abstracts, which includes some 3000 items in each printed monthly issue. Scope includes all aspects of metal science technology beginning with ore preparation and all the way through to and including the properties and applications of finished products. Powder technology and matrix composites are also covered. The online counterpart, *Metadex,* provides coverage beginning with 1966. Types of materials covered include patents, journal papers, books, conference proceedings, technical reports, and reviews.

Engineered Materials Abstracts, which covers the technical literature on the uses of polymers, ceramics, and composites in engineering applications. Coverage focuses on structural materials, broadly defined, but important nonstructural materials are included, such as optical fibers, nuclear fuel rods, piezoelectric materials, and superconducting ceramics. *Engineered Materials Abstracts* coverage begins with 1986. Types of materials covered include patents, journal papers, books, conference proceedings, technical reports, and reviews (European Space Agency). *EMA* is the database name for *Engineered Materials Abstracts.*

Aluminum Industry Abstracts (published in association with the Aluminum Association, Washington, DC, and the European Aluminum Association) coverage begins in 1968.

Steels Alert, Nonferrous Metals Alert, and *Polymers/Ceramics/Composites Alert* provide technocommercial information in the form of monthly abstracts journals. Topics covered include trade, plant and product developments, health and safety, production and price trends, competitive materials, environmental issues, legislation, and company news. *Materials Business File* database embraces the three *Alerts.*

Metals Datafile, which contained numeric property information on nearly all ferrous and nonferrous alloy systems, has been discontinued.*

Most of the previous ASM databases are accessible online through Cambridge Scientific's *Internet Database Service.* In addition, some of the files are available through DIALOG, STN International, Orbit, and ESA (European Space Agency).

In addition, all of the material abstracted and indexed by MI since 1966 is available on CD-ROM as the *Metadex/Materials Collection.* Updating is quarterly.

Other products and services that MI has offered in the past, and that are apparently to be continued by CSA, include

World Calendar, a quarterly diary of forthcoming meetings that are materials-related

Source Journals in Metals and Materials, a list of journals in the field

Bibliographies, of which there are now 250 (Search-In-Print Series)

*CSA now produces *Ceramic Abstracts,* previously issued by the American Ceramics Society.

Industry Reports, each of which is a major study and analysis of an important industrial sector or process, such as *Non-Ferrous Metal Heat Treating* (now discontinued)

Custom searching of the databases on a "bureau" basis (now discontinued)

The Materials Information Translations Service, for items not in English, is now handled through the Institute of Materials in the United Kingdom. The document delivery service is to be continued by the ASM library.

8.13. GOVERNMENT SERVICES

The U.S. Government is an important publisher of abstracting and indexing services. Some of these are summarized in Chapter 9.

8.14. PATENTS; ENVIRONMENT, SAFETY, AND HEALTH

A number of specialized services have been developed to cover information found in patents and on environment, safety, and health. These are discussed separately in Chapters 13 and 14.

REFERENCES

1. H. Skolnik, *The Literature Matrix of Chemistry,* Wiley, New York, 1982.
2. E. J. Crane, A. M. Patterson, and E. B. Marr, *A Guide to the Literature of Chemistry,* 2nd ed., Wiley, New York, 1957.
3. C. E. Huber, "SciSearch on STN—Unique Features for Sophisticated Searching," *Database,* 52–62 (April/May 1995).
4. A. R. Flambard and Christian M. Weiske, "Fachinformationszentrum Chemie GmbH, Berlin—A Decade in the Service of Chemistry Information," *Chem. Ber.* **123,** XXV–XXXI (1992).

List of Addresses

BIOSIS
2100 Arch Street
Philadelphia, PA 19103-1399
(800-523-4806)

CAB International
Wallingford
Oxon OX10 8DE
United Kingdom

Cambridge Scientific Abstracts
5161 River Road
Bethesda, MD 20816

Engineering Information, Inc.
Castle Point on the Hudson
Hoboken, NJ 07030
(800-221-1044)

ISI
3501 Market Street
Philadelphia, PA 19104
(800-523-1850)

IEEE Operations Center
IEEE/INSPEC Department
445 Hoes Lane
Piscataway, NJ 08855-1331
(732-562-5548)

Institute of Paper Science and Technology
500 10th Street, NW
Atlanta, GA 30318
(800-558-6611)

Rapra Technology Ltd.
Shawbury, Shrewsbury
Shropshire SY4 4NR
United Kingdom
or
c/o Plastics Design Library
13 Eaton Avenue
Norwich, NY 13815
(607-337-5080)

Rapra now has a U.S. office in Charlotte, NC.

9 Some United States Government Technical Information Centers and Sources

9.1. INTRODUCTION

The research and development effort of the U.S. government is huge. There are approximately 500 federal laboratories, and the total annual budget is estimated at $70–75 billion; of the total, only about one-third is actually expended on site at the federal laboratories, whereas the bulk goes to federally funded contract and grant research in industry, universities, and other organizations. There is at least one federal laboratory in every state; California, with 46, has the most. A large fraction of the work is of direct interest to chemists and chemical engineers. Examples of federal laboratories that do work of chemical interest include the National Institute of Standards and Technology (see Chapter 15), the National Library of Medicine and other agencies of the National Institutes of Health (see Section 9.5), the Department of Energy (see Section 9.4), the Department of Agriculture, the Environmental Protection Agency, and the Defense Department (see Section 9.3), to name some of the most pertinent. The massive government research effort yields thousands of reports each year and a large number of potentially licensable inventions, and other technologies that can be transferred in other ways to industry. In addition, thousands of scientists are employed in government laboratories, and many are willing and able to talk about their expertise and capabilities with individuals in industry and universities.

Budget reduction efforts affected the scope of the U.S. government research effort in 1996–1997, and further reductions can be expected. One example that impacted chemists was the closing, as such, of the U.S. Bureau of Mines and its laboratories. A publication that has been similarly affected was the discontinuance in 1996 of the compilation (by the U.S. International Trade Commission) that recorded the production of U.S. synthetic organic chemicals—a void that the chemical industry economists say they will sorely miss.

9.2. NATIONAL TECHNICAL INFORMATION SERVICE

The National Technical Information Service (NTIS), Springfield, VA, is a central resource for documents in a variety of formats (printed, electronic, audio, etc.) relating to government-sponsored efforts in scientific, technical, and business fields. NTIS is also a leading U.S. government agency in international technical and business information exchange, and it actively acquires and distributes information from foreign governments and other organizations. Worldwide sources include government and industry groups in 24 different countries. It is a self-sustaining agency of the U.S. Department of Commerce's Technology Administration; the sale of its products and services cover its expenses. Donald R. Johnson is the current director, and there is a staff of some 380 persons. Estimates are that the document collection exceeds 2.5 million titles. The Internet location is www.ntis.gov., where products, catalogs, services, and so on can be read, searched, downloaded, or printed.

Some of the functions of NTIS include

1. Providing copies of government reports at reasonable cost. NTIS receives over 1600 new titles each week. Many, but by no means all, government reports are available from NTIS.
2. Providing online and other search tools to identify government documents of interest within NTIS holdings. The NTIS database of holdings is available for searching on a number of systems, as, for example, STN International and DIALOG. The Internet site contains a 30-day rolling window subset known as *NTIS OrderNow;* this is also available in CD-ROM format with over 2 years of searchable information.
3. Compiling and publishing searches on topics (now over 3500) deemed important by customers and NTIS.
4. Semimonthly bulletins (*NTIS Alerts*) with summaries of new reports categorized into separate topics such as Materials Sciences, Environment Pollution and Control, and Government Inventions for Licensing. Chemistry as such is not an available category.
5. Home of the Federal Computer Products Center, which provides access to information in electronic formats. It includes software and datafiles from 1990, primarily as sponsored by the federal government, but also from some state governments and even a few from the private sector.
6. Operation of the FedWorld system on the Internet, a major source for identifying and linking into government electronic files, as described further in Section 9.7.

NTIS published the fifth edition of its valuable *Directory of Federal Laboratory Resources and Technology* (PB 93100097) in 1993. This thick compilation is considerably out of date now, but can still be useful in some cases because of the rather extensive abstracts and the contact names given. NTIS says it has no plans to update this directory, but the government-funded National Technology Transfer Center, Wheeling, WV, has begun work to update it on the Internet (www.nttc.edu/brs/update.html). (The telephone contact is Judy Kirker at 304-243-2591.)

Search of the online NTIS databank is recommended because it includes some information of interest to chemists that is not readily found elsewhere. It is an especially simple file to use, and copies of almost all documents listed can be obtained from either NTIS itself or a variety of private document delivery services. Cost of use online is relatively low.

Despite all of its services and important advantages, users of NTIS should be aware that its document operation can be likened to that of a large, well-organized, and intelligently managed mail order facility, except that it is not for profit. NTIS staff have neither the time nor the background required to evaluate the novelty or quality of all the many thousands of documents they handle nor to give expert referrals to the best sources of information. The quality of NTIS abstracts and indexes is much inferior to that of CAS or Derwent, and chemists using NTIS indexes must exercise ingenuity and creativity in finding what is needed.

9.3. DEFENSE TECHNICAL INFORMATION CENTER AND OTHER DOD SOURCES

The Defense Technical Information Center (DTIC) is operated by the U.S. Department of Defense (DOD) to serve the information needs of DOD personnel and DOD contractors and potential contractors, and of other U.S. Government agency personnel and their contractors. DTIC users include universities and industries throughout the United States. As noted above, the primary focus of DTIC is the defense community, and for most of its services, users must be appropriately registered with DTIC. However, others may access some of its resources, most of which are described at the Internet address www.dtic.mil. For example, a database of unclassified reports is available to the general public on the Internet at the address www.dtic.mil/STINET/Public-STINET; this now embraces over 150,000 reports. In addition, small businesses seeking DOD SBIR (Small Business Innovation Research) grants may contact DTIC to obtain free Technical Information Packages listing pertinent unclassified DTIC documents on topics of interest to DOD. DTIC also can provide copies of reports identified through these background searches; the first 10 reports are free.

Because DOD is a huge generator of technical information, much of which is of interest to chemists, access to its data and documents through DTIC is important. DTIC headquarters is located in Fort Belvoir, VA. In addition, there are regional offices in Albuquerque, NM; Boston, MA; Dayton, OH; and Los Angeles, CA. The services of DTIC include providing full-text copies of reports, bibliographies, and current awareness and retrospective searches. Resources of DTIC are also searchable online. The principal DTIC online system is known as DROLS (Defense Research Development Test & Evaluation Online System), and it is available to properly registered users only. The total file at DTIC consists of approximately 2 million reports and documents. The categories of documents includes those on (1) completed research, (2) research in progress, (3) planned research, and (4) independent R&D.

Although DTIC could logically be expected to get copies of all technical reports sponsored by DOD, in fact it does not. For one reason or another, thousands of reports are not available through DTIC, but rather only through the specific Defense Department agency or office that funded the work. In addition, DTIC is not well equipped to welcome visitors who may desire to do background reading on site within its specialized files. This is because DTIC headquarters houses almost all of its reports only in microfilm or microfiche form. DTIC is geared to serving its users on a remote basis, both electronically and through the mail.

However, many unclassified reports originating from work sponsored by the Defense Department, are made widely and readily available by DTIC through the NTIS, which is described in Section 9.2. DTIC also sponsors 13 Information Analysis Centers (IACs), which provide specialized and expert services and products such as state-of-the-art reports, handbooks, database compilations, and so forth.

DTIC IACs which may be of particular interest to chemists include these examples:

1. AMPTIAC (Advanced Materials and Processes Technology Information Analysis Center), Rome, NY, which is discussed in Chapter 15. Founded in 1996. (http://rome.iitri.com/amptiac)
2. CPIA (Chemical Propulsion Information Agency), the Johns Hopkins University, Columbia, MD. In existence since 1946. (http://www.jhu.edu/~cpia/)
3. CBIAC (Chemical Warfare/Chemical & Biological Defense), Edgewood, MD. (http://www.cbiac.apgea.army.mil)

Some IAC services to the public may be free (simple inquiries), and for other services there is a fee. In addition to the 13 DTIC-sponsored IACs, there are 10 other Information Centers operated by other DOD components. Further, the Defense Logistics Agency has established a Hazardous Technical Information Services operation at its Richmond, VA location (804-279-5168); this is not generally open to the public, but the Internet address can be found useful (www.dscr.dla.mil/htis/htis.htm). This Web site contains selected news developments relating to hazardous materials as well as helpful links.

Like DTIC, the services of the IACs are aimed primarily at government agencies and government contractors, but on occasion, the user community is more broadly defined. Services of the IACs typically include analysis and evaluation of reports and other literature, abstracting and indexing, bibliographies, issuing of specialized compilations, and copies of documents. Some IACs are not directly funded and managed by DTIC but are operated in cooperation with DTIC.

Except for the IACs, which have some evaluative capabilities, most users can expect warehouse type of service from DTIC. This means that users are left with the task of pinpointing what is important and with interpreting results. Precise chemical nomenclature, or precise chemical abstracting and indexing, are not offered by DTIC.

DTIC and IAC services can be extremely valuable, but because much of the content of these centers is government classified, there is a complex procedure for use. This procedure requires that potential users fill out the appropriate registration forms that must then be approved by the government. Once approved, some DTIC services are free.

9.4. ENERGY

A major source of energy information is the U.S. Department of Energy's database *Energy Science and Technology.* The editorial staff is located at Oak Ridge, TN, which reflects the original nuclear science orientation of this database, but the current scope covers energy very broadly. This file can best be searched online through DIALOG. It begins coverage with the year 1974.

Other major sources of energy information include *Chemical Abstracts* and *Ei Compendex Plus.* The American Petroleum Institute's online literature and patent files (*APILIT* and *APIPAT*) can also be found helpful.

9.5. NATIONAL LIBRARY OF MEDICINE

The National Library of Medicine (NLM) is the strongest library of its type in the United States and possibly in the world. A complete range of information services is offered. Facilities are located in Bethesda, MD. Several very important online databases of interest to chemists are offered by NLM under the leadership of Associate Director Melvin Spann. These are described in the following sections.

The principal NLM file is, of course, *MEDLINE,* noted in Chapter 8. The worldwide journal literature is covered from 1966 in this excellent abstracting and indexing service. In 1996–1997, *MEDLINE* experienced serious time lags, but the problems have now been completely resolved and the product is on a current basis.

During recent years, several major enhancements have been made to facilitate the availability of *MEDLINE.* It has long been available on virtually all major commercial hosts, of the type described in Chapter 10, and also continues to be searchable directly through the NLM *MEDLARS* system on a modest fee basis. In mid-1997, it additionally became available, on a totally free basis, through the new, experimental PubMed Internet site of NLM (best accessed through www.nlm.nih.gov, which also lists other options). The PubMed system was developed by NLM's National Center for Biotechnology Information (NCBI). In a landmark development, PubMed includes also links to the full text of retrieved articles for some journals; in some cases, there is a fee for the full text, or the user must register, depending on the publisher, but in any event, this is a major step forward. *MEDLINE* is also available on the Internet through several other sites such as that of *BioMedNet* (biomednet.com.gateways/db/medline). This latter site includes an "evaluation" feature— that is, recommended articles "selected and annotated by experts," as well as links to some full-text articles. In addition, *PREMEDLINE,* updated daily, has been introduced by NLM. It is available without charge on the Internet, and it may also be searched directly through NLM. *PREMEDLINE,* as the name implies, lists citations of materials to be later indexed and abstracted in *MEDLINE.* Another change announced is that *MEDLINE* coverage is to be extended back through 1962 in the new database known as *OLDMEDLINE.*

Special note should be made of the *Grateful Med* software that facilitates searching the NLM files either on a direct dialup basis or through the Internet. This is an extremely user-friendly product intended for direct use by the end user, specifically the health professional. The *"MESH"* thesaurus (standard search term vocabulary) is included. Use of this software is not required, but it can help considerably. This is the first of the major end-user products for searching of databases, and is probably the most widely utilized. See also Chapter 8 and Chapter 10.

Other sources of information produced by, or associated with NLM, include the following.

A. *TOXLINE*

This is the NLM's extensive collection of computerized bibliographic information covering the pharmacological, biochemical, physiological, and toxicological effects of drugs and other chemicals. The *TOXLINE* family of databases contains over 4 million citations, almost all with abstracts and/or indexing terms and CAS Registry Numbers. Records in the *TOXLINE* file, and in the companion file for royalty information, *TOXLIT,* cover the literature from 1981 to the present. Older information is found in the *TOXLINE 65* and *TOXLIT 65* backfiles. The information in *TOXLINE* is taken from 16 secondary sources that formulate the following subfiles:

Aneuploidy (*ANEUPL*)
Chemical Biological Activities (*CA*)
Developmental and Reproductive Toxicology (*DART*)
Environmental Mutagen Information Center File (*EMIC*)
Environmental Teratology Information Center File (*ETIC*)
Epidemiology Information System (*EPIDEM*)
Federal Research in Progress (*FEDRIP*)
Hazardous Materials Technical Center (*HMTC*)
International Labor Office (*CIS*)
International Pharmaceutical Abstracts (*IPA*)
NIOSHTIC (*NIOSH*)
Pesticides Abstracts (*PESTAB*)
Poisonous Plants Bibliography (*PPB*)
Toxic Substances Control Act Submissions (*TSCATS*)
Toxicity Bibliography (*TOXBIB*)
Toxicological Aspects of Environmental Health (*BIOSIS*)
Toxicology Document and Data Depository (*NTIS*)
Toxicology Research Projects (*CRISP*)

Note: Some of the above sources are no longer published.

B. *Registry of Toxic Effects of Chemical Substances*

The *Registry of Toxic Effects of Chemical Substances* (*RTECS*) is the online version of the National Institute for Occupational Safety and Health's (NIOSH) annual compilation of substances with potential "toxic activity." The original collection of data that makes up *RTECS* was known as the *Toxic Substances List,* compiled in 1971 by NIOSH in response to the Occupational Safety and Health Act of 1970. NIOSH is responsible for the file content in *RTECS,* and for providing the quarterly updates to NLM and other vendors. *RTECS* currently contains toxicity data for some 130,000 substances. The information in *RTECS* is structured around chemical substances with potential toxic action, and thus provides a single source for basic toxicity information. *RTECS* includes some listings of basic toxicity data and specific toxicological effects that are searchable. The sources of toxicity data are identified, with the names of the journals, volumes, pages, and years given. Also included in *RTECS* are threshold limit values, air standards, NTP (National Toxicology Program) carcinogenesis bioassay information, toxicology–carcinogenic review information, status under various federal regulations, compound classification, and NIOSH Criteria Document availability. See also pages 380–384.

C. *Chemical Dictionary Online* and *ChemID*

The *Chemical Dictionary Online* (*CHEMLINE*) is the NLM's online, interactive chemical dictionary file developed under contract with Chemical Abstracts Service (CAS). It provides a mechanism whereby over 1.4 million chemical substances can be searched and retrieved online. This file contains CAS Registry Numbers, molecular formulas, *CA* chemical index nomenclature, generic and trivial names, and a locator designation that points to other files in the NLM system and the TSCA (Toxic Substances Control Act) Inventory. In addition, where applicable, each record contains ring information including: number of component rings within a ring system, ring sizes, ring elemental compositions, and component line formulas. *ChemID* was developed more recently by NLM to provide users with an inexpensive, nonroyalty source for CAS Registry Number and locator information. This file contains nearly 300,000 chemical records and also includes *SUPERLIST.* The latter is the term used in *ChemID* to designate a collection of some 34 lists of chemical substances, as maintained by key U.S. federal and some state regulatory agencies, as well as a few from Canada and Europe, and by a few selected scientific organizations concerned with health and environmental hazards of chemical substances. *ChemID* provides "directory assistance" to these lists, specifically giving the name of the list in which the substance appears and the chemical names utilized in the specific lists.

D. *Directory of Information Resources Online* and *TIRC*

NLM offers a free online database *Directory of Information Resources Online* (*DIRLINE*), which contains descriptive information on a wide variety of information resources concerned with health and biomedicine. Over 15,000 records are contained in this important file.

In addition, in 1971, the NLM initiated the Toxicology Information Response Center (TIRC) at the Oak Ridge National Laboratory (ORNL) to serve as an international center for toxicology and related information. It provides information on individual chemicals, chemical classes, and toxicology related topics to the scientific, administrative, and public communities. As an information analysis center, TIRC synthesizes comprehensive literature packages according to a user's specific request. Formats may include, but are not limited to, custom searches of computerized databases, individualized literature searches, annotated and/or keyword bibliographies,

SDI (selective dissemination of information) service, or written summaries of the material obtained. Published bibliographies are available from NTIS. Most services are provided on a fee (cost-recovery) basis. For more information, the contact is Toxicology Information Response Center, Oak Ridge National Laboratory. This Center is now operated independently of NLM.

E. *TOXNET*

Established in 1985, this powerful and user-friendly array of files is also known as the *Toxicology Data Network.* It is managed by NLM and is available through its MEDLARS system (and soon through the Internet). This includes the following files:

HSDB (*Hazardous Substances Data Bank*). A continuously updated and peer-reviewed file covering over 4500 potentially hazardous chemicals. Focus is on toxicology, but environmental, handling, detection, and regulatory aspects are also included. This is a very significant file.

TRI (*Toxic Chemical Release Inventory*). (See discussion in Chapter 14.)

IRIS (*Integrated Risk Information System*). This EPA file contains EPA carcinogenic and noncarcinogenic health risk and regulatory information on over 600 chemicals. The risk assessment data are scientifically reviewed. Also contains drinking water health advisories.

RTECS. (See description earlier in this chapter.)

CCRIS (*Chemical Carcinogenesis Research Information System*). National Cancer Institute sponsors this file, which contains test results on almost 7,000 chemicals.

GENE-TOX (*Genetic Toxicology*). Created by EPA, this includes test results on about 3000 chemicals.

DART (*Development and Reproductive Toxicology*). A bibliographic database that covers literature on toxicology and related topics. This is a continuation of *ETICBACK* (*Environmental Teratology Information Center Backfile*).

EMICBACK (*Environmental Mutagen Information Center Backfile*). Bibliographic database (1950–1990) on chemical, biological, and physical agents tested for genotoxic activity.

9.6. FEDERAL LABORATORY CONSORTIUM FOR TECHNOLOGY TRANSFER

Another significant way to access the know-how of the federal government in the United States is through appropriate contacts with the Federal Laboratory Consortium for Technology Transfer. The Consortium was established to help scientists in industry, and others in the United States, identify and use the results of government research.

The expertise, and sometimes the associated facilities, of thousands of scientists and engineers at hundreds of government laboratories are potentially accessible to the private sector through the offices of the Consortium, most notably the FLC Management Support Office as described below. The chairman of the Consortium is C. Dan Brand, who is located at the National Toxicology Research Center, Jefferson, AR.

FLC has, for some years, engaged a private firm (DelaBarre and Associates) to establish and maintain a Management Support Office. This office is located at 224 West Washington Street, Suite 3, PO Box 545, Sequim, WA, 98382 (360-683-1005). One of its functions is a locator service that responds to inquiries from the public regarding which federal laboratories may have expertise or facilities in specific areas of science and technology. The response will usually include the locations of any such laboratories and the names and telephone numbers of the FLC representatives (the technology transfer persons) at these laboratories. Andrew Cowan is manager of the locator function. The FLC Internet location is www.zyn.com/flc, and this contains some excellent links to federal laboratory servers and other government sites.

Each federal agency that is a member of the Consortium has one or more FLC representatives one of whose major functions is to work to help transfer federal technology developed in that agency to the private sector. Another service offered by the representatives is to direct or refer those private sector scientists who may inquire to appropriate federal scientists and engineers in the agency who may be available to discuss specific research projects and technical problems of mutual interest.

The FLC Management Support Office periodically publishes a list (*Participating Representatives and Contacts*) of all the officially designated representatives, with their addresses and telephone numbers.

FLC representatives hold regular national and regional meetings, and most of these meetings are open to the public. There are six regional components of FLC. Each publishes regional directories, and some of these directories contain descriptions of the laboratories in the region. An effort is underway to attempt to place the content of these directories on the Internet.

Unfortunately, at this writing, the government has not made available a good, recent, overall guide summarizing the activities of all government laboratories.

9.7. THE INTERNET AND OTHER SOURCES OF GOVERNMENT INFORMATION

U.S. government agencies have made a huge amount of information available on the Internet. Most, but not all, of these files are available without charge. Assistance in identifying and linking to government Web sites can be obtained through the FedWorld location, which is maintained by NTIS at www.fedworld.gov. Another good such source is maintained by the National Technology Transfer Center (Wheeling, WVA) at www.nttc.edu./gov_res.html. The FLC site contains some excellent links to federal laboratory servers, as mentioned above.

At a minimum, most government agency sites on the Internet provide the names and telephone numbers of key officials. Key programs are also described in many cases. Some of these government sites are described elsewhere in this book, as, for example

U.S. Patent and Trademark Office—tools to search U.S. patents. (See Chapter 13.)
U.S. National Institutes of Health—CRISP files to search funded research projects. (See Chapter 10.)
U.S. National Institute of Science and Technology—physical property information and other output from the Standard Reference Data Program. (See Chapter 15.)

A good example of another government agency that has an extensive presence on the Web is the U.S. Environmental Protection Agency (www.epa.gov). This site includes, for example

1. *Rules and Regulations*
2. *Envirofacts.* Provides detailed factual information regarding the EPA Toxic Release Inventory System and several other EPA program system databases, with monthly updating. These contain data about many thousands of industry facilities insofar as they are subject to these programs. In addition, there is a grant information database, three integrating databases, and mapping applications. The specific Internet site is www.epa.gov/enviro/index.html.

The Technology Transfer Information Center at the U.S. Department of Agriculture's National Agricultural Library maintains a useful site (www.nal.usda.gov/ttic) that permits searching of Agricultural Research Service (ARS) research, new technologies from the Department, patents, and other useful information. Thus, their TEKTRAN database permits searching of prepublication notices of ARS research results.

In addition, the FLC Management Support Office is in the process of working on an extensive federal laboratory resource directory. Unfortunately, this directory will not, at least initially, contain narrative descriptions of each of the laboratories. However, a goal is to ultimately include lists of technologies available for licensing and information about CRADAs (Cooperative Research and Development Agreements) that industry can develop with the federal laboratories.

Additionally, members of Congress can often make and expedite contacts with the most appropriate federal officials. Further, many industry (trade), scientific, and professional associations are located in or near Washington, DC; some of these have staff who are skilled in advising as to the status of both proposed and introduced government actions, especially in the regulatory arena. They also have know-how as to the best government officials to contact on specific matters. In addition, there are several publishers who specialize in government information. One of the best known is the Bureau of National Affairs (BNA), Washington, DC, one of whose specialties is government actions relating to environmental matters (see Chapter 14). Private firms also publish directories of federal staff. It is often advisable to communicate directly with these staff members. Since there are frequent government reorganizations and personnel shifts, it is important that any directory purchased be updated on a frequent basis. One of the best such sources is *Federal Directory,* which is a product of Carroll Publishing Co., Washington, DC. This is updated six times a year and is available in printed and CD-ROM formats. Carroll's Internet site (www.carrollpub.com) lists its products and includes useful links to other sources of government information.

In summary, some of the principal sources of government information of interest to chemists have been described in this chapter. Examples of other important agencies discussed more fully elsewhere in this book include the U.S. Patent and Trademark Office (see Chapter 13); the National Institute of Standards and Technology, also known as NIST (see Chapter 15); and the regulatory agencies such as the Environmental Protection Agency (see Chapter 14). Another excellent resource is the Library of Congress in Washington, DC. Private-sector publishers also produce helpful resources.

List of Addresses

Carroll Publishing Co.
1058 Thomas Jefferson Street NW
Washington, DC 20007

Library of Congress
Washington, DC 20540
(202-287-5680)

National Library of Medicine
8600 Rockville Pike
Bethesda, MD 20894

National Technical Information Service
Springfield, VA 22161
(800-553-NTIS)

Toxicology Information Response Center
Oak Ridge National Laboratory
1060 Commerce Park
Mail Stop 6480
Oak Ridge, TN 37830
(423-576-1746)

10 Online Systems and Databases, the Internet, CD-ROMs, and Related Topics

10.1 INTRODUCTION

The purposes of this chapter are to give chemists and chemical engineers an enhanced appreciation of how online systems, databases, and the Internet function; to review some of the advantages and limitations of online and Internet searching; to describe and explain some of the leading sources; and to suggest additional strategic concepts (beyond those mentioned in Chapter 3) that need to be considered in searching of online systems and databases, the Internet, and other tools. Discussions on selecting tools and sources are interspersed, as appropriate, with pertinent, selected lists of systems, vendors, and so on. Accordingly the reader may find it helpful to quickly scan the table of contents for this chapter before reading it further. Note that topics pertinent to online systems and databases, the Internet, and CDs are discussed throughout most of the chapters of this book. For example, current awareness mechanisms are discussed in Chapter 4; Chapters 6–8 discuss Chemical Abstracts Service, ISI, and other major producers of online products; Beilstein's computerized systems are described in Chapter 12; Chapter 13 explains computerized systems for searching patents; material safety data sheets online and other environmental–safety systems are discussed in Chapter 14; and Chapter 15 discusses physical property and other numeric databases.

10.2. SOME RECENT DEVELOPMENTS

Since the publication of the second edition of this book, the use of personal computers and their desktop availability to most chemists has become commonplace, and the availability and searching of online databases and other files has greatly improved and expanded. A few of the most significant developments include the following:

1. Search methods have become both more powerful and more simple. For example, such features as detection of duplicates (when more than one database is searched at the same time), and ranking and other related statistical capabilities, are now available on a routine basis as needed by the user.

2. Full-text databases and files have become more significant and numerous, especially for important chemical journals.
3. More numeric databases are now available and can be searched more efficiently than ever before.
4. System speeds, based on baud rate, are now faster than ever; 9600 is the usual minimum, but frequently extends to more rapid speeds by factors of 2 or 3 and sometimes more. This has the effects of saving search time and of lowering certain costs.
5. It is now possible to display images, such as chemical structures and diagrams, online for a number of databases. This feature is especially important with patents and trademarks.
6. The Internet is in process of emerging as a major factor in availability of information of interest to chemists and chemical engineers. However, important questions need to be resolved as to its reliability, security, and frequently undisciplined nature.
7. There are now so many databases and other files available online (through commercial and proprietary sources or the Internet), and/or on CD-ROM, that selecting the best databases to use is an important issue. Similarly, selection of which host system (online service) to use is also a significant matter.
8. Relatively few databases are now exclusive to a single online system. Most are available on several systems. However, there are a number of significant exceptions.
9. The premier provider of information products to chemists, Chemical Abstracts Service, has significantly strengthened and diversified its armamentarium of products and services. Its competitors have also strengthened their offerings.
10. The accelerated interest in combinatorial chemistry approaches as utilized by the pharmaceutical industry has presented a significant new challenge to chemical information management. A number of innovative software products are available to help drug discovery chemists optimize work in this field.
11. The exploding growth of CD-ROM technology (many databases and other files are available on CD-ROM) offer challenges to online searching in many situations, especially where the frequent use of CDs makes them very cost-competitive.
12. The online products industry (including both online services and online databases) has matured considerably since the previous edition of this book, and there have been some major acquisitions and mergers. Many online products are offered by multi-billion-dollar publishing industry giants. But, at the other end of the spectrum, there are a number of freestanding information companies, some of them small and relatively new to the scene, that show their presence by introducing new products, especially on CD-ROMs, diskettes, client/server packages, and on the Internet.
13. The online industry is more competitive than ever, and advertising to catch the eye and interest of the chemist and chemical engineer is extensive and

flashy. Diligent comparison-shopping by the user of online products is essential not only from the perspective of cost, but also to achieve the best results as efficiently as possible.

14. Concurrent with the increasing number of online and other electronic product delivery options, there appears to be a very slow decline in sales of the printed counterparts of such electronic products. But the market for printed resources continues to be significant, especially in universities.

In all of the excitement about the Internet, it is sometimes forgotten that "conventional" online services (such as the DIALOG system and the system now operated by Questel•Orbit) have been operating very efficiently on a large scale since the early 1970s. These services, later joined by STN International and others, have been delivering substantial amounts of reliable chemical information ever since that time. In any event, every chemist and chemical engineer knows, or should know, that many important chemical information sources are available online, either through one or more commercial or proprietary online services, the Internet, or both. CD-ROM availability and databases on diskette are also commonplace, although not nearly as widespread as is online or Internet availability.

However, by no means is all significant chemical information online or in other electronic form. Furthermore, many online sources do not begin their coverage until the late 1960s (or subsequent to that), and a vast amount of useful data appear in the earlier chemical literature. At the same time, not all chemical information is available through conventional printed journals and books. Rather, a combination of media and sources is frequently desirable to locate what is needed.

10.3. THE BASICS

An online information retrieval system is one in which a user can, via computer, directly interrogate a machine-readable database such as *Chemical Abstracts.*

There is two-way (interactive) communication between user and computer through local input/output devices, such as a suitably equipped typewriter-like device and/or a cathode ray tube (CRT) display (or monitor), connected to a main-frame computer center by a telephone line. The communication (input) from the user is the query. The communication from the computer (output) is the system's response to the query, or a prompt for the next query.

Input devices may range from "dumb" or nonintelligent terminals (no memory or logic), to personal computers. In all cases, a device called a modem (modulator-demodulator) is required for telephone communication unless the terminal is directly connected to the computer. These devices may be either separate or already built in. Certain combinations of modems and computer terminals along with the appropriate software, permit users to automatically dial and log into pre-selected online systems. A telephone instrument is not needed; a telephone line will suffice. Modems convert digital signals into analog signals in sending and analog into digital while receiving.

Online output usually consists of references, including full bibliographic citations, abstracts, and indexing terms, ideally including also CAS Registry Numbers. In some cases, for example the *Registry File* on STN International, structure diagrams for substances retrieved are offered as answers, along with other information about the substances. Full text (full journal paper or full patent) is increasingly available, as are images (diagrams and related graphics). Output format can usually be tailored to the needs of the individual chemist, and the content of the output can be limited to what is of primary interest. Results can be sorted (by date, company, and in other ways) as needed, or ranked by frequency. Full copies of documents can be ordered online, and, in some cases, may be delivered electronically. Tables of contents for journals are available from several sources online, including DIALOG's UnCover affiliate, the British Library Document Supply Center, and the Institute for Scientific Information, the latter with abstracts. Such services alert chemists and chemical engineers to what is being published on a very timely basis through the Internet or traditional online services.

The online concept, as applied to chemistry, did not win truly widespread acceptance in the chemical community until the late 1970s and early 1980s. Online databases continue to proliferate at a relatively modest pace, but new Internet files are rapidly multiplying. Many chemists applaud the introduction of new databases and systems that fill an important void or that offer other benefits. On the other hand, some feel that they must now learn the details of too many databases and systems, and that this dilutes the opportunity to develop full expertise in methods of use. Although the "conventional" online industry clearly shows signs of maturity, the Internet is still in a developmental stage from the perspective of most chemists. Those databases, files, and systems that meet real needs, are user-friendly, and are profitable, will survive, while others will fall by the wayside or be merged into other products and vendors.

Several directories of online databases are available, and there are a number of English-language journals in this field (1). None of the directories, valuable though they are, make it significantly easier for chemists to learn details of how to use and evaluate competing systems and databases. See also Section 10.5B.

10.4. WHY USE ONLINE SERVICES?

Online information retrieval offers chemists important advantages in searching of patent, journal, and report literature. Rapid, convenient, and fingertip access is made possible to millions of references and chemical substance structures as well as hard data and full text of journal articles, patents, and so on. The systems available often permit users to identify recent pertinent references more completely, quickly, and effectively than is possible by using other techniques.

Access to printed indexes—whether these are alphabetical, classified, numerical, or specially organized—is inherently limited by index arrangement. In addition, users can find information in printed indexes only through a limited number of primary access points, and these are usually in a predetermined order. (More can be done if the user wants to spend the time.) In contrast, online services often afford many more access points and have much increased flexibility. The following are examples.

If the index heading is **Polymerization** and indexing policy is to index *oligomerization* under the more general heading, the user must turn to the heading **Polymerization** and look for *oligomerization* as a potential indenture or a term somewhere in a phrase (or modification) under that heading. Yet, in online access, *oligomerization* can directly be accessed no matter where it occurs in the file.

If the index heading is **Benzoic acid, 4-(1*H*-tetrazol-5-yl)-,** the user cannot usually find from the printed index, for instance, all (or certain) derivatives of *1H-tetrazole,* because the alphabetically arranged chemical substance indexes contain unique names constructed according to an order of precedence of chemical functions and compound classes.

If the index headings in the Formula Index are $\mathbf{C_{17}H_{18}IN_5O_4S}$ and $\mathbf{C_{24}H_{20}I_6N_4O_8}$, the user cannot usually find [unless the whole index is scanned (an impossible task), or unless it is a permuted index] other organic compounds each containing I, N, and O. Online access allows the user to go into *the middle* of a chemical substance name or formula, and then use combinations of terms such as described later.

The online approach has the capability to search for individual (or combinations of) access points that are not usually easily available in printed indexes. These may include, for example, not only index terms, keywords, authors, formulas, and patent numbers (these are readily available now in printed indexes), but also parts of chemical substance names, formulas, and notations; generic features; periodic groups; classes of substances; and formula weights. These are all relatively difficult to locate in most printed indexes, and combinations with one another, as well as concurrent limiting by language or type of document, are even more difficult or impossible in traditional indexes.

The online approach permits users to create and use, as necessary, search logic that is much more complex than can usually be handled manually. This is done by combining key terms (such as names of substances and other subjects) used in searching, as shown in the following examples:

A or B or C or D

A & B & C & D

(A & B) or (C & D) not (E & F)

Similarly, users can specify that only certain types of publications are wanted, such as patents issued in a specific country, language within a given time, articles that contain one or more CAS Registry Numbers in combination with a specific end use, or review articles and books on a combination of topics. All other materials can be automatically excluded. Doing this kind of specification in a conventional manual search may sometimes be possible, but it is much more time consuming. In addition, online systems facilitate ranking and sorting of results.

Online systems also make it possible for the chemist to search a wide variety of sources and tools that might otherwise not be conveniently available because of budgetary, geographic, or other limitations.

10.5. BASIC STRATEGIES FOR GETTING STARTED IN ONLINE AND CD-ROM USE AND KEEPING UP-TO-DATE

Getting started with online use is fairly straightforward for most chemists and can be achieved through methods such as described below. Keeping up with new databases, other new products, and system developments is a more daunting task. Hardly a month goes by without several new chemical information products and/or online system improvements being introduced. It is no wonder that both full-time chemical information professionals and (especially) laboratory chemists and engineers are truly challenged to keep up with the numerous, rapidly moving developments that often border on "overload." The suggestions given below present a systematic strategy for an approach.

A. Getting Started: The Fundamentals

1. Read this chapter and the other chapters in this book.

2. Consider and evaluate the basic features of a few of the principal online services that are available, and attempt to identify one or two that seem to offer the closest fit with professional interests and preferences. Good initial choices for most chemists and engineers include STN International, Columbus, OH, and either DIALOG, Mountain View, CA, or Questel•Orbit (the U.S. office is in McLean, VA). Ideally, one might at least consider for future use one or more of the more specialized sources such as MEDLARS, Technical Database Services, or Chemical Information System, depending on individual needs and interests. To conduct this selection process, obtain and review copies of the free advertising literature and catalogs from the online services, and discuss the main features and pricing structure with the sales representatives of the services of most interest. Then obtain and review the basic user guides. Later, take at least one basic training course with hands-on online practice, if possible; good choices include the introductory courses offered regularly by STN at American Chemical Society national meetings and elsewhere; independent-study materials are also available. The choices for Internet access include local or national Internet service providers and such services as CompuServe and America Online. This situation is still evolving and developing, and additional options can be expected through such firms as Microsoft.

3. Identify and become familiar with the core databases most pertinent to specific projects of interest. This will usually include, at a minimum, the powerful *Chemical Abstracts* and *Registry* files on STN. Selection of databases is usually based on content, timeliness, quality, affordability, and user-friendliness. Obtain and study the summary sheets and (time permitting) the detailed manuals dealing with those databases most critical to one's current interests. As mentioned above, if possible, take a basic training course, and then practice in the low-cost "learning" or training online files available on major systems. Those who are not information professionals should try to investigate, and consider using, products intended for direct use by the laboratory scientist or other end-user. Examples include the "Grateful Med" software of the National Library of Medicine, the in-house *SciFinder* product of CAS and the *STN Easy* Internet product CAS (Internet address www.cas.org), and the *DIALOG ScienceBase*

and *DIALOG Select* Internet products (Internet address www.krinfo.com). Related products are *Custom DIALOG, DIALOG QuickStart,* and a "Quick Search" approach through DataStar Web. Many CD-ROM products are also intended for end users.

4. As a guide to selecting both hardware and software, it is prudent to ask the host (online) systems and the producers of databases one plans to use (see sections on these below) what hardware and software configurations they require and recommend for optimum results. This is especially important with databases and files that are provided on CD-ROM or diskette. The type of information that is needed before a final decision is made typically includes processing–operating system; minimum and suggested RAM (random-access memory); hard-disk space required; type of videodisplay (monitor) equipment required; peripherals and output devices supported or recommended, including printers and modems; and special software requirements.

5. When talking with potential vendors of hardware and software, it is also desirable to determine the nature and details of the technical support offered. Is there any charge for this support? Is there a toll-free phone number? Are there return privileges? Does leasing make sense? Are instruction manuals and training courses offered? At what cost? Version ".0" such as 1.0 or even 5.0 of any product is frequently not a good buy; it is usually better to wait until any deficiencies are identified and eliminated as reflected in a later version number such as 5.1.

6. Review, select, and install the proper equipment. Consider such features as (a) the power and speed of the computer including memory options, (b) CD-ROM drive speed and type, (c) monitor size and pitch, (d) modem speed, and (e) type and speed of printer. The most crucial factor in online–Internet communications, in addition to computer speed and proper software, is modem speed; this should be at least 28.8 baud. When buying hardware, it is usually better to "overpower" somewhat at the beginning, if possible, rather than be forced to upgrade a system as the need surfaces.

7. Concurrent with step 6, review, select, and install the proper software, including good communications software (such as DIALOGLINK and/or STN Express), and CD-ROM software as required. Also, for Internet use, consider *Netscape Navigator, Microsoft Internet Explorer,* or a comparable Internet World Wide Web browser. E-mail software is typically included in major browser products. Of course, today's computers often come with much of the desired hardware and software already installed and bundled in the price of the original equipment.

B. Online Databases of Interest to Chemists and Chemical Engineers

The brief list of online sources of interest to chemists and chemical engineers in Table 10.1 is merely representative. A recapitulation of databases that emphasize patents is presented in Chapter 13, and other online sources are discussed throughout this book, especially in the chapters on physical properties (Chapter 15) and environment and safety (Chapter 14). New databases (sometimes also called databanks or files) are added and others are dropped periodically, depending on availability, demand, economics, and interest or need. The principal criterion as to the availability of a database to the public is usually financial (profit or loss).

TABLE 10.1. Some Online Databases of Special Interest to Chemists and Chemical Engineers (see also Appendices C and D)

Beilstein (*Beilstein Handbook of Organic Chemistry*)
Biosis Previews/RN
Chemical Abstracts Service: *Registry; Chemical Abstracts; Chemical Industry Notes*
Compendex Plus (*Engineering Index*)
Derwent World Patents Index
Gmelin (*Gmelin Handbook of Inorganic and Organometallic Chemistry*)
CLAIMS/U.S. Patents Abstracts (*IFI*)
Medline (National Library of Medicine)
NTIS (National Technical Information Service)
SciSearch (*Science Citation Index*)-*ISI*

A complete list of online files of potential chemical interest might exceed 200–300 databases. Chemists and engineers can consult an information professional for full details on what is available or one of the several good lists of online databases, as, for example

1. The major online systems, such as STN International, DIALOG and Questel•Orbit, each publish catalogs describing the databases that are available on their systems. These valuable resources are usually free, but are frequently overlooked.

2. *CEP SOFTWARE DIRECTORY.* Karen E. Simpson, editor. Published annually by the American Institute of Chemical Engineers, New York, NY, with the December issue of *Chemical Engineering Progress.* Also available separately and very inexpensively from American Institute of Chemical Engineers, New York. This product is not nearly as well known as it should be, but it is quite useful. The 1996 edition covers over 1600 computer programs and online databases from 510 vendors in the United States, Canada, United Kingdom, Europe, and elsewhere. The focus is on programs and databases of special interest to chemical engineers, but the scope is very broad, so that chemists and other scientists will find it of great value. Listings are presented in 32 broad subject categories such as physical and chemical properties, safety and material safety data sheets, thermodynamics, scientific word processing, and environmental control. The information about each program or online database is the name and city of the producer (but no phones or addresses) and a succinct abstract. Also supplied is the price and the basic features of the computer system required to run the software. The reader must fill out enclosed inquiry cards for further details on all products listed. This is one of the most useful listings of its type because of its scope and organization.

3. *Gale Directory of Databases,* Gale Research, Detroit, MI (800-877-4253). Published annually. This powerful tool is the most extensive publication of its kind. It includes excellent indexes. The book is available in printed form (two volumes: one for online databases and one for "portable" databases such as CD-ROMs, diskette, magnetic tape, hand-held, and batch access products), on CD-ROM, and online through Galenet, Questel•Orbit, and DataStar. In the volume on online databases, there are

listings in the following categories: online databases, including description of scope and coverage, producer, and online availability; database producers; and online services, including a list of each of the databases available on each service. See also the related Gale product available on the Internet at the address http://www.cyberhound.com; emphasis of this unique search engine is on products that are available on the Internet. A special feature is a rating or evaluation system. Gale also publishes *CyberHound's Guide to Internet Databases,* 2nd ed., Gwen Turecki, Ed., 1996. This guide includes approximately 2750 databases. A description and rating for each of the databases is provided. As mentioned, an online version of this is available on the Internet.*

4. *DataBase Directory,* Knowledge Industry Publications, White Plains, NY (914-328-9157). Revised every 2 years. This is part of the DataBase Directory Service that includes semiannual supplement volumes and a monthly *DATABASE ALERT* newsletter. Approximately 2000 databases from over 1200 producers in all disciplines (extending far beyond chemistry) are included. About one-third of the databases are produced outside the United States. The files included are those that are technically accessible in North America.

5. *Ulrich's International Periodicals Directory,* published annually by Reed Reference Publishing, New Providence, NJ. This widely available reference source indicates which periodicals included are available online and on which systems. This is also available online through such sources as DIALOG.

6. *World Databases in Chemistry,* published by Reed Elsevier, New Providence, NJ. This is a relatively new, very extensive source with a convenient arrangement by type of database and a subject index.

Appendices C and D tabulate selected online databases of interest to chemists and chemical engineers.

C. Evaluation of Databases and Other Electronic Files

There are so many chemical information tools, many very new, and many very expensive. Increasingly, chemists are the targets of ads for chemical information products in chemical magazines and through the mail, and many of these ads are large, elaborate, and expensively produced. For this reason, careful evaluation of the claims is needed. It is important to cut behind the wording of the ads and to understand both the benefits and the shortcomings in specific situations.

By far the most straightforward and most logical solution for most chemists and chemical engineers is to rely on the *Chemical Abstracts* and *Registry* files as available on STN International as the principal day-in, day-out source. These are powerful and reliable products with excellent quality, timeliness, and comprehensiveness of coverage. However, in addition, there are other highly significant tools that the chemist will often wish to consider in specific situations; a number of these are dis-

*Since these words were written, the CyberHound Internet site has been "unplugged."

cussed in this chapter and in other chapters in this book. As noted elsewhere in this chapter, it is frequently a very good idea to search more than one source.

To proceed beyond *Chemical Abstracts,* several proven methods can be used to assist in one's evaluation of other sources. For example, many products offer demonstration floppy disks or CDs. Others, especially those online, or available on the Internet, will offer a trial period without charge. Still others, shortly after introduction of a new product, will offer free search and display time for a specified number of days. Finally, many online services offer "learning files." Free, no-obligation 30-day inspection periods should be requested by anyone who is seriously considering purchase or subscription to an electronic tool.

To start with, the database evaluation process should take into account the user's situation as follows:

- What is needed to meet current and future information needs, and in the case of industry, what is needed to help increase profits and/or reduce costs?
- Are present resources sufficient to do all of what is needed?
- Do new products being considered offer true advantages over what is already available?
- What would a new product add in the way of capability?
- What can be afforded?

Regarding specific databases, some methods for evaluation are discussed below.

1. Inherent Quality. In considering any new database product (new to the organization), one should determine the inherent quality of the product. Consider such factors as quality, timeliness, and coverage.

As to quality, what is the reputation of the producing organization? Is it known to be fully qualified, reliable, and accurate? What is the quality of the product in terms of content, arrangement, and understandability? Who does the indexing, and how deep is it (how many index entries per document)? Is the indexing vocabulary controlled or uncontrolled? How lengthy are the abstracts, and how are they written or derived?

> Indexing should be done by qualified scientists as based on the full original document. The deeper the indexing, the better. The acceptable number of subject entries per document is in the range of 6 or more, but some sources index more complex documents as deeply as 50 or more subject entries or chemical names per document, thus increasing the searcher's opportunities to find what is wanted. Controlled vocabulary and/or the equivalent coding is preferred in most cases, rather than uncontrolled vocabulary. Intellectually written abstracts of 100 words or more are preferred over author abstracts in many cases, especially for patent documents. Original document titles that are cryptic or brief should be rewritten or augmented to make them meaningful.

By *timeliness* is meant how much time elapses between appearance of the original document and its online accessibility, including both abstract and full indexing. What is the time period covered, how up to date is the coverage, and what are the provisions and schedules for updating? Specifically, in the case of an abstracting and indexing service, what is the average time lag between publication of the original document and its appearance in the abstracting and indexing service? What is the frequency of updating (e.g., daily, weekly, monthly)?

> Citations without indexing, but possibly including abstracts, may appear within approximately a week or less in some top sources, and a number of the top sources provide complete indexing and abstracting within approximately 6–12 weeks after the appearance of the original.

As to coverage, what types of documents are covered or not covered (e.g., patents, books, journal papers, reports)? What is the subject matter that is covered or not covered? What countries and languages are covered or not covered?

The broader the scope and the fewer the exclusions, the better. Databases that include all types of literature (journal papers, patents, etc.) are generally the most useful, but some excellent files are limited to a certain type of literature, such as patents, only. In most cases, but not all, essentially worldwide coverage is expected. Any limitations of coverage should be clearly specified by the database producer.

2. Documentation, Training, and Related Topics. What kind of printed and/or online documentation is offered? How much does it cost? (Basic documentation should be free.) Is the documentation easy to read and use? Are free or low-cost training seminars offered nearby? What kind of free user support is offered? Is there a "learning" or training file available on which to practice? Are free demonstration disks and/or initial free online time available for evaluation? In the case of a disk, are there return privileges? Is there a toll-free help desk? What type of help is given by the help desk, and what are the hours? An example of an outstanding help desk is that of Chemical Abstracts Service, whose help desk is staffed by qualified chemists and other scientists who can review proposed search strategies and who may do some online testing of strategies while the customer waits on the phone. In some cases, the database producers or online services are willing to conduct the complete search for the user on a fee basis.

3. User-Friendliness. On the basis of actual hands-on testing, are the basics of the new product easy to use? (Ideally, the basics will be virtually intuitive.) Is the presentation acceptable from a visual or graphics perspective? How long does it take to learn how to use the basics of the product? Are there a variety of access points, such as subject, author, CAS Registry Numbers, and molecular formulas? Does the product seem to be geared to the information professional or the end user? (Bear in mind that it may take many hours of study and considerable practice and experience to become a fully effective user of a sophisticated information source such as *Chemical Abstracts.*)

4. Affordability. Each individual chemist needs to determine on a case-by-case basis what can be afforded in terms of budget and in terms of value delivered. In many cases, a chemist's organization will have made the basic decisions and set the framework in which the chemist must operate. However, there are some specific issues that can help make the choices more clear. For example, are the databases and/or systems priced on a pay-as-you-go basis (this alternative is usually far preferable to up-front or subscription fees)? Does online pricing offer discounts for those who also subscribe to the printed or other electronic versions of the product? (In some instances, only those who subscribe to other versions, such as the print versions, are permitted to use all of the features of the online file.) Are there discounts, as, for example, to chemists in colleges, universities, and government; for certain levels or volumes of use (previously agreed on); or for off-peak hours? What is the cost in terms of connect time, search term charges, if any, and print or display charges—both on- and offline?

*5. **Reviews.*** Important new chemical information products in electronic form are reviewed and critiqued in the *Journal of Chemical Information and Computer Sciences,* which is edited by George W. A. Milne, National Institutes of Health. An example of another excellent source is *Database,* a magazine that is published by Online, Inc., in Wilton, CT, and edited by Judi Copler. Online, Inc. presents some of the material from its several magazines on the Internet as the address http:/www.onlineinc.com. For other valuable information and sources of reviews, see also the journals mentioned in the following section.

D. Keeping up with New Databases and with Changes in Databases

The following is a strategy for dealing with the proliferation of new databases of chemical and chemical engineering interest:

1. Accept the fact that keeping up with the new products and capabilities of online or Internet hosts and of databases of chemical interest *is* a significant challenge, and that it is realistically all but impossible to keep up with *all* new developments, except in a limited specialty.
2. If a chemist's organization has an information professional such as a chemical information analyst or librarian, that person can help the chemists in that organization become aware of new developments that are important to specific projects and interests.
3. For those who belong to a project group or team, it may help if a member of the team can take on the specialized task of informing the others as to what is new and important.
4. One of the best ways to keep up is to read the column on the new software and databases that appears periodically in *C&E News;* this includes brief squibs, with a postcard opportunity to request additional detail from the manufacturer. For the most recent Internet offerings, read the regular Internet columns in *Chemtech, Chemical Engineering Progress,* and *Environmental Science and Technology,* as well as the occasional column in *C&E News.* The Internet files that list Internet databases of interest to chemists should be especially helpful in this regard. A valuable column that discusses and evaluates Internet tools and approaches, edited by Stephen R. Heller, U.S. Department of Agriculture, appears in the journal *Trends in Analytical Chemistry* (*TrAC*), published by Elsevier; the full text of this column is available on the Internet at the address http://www.elsevier.com:80/inca/homepage/saa/trac/intntcol.htm.*
5. Scan the monthly customer newsletters issued by the various online (host) systems. An excellent example is DIALOG's *Chronolog.*

*The CambridgeSoft Corporation's magazine *CS Catalyst* contains brief articles and ads of interest.

6. For more in-depth coverage, read *Journal of Chemical Information and Computer Sciences* and skim one or more magazines such as *Online, User, Database, Online, Searcher,* and *Information Today* that deal with new developments in online technology. (It may be useful to mention at this point some other sources of current information for those who are interested in computers and chemistry. These include the following examples: *Computers in Biology and Medicine, Computers in Chemistry, Information Processing, Information Sciences, Journal of the American Society for Information Science, Journal of Computational Chemistry, Journal of Computer-Aided Molecular Design, Journal of Mathematical Chemistry, Journal of Molecular Graphics,* and *Journal of Molecular Modeling*). To keep up to date very broadly with developments relating to computational and modeling methods utilized in the pharmaceutical and agrochemicals industries, in addition to the specialized journals mentioned, one good source readily available to almost all chemists and engineers are the periodic articles on computational chemistry that appear in *C&EN*. These are based primarily on the national ACS meetings. Other good sources include the product newsletters and bulletins of the companies in the field. Chemists and engineers who are interested should get on the mailing lists for this material. See also chapters in the well-known series of books on medicinal chemistry originally edited by Burger (14).

7. In addition, consider that STN and other online systems offer user meetings at national American Chemical Society meetings, and that other systems offer regional updates from time to time. For those who wish to explore in even more detail, there are major trade shows presented annually that deal with online and/or CD-ROM technology; an example is the National Online show (sponsored by Learned Information, Medford, NJ, and usually held in New York City in May of each year). In addition, there are several other online shows in the United States and in Europe.

10.6. REPRESENTATIVE ONLINE SERVICES (HOST SYSTEMS)

Fortunately for the chemist, most major databases of chemical interest are available through a choice of several suppliers, most of whom are readily accessible throughout the United States and virtually all other significant industrialized nations.

An online service (which may also be called *host system* or *vendor*) is a service that, for a fee (typically based on connect time and number of hits viewed), provides online public access to a large number of databases usually, but not always, from a number of different database producers. (Some systems, such as the MEDLARS system of the National Library of Medicine, offer primarily their own databases.) Hosts typically offer a search system that permits users to search the databases and display results with any or all of the databases on the system, "learning files" that permit users to practice with specific files at very low cost, communications software, a help desk with a toll-free phone number, training courses, a catalog describing each database briefly, and extensive documentation on how to use each of the databases. The search system utilized varies from service to service, but the general types of commands are fairly similar.* Users typically are given such options as ranking and sorting search results and the capabilities to search more than one database concurrently. An example of another capa-

*The list of commands for STN International is shown on page 124.

bility usually offered is the "crossover" of results from a search of one database to a search of another that may contain additional information on these results or may permit different kinds of searching. Access to the host is usually through a local telephone number or direct dial, and Internet access to hosts is now becoming available as well. Host systems may also offer other products such as current alerting products, document delivery services, and compact disk counterparts of some of the online databases. Some will offer to search their databases on a fee basis for customers who want this.

The *Gale Directory of Databases* (2) states that the total number of online systems (hosts) and vendors is nearly 1100, the total number of databases is nearly 6500, and there are over 3700 CD-ROM products. Below are listed some representative services (hosts) that should be of special interest to chemists.

Cambridge Scientific Abstracts
(offers Internet Database Service)
7200 Wisconsin Avenue
Bethesda, MD 20814
(301-961-6750); (800-843-7751)

Chemical Information System
(part of Oxford Molecular Group)
810 Gleneagles Court, Suite 300
Towson, MD 21286
(410-321-8440); (800-CIS-User)

European Space Agency (ESA) Information Retrieval Service (IRS)
ESRIN, Via Galileo Galilei
1-00044 Frascati (Rome, Italy)
(39-6-94108726)

(DIALOG, DataStar)
2440 El Camino Real
Mountain View, CA 94040
415-254-7000 or 1-800-3-Dialog.

LEXIS-NEXIS
(formerly Mead Data Central)
9443 Springboro Pike
Dayton, Ohio 45401
513-865-6800; 800-227-4908

National Library of Medicine, Specialized Information Service
(MEDLARS)
8600 Rockville Pike
Gaithersburg, MD 20899
(301-975-6776)

Ovid Technologies
333 7th Avenue
New York, NY 10001
(212-563-3006); (800-950-2035)

PLASPEC
D & S Data Resources
218 East Bridge Street
Morrisville, PA 19067
888-752-7732

Questel•Orbit
U.S. address: 8000 Westpark Drive
McLean, VA
(703-442-0900); (800-45-Orbit)
Headquarters address: Le Capitole,
55 avenue des Champs Pierreux,
F-92029
Nanterre, Cedex, France

STN International
Chemical Abstracts Service
2540 Olentangy River Road;
PO Box 3012
Columbus, OH 43210-0012
(614-447-3731); (800-848-6538)

Technical Database Services
135 West 50th Street
New York, NY 10020-1201
(212-245-0044)

For some chemists, their preferred online system is frequently the one that they learned first—in other words, the one that they "grew up with." This is human nature. However, it is desirable to try to find the time to at least explore a few of the other important systems that could potentially be useful. If funds and time permit, it also makes good sense for most organizations to become a user of at least two host systems in order to have available as many different search features and databases are possible.

Many hosts are usually available essentially around the clock (except for maintenance during some weekend hours), and, as previously noted, most are available in all of the major industrialized nations of the world. Some of the hosts offer unique, exclusive databases, although there is an increasing trend away from exclusivity. The basic searching features and power are broadly similar, but there are differences, and some hosts offer more searching power and more special features than others. The commands utilized to navigate around the various online systems vary sufficiently so that many searchers find it useful to keep on hand a chart that outlines and compares the commands in each of the major systems. Besides differences in features and commands, there are also variations in pricing.

In this book, conventional online services (those that use commercial telecommunications networks such as SprintNet) are treated separately from Internet sources, which are discussed later in this chapter. Some online services, such as STN International and DIALOG may be accessed either conventionally or via the Internet, and so some of the distinctions between the Internet approach and more "traditional" approaches may be beginning to blur somewhat.

The major online systems have excellent "help desks" accessible by toll-free phone numbers in the United States. In the case of STN, their help-desk personnel have degrees primarily in chemistry or in other scientific fields, and some are PhD-level chemists who can handle highly technical inquiries. If necessary, they may call upon the advice of 12 consultants in the editorial division, many of whom are PhDs in their fields. In addition, hundreds of other scientists at CAS can be consulted for specialized assistance. All of the major hosts and some of the database vendors offer training classes, many of which are offered without charge. Extensive printed documentation is offered by all major hosts.

Probably the best known major hosts of principal interest to chemists are STN International, Knight-Ridder Information, Inc. (including DIALOG and DataStart), and Questel•Orbit.

Both STN and Questel•Orbit permit structure and substructure searching utilizing a topological approach. Examples of other systems that permit this include Beilstein's *CrossFire* and *Current Facts* products, described in Chapter 12; Chemical Information System, described in Chapter 15; and MDL's in-house systems, as described on page 240 and following.

STN also has the advantage of being the only service that contains *Chemical Abstracts* abstracts (from 1967 to date) and that offers *CAplus* (daily updating of *CA* and other unique features).

Among the various host systems, STN offers the strongest combination of features, databases, and support services that are important to chemists, but other major hosts such as DIALOG and Questel•Orbit still deserve important consideration by chemists since each has unique files and features of significance. DIALOG is the most user-friendly of the online services.

STN has a number of powerful features of special interest to chemists (see Chapters 6 and 7 for additional background). As noted, STN is the only host that permits the searching and display of abstracts from *Chemical Abstracts* from 1967 to date. Another example is the extensive carry-through of CAS Registry Numbers into non-*CA*-produced files. All of the Chemical Abstracts Service databases are available on STN with full searching power, including the capability to draw and search chemical structures in the *Registry File* and in certain other files. Searching using CAS Registry Numbers is possible on a number of databases within STN. Other databases on STN include the full text of chemical journals published by the American Chemical Society and certain other major publishers, and there a number of exclusive databases. Since 1994, CAS has begun to offer innovative online and other electronic tools aimed at the end user, that is, the laboratory chemist or engineer, rather than the information professional, in the form of its *STN Easy**, *Surveyor*, and *SciFinder* products. As noted elsewhere in this chapter, STN's skilled help-desk personnel can, if appropriate, test specific potential search strategies online for the user and report back on results with their advice.

One of the system features that STN offers is the option of both left- and right-hand truncation. Further, strategy and answers may each be saved separately for an indefinite period of time, thereby saving both time and costs by avoiding having to reexecute a search.

A related system feature that searchers find helpful is that STN permits the user to "hold" a search session for one full hour. This means that if the searcher enters the command "log hold," the entire session is automatically held for a full hour.** This saves the time and expense of repeating work already done. It gives the searcher time to consider the merits of the strategy utilized and the results to date before either resuming the search, revising tactics, or discontinuing the effort for the present. (Some other systems do not permit this long a hold time, but Questel•Orbit does even better by offering a very generous 2-hour hold time.)

The file *USPATFULL* on STN has the capability to display images (graphics) online with the STN Express communications software package when there are images in the original patent from 1993. A similar capability (chemical structure drawings—one representative drawing for patents since 1992) is offered with *WPINDEX* (*Derwent World Patents Index*). This is an especially important feature in patent searching. (All other online services that offer the Derwent database offer the same capability.) Another example is found in the full-text files of journals of the American Chemical Society; images may be printed from 1992.

STN now offers over 200 online databases.

A new (Dec. 1996) product, *STN Easy*, available on the Internet only, offers a simple approach to searching a selected number of the databases available on the regular STN. There is no monthly fee or hourly connect charge; users are charged a flat search fee and pay for each answer displayed.

*These products include *CA* abstracting and indexing.

**Experienced searchers frequently use this command (or equivalent) when logging off.

Headquartered in Mountain View, CA, DIALOG began commercial operations in 1972 as part of the well-known Lockheed aerospace firm.* DIALOG has more total databases of all types (over 450) than does any other online system of major interest to chemists and engineers. It is also distinguished by a highly user-friendly architecture and powerful search features. The DIALOG help desk has been excellent over the years. Some of the databases of special interest to chemists include the full texts of the *Kirk-Othmer Encyclopedia of Chemical Technology* and the *Encyclopedia of Polymer Science,* as well as of U.S. patents since the 1970s. Additional databases are offered through a sister service, DataStar, whose computers are located in Berne, Switzerland. The two services have some overlap with one another, and each has some unique files.

As mentioned, the DIALOG online search system offers many special features. For example, to name just one, a special low-cost "pause" feature permits searchers to survey their results without logging off.

DIALOG offers a number of its online products in CD-ROM form, and it also offers DIALOG ScienceBase as an end-user product.

As the name of the online service Questel•Orbit indicates, this online service was formed by the combination of two long-established and well-known online entities, that is, Questel (a French firm) and Orbit (a U.S. firm). Both systems may now be searched through a single user contract, but two separate user IDs are given out—one for Questel and one for Orbit—and there are two separate sets of search commands. One phone number will provide access to either service; this feature is known as *common access.* The combined firm is owned by France Telecom Group, the French telecommunications giant. The principal computer facility is located in France, but, of course, this location, as with all online hosts, is seamless to the user. A total of approximately 300 databases are offered; a number of these databases have a European orientation and are in European languages such as French, German, and Italian. The service may be accessed directly, and, in addition, searching of Markush DARC files (see Chapter 13) on Questel is also possible through a "gateway" in *DIALOG.* This is available only to Derwent subscribers.

One of the main features of Questel•Orbit is the capability to search chemical structures graphically utilizing the unique Markush DARC approach (3). Questel•Orbit says that the Markush DARC approach permits a higher degree of specificity than with other systems. Markush DARC searching is possible in the unique *Derwent World Patents Index Markush* file, but this is for Derwent subscribers only, and in the *Pharmsearch* file. The *Pharmsearch* file has certain enhancements not yet available on the full system. See also page 213 and Chapter 13.

Generic DARC searching is possible in the *Chemical Abstracts* Registry File designated as the *EURECAS* file. Text input (molecular formula, atoms, etc.) is preferred; graphic input is also possible, but it is time-consuming and not widely used. Such generic searches may yield some of the same results as a comparable search on STN, but may also yield some different results, thereby helping ensure completeness of a

*Later, DIALOG was acquired by Knight-Ridder, Inc., and soon became a key part of Knight-Ridder Information (KRI). In late 1997, M.A.I.D., a UK online supplier of business intelligence, acquired KRI.

search, according to Questel•Orbit. Results from a *EURECAS* search may be crossed over into Questel for a search of the bibliographic files there.

The Questel DARC system was first conceived of by Jacques-Emile Dubois (University of Paris) in 1963, and was fully commercialized in 1980 with a portion of the CAS files. DARC is believed to be the first commercially available structure search system, if the Chemical Information System can be regarded as originally a public-sector operation.

One of Questel•Orbit's particular strengths is its coverage of databases pertaining to intellectual property (patents and trademarks). The patent files are listed in Table 13.6. One of the newest products offered is an Internet product *QPAT-US*. Another noteworthy and unique patent file is *Derwent World Products Index with API Indexing,* but this is available only to Derwent and API subscribers.

Ovid Technologies, headquartered in New York City, embraces the online service known as *BRS,* (Bibliographic Retrieval Services) and was formerly known as *CD Plus*. Approximately 80 databases are offered. Ovid is not strong in chemistry as compared to the other hosts mentioned above, but is very strong in biomedical databases and includes the full text of a number of medical journals. Ovid also includes the H. W. Wilson Co. (Bronx, NY) series of indexes in applied science, biology, and related subjects. These tools are economical and convenient to use as a first approach to many topics, and they are also available through other online services. Other hosts are described elsewhere in this book.

10.7. TELECOMMUNICATIONS AND RELATED TOPICS

In most metropolitan areas, there is a choice between several telecommunications networks, each of which permits the searcher to utilize a local phone call to access a number of online services.

The leading such networks in the United States, and the approximate per-hour costs, are as follows: CompuServe, $11.00; SprintNet, $12.00; and BT Tymnet, $12.00. Alternatively, the Internet may be utilized at a cost of only $5.00. (Costs are as charged by STN International.)

Of course, users also have the option of dialing directly into the host computer, but this usually involves a long-distance phone call unless there is a dedicated connection or a similar arrangement.

A modem or modemlike device is usually the required interface between online services and a local computer. Operating speed for modems is typically expressed in baud rates. Another form of expression of these same speed rates is *kbps* (kilobits per second). Baud or kbps rates for some online services have increased in recent years to 28.8, with further increases in speed anticipated in the future. Internet connections are typically much faster. This means that the chemist can expect to see higher-speed transmission of the searching process, including results of searches, and in some cases, of copies of full documents such as patents, for example.

There are special communications links that are even faster than the modem speed described above. One approach to high-speed communications is to lease an ISDN (Integrated Services Digital Network) line from the local phone company. This typically yields speeds of 64–128 Kbps (kilobits per second). The most widely used very high-speed commercial technology option at present is probably a "T1" line (developed by AT&T; a time-division multiplexed digital transmission facility), which offers speeds of 1.54 mbps (megabits per second). In addition, other forms of very high-speed connections are available. Very high-speed connections are expensive and are most suitable for large-volume users of online services and/or the Internet.

The principal advantages of all this speed include faster searching and lower cost (based on lower actual connect time.) However, the higher transmission speed also means that the searcher cannot scan the results as they are displayed on the screen. Thus, it may be more difficult to determine if the results are indeed those desired. There are several ways to cope with this potential "disadvantage":

View only a few answers at a time, especially at the outset, to determine whether the types of results being received are those desired. This is always a good idea as a method to test the validity of a search strategy.

Have the results transmitted offline via conventional mail or electronically.

After an initial group of answers is displayed, utilize "log hold" commands that permit one to log off temporarily (typically from 30 minutes to 2 hours depending on the host), and then to log back on after examining initial results. (With some hosts, the same telecommunications system must be utilized when logging back on.) The user reenters the search at the point it was left in the same database. Or, after the initial group of answers is displayed, utilize the "pause" command (at 75 cents per minute) for a similar reason as the above. This permits one to look at results while incurring only an economical connect charge. This command is currently available on DIALOG as mentioned earlier.

"Save" commands can alternatively be specified for much longer hold-times, but the database must be reentered before the saved strategy can be executed and viewed. *In the case of CA on STN, both answers and strategies may be saved.*

Other ways to speed up viewing time include specifying that the terms searched for be highlighted in the results, and that results be displayed in a brief format at first. The KWIC (Keyword-in-Context) format can be especially helpful in evaluating pertinence. In this format, a user-specified number of words is shown on either side of the "hit" term.

Of course, baud rate can always be slowed to lower rates of transmission that do permit viewing of results as they are received. This is not recommended because it results in increased costs (increased connect time).

A. Communications Software

A number of excellent communications software packages are available. These are essential for utilizing online services. Two of the most widely used are STN Express, a product of Chemical Abstracts Service, and DIALOGLINK. These, and other competitive packages, will provide basic operational capability with any number of online services, not just those of the producer.

Most software packages of this type permit automatic dial and logon so that the searcher does not need to dial and enter passwords every time a connection is made with an online service. Other features usually include type-ahead buffers that permit the searcher to type search strategies before going online and then to load them automatically when logon is established. This saves time and cost. In addition, full results can be saved after logoff for further examination.

The most outstanding feature of STN Express is that it permits drawing of chemical structures, an invaluable capability for searching the *Registry* file of CAS.

DIALOGLINK is probably the most user-friendly communications software, but it does not have the same powerful structure-drawing capabilities offered by STN Express. Some of the attractive features of DIALOGLINK include

- Very easy-to-use type-ahead buffer that permits entry of search strategies before going online. This is an important time- and cost-saving feature.
- Easy selection of any one of a number of host systems.
- Easy modification of host parameters if this is necessary. For example, the type-ahead buffer may be easily turned on or off.
- Easy saving of results (retrieve buffer) and of search strategies.
- Excellent user support through an 800 (toll-free) phone number.

These key features mean that the chemist can concentrate on the search and need not be slowed or hampered in any way by the intricacies of communications software.

One important caveat regarding software packages is that if the user is interested in displaying images, the package sold by a specific host for its system will usually need to be used when images are to be displayed with files on that host.

10.8. USE OF ONLINE DATABASES: SOME IMPORTANT CAVEATS

The online database concept is not a panacea. For many online databases, coverage of the literature and patents does not begin until the late 1960s, although, for some, coverage begins much earlier, and for others, much later. For example, although the printed *CA* begins in 1907, the online version of *CA*'s bibliographic files does not begin coverage until the 1967 volumes.

Most major online systems that offer online databases of interest to chemists did not start on a full commercial or public basis to the chemical community until the early 1970s or later. This means that for complete investigations—especially those that require going back to the *beginning*—the chemist also needs to conduct a conventional manual search. Accordingly, online services are often complementary to printed sources and manual search procedures. In addition, printed tools can have such important potential advantages over their online counterparts as: unlimited browsability, ease of use (no online instruction or equipment needed), and continuous availability (condition or availability of equipment, or online system schedules are not factors when using conventional printed sources). Finally, some important printed sources are not yet available online, and to ignore such sources, if they are pertinent, can be a significant error.

Availability of data through a computer, however sophisticated the hardware or software, does not in itself make that data accurate or reliable. This is a function of the original research that produced the data, the care with which these data are abstracted and indexed by secondary sources, and the care with which the data are entered into online files and systems.

The completeness that some users may attribute to online output per se can frequently be a myth. What count most are the intrinsic completeness and quality of the original sources, validity of search strategies, and the skill with which these are constructed and executed. Strategy is important because of the considerable chance for information loss due to such reasons as variations in: chemical names for the same substance, expression of concepts, and spelling of the names of organizations and people. Another example of the importance of strategy is in searching for patents assigned to a specific company. In such instances, it is important to look for patents to subsidiaries and affiliates in addition to simply looking under the name of the *parent* company and variations of that name.

As an example of skillful implementation of strategy, it is essential to understand and follow the protocol and logic of the database system. For example, a "minor" error such as a misplaced space or punctuation mark, or failure to use one when it is called for, can lead to a significant loss of information. An error in logic can be even more serious. The system may not signal when an error occurs and so the mistake can go undetected. A computer answer (output) showing zero "hits" always deserves especially careful scrutiny for this and related reasons.

A thorough knowledge of the database being searched is as essential for online searching as it is for manual searching. The user needs to know as much as possible about the database, including scope, kind of indexing and abstracting used, and any special features or limitations. In addition, the user needs a thorough knowledge of the online system being used. This does not mean becoming a computer programmer, but it does mean study and knowledge of the manuals that suppliers of online systems make available. Further competence can be gained by attending training sessions held by suppliers at central locations, or as an option, at one's own facility. Expertise in use of systems can be further enhanced by online experience in real-life situations. A number of months may elapse before a user can be considered expert, although the key is regular and frequent use. Furthermore, online suppliers make frequent improvements and other changes in details of operation. Sources (databanks) that constitute input to the systems may also frequently change scope and other policies. The expert user needs to try to keep up with these changes by perusal of the newsletters that systems issue periodically, the notices that most systems supply online and on the Internet, and study of the sources in both printed and online forms.

In this connection, it is important to note that there may or may not be one-to-one correspondence between the printed and online forms of the same database. One may contain information or approaches that the other lacks. Accordingly, both may sometimes need to be consulted. It is often desirable to scan the printed form before going online to help develop the best strategy and for other similar reasons.

Expertise with any one online system does not assure competence in use of other systems, although all have fundamental similarities. For this and other reasons, user organizations should train more than one person, each of whom specializes in one or two systems. Basic typing skill is useful but not essential. This helps because communication with the computer (input) is by a typewriterlike keyboard. Ability to operate this equipment quickly and accurately saves times and money. From this, it should be clear that special training, and considerable study and effort are prerequisites to optimum use of online systems. Many laboratory chemists have the time and interest required, but many others have their online searches conducted for them by information professionals. The laboratory chemist should provide the person who does the searching with a complete description of what is wanted and should help evaluate strategy and output. Best results come from close partnership.

With the preceding qualifications and other limitations in mind, and especially if an important decision of any kind is to be made on the basis of a literature or patent search, it makes good sense when evaluating search results to examine those full, original documents that appear pertinent, rather than rely on what is reported about those documents in a computerized (or printed) abstracting and indexing service or other such source.

10.9. COSTS AND PRICING

The usual practice among the most popular online services is that users are charged a relatively modest annual fee (up to about $150) that permits them to utilize the system, and that provides extensive documentation and other support services, but that does not include actual use. (However, STN International charges no annual fee, but does have a very small, one-time, start-up fee.) In addition, under these systems, users are charged for (a) connect time (the time the user is connected to the central computer), (b) the number and kind of hits displayed or printed, and (c) telecommunications time. Each database has its own clearly identified rate (connect time) on each online service. Connect hour costs can vary from $30 per hour to $300 per hour or more (prorated, of course, to actual use), and there are often some very complicated pricing arrangements, depending on the database, type of use, and the online service. In some cases, there are other charges as well. Thus, for example, when searching *Chemical Abstracts,* there is a charge for search terms, with reduced connect hour charges, contingent on the user's choice from among the several pricing plans that are offered. (One of these plans has no charge for search terms but higher connect charges; another offers higher search term fees, but no connect hour charges.)

Other charges can be incurred when a chemist specifies during an online session that a search or its results be saved for future use beyond the free "log hold" period. Typical costs per hit displayed or printed depend on how much of the hit is displayed (ranging from title only to full reference, abstract, and indexing). If a full record from *Chemical Abstracts*—including citation, abstract, and index terms—is displayed on STN, the current cost per record is $1.39 for subscribers to the printed *CA* and $1.83 for nonsubscribers. The chemist should check carefully with each system utilized to determine what other special charges may apply; this should, of course, be done before the search is initiated.

Connect time costs can be reduced or controlled if the chemist specifies that the results are to be printed offline (i.e., not during the online session) and then mailed, either through conventional mail, or electronically. Other cost-saving techniques include scanning of titles before printing the full records, use of the "expand" command to determine the likelihood of hits, and careful development and testing of strategies before full execution, as described in this Chapter. Examination of titles only, at first, may indicate in which instances a more complete display may be needed, although titles can be very misleading. In many cases, the user may not truly need the full record; a partial display such as evoked by a "HIT" or "KWIC" display, along with the bibliographic reference, may prove adequate to help confirm a search strategy (although in many other cases, study of the full record can be very helpful). In these ways and through a number of other options, the user can exercise some degree of control over costs.

The pay-as-you-go approach as described above may appear a little complicated, but it gives the searcher a flexibility that is usually needed, and it permits the user to control search costs.

In some cases, especially in the case of some newer end-user products, a substantial up-front minimum annual contract or subscription is required. This is a more rigid, and hence less desirable, situation from the perspective of many users, especially in smaller organizations, assuming that everything else is equal. It also effectively limits use of such databases or systems to larger organizations that can afford the up-front fee, and who can forecast a year in advance what their requirements will be.

Chemists and engineers at smaller firms and at smaller colleges and universities that cannot afford the up-front fee are thus at a clear disadvantage in their searches for chemical information. (In some cases, academic discounts may be offered.) Shopping around among the various systems and plans available can sometimes identify suitable alternatives.

Examples of other cost factors the user may want to take into account include

1. Any charges for training or documentation.
2. Cost of required software.
3. Cost of required hardware.
4. Supplies, including printer ink cartridges, paper, and so forth.
5. Cost of authorized downloading (electronic capture of results as permissible).
6. Savings from volume or other discount plans or from use of offline prints (especially advantageous when there are a large number of references to be studied). In some cases, subscribers to the print version of a database, or to other products from the same producer get a discount or other benefits.

Chemists and engineers should consider maintaining a simple log book of online time and expenses to sharpen awareness of these aspects.

10.10. ONLINE SEARCH STRATEGY

This section further develops some of the material noted in previous sections and introduces additional concepts as well.

For most effective search results, strategies should be developed on computer, blackboard, or paper before going online. Search objectives, basic logic, and search terms should be clearly understood and specified. Certain search aids, such as thesauri and other documentation, are available to help online searchers in their planning. Once online, users have available additional searching aids, some of which are not available manually.

A good first rule of thumb is to *always* consider use of more than one database, preferably databases from different producers. The principal advantage of this is that there are often important differences both in scope of coverage and in abstracting and indexing, and this means that additional information can frequently be gleaned from such an approach. Thus, for example, *Chemical Abstracts* and *Derwent World Patents Index* typically abstract and index the same patents differently. Each frequently has at least some details not in the other. For example, *CA* is especially noted for its inclusion of CAS Registry Numbers as appropriate. On the other hand, the *Derwent* record includes the numbers of other patent documents in the same patent family (see Chapter 13). In addition, the abstracts in each service will typically differ somewhat in content and length, and thus it often helps to have both in hand. Further, many documents found in one service may not be found in the other, and even when the same documents are found in both services, timeliness of coverage may vary from one service to another.

Most of the online systems offer a convenient feature that permits the user to search several databases concurrently and then to identify and eliminate any duplicate hits. However, this feature does not usually permit the types of comparisons described above.

In addition, because the various online services (hosts) often offer search capabilities that are different or unique, it makes good sense to consider use of more than one host as well for that reason as well.

The simplest approach to selecting pertinent databases is to simply study the previously mentioned free database catalogs provided by the host systems and to make a selection on that basis. A telephone call to the help desk of the host system may also be a good approach to obtain some initial suggestions. Another way to select pertinent databases is to utilize features such as DIALOG's *DIALINDEX/One Search* categories. (Similar arrangements are available on STN International through the "clusters" that are available for *STNindex* searching.) A quick visual review of these categories alone suggests a range of databases for the searcher to consider. But even more to the point, *DIALINDEX* and *STNindex* features permit the searcher to submit a search strategy online to broad categories of related databases at the same time and, as a result to quickly determine those files in which most of the hits are to be found. Categories in *DIALINDEX* include CAS Registry Numbers (all files that contain CAS Registry Numbers); Chemical Business News; Chemical Properties; Chemical Regulations; Plastics, Polymers, and Composites; and many other categories. If needed, even broader categories, such as "ALLSCIENCE," are available. In the next step, these databases can then be immediately and easily ranked as to frequency of hits, and then selected and searched. As mentioned, an excellent first choice for most chemical searches is *Chemical Abstracts* and/or *Registry* on STN International, but, in practice, search requirements often require use of other databases as well. As noted elsewhere, it frequently makes good sense to search more than one database, especially when important decisions hinge on the outcome.

Once one or more databases are selected, a good way to get started with any particular database, is to attempt to approach each database with at least one example of a pertinent reference bearing directly on or very close to the topic and that has been previously located. This permits the searcher to enter the database and to examine how any such pertinent references are treated with respect to abstracting, type of indexing, placement as to *Chemical Abstracts* section, and other such clues. Of course, this idealized approach is not always possible. In such cases, an attempt may be made to identify the closest approaches to the desired types of references and to observe how these are treated by the abstracting and indexing service.

When the search deals with specific chemicals, the optimum starting points are, of course, the CAS Registry Numbers, when these are available, as is the case with *Chemical Abstracts* and certain other online databases. The Registry Number system is straightforward and easy to use, but it is a tool of great power, especially in searches of *CA*.

When working with Registry Numbers in Chemical Abstracts Service files, the best approach is to start by specifying the chemical in *File Registry* (to identify both current and discontinued Registry Numbers) and then to "cross over" to the *Chemical Abstracts* File. If preparation is what is of interest, the searcher can be more specific by following the resulting "L" number with a slash and a suffix as appropriate, thus: L1/P for preparation of that chemical; or L1/DP to indicate preparation of a nonspecific or generic derivative of that chemical.

It is usually wise to have as broad an initial strategy as makes good sense, consistent with time or budgetary restraints dictated by the situation, because any broad initial strategy can be easily limited subsequently to control the number of hits as is described later below. The advantage of a broad approach at first is that the testing process frequently identifies alternatives and approaches to a research problem that may not have been envisioned at the outset. One way to achieve breadth of search strategy is by use of appropriate related terms for the same concept; other ways are described in some of the following paragraphs.

The strategy should be developed with printed documentation relating the database(s) to be utilized immediately at hand. Most major online hosts and database producers produce good documentation usually in both summary form and in detailed chapters or manuals. Documentation to be considered should include thesauri (controlled terms utilized for indexing, as e.g., the *CA Index Guide*. These user aids may be available both in printed form and online, depending on the database and the online service provider, and may also be accessible through the Internet.

It often makes good sense to consider talking with the database producers and/or host system experts to obtain assistance with optimizing search strategies, especially those that may be especially important, complex, or difficult. Toll-free telephone numbers for help desks at both the database producers and the online services are usually available, and most such consultations are typically without charge. However, the quality of help-desk personnel can vary widely. Some of these people are considerably more expert and generally helpful than others; after a little experience, the user can get to recognize the difference. Consulting firms can also provide assistance.

It is prudent to test the logic and suitability of a search strategy before it is fully implemented. In general, these ways utilize just a portion of the file. There are several ways to do this.

For example, one approach is to experiment with a strategy in one of the "learning" (STN) files or the comparable ONTAP (DIALOG) files that are available for some of the key databases. These are intended primarily for practice and training use to help beginning searchers learn how to utilize these files, but they can also be utilized by more experienced searchers as one way to test search strategies. The principal advantage of this approach is that the learning files are considerably less expensive than the regular files. The principal disadvantage is that the files are quite small and are probably not sufficiently representative in scope to fully test all aspects of a given strategy. Furthermore, the training files are usually several years old and may not properly represent the most recent developments. But at a minimum, this approach should be sufficient to detect and highlight basic errors in logical construction, search terms, and spelling and to determine whether the system will accept the proposed logical statements.

Another approach is to utilize the regular file (database) but to delimit time of coverage. For example, the strategy could be tested against the latest update only or perhaps for the most recent 3–12 months only. This will not only test logic, but may also give some idea of the approximate number of hits and the kinds of results that can probably be expected when the full file is utilized.

Since current alerting profiles [also called SDIs or selective dissemination of information) profiles, also discussed in Section 4.3] may involve consistent execution of a strategy over a period of many weeks, months, or even years, it is especially important that strategies for any SDI be as carefully constructed as possible. A first test should be based on the most recent 6–12 months. Initial output may be either too broad or too narrow. Before "finalizing" these, it is a good idea to go through three or four iterations so that the content, size, and cost of the profiles will be as completely satisfactory as possible. In addition, strategy and output should be regularly scanned for potential additions and deletions to the search strategy.

The number and pertinence of the hits resulting from a test search is frequently a good indicator of the soundness of a strategy.

One concern is if there are too many hits, beyond those that can be comfortably managed if the initial test results are extrapolated to the full file; in such cases the chemist has several options when implementing the strategy:

1. Limit time coverage as, for example, to the most recent few years or even few months.

2. Limit language, as, for example, English only.

3. Limit types of documents covered as, for example (a) books and reviews in English only, (b) U.S. or U.K. patents only, or (c) nonpatents only.

4. In the case of *CA,* limit the search to certain sections of *CA* only, such as to the "BIO" sections only, or even more narrowly, to certain specific sections or even sub-sections within the BIO category. (Bear in mind, however, that assignment by *CA* document analysts to sections is somewhat arbitrary and that limiting a search to one or more sections or especially sub-sections may result in missing some pertinent results.)

5. In the case of *MEDLINE,* limit output to those cites dealing with humans only, thus eliminating any unwanted hits pertaining to plants or animals.

6. Tighten adjacency requirements as appropriate. For example, require that two words be directly adjacent to one another rather than, say, within two words of one another. And it can be specified that the words will be in a specific order or sequence. This can be done very simply, as, for example A(W)B; this indicates that A must be within one word of B and in that order. Other limitations that may be possible include, for example, limiting to the same sentence or to the same part of the record, but these are less restrictive than requiring immediate adjacency and in a specified order. In STN, the searcher can "link" search terms, thus requiring that the terms be in the same part of the record; this is a simple but powerful approach that can increase relevancy and cut down on "false drops." A very simple hypothetical illustration may clarify this strategy. Let us assume that the searcher is interested in catalytic hydrogenation. An initial approach might specify the following: cataly? and hydrogen?.* (The question marks simply indicate truncation on the right-hand sides so that any subsequent variations of these words are acceptable, such as catalyst, catalysts, catalytic, hydrogen, and hydrogenation.) The use of "and" logic here is a very broad-brush approach that merely requires that the two search terms be anywhere in the same record. Such a search is likely to yield a large number of nonpertinent results as well as pertinent results. But, if instead, the searcher specifies cataly?(W)hydrogen?, this much tighter logic requires that the two words be adjacent (within one word) and in the specified order. [Alternatively, in a slightly less specific approach, the searcher could have specified, e.g., 2W (within two words) or A (within one word but adjacent in any order).] There are, of course, other ways, some of which are conceptual, and others more specific, to search for the desired technology. This is merely a very simple illustration of one possible approach. The search could be further tightened in such ways as, for example, specifying certain catalysts by using their CAS Registry Numbers; specifying certain chemicals that are to be hydrogenated by using their Registry Numbers; limiting to certain types of documents (e.g., reviews, books, or patents), certain years of coverage, and certain languages only; and in several other ways.

7. Limit output to those cites dealing with preparation or manufacture only if that is the only aspect of interest. Similarly, other roles or functions may be specified. The

*Thus, ((cataly? (l) hydrogen?))/It will retrieve only documents so indexed in the same information unit in *CA* on STN. It = index term.

use of roles in *Chemical Abstracts* (see page 127) online facilitates this kind of limiting.

8. Limit results to a particular country of origin or to a particular university, company or other originating organization. The latter limitation requires careful listing of all potential variations of corporate name since these are seldom standardized. In the case of corporations, identification of all likely subsidiaries, affiliated companies, and simply minor variations in expression of the company name, can be difficult, yet important. Some database producers do attempt to standardize company names; these include IFI and Derwent (see Chapter 13). A related approach is to limit output to only those documents written by specific authors or inventors. Another option is to select work done in universities only, or to work not done in universities; this can be useful in certain special situations.

9. Utilize "not" logic, specifying that documents containing certain search terms are not wanted. But see the important caution below.

The approaches just described all run the risk of missing something crucially important. For example, use of "not" logic, although a powerful tool, is particularly dangerous and can result in exclusion of significant references. Hence, use of "not" logic is best minimized.

Limitation to a *CA* section is also risky. This is because assignment of a document to a given Section is relatively arbitrary and subjective.

To increase the pertinence of results, and at the same time control the number of hits, use of CAS Registry Numbers for specific compounds usually helps as previously noted. In addition, the chemist can require that all search terms selected are *CA* index terms and hence controlled vocabulary, not just anywhere in the record (title or abstract). Consultation of the *CA Index Guide* can help in this regard. Selection of an index term can be done by following that term with a slash and the letters IT, thus: benzene/IT. This means that this term must have been selected by *CA* staff in the indexing of the document. If this is the case, then it is highly pertinent to the document, not just mentioned casually in the abstract. Most abstracting and indexing databases have a published and/or online thesaurus of authorized indexing terms (controlled vocabulary) to guide users of that database.

In some cases, a search strategy will yield no hits, which is almost always a suspicious result. If the results show no hits or fewer hits that could reasonably be expected, and more would be expected or desired, once again there are several remedies that permit expansion of the scope of a strategy.

1. Expand the search strategy so that it is broader or more generic in coverage. For example, rather than limit a search to a specific halogen substituent, consider other halogens as appropriate. This case and many other variations are facilitated by the structure and substructure searching capabilities of the *Chemical Abstracts* Registry file on STN International.
2. Remove or soften previously imposed limits as word adjacency, years covered, language, country, or type of document.

3. Search the entire record, including the full abstract, and forego the option of searching just the indexing terms.
4. Explore utilizing other databases and other host systems beyond the ones originally selected.
5. Consider full-text files as appropriate.
6. Consider alternative spellings, especially British spellings (see Table 8.1).
7. Utilize citation searching.
8. Use "expand" commands to help identify search term variations.
9. Truncate as appropriate. For example, "cataly?" will include such variations as, for example, catalyst, catalysts, catalytic, and catalysis. Variations are discussed further in the section that follows.
10. Try browsing for additional ways to approach the project. This can be done in the file presently being worked with, in many cases, or in the printed version of that file (e.g., browse a few of the recent printed semiannual indexes to *CA*), or it can be done in an encyclopedia such as the *Kirk-Othmer Encyclopedia of Chemical Technology* which very often acts as a "mind stretcher" or creative source.

In some cases, such as certain patent searches, the chemist wants everything that can be found that is pertinent; in fact, limiting can be disadvantageous.

A. Use of Variations of Search Terms

Especially in databases that do not employ tightly controlled vocabulary, but in other databases as well, the searcher should at least consider utilizing as many term variations as seems reasonable, and within time or economic constraints. As mentioned above, the use of the "expand" command will often indicate useful variations.

Another obvious case relates to consideration of both singular and plural search terms that can be easily specified by use of a simple truncation symbol. This can be done automatically in STN by designating this as a choice.

> Truncation is usually indicated by a symbol such as a questionmark, asterisk, or colon, depending on the online service. Truncation may be limited to just one character, or it can be unlimited. In some systems (such as STN International and Questel), left-hand truncation is also possible. This can help locate variations such as "metallurgical" or "electrometallurgical." Most systems also permit truncation or "wildcards" in the middle of a word. Thus, the expression "anodi?ed" will allow for both "anodized" (American spelling) and the corresponding British spelling "anodised." (These differences in spelling would be unlikely in a database such as *Chemical Abstracts,* but could be encountered in databases produced outside the United States or in full-text databases.) There are other occasions, as well, in which wildcard truncation can be useful.

As another example, consider the case of a chemical engineer concerned with selection of materials of construction to be used with a specific corrosive chemical that is a process by-product. The search terms might include the name or CAS Registry Number of the corrosive chemical together with such terms as *materials of construction* or *building materials,* depending on the database. In addition, other entries in an online search might include

acid- or alkali-resistant materials
anticorrosive (and variations)
compatibility (and variations)
corrosion (and variations)
inhibition (and variations)
inhibitors
noncorrosive (and variations)
prevention (and variations)
protection (and variations)
resistance (and variations)
rust
specific materials, such as type of steel
stabilization (and variations)

Or consider the case of an analytical chemist searching for references on gases evolved on decomposition of a polymer. If this search were conducted online, an array such as listed in Table 10.2 could be constructed. Note that this array takes into account variant terms as well as abbreviations, all of which might be needed to ensure that the search is as complete as possible. Of course, considerable time and effort can be saved by truncation. Thus, for example, "decomp?" and "dissoc?" will embrace the abbreviated variations of these words. Additional variations for the same concept may also be possible; the *Chemical Abstracts Index Guide* can be utilized to suggest some of these. Of course, use of multiple term variations is less important in an expertly indexed source such as *CA*.

If a database does not utilize CAS Registry Numbers, controlled vocabulary, or systematic nomenclature, variations in chemical names must often be included. A

TABLE 10.2. Examples of Term Variations Used in Online Searching[a]

decomp.	decomposed	dissoc.	dissociation	oxidation
decompd.	decomposing	dissocd.	dissociating	oxidn.
decompg.	decomposition	dissocg.	dissocn.	pyrolysis
decompn.	degradation	dissociate	heat	temp.
decompose	degrdn.	dissociated	hot	temperature
				thermal

[a]In *CA* on STN, abbreviations and plurals may be automatically "turned on" by users.

simple case is polyvinyl chloride, which might be found entered as a trade name or, more likely, as any of the following examples:

PVC
polyvinylchloride (one word)
poly(vinyl chloride)
polyvinyl chloride (two words)
vinyl
vinyl compounds, polymers
vinyls

Note inclusion of both singular and plural variations of the same word. This list is not necessarily all-inclusive; other variants may need to be used.

Another variation is due to changes in company or organization names over the years as shown in the example in Section 13.10. Failure to take such changes into account can result in loss of information. Fortunately, some database producers such as Derwent, IFI, and EPIDOS-INPADOC (see Chapter 13), have attempted to standardize many company names. Such standardization is, however, incomplete and otherwise imperfect and may not take into account names of subsidiaries and affiliated companies.

In some cases, companies choose not to have patents assigned immediately after issue. Such patents may be found only under names of individual inventors; none of the strategies described previously will help locate such patents quickly.

B. Commercial Aspects in Scientific Databases

A frequently occurring issue when conducting a search in an online database such as *Chemical Abstracts* is how to identify technology that is commercial or potentially commercial in the opinion of the author of the document.

One obvious way to handle this in an online search is to limit the search to patent documents. If a technology is patented, chances are excellent that it is at least potentially commercial; otherwise the inventor or the inventor's organization would not have spent the funds and time required to file for a patent. If a patent is filed in a number of major industrialized nations, that is usually a particularly strong indicator of intended commercialization in those nations.

In addition, certain words may be specified in nonpatent literature in order to identify technology that is potentially commercial. Examples of such words are

commercial/business
large-scale
industrial/industry
scale-up
pilot plant

economics/economical
capital investment
cost or cost-effective

The inclusion of one or more of these words, and other such, in any document is sometimes a good indicator of commercial intent or application. Such words may be utilized in combination with the name or CAS Registry Number of the chemical of interest. The searcher should consider the results carefully, bearing in mind that "false drops" are always possible. There are, of course, online databases dedicated to the commercial aspects, such as *CIN* (*Chemical Industry Notes*). See also Chapter 16.

10.11. CHEMICAL STRUCTURE AND SUBSTRUCTURE SEARCHING USING COMPUTERIZED SYSTEMS

The quickest and most unambiguous way to search for a specific chemical structure is, of course, to have readily at hand a CAS Registry Number or a systematic name, preferably the official CA Index Name. CAS Registry Numbers are widely utilized in a number of online databases and in printed sources. However, in many cases, this information may not be readily available, or the chemist may wish to search for certain relatively broad types of structures or substructures on a nonspecific or generic basis.

A number of approaches to this problem are possible as summarized some years ago by Heller and Milne (4) and Witiak (5). The most efficient of these is graphical or topological searching (actually "drawing" the structure or substructure of interest on one's computer using either the mouse or the keyboard for text input); this approach is typically based on the use of connection tables.

Connection tables can be compiled by humans or computers. For any given structure, the table shows numerically, and usually unambiguously, the sequence and ways in which specific atoms are connected to one another. Some systems include relative stereochemistry.

The advantage of this approach is that it is independent of any vagaries or artificialities of nomenclature, which, on occasion, can be rather arbitrary and are frequently inconsistent with names utilized in commerce. Graphical or topological searching provides the potential for effective searching for less-well-defined structures including Markush and generic structures. The best known and the biggest single group of connection tables for chemical structures is the *Registry File* of Chemical Abstracts Service, which included over 16 million structures as of 1997.

The structure search capabilities offered by Chemical Abstracts Service on STN International and by Questel DARC on the Questel•Orbit system represent the largest and most important online structure–substructure search systems. Both permit structure searching of the CAS *Registry File.* In addition, Markush DARC on Questel permits unique Markush searching (see box below) of the *Derwent World Patents Index Markush* (for Derwent subscribers only) and of the *Pharmsearch* patents databases. The CAS system also permits Markush searching of its unique *MARPAT* file of international patents and structure searching of certain other files on the STN International system such as the *Beilstein* and *Gmelin* databases (see Chapter 12). (Chemical Abstracts Service products and STN are discussed in Chapters 6 and 7.) Both the CAS (STN International) and Questel•Orbit DARC systems are available on a worldwide basis. DARC employs computers located in France, and STN is made available through computers located in the United States, Germany, and Japan.

Note that the CAS Registry system has been utilized not only for public access but also for some private structure files of individual companies. Similarly, the DARC system may be purchased for special private applications.

> The designation *Markush* comes from the name Eugene A. Markush whose U.S. patent (1,506,316) introduced this concept in 1925, although some may also use the term loosely to designate any type of generic claim or generic searching. Markush language in patent claims designates artificial groupings by wordings such as "material selected from a group consisting of A, B, C, or D," where A, B, C, and D are different but can serve the same function. Possible variations in patent claims are almost infinite, and the example just given is merely illustrative. According to CAS, a typical Markush structure represents more than 250 hypothesized structures.

In graphical searching, there is no need to know systematic nomenclature rules or special chemical codes. Chemists or engineers need know only how to use the computerized systems to draw chemical structures or substructures of interest. Ambiguity is essentially eliminated for most chemical structures.

Structure searching is not, however, a completely simple matter. Some study and practice is required. Details of structure searching with CAS products are discussed in Chapters 14 and 15 of Ridley's excellent book (6), in Chapter 7 of Schulz and Georgy's volume (7), and in instructional materials available through CAS. Even more to the point, CAS offers 1–2-day training sessions on structure searching, and these are highly recommended.

To recapitulate, the chemist can search for structures and substructures in the CAS *Registry File* and in other databases suited for graphical searching, in a number of useful ways. As mentioned above, for complete structures of known identity, the best way is still to use the CAS Registry Number, or *Chemical Abstracts* Index Name, if available. Alternatively, other names, including trade names, can be entered, and these often yield good results by leading quickly to the Registry Number and the CA Index Name. However, when searching for structures in a generic way or for substructures, the most efficient way is usually to "draw" the structure or substructure utilizing a mouse as the "pen" and STN Express or other compatible software. Use of the mouse as described is possible on the STN system. In the case of the use of Questel to search the *Registry File* (termed by them *EURECAS*) or *Pharmsearch,* generic DARC access is best achieved by text input (using the keyboard to specify the atoms and their positions). Text input (or keyboard-input) is also possible with STN. In both cases, "drawing" of the structure should be done offline when possible so as to minimize connect time.

There are other important systems that permit connection table, graphical, or topological searching of chemical substructures. Examples are the products of MDL Information Systems and Daylight Chemical Information Systems, Mission Viejo, CA (http://www.daylight.com), which are intended primarily for in-house use; Beilstein's new *CrossFire* product; and the SANSS component of the Chemical Information System.

In addition to topological searching, other sell-accepted approaches to searching chemical structures are in use. These include fragmentation or chemical coding systems, notation systems, classification approaches, and dictionary or nomenclature approaches. Fragmentation code systems (used to describe structural features) are extensively and successfully employed by IFI/Plenum Data Corp. in its *Comprehensive Data Base* and by Derwent Information Limited in the *Derwent World Patents Index* (see Chapter 13). Fragmentation codes are alphanumeric systems that have been found especially useful in patent searching, as in the two databases noted above.

Notation systems are another approach to structure storage and searching. These also employ combinations of letters and numbers, and may use punctuation marks as well. IFI/Plenum Data Corp. utilizes a searchable line notation system in its CLAIMS patent databases (Chapter 13).

Classification approaches are hierarchical in nature and can be utilized to group together chemicals that have similar structural features into relatively manageable classes. Among the best examples that lend themselves to computerized searching are the U.S. Patent and Trademark Office's classification system and the International Patent Classification or IPC (see Chapter 13), which is maintained by the World Intellectual Property Organization. In addition, the European Patent Office Classification System offers a further extension of the IPC that offers the advantage of additional detail. Another example is the Derwent classification system devised by Derwent Information Limited (Chapter 13); Derwent classes are relatively broad in scope and can be utilized in patent searching. Note that patent issuing authorities frequently revise their classification systems and that therefore, depending on subject matter, several different classification numbers may need to be utilized to express the same concept or structure; thus, for example, several editions of the International Patent Classification are available.

A "dictionary" or nomenclature (full or partial name match) approach to searching for chemical structures can employ any or all of the following examples, which are possible when searching the CAS *File Registry:* chemical name segment, heading parent, index name segment, other name segment, molecular formulas and weights, atom counts, and/or elements present. Note that structure drawing and dictionary approaches may sometimes be used effectively in combination.

Also of interest from a structure searching perspective are a number of reaction searching files. These include the following examples: Beilstein's large and powerful *CrossFire* product (see Chapter 12); *CASREACT,* available on STN only, which covers organic chemical reactions as reported in *CA* since 1985 for journals and from 1991 for patents; MDL Information Systems products for in-house use; Derwent's *Journal of Synthetic Methods* and related products (see Chapter 13); INPI's (French Patent Office) new *Rxn Core Reactions* product; and ISI's new *ChemPrep* products available on CD-ROM and their in-house reaction searching system.

10.12. NOTATION SYSTEMS

Notation systems are typically alphabetical and numeric representations of chemical structures. When the first two editions of this book were written, chemical notation systems were of very broad interest. At that time, notation systems were a vehicle for chemist interpretation (could be read visually by the chemist quite readily) and also a means to facilitate computer manipulation such as input and communication of structures and searching for structures and substructures, all in a concise manner. Because of the development of technologies that permit today's chemist to draw chemical structures on a computer, along with the continued advances in computer power and storage efficiency, the interest in notation systems is now considerably less than it was 10 years ago. More recently, however, new notation systems have been developed that address the important special cases that can still require concise handling of structures in computer systems, especially network computer systems, and for other practical reasons. These more recent schemes provide a good balance between conciseness and extensibility and cope well with new developments such as combinatorial libraries, which are crucial to pharmaceutical research today.

One example of a notation system in current use is the SMILES (Simplified Molecular Input Line Entry Specification) notation system. This was developed by Weininger (8) while he was with a government agency, and became a basis for the formation of Daylight Chemical Information Systems, Inc., Mission Viejo, CA and Santa Fe, NM. Daylight was a spinoff from the MedChem Project at Pomona College.

As examples of SMILES notation, benzene is represented as

clccccl

Methane is represented as

C

Ethanol is represented as

CCO

Another line notation system is the SYBYL Line Notation (SLN), which has been developed at TRIPOS, Inc., St. Louis, MO (9). This "was inspired" by the SMILES notation described by Weininger. Among its other capabilities, SYBYL is said to permit specification of full substructure queries, including Markush structures and queries that are especially important when dealing with patents, and when representing and searching combinatorial libraries.

As an example of SYBYL notation, benzene is represented as

C[1]H:CH:CH:CH:CH:CH:@1

Ethane is represented as

CH3CH3 or CH3-CH3

Both Daylight and TRIPOS provide sophisticated and powerful chemical information processing tools to chemists. Both have a special focus on the needs of research workers in the pharmaceutical industry and the current high interest in combinatorial chemistry.

Another example of a line notation system is that utilized by IFI/Plenum Data Corp. to describe two-dimensional chemical structures that appear in the patent claims listed in the U.S. Patent and Trademark Office's *Official Gazette*. The notations are available for all chemical *Gazette* claims in the online *CLAIMS* database, from 1950 onward. This system was designed to be easily understood by anyone with a basic knowledge of chemical nomenclature, and enables searchers to review claims containing chemical structures without the need to retrieve a copy of the patent. Structures are employed for the acyclic portions of compounds and names are used for the cyclic portions. There is a very simple set of rules, and these are available from IFI. See also Chapter 13.

10.13. PRODUCTS INTENDED FOR DIRECT USE BY THE END USER

In recent years, there has been considerable discussion of so-called end-user chemical information products (databases, other files, and search software) that are intended to be especially user-friendly. The *end user* is usually defined as the laboratory scientist or engineer (as distinct from the information professional). Thus, these are usually products designed specifically for direct use by the laboratory scientist or engineer on that person's individual laboratory or office computer. With the emergence of the Internet as a major factor, and the trimming of certain parts of research and development budgets, this trend is expected to strengthen.

Sometimes, but by no means always, the producers of end-user information products prefer to market these products directly, without the involvement of an information professional who may be perceived by some producers as a marketing barrier rather than as an ally. However, Jan Williams, formerly of the Monsanto Company (see Ref. 1 in Chapter 7) has written about the introduction of a major end-user product into a leading company. She explains the positive role of the information professional in the process of introducing end users to such products and in helping ensure optimum and appropriate use.

Paradoxically, end-user products are frequently priced at the budgetary level of an entire organization (company) or an entire research laboratory, and hence, are frequently, but not always, expensive. Because of the cost, the decision on whether to obtain any of these products is, with some exceptions, usually at the level of a senior administrator rather than the laboratory chemist or engineer. The same caveat applies to some CD-ROM products that are also usually oriented to the end-user. As things stand now, chemists and engineers affiliated with smaller colleges or companies) may be very limited in their access to many of the more attractive end-user products because of the costs of many of these products. As mentioned before, however, there may be academic discounts in some cases. In addition, *STN Easy,* a new end-user Internet product from CAS, has an interesting, although not cheap, cost structure aimed at individual end users. (There is a flat fee of $2.00 per click of the "search" button, and, in addition, each display from the powerful *CAplus* database costs $3.50. Display of titles only is free. Other databases are also available on *STN Easy.*)

Some of the other leading products specifically designed for the end user include

Grateful Med software for searching all of the online databases offered by the National Library of Medicine. Introduced in 1986, this inexpensive software was the first of the major new end-user products for searching of electronic databases. It now accounts for 80% of all of the searching on the National Library of Medicine Medlars online system.

SciFinder (a client/server product) and *CASurveyor* (a CD-ROM product series), both landmark products from Chemical Abstracts Service. In 1996, it was announced that *SciFinder* won an "R&D-100" award for one of the top 100 new research products of the year.

CASurveyor is an especially attractive CAS product. Each CD provides searchable "slices" of *CA* coverage (with citations, abstracts, displayable structures, and index entries) for the current month and the previous 36 months in each of some 15 categories of chemistry preselected by *CA*. The affordable license fee (about $1000/year) can be very appealing to organizations with highly focused and intensive fields of interests, such as polymer coatings and adhesives, and that wish to have the advantage of unlimited search time in these areas. Results from searches can be grouped on the basis of number of hits per search term, hence providing listings automatically ranked by relevance. Updates are provided on a monthly basis.

> Note that ranking of search results is also possible with other systems noted in this chapter. For example, this is possible with several Internet search engines. As another example, ranking is also now possible with STN databases using a special command; this is achieved by a special algorithm that (1) counts occurrences of terms weighted by field, (2) measures proximity of terms to one another, and (3) determines where the terms appear in the field. Similar approaches are offered by the rank commands and the statistical analysis of search results that are possible with most other major online systems, and in some cases special software is offered. These approaches may be useful not only in connection with technical matters but also for other matters such as developing contact lists or customer lists. Although rankings provide valuable information, they are not intended as a substitute for professional judgment, but rather as one basis for such judgments.

DIALOG *ScienceBase,* an Internet product of DIALOG, Mountain View, CA, was introduced in 1996 as a product intended primarily for direct use by the end user. The Internet address is http://krscience.dialog.com. Users must register first; there is a monthly access fee and a price is charged for each database record viewed. Content includes chemical, pharmaceutical, and biological research as well as other topics. The sources utilized are selected from the DIALOG array of online databases. References may be searched for by author, corporate source, or table of contents from journals; chemical substances may be searched by name, physical properties, uses, separation, derivatives, and polymers; and other chemical topics may be searched for by substance preparation, reactions, analytical methods, toxicology, and so forth.

QPAT-US, an Internet-available product of Questel•Orbit, offers powerful, full-text searching of U.S. patents since the 1970s (see Chapter 13). A competing product, very powerful, and easy to use, is CAS's *Chemical Patents Plus* (despite its name, this file is not limited to chemical patents). These two Internet products have different capabilities. There are other patent search files on the Internet such as those of MicroPatent, East Haven, CT, and of the U.S. Patent and Trademark Office. See Chapter 13.

CrossFire, offered by Beilstein Informationssysteme GmbH, Frankfurt, Germany, is a unique and useful client/server product that includes a very large amount of information from both the *Beilstein* and *Gmelin* handbook series (see Chapter 12).

Virtually the complete family of client/server and other files offered by MDL Information Systems, Inc. for searching of both in-house data and certain other files and databases can be described as end-user products. In addition, as mentioned, many CD-ROM products (see Section 10.16) are intended for direct use by the end user. Most chemical files on the Internet are, of course, aimed at the end user. A number of the current alerting products mentioned in Chapter 4 are targeted for direct use by the laboratory scientist. In addition to electronic products, such as those just mentioned, there are, of course, a number of printed chemical information products that are end user in pricing and orientation. One outstanding example is the *CA Selects* series noted in Chapter 4.

10.14. FULL-TEXT DATABASES

An increasing number of important journals, two major chemical encyclopedias, and many patents are now available online for full-text searching. Examples include

1. Databases covering all U.S. patents since the 1970s (available on several hosts and the Internet). See Chapter 13.
2. ACS journals, chemical journals of The Royal Society of Chemistry, chemical journals of Elsevier Science Publishers, chemical journals of Verlag Gesellschaft, chemical journals of the AOAC (oil chemists association), and the chemical journals of John Wiley and Sons (all available on STN International).

Furthermore, many journals are now available in full-text form on the Internet, usually on a subscription or other fee basis.

3. Encyclopedias, such as the *Kirk-Othmer Encyclopedia of Chemical Technology* and *Polymer Online* (*Encyclopedia of Polymer Science and Technology*). Both of these John Wiley and Sons products are available online in DIALOG and in CD-ROM form).

Full-text searching is important in helping ensure a higher degree of completeness of some searches and in searching for unusual words or phrases that may be less likely to be indexed, as, for example, trade names that are frequently encountered, but not always indexed and abstracted. Similarly, "trivial" or "ordinary" aspects of reports on reactions or other types of research are less likely to be indexed in conventional databases, but would be included in full-text databases; these might include, for example, certain solvents utilized in reactions.

The principal advantage of a searchable full-text database is that essentially every word in the documents covered is searchable (the only usual exceptions are common "stopwords" such as articles and prepositions, and even these may be searchable as parts of phrases). This compares with usual bibliographic database, in which the "only" elements searchable are the bibliographic elements, the abstracts, the keywords, and the indexing, including indexing of concepts. A full-text search—in theory, at least—helps reduce chances of missing any information or detail, no matter how brief or casual the mention.

Thus, in the cases of full-text databases covering patents, encyclopedias, or journals, every significant word in the patents, encyclopedias, or journal articles included is searchable. This has the advantage of minimizing the chance of missing something that is not included in the bibliographic elements, the abstracts, or the indexing.

Another advantage of full-text databases is that the user can, in fact, achieve "instant" document delivery of the desired journal paper or patent; there is no need to place a special order.

On the other hand, full-text searching frequently yields far too many hits to be useful—in other words, the precision of recall is not good. This, in turn, can lead to time-consuming searching. (Both of these matters can be addressed by more careful planning of full-text searches; for example, in multiterm searches, by requiring that the search terms be adjacent or at least in the same sentence or paragraph.)

Further, full-text searching may be deficient in the searching of concepts such as reaction types or types of compounds. There are no indexers to add CAS Registry Numbers or to ensure systematic nomenclature. Reliability is not good because the vocabulary is not controlled. Different authors frequently use different words to express the same thought or to describe the same chemical. This means that the searcher needs to attempt to imagine all the different ways in which a thought or a chemical name might be expressed. Full-text searching of patent databases adds another complicating factor in that most patent documents are written by patent attorneys with the legal aspect in mind. Clearly, then, full-text searching is not a panacea and should not give the impression of all-inclusive searching.

For the most comprehensive searching, the searcher should consider utilizing one or more conventional (bibliographic) databases and, in addition, one or more searchable full-text files, if needed and if these exist in the field of interest. For most searches, however, search of several conventional databases is usually sufficient. If such a search yields unacceptable results, or if more details or more completeness are required, then a search of a full-text file and of other databases is indicated.

There is at least one database, *USPATFULL* (available on STN and also known as *Chemical Patents Plus* on the Internet), which combines the advantages of both full-text searching and availability (for chemical patents only) of indexing based on systematic nomenclature and controlled vocabulary. In this "hybrid" database, a product of Chemical Abstracts Service, the searcher may utilize any combination of full-text and controlled vocabulary search methods.

It is important to note that not all full-text databases are searchable; some provide only the page images for viewing.

10.15. INTERNET ISSUES AND TOOLS

A number of specific information products available on the Internet are discussed in appropriate chapters throughout this book. In this section, special points are discussed and some of the other files available on the Internet are discussed and reviewed.

Government agencies and the academic community were the first to establish and utilize the Internet. It was not until approximately 1994 that the chemical industry began to recognize the opportunity on a large scale, although some companies were very active users several years earlier. The part of the Internet that is of most interest to most chemists and chemical engineers is the World Wide Web, often referred to simply as the "Web." The Web emphasizes both words and graphics. In this book, the use of the word Internet is generally meant to imply the Web in most cases.

Any consideration of the Internet should also take into account the relative pros and cons of commercial–proprietary online services and systems, as described earlier in this chapter. As compared to the Internet, traditional online systems are mature (began in 1970s) and well-developed, yet still capable of new and innovative approaches. The traditional systems offer an extremely broad array of well-organized and well-documented databases and files produced by well-known organizations, for the most part. The commercial–proprietary services are very fast, offer extensive search power as desired, and are crisp and responsive in performance. Searches can usually be pinpointed to exactly what is needed, and the type of response desired can also be made highly specific. In almost all cases, help-desk assistance is readily available. The systems take time to learn well, but the excellent documentation, training classes, and help-desk assistance that are available facilitate this process. However, some end users find that commercial–proprietary systems are not sufficiently user-friendly. There may be a modest annual use fee, but usually, fees are on a pay-for-what-you-use basis. Nevertheless, use of certain databases can be expensive.

The Internet, although still in its relative infancy (some would say adolescence) from the perspective of chemical information, offers a number of important features and benefits. Some examples of these are

It includes a vast number of files and databases, many of which cannot be readily found in any other way. Thus, it can be a rich source for browsing through previously unknown or unexpected resources.

With the important exception of those files that charge a fee for use, searching can be "free," especially if one's organization offers "free" access, as in the case of many universities and colleges and government agencies and an increasing number of industrial firms. An outstanding example of an Internet file on which searching is free is the database of U.S. patents offered by International Business Machines Corp. (see Chapter 13).

A convenient, high-speed way to deliver copies of certain published documents, such as patents, and of search results with commercial/proprietary online search systems is possible through the Internet in some cases.

The Internet provides a very large amount of "directory" information that is useful in learning about chemists and their research interests. For example, many chemistry departments in colleges and universities list their faculty with background about each, telephone numbers, fax numbers, and e-mail addresses.

The Internet is a good place to learn about scheduled professional meetings and, in some cases, to review titles and abstracts of papers.

It offers a vehicle for electronic conferences. Several of these involving chemists have already been held.

A large number of chemical and other firms have sites on the World Wide Web on which descriptions and properties of their products are included.

All major publishers, and many lesser known publishers, have home pages that advertise and describe their products, thus offering chemists and chemical engineers an excellent means for determining availability, prices, and contents of journals, books, and other publications. Many publishers also offer the tables of contents and full texts of their journals, usually, but not always, on a fee basis.

The e-mail capabilities of the Internet facilitate ready communication and exchange with colleagues, suppliers, and customers around the world. A related benefit is offered by the discussion lists and newsgroups in various areas of interest to chemists.

Certain Internet files have unique display features. Thus, *Chemical Patents Plus* permits 2D and 3D display and rotation of chemical structures, if the user has the proper software. If the same or similar source is available both on the Internet and on a "conventional" online service such as STN International, it is worthwhile to compare coverages and search and display capabilities.

A search of the Internet may often yield data and clues that can be useful in conducting a more effective search of "traditional" sources.

On the other hand, chemists should be aware of potential disadvantages:

A major concern is that, although the Internet is a factor in chemical information, it is still emerging and developing, and important questions need to be resolved as to its reliability and its frequently undisciplined nature. The fact that a file appears on the Internet offers no guarantee as to the accuracy, currency, and completeness of the data. Accuracy is contingent on the reliability and diligence of the submitting person or organization, as in the case of "conventional" online services. (With "conventional" online services, there is extensive documentation available so that the user knows exactly what can be counted on or not counted on and how to optimize a search.) Another example of the question of reliability occurs when a research worker posts an unrefereed piece of research on the Internet; compare this with the reliability of a refereed article appearing in an ACS or other reputable journal. The problem is separating the wheat from the chaff.

There is no single source such as a central help desk to answer questions as to file content, timeliness, and reliability of all the files and databases. However, some few of the individual files and databases do have help desks that are identified.

Although there are many books describing the Internet as a whole, and at least one book dedicated to chemistry on the Internet (10), for most of the Internet files, there are few detailed printed descriptions of each individual database or file and how best to utilize it on the Internet.

Most files for which a search charge is imposed in the conventional online hosts, such DIALOG also impose a search charge in the Internet, so there may be little or no cost advantage in such cases. However, telecommunications costs and connect time may be less that with commercial hosts. T. H. Pierce and T. J. Cozzolino of Rohm and Haas report (11) that, of 65 respondents to their survey, a large number used the Internet to connect with online services such as DIALOG and STN International; the connections were described as "faster, more robust, and easier to use than modems."

Security is still not good. This allows for the possibility of intrusion by outsiders. Confidential data may be accessed by unwelcome strangers under certain circumstances. Encryption and "firewall" approaches can help.

The Internet is often very slow, especially during the middle of the business day, even when the user employs high-speed modems and top-grade computers. (Users report that a quick phone call, fax, or lookup in a nearby desk handbook can often get the information desired more quickly and economically.) Even commercial–proprietary online system users may experience some occasionally slow results, but the problem is much more pronounced with the Internet. The problem is not without its solutions.

One good example of the advantages and disadvantages of the Internet is found in the file covering the research projects of the U.S. Public Health Service, an agency that includes the National Institutes of Health. This database is known as *CRISP*

(Computerized Retrieval of Information on Scientific Projects). It covers, on a year-by-year basis, all the research projects in universities, hospitals, and industry that are funded by the Public Health Service. For most of the projects, in addition to the names and addresses of the investigators, abstracts are provided.

The Internet file of *CRISP* (gopher://gopher.nih.gov:70/11/res/crisp) is updated on a weekly basis and is more up-to-date than the version of the file found in any other source. On the other hand, the mechanism provided for searching is much weaker and less versatile than would be desired. Thus, searching by field and control of output by field is not possible on the Internet. On the other hand, a similar version of the same file, although updated only quarterly, and that is available on CD-ROM, provides much more powerful searching capability and flexibility.

Chemical Abstracts started to cover electronic (Internet) publications in April 1995, but during the first year of such coverage, only a relative handful of such publications (197 papers from 14 online journals and 45 papers from two online conferences) met their minimum criteria for inclusion, which are as follows:

1. Report new information of chemical or chemical engineering interest.
2. Present information in a form similar to that of a scientific papers (i.e., reporting novel findings with some discussion of experimental detail).
3. Identify authorship.
4. Be publicly available.
5. Have some relative permanence or continued accessibility.
6. Be original publications, not issued previously in another form or medium.

The 14 Internet journals (online only) monitored by *CA* to date are

Title/URL (URL = Uniform Resource Locator, or "address" on the Web)

Chemical Educator (http://journals.springer-ny.com/chedr/)

Complexity International (http://www.csu.edu.au/ci/)

Electronic Journal of Theoretical Chemistry (http://ejtc.wiley.co.uk/)

Experimental Biology Online (http://science.springer.de/ebo/ebo-main.htm)

Frontiers in Bioscience (http://bioscience.org)

Journal of Corrosion Science and Engineering (http://www.cp.umist.ac.uk/JCSE/JCSE.htm)

Journal of Molecular Modeling (http://science.springer.de/jmm/jmm.htm)

Journal of Quantitative Trait Loci (http://probe.nalusda.gov:8000/otherdocs/jqtl/index.html)

Molecular Vision (http://www.emory.edu/MOLECULAR VISION/index.html)

Molecules (http://link.springer.de/link/service/journals/00783/index.htm)

MRS Internet Journal of Nitride Semiconductor Research (http://nsr.mij.mrs.org/)

Network Science (*NetSci*) (http://www.awod.com/netsci/)

Neuroscience-Net (http://www.neuroscience.com/)

World Wide Web Journal of Biology (http://epress.com/w3jbio/)

The online conferences monitored by *CA* to date are

Title/URL

Applied Technology of Teaching Chemistry, On-Line Computing Conference (gopher://gopher.inform.umd.edu/11/EdRes/)

InCINC'94 International Chemometrics InterNet Conference (http://www.emsl.pnl.gov:2080/docs/incinc/homepage.html)*

CA editor David W. Weisgerber has noted that authors prefer to publish in conventional journals. He cites such factors as lack of awareness of electronic journals, a general perception about the lack of recognized credentials for online publications, and a concern about visibility and readership among one's peers. Online journals can have irregular publication schedules, and their very existence can be difficult to learn about. Individual chemists using the Internet would do well to keep in mind *CA's* six minimum criteria, noted above, as they browse the Internet and seek to determine what is and what is not reliable and potentially worth spending time perusing.

Major online services such as STN International and DIALOG are also available on the Internet. In these cases, all the good features of these hosts and their databases are fully available. There is, however, little or no cost advantage to utilizing these hosts on the Internet as connect and search charges are still incurred. However, telecommunications charges may be less.

A. Searching on the Internet

Whenever possible, strategic approaches as described in Section 10.10 should be employed when searching Internet files. However, the capability to search many Internet files is often less powerful than for "conventional" online files; this can mean that important information may be lost or that certain more complex queries may not be possible on the Internet. Furthermore, refinements such as ranking and sorting are frequently lacking in many Internet files.

Fortunately, a number of "search engines" are available to help locate information on the Internet. As with searching in "traditional" online databases, it makes good sense to utilize several search strategies and several sources when searching the Internet. Thus, for example, several (perhaps all) search engines may need to be utilized in order to locate what is needed; a search using just one engine is probably not complete. With the exception of traditional databases and host systems that are on the Internet (e.g., STN International and DIALOG), searching of the Internet is often based on keywords, and this is not a reliable or complete way to search for chemical information and concepts. Alternatives are tools such as *ChemFinder* from CambridgeSoft, which permits the user to search more precisely for certain chemical information. In addition, the various lists describing chemistry resources on the Internet can be very helpful.

Some of the leading Internet search engines include *Alta Vista, Excite, HotBot, In-*

*For updates, please see www.cas.org/EO/ejournz.html.

foseek, Lycos, Magellan, and *Yahoo*.* (All of these are proprietary names.) A number of search engines will group results by "relevance," but this may or may not be a reliable indicator.

As to the question of slow speed, in addition to upgrading hardware and software at one's desk, the solution is sometimes simple—for best results, some users try to utilize *any* online system or the Internet during off-peak hours. Optimum hours will, of course, vary with the time-zone location of the user. High-speed telephone connections are available today, but this can be expensive. (An important recent development is the plan to develop an "Internet II" that would be much faster than the current Internet. Access would be limited to participating institutions of higher learning. Similar high-speed Internet networks for commercial use exist or are planned; payments of a premium fee may be required.)

Some of the most important issues relating to searching for chemical information on the Internet are thoughtfully discussed in the excellent review article by Murray-Rust and colleagues (12).

B. Tools to Help Chemists Keep Up-to-Date with and Evaluate What's New on the Internet

Keeping up-to-date with new Internet files of interest to chemists is a particularly difficult and often frustrating task.

One way is through columns that are published in American Chemical Society and other journals. For example, columns on the Internet appear in *ChemTech, C&ENews, Environment Science and Technology,* and *Chemical Engineering Progress.*

Another excellent approach is on the Internet itself. There are a number of useful compilations that list and briefly describe chemistry files and databases from a broad variety of sources that are on the Internet. Examples of such lists that may be accessed on the Internet include those prepared by Steven M. Bachrach, Northern Illinois University; Max Kopelevich, University of California at Los Angeles (UCLA); Joseph Warden, RPI; Gary D. Wiggins, University of Indiana; and Mark Winter, University of Sheffield. Other tools that can be helpful are Heller's Internet column, mentioned earlier in this chapter, and some of the sources and methods listed in Section 10.5.C and 10.5.D.

Bachrach's excellent list is intended to be selective and descriptive; in addition, Bachrach and colleagues list the 25 "best" files on the Internet. The UCLA list is intended to be comprehensive. Warden's list is both well organized and comprehensive. The Wiggins list is superb in its coverage of chemical information sources.

Other sources that can help the user sort out and evaluate Internet files are starting to appear. Examples include the Cambridge Scientific Abstracts product *Environmental RouteNet* and Engineering Information's *Engineering Information Village.*

Other key Internet sites that can help in selection are those offered by the leading professional societies and publishers. The ChemCenter site of the American Chemical Society (see below) and the home page of The Royal Society of Chemistry are outstanding starting points. The ChemWeb location operated by MDL Information

*The Internet tool *MetaCrawler* permits use of several search engines concurrently. It is recommended.

and another company is also important. The John Wiley & Sons location, especially the location for its *Journal of Computational Chemistry,* is excellent in terms of identifying and providing links to some key sites.

One of the newest examples of Internet user aids is *CyberHound Online,** an Internet product of Gale Research, a publisher located in Detroit, MI (Internet address http://www.cyberhound.com). This is a sophisticated search engine that currently both categorizes and evaluates some 30,000 Internet sites, with new sites added at the rate of approximately 2000 per month. A variety of fields can be utilized to limit and fine-tune an Internet search including, for example, full text, time span, language, frequency of updating, site name and organizer, geographic coverage, ratings (Gale evaluations), fees, and many other parameters. There are also separate categories from which the user can select, such as journals, databases, and physical properties. Once the search engine has identified the sites of interest, users can read the CyberHound reviews of the site, and finally can look at the sites themselves. Internet files are evaluated utilizing such criteria as content, "technical merit," and design (see Fig. 10.1). The net benefit to the user intended is to help select the most pertinent sites. A current major limitation for chemists is that coverage of chemistry is, at present, not extensive, but considerable improvement is anticipated. Gale's printed volume was mentioned earlier.

C. Internet Costs

Basic access to the Internet is available directly to individual users on a personal basis from a number of Internet service providers, and/or, more typically, the company or university where the chemist or engineer works provides access. Internet access to specific data files may be free, or there may be a surcharge based on time or units utilized (as in conventional online services). In the case of files put up on the Internet by universities and some branches of government, chances are excellent that the access will be free. In the case of files put up by commercial sources, these may sometimes be free, if the sponsor elects to derive income or other benefits through advertising. For a "conventional" online system such as DIALOG, however, use of the databases (in terms of per-unit connect and display charges) costs essentially the same on the Internet as through "commercial" telecommunications lines, except that the telecommunications charges will usually be less. The full texts of many journals are now on the Internet and are typically available either to subscribers to the print edition (perhaps with a surcharge) or to those who wish to subscribe to the Internet version.

D. Cambridge Scientific Abstracts—Internet and Other Products

It is interesting to note that one of the newest online service providers, Cambridge Scientific Abstracts (CSA) offers its service on the Internet. CSA is a privately held firm with a staff of more than 100 persons that is located in Bethesda, MD (800-843-7751). CSA, founded in 1957, had long been a provider of printed abstracting and indexing services. These were, and still are, available in printed form and through major online services such as DIALOG.

*This product was just recently discontinued.

CyberHound Rating System http://www.thomson.com/cyberhound/ratings.html

[Back to Review]

Content
Design
Technical Merit
Entertainment

Rating Criteria

We could have rated sites by our mood of the moment. But we didn't. You don't even know us--how could our subjective opinions be valuable to you? So, to provide a comprehensive (not to mention fair) rating system, we created a checklist for each site that our in-house and freelance experts used to determine whether an entry would "wipe out" or rate as a tsunami.

For each site, you will find an overall rating, as well as individual ratings for content, design and technical merit. For each of the individual ratings, there are four criteria. We asked our surfers to ponder the questions under each of the four criteria, thereby determining if it gets a star or not. A site can receive one to four stars in each of the individual ratings, depending on how many of the criteria they meet. For instance, a site would get four stars for content if it is: comprehensive, up-to-date, unique and authoritative. If it's only unique, then it will get one star for content.

Once the individual ratings were compiled, we averaged them out to calculate the overall rating. If the average came out to 3 1/4, we rounded down to three stars. If the average came out to 3 1/2, we rounded up to four. And although the overall rating is just an average of the individual ratings, if a site receives four stars, you can be sure that:

- It is a site you will remember
- It is a site you'll go back to
- When you talk to friends about web sites, you'll mention this site
- It is a valuable research tool, or provides comprehensive information on the topic
- You will get the information you need

The following are the individual rankings, and how they break down into four criteria. You'll also see the "questions to ponder" that our surfers used to rate the site.

Return to Index

Content:

1. **Comprehensiveness:** Does the site cover all the facets of a particular topic? Is everything you expect to find in the site there? Are there any logical gaps of information?

2. **Up-to-date/trendiness/timeliness of data:** Is this a popular topic that people will want to know

Figure 10.1. CyberHound rating system. (Reprinted with permission of Gale's CyberHound Online Internet directory Service.) Product discontinued.

In 1995, CSA began to establish a strong position as a database provider on the Internet. This service, designated as *Internet Database Service,* embraces the following principal areas, many of which are of interest to chemists: materials information (based largely on the 1996 acquisition of ASM's suite of abstracting and indexing products); environmental sciences and pollution management; aquatic sciences and

CyberHound Rating System http://www.thomson.com/cyberhound/ratings.html

about? Has the data been updated recently?

3. **Unique or hard-to-find information:** Does the site contain information that is not readily available? Would you have to look long and hard to find similar data, or is it a dime a dozen?

4. **Authoritative/reliable information:** Who's providing this information? Do you trust them?

Return to Index

Design:

1. **Good use of graphics:** Are graphics small enough so that the site loads quickly? Are graphics large enough so that you know what they are?

2. **Aesthetic layout:** Does the site have an interesting design? Does the site try too hard to be cute? Do you drown in cuteness and lose the content?

3. **Use of audio and video:** Does the audio/video add to the data content?

4. **Catchy concept:** Will you remember this site because of its creative, imaginative approach?

Return to Index

Technical Merit:

1. **Ease of navigation/dummy proof/good online documentation:** Can a beginner figure out this site without too much trouble? If you get stuck, is there a way to get help?

2. **Searchability:** Do you get lots of false hits? Do your searches return valuable resources?

3. **Logical presentation of information:** Is there a logical progression to the information?

4. **Put together well:** Do the links work? Do the online forms work? Are there a lot of hypertext links to other resources? Any additional software provided? Does the site have any transactional functions (i.e. questionnaires, games)?

Return to Index

Entertainment:

1. **Editorial style:** Does the site make good use of humor? Is it creative in its approach to whatever subject matter is covered? Is the writing clever, witty or intelligent (all three if you're lucky)?

2. **Subject matter:** Does the site cover a popular subject of high interest to many Internet users? Does it cover something that has cult appeal? Is the site controversial?

3. **Lasting impression:** Is the site memorable? Would you recommend the site to friends? Would you bookmark the site to revisit frequently?

4. **Educational value:** Is it educational in the traditional sense (for example, a science page) or the non-traditional sense (bios on TV stars)? Does the site contain reviews/critiques of movies, books, music, computer software--anything that teaches you about a product or service?

Return to Index

Figure 10.1. Continued.

fisheries; biological sciences, biotechnology, and bioengineering, including AIDS and cancer research; and safety science and risk-related topics.

CSA's Internet offerings include not only its own databases but also some of the files of 11 other information providers: American Society of Health System Professionals, U.S. Department of Education, Elsevier Science/Excerpta Medica, Informa-

CyberHound Rating System http://www.thomson.com/cyberhound/ratings.html

[Back to Review]

[About CyberHound] [Search] [Homepage]
[About Gale] [Write to Us] [Help] [Register] [Bulletin Board]

Figure 10.1. Continued.

tion Today, International Academy at Santa Barbara, MDL Information Systems, U.S. National Agricultural Library, U.S. National Library of Medicine, National Information Services Corporation, Petroleum Abstracts/University of Tulsa, and Research Information Systems. At this writing, CSA offers numerous databases on the Internet in addition to print and CD-ROM versions of some of its own products. Each database that Cambridge offers is available on an annual subscription that permits unlimited use for the year; rates start at about $850 for an individual abstracting journal. The price includes several years of backfiles.

CSA's *Environmental RouteNet* product (described more fully in Chapter 14) and available on the Internet only, is of special interest. Included are 11 files from CSA, files from seven other publishers, and pointers to hundreds of to other Internet sites that CSA claims to have carefully selected on the basis of their evaluation. The annual subscription rate is approximately $1500.

The Internet address is http://www.csa.com.

E. Knowledge Express Data Systems on the Internet

An example of another interesting online service provider on the Internet is Knowledge Express Data Systems ("KEDS"), Berwyn, PA 19312 (800-529-KEDS). "KEDS" offers unique files such as lists of technologies that are available for license or joint venture from over 150 universities in the United States and abroad. There is also a unique file of company needs and capabilities, including company technologies available for licensing or joint ventures. Other databases include a number of files that list and abstract government inventions that are available for license.

With such a background, it should be no surprise that Knowledge Express provides special access to members of the Licensing Executives Society (LES), including access to the group's membership directory and to its journal. Others pay a standard fee. The Internet address is http//www.knowledgeexpress.com.

F. Selected Internet World Wide Web Addresses of Chemistry Interest

There are, of course, thousands of addresses on the Internet (note that each address is to be preceded by http://) that may be of interest to chemists and chemical engineers. The author has found the following to be especially interesting or helpful:

1. Aldrich Chemical Company (www.sigald.sial.com/aldrich/al_e.htm). Aldrich browsable on-line catalog.
2. American Chemical Society ChemCenter (www.chemcenter.org). This has become the primary ACS site, along with the location for CAS. It includes pointers to related sites such as the Publications Division (pubs.acs.org) and the Society as a whole (www.acs.org). The ChemCenter site is to include all available ACS electronic editions of journals and other tools such as certain ACS directories. There are also extensive links to other chemistry sites on the Internet. The components at this time include ACS services; chemistry resources; conferences, newsgroups, and similar; electronic publications; publications of ACS; ACS/CAS database searching (includes links to the STN International database site, to the *Chemical Patents Plus* file of Chemical Abstracts Service, and to ACS journals); searching (people, Yellow Pages, etc.); "shopping center": and "education center," including electronic courses and other.
3. American Chemical Society Publications Division (pubs.acs.org). Broad information pertinent to subscribing to and writing for ACS publications. Internet journal issues. ASAP service.
4. Steven M. Bachrach. Northern Illinois University.
   a. hackberry.chem.niu.edu/cheminf/html. A compilation of Internet sources of specific interest to chemists.
   b. hackberry.chem.niu.edu/Infobahn/Paper38.bow_notable.html. A compilation of "best" chemistry sites on the Internet in 1995 by Bachrach, Thomas Pierce, and Henry Rzepa; it is believed that this will be updated on annual basis.
5. Beilstein Informationssysteme GmbH (www.beilstein.com). Includes a searchable database with data since 1980 (subscription basis) and descriptions of electronic and printed products. Excellent set of links to other files and databases of interest to chemists.
6. BIOSIS (www.biosis.org). This major abstracting and indexing service covering biological and biomedical topics is described here, and search strategy information is given.
7. Cambridge Scientific Abstracts (www.csa.com). Includes descriptions of their abstracting and indexing products, as well as access to their fee-based *Internet Database Service* and *Environmental RouteNet* products.
8. CambridgeSoft Corp. (www.camsoft.com). Among other products, they offer a free product that is resident on the Internet (http://chemfinder.camsoft.com) that searches over 200 sites and over 22,000 chemicals by chemical name, formula, molecular weight, CAS Registry Number, and molecular structure. Re-

sults are basic physical property information about the compounds searched, along with links to other sites that contain additional information.

9. ChemPort (www.chemport.org). This location provides fee-based full-text access to chemistry and other scientific journals not only from ACS, but also from other publishers. Some 200 journals now included.
10. ChemWeb (www.chemweb.com). Joint venture of MDL and Current Science Group.
11. Chemical Abstracts Service (www.cas.org). Includes, for example, list and description of products, basic STN documentation, and link to Telnet access to STN. Also link to ChemCenter, the principal ACS Web site, and to fee-based *STN Easy.*
12. Chemical Patents Plus [casweb.cas.org/chempatplus (do not use www)]. Full-text searching of U.S. patents (including Chemical Abstracts Registry Numbers and subject indexing for chemical patents) since 1974. Display of most results is on a fee basis. Content the same as the comparable file on STN International, but has some unique features. A CAS product.
13. CISTI (Canada Institute for Scientific and Technical Information) (cisti.nrc.ca/cisti/). Among other components, includes free access to their online catalog covering books, serials, reports, and conference proceedings; registered users may order document copies. Also includes access to fee-based *SwetScan,* a table-of-contents service.
14. Derwent (www.derwent.co.uk). Includes general data about Derwent and its products. Text of *Derwent Information Newsletter.* Linkages to Derwent online hosts. Derwent selected Usenet Newsgroups by industry sector. Links to other scientific and patent Internet sites.
15. DIALOG (www.dialog.com; or dialog.krinfo.com). Includes Telnet or browser-based access to DIALOG and DataStar databases. Includes access to instructional sheets that describe the databases. Link to DIALOG ScienceBase—see item 22 below. Link to document copy ordering capability.
16. DIMDI- Deutsches Institut für Medizinische Dokumentation und Information (www.dimdi.de/homeeng.htm). A major database host, located in Cologne, Germany, that emphasizes biomedical information.
17. Engineering Information Inc. (www.ei.org). Includes access to *Engineering Information Village,* a fee-based service that includes extensive evaluated links to other Web sites.
18. Elsevier Science (major publisher) (www.elsevier.com). Includes catalogs for the world's largest scientific publisher, among other information. Also offers access to valuable table-of-contents service (ESTOC) for nearly all 1100 Elsevier Science journals and other alerting services, mostly free, others to those at subscribing institutions only. See also www.scidirect.com. This new (1997) site is to contain the full texts of all Elsevier journals in the ScienceDirect collection, and, in addition, those of other participating publishers. Authorized subscribers can search and download, as well as navigate to other referenced articles within the collection.

19. Gale Research (major publisher) [galenet.gale.com (do not use www)]. Offers fee-based access to many of Gale's directory-type databases.
20. (www.cyberhound.com). A search engine from Gale Research that helps users search for evaluate sites on the Web.
21. Institute for Scientific Information (www.isinet.com). Home page for major scientific information publisher. Among many products and services, ISI offers inexpensive automatic journal table-of-contents service with abstracts.
22. DIALOG ScienceBase (www.krscience.dialog.com). Intended for direct end-user searching of databases that are also available on DIALOG.
23. M.Kopelevitch (UCLA) (www.chem.ucla.edu/chempointers.html). A very useful compilation of Internet sources of specific interest to chemists.
24. LEXIS-NEXIS (www.lexis.nexis.com). This is a major online service provider with many databases, offered on a fee basis. Emphasis is on legal and business areas.
25. MDL Information Systems (www.mdli.com). Home page for a major software company that specializes in in-house systems of particular interest to companies focusing on pharmaceutical and agrochemical applications.
26. MicroPatent (www.micropat.com). Patent and trademark searching and downloading for a fee, but some searching is free.
27. Ovid Technologies (www.ovid.com). Major online service provider.
28. Questel•Orbit (www.questel.orbit.com). Includes Telnet access to Questel•Orbit databases. Includes link to their *QPAT-US* site (see item 29 below). Includes useful links to other Internet sites.
29. Questel•Orbit Full Text U.S. Patents (*QPAT-US*) (www.qpat.com). Subscription service that provides unlimited full-text access to all U.S. patents since the early 1970s. In addition, Questel•Orbit offers free searching of front-page information of U.S. patents since 1974 (for registered users).
30. The Royal Society of Chemistry (chemistry.rsc.org/rsc/). An especially valuable location for chemists. Includes RSC catalog and the library and information center's holdings catalog. Also includes new developments such as press releases on hot topics and information about conferences. Includes links to other chemistry sites. All 12 RSC primary journals and their *Analytical Abstracts* publication and database are also on the Internet on a fee basis.
31. U.S. Patent and Trademark Office (patent searching site; there are other U.S. Patent Office sites) (www.uspto.gov). Free searching capability for the most recent 20 years of U.S. patents. Not full-text and not indexed. System operated by CNIDR (Center for Networked Information Discovery and Retrieval of MCNC, Research Triangle Park, NC).
32. John Warden (RPI) (www.rpi.edu/dept/chem/cheminfo/chemres.html). An outstanding compilation of Internet sources of specific interest to chemists.
33. Gary D. Wiggins (University of Indiana) (www.indiana.edu/~cheminfo/). Chemical information sources on the Internet. Superb.
34. John Wiley & Sons, Publishers (www.wiley.com). Includes complete catalog

as well as valuable links. One link of special value is to *Journal of Computational Chemistry,* which, in turn, has links to Usenet news groups and to chemistry resources on the Internet.

35. Mark Winter (University of Sheffield) (www.shef.ac.uk/chemistry/chemdex/welcome.html). Chemistry resources on the Internet.
36. H.W. Wilson Co. (www.hwwilson.com). Home page for an important publisher of economical, reliable indexing and abstracting services.

In addition to these, there is, as mentioned earlier, an important printed book edited by Steven M. Bachrach, Northern Illinois University: *The Internet—A Guide for Chemists,* American Chemical Society, Washington, DC, 1996. This contains descriptive information about databases, electronic discussion groups, and other sources on the Internet that are of direct interest to chemists. See also: H. N. Tillman, Ed., *Internet Tools of the Profession: A Guide for Information Professionals,* 2nd ed., Special Libraries Asso., NY, NY. 1997.

New Internet addresses are, of course, added on a regular basis, and others change or are discontinued.

G. Evaluation of Internet Files (see also pp. 223–224)

Because of the unclear or even dubious nature of some Internet files, evaluation criteria such as the following may be worthwhile (in addition to suggestions for evaluation of electronic files made earlier in this chapter):

1. What is the source of the data? Is it a reliable, known source? The organization's name should be given, as should the name and location of the editor or author. Is the physical address of the source given, or can it be readily found? Ways to contact the editors or authors for questions should be provided. Sources that are unknown to the user may still be reliable, but this must be demonstrated or proved somehow to the user's complete satisfaction.
2. Is the scope (what is included and what is excluded) of the file clearly and understandably defined? This should include data of issue or date of most recent update, and, as appropriate, scope of coverage and indication of time period covered. A fuzzy scope or one with sweeping generalizations is usually suspect. So is a database or file that seems to change without a systematic rationale being clearly evident.
3. In the case of research work, are sufficient details given as to methodology? Has the work been subjected to a peer review process?
4. Is the database or file easy to use? Are instructions for use provided, and/or is a user manual available?
5. Are the graphics and presentation clear and unambiguous?
6. Are references (citations) given, as appropriate?

H. Listservs, Newsgroups

The Internet is distinguished by having a number of special-interest or user groups that, in effect, maintain electronic forums. Many forums are listed in the various compilations of Internet resources in chemistry that were noted earlier. The Internet address, www.liszt.com, includes a listing of thousands of listservs. *Listservs* (list servers) are mailing lists that provide for automatic distribution of submissions via e-mail to all those who subscribe to the particular list. They are, in effect, discussion groups and provide one way of keeping up with developments and current issues in a field. However, the messages may sometimes be too numerous, and the content may not be reliable since, for many lists, there is no central monitor. Of course, there is no referee process. To get on a listserv mailing list, the user must first "subscribe" electronically; thereafter, everything is automatic until the user wishes to "de-subscribe."

A few examples of listservs pertinent to chemical information include

CHEMIND-L. Discussions on the subject of chemical structure indexing

CHEMINF-L. Discussions of chemical information sources*

PIUG-L. Discussions between users of patent information sources; the Patent and Trademark Group of the UK participates.

Newsgroups are similar to listservs except that distribution of messages is not automatic; the user must initiate the action of accessing the group on Usenet. A few examples of newsgroups are

sci.chem

sci.chem.analytical

sci.chem.electrochem

This is all in addition to the well-known e-mail capability of Internet in which individuals correspond with one another through the Internet, but do so in a more private manner by using one another's e-mail addresses.

For an extensive list of e-mail servers, listservs, newsgroups, and similar, see the excellent list compiled by Professor Joseph Warden at

http://www.rpi.edu./dept/chem/cheminfo/chemres/chemres_05.html

I. Electronic Conferences

Conferences of interest to chemists are now being held electronically on the Internet (13). ChemConf'93, the first electronic chemistry conference, took place on June 14–August 20, 1993.

The First Electronic Computational Chemistry Conference was held on the Internet on November 7–18, 1994. This attracted a total of 321 subscribers and 73 final papers. The second such conference took place November 6–December 1, 1995, dedicated to all aspects of computational chemistry. It included the capability to comment on and discuss the papers presented.

Presentation of papers on the Internet allows transmittal and discussion of the most

*Henry S. Rzepa coordinates an important list for chemical applications of the Internet. This is archived as www.lists.ic.ac.uk./hypermail/chemweb/.

recent results. It permits low-cost participation by many who could not otherwise attend a conventional conference.

The version of HTML (HyperText Markup Language) available at the time of the First Electronic Computational Chemistry Conference did not support superscripts, subscripts, tables, or the Greek alphabet; most of these obstacles have already, or will be, overcome.

The conferences described above were held entirely on the Internet, and have been available on the Internet, at least temporarily. The First Electronic Computational Conference is to be available on CD-ROM and is to then be removed from the Internet.

10.16. "PORTABLE DATABASES": CD-ROM PRODUCTS

The phrase "portable databases" was popularized by Carlos Cuadra, a well-known pioneer in information systems, as a designation for CD-ROMs and similar products such as diskettes that can be easily moved from computer to computer and that, therefore, are portable. CD-ROM products require a special drive that can either be connected externally to a computer or is an integral part of a computer system. CDs are now by far the most popular of the portable databases.

A CD-ROM is a very efficient storage medium. Thus, a typical disk will hold the contents of 1500 floppy diskettes or approximately 250,000 pages. Each disk producer usually provides complete search software and an instructional manual along with the purchase or license of the disk.

As mentioned, a CD-ROM drive is required to read the disks. This may be built into the computer at time of purchase, or it may be added at any time. Some compact disk products, but not all, are available for either DOS, Microsoft Windows, or Macintosh systems, but others are limited to only one or two of those systems. In considering a compact disk, it is wise to determine from the producer of that disk what the requirements are in terms of hardware (operating system, RAM, hard-disk storage capacity, and CD-ROM drive, including speed) and software. It is also prudent to ascertain the extent and type of documentation and user support that is available. Availability and compatibility for potential networking should also be determined, as should scheduling of any updates, and future software enhancements.

CD products have a fixed price, unlike the "open" pricing structure of online systems for which prices depend on extent and type of use. But pricing of most CDs is high, frequently $1000 or more, and marketing efforts are often aimed at the large organization (industrial firm or university) rather than the individual chemist, or even the small firm or small college, as the ultimate purchaser.

From a searching perspective, the principal advantage of the CD is that because there are no online charges, the user is freed from any pressure that this might create. The fact that the price is "fixed" at time of initial purchase encourages the user to browse at will, with unlimited access as needed. Once the disk has been purchased or licensed, there are no time or cost constraints or pressures.

CD products also offer the advantage of excellent graphics in many cases; for example, a Greek symbol can be shown as such rather than in a representation of the symbol. Quality is similar to print versions of the same materials. Good chemical substance images can often be displayed and printed.

For libraries, CDs offer important space-saving advantages. Interestingly, however, at least one organization that produces CDs has found that most of the interest in CD-ROM versions of journals is with individuals rather than libraries. Libraries are concerned about CDs for several reasons: (1) the jukeboxes utilized by libraries for CDs can be expensive to purchase and maintain, (2) CDs from a variety of sources often have different interfaces that can cause problems, and (3) some users may either inadvertently or otherwise tend to make off with CDs.

Some producers of CD products believe that CD technology is particularly well suited for the end user (i.e., the bench chemist or engineer) rather than the information professional. They believe that their CD products are particularly user-friendly since many of the products require merely simple "point and click" operations in a Windows environment.

The CD concept makes for more portability than does the online system, in that neither a telephone line nor a modem is required. On the other hand, as previously noted, a CD reader and a computer are, of course, required.

Some CD products lack the full power and sophistication of their online counterparts. For example, with CD products from CAS, such operations as ranking and sorting of results, and searching by drawing of a structure—possible in major online files generated at CAS—are not possible with the CD products. In addition, searching more than one file (database) concurrently is obviously not possible with CDs. On the other hand, CDs from DIALOG are said to have essentially all the power of their online counterparts with the exception of the multiple file limitation just mentioned.

CDs are frequently searched one disk at a time. When dealing with a large number of disks, this can be inconvenient and time-consuming. Special technology is available to help ameliorate this challenge. When using DIALOG CD products, if a user changes disks, the system asks if the user would like the search repeated; this feature eliminates the task of rekeying search statements when using subject-related products or when using a multidisk product. In addition, for some DIALOG CD products, DIALOG personnel have created software, known as *Site-Enhanced* version, that—with the use of a tower or daisy-chained CD-ROM drive—will permit the user to conduct a single search of all the disks for that product e.g., *Ei Compendex.* "Jukeboxes" are available that will hold up to 100 disks at a time, but these rarely permit the user to search all disks as if they were one.

Timeliness can also be a special issue with most CD products, and this should be considered at time of purchase or license. (Of course, timeliness must be considered with any chemical information product.) Disks are typically updated at set intervals, such as quarterly (or every 90 days), although some are updated less frequently, some are updated more frequently (weekly in some few cases), and some may not be updated at all. (Some products, such as encyclopedias, are not updated in either printed or electronic form, until new editions are published.) Thus, CD-ROM products are seldom as current as their online counterparts. A search with a CD-ROM product usually needs to be augmented by an online search if the user is interested in the more recent data. However, Intel Corporation ads have touted CD-ROMs that are described as hybrid or "connected." These disks offer built-in access to relevant information on the Internet.

In summary, the principal advantages of portable databases are that

- They can be utilized by the searcher without need to access an online system. Thus, there are no connect time or other charges after purchase or license.
- Since connect charges and time are not a factor, browsing at leisure is greatly facilitated. There is no pressure. Prices are fixed.
- CDs do not require use of a modem or access to an online system of any kind.
- They can be taken home or to any other computer equipped with an appropriate drive.
- Most CDs have very friendly user interfaces.

The usual disadvantages are as follows:

- If a CD is involved, the searcher's computer must be upgraded to read CD-ROM products if it is not already so equipped.
- CD-ROM pricing structures are frequently based on potential sales or license to organizations such as larger universities and corporations. This often puts CDs beyond the budgets of individual chemists and many smaller organizations.
- The products are out of date as soon as they are produced. They need to be updated by use of online or Internet databases or by obtaining any updates or supplements that may be made available later. But even frequent updates, as often found with CD-ROM files, don't compete effectively with the timeliness of the many online files that feature monthly, weekly, or even daily updates in some cases. (For some users, timeliness is not an issue and may not be viewed as a disadvantage.)
- Some of the more useful searching features of commercial–proprietary online files are not always available on many CD-ROM and other portable products.
- If CDs are searched one at a time, searching can be time-consuming and inconvenient. However, as mentioned, special technology to overcome this problem is now available.

Typical CD-ROM products of interest to chemists are based on products that are already available online and/or in printed form, although some CD-ROM products are available in that form only. Some representative CD products of interest to chemists and chemical engineers include those products listed below. Names of information providers are shown in parentheses. DIALOG OnDisc products are available from the DIALOG Corporation.

1. *DIALOG OnDisc Academic Press Encyclopedia of Physical Science and Technology* (Academic Press, Inc.).
2. *DIALOG OnDisc Advanced Materials* (polymers, rubbers, ceramics, nonmetallic composites).

3. American Chemical Society: *Directories on Disc,* which includes *Directory of Graduate Research, College Chemistry Faculties,* and *Graduate School Finder.*
4. Beilstein Institute: *Current Facts in Chemistry.*
5. *DIALOG OnDisc Biotechnology and BioEngineering* (Cambridge Scientific Abstracts).
6. Chemical Abstracts Service: *CA 12th Collective Index,* 1987–1991; *CASurveyor* series; *CASSI; Chemical Abstracts,* 1987–1991 and 1996 to date; 13th *Collective Index* is also available. See also Chapters 6 and 7.
7. *DIALOG OnDisc Chemical Business NewsBase* (The Royal Society of Chemistry).
8. *DIALOG OnDisc Chemical Engineering and Biotechnology Abstracts* (The Royal Society of Chemistry).
9. *DIALOG OnDisc Derwent Petroleum and Power Engineering* (from Derwent *World Patents Index* database); Derwent offers other CD products directly. See Chapter 13.
10. *DIALOG OnDisc DOE Energy Science and Technology* (U.S. Department of Energy).
11. *DIALOG OnDisc Ei ChemDisc* (Engineering Information, Inc.).
12. *DIALOG OnDisc Ei Compendex* (Engineering Information, Inc.).
13. *DIALOG OnDisc Environmental Management* (Cambridge Scientific Abstracts).
14. *DIALOG OnDisc Kirk-Othmer Encyclopedia of Chemical Technology,* 3rd ed. complete; and 4th ed. in progress with images (John Wiley & Sons, Inc.).
15. *DIALOG OnDisc MEDLINE* (National Library of Medicine).
16. *DIALOG OnDisc Metadex/Materials Collection* (Cambridge Scientific Abstracts).
17. Micropatent: *RetroChem* (covers U.S. chemical patents) and other intellectual property products. See Chapter 13.
18. *DIALOG OnDisc NTIS* (U.S. National Technical Information Service).
19. *DIALOG OnDisc Nuclear Science Abstracts* (U.S. Department of Energy).
20. *DIALOG OnDisc Paper, Printing & Packaging Database* (Pira International, UK).
21. *DIALOG OnDisc Petroleum Abstracts.*
22. *DIALOG OnDisc Polymer Encyclopedia* (*Encyclopedia of Polymer Science and Technology,* John Wiley & Sons).

An example of another major vendor of CD-ROM and other computer software products of interest to chemists and that are made by a number of producers is Sigma-Aldrich Chemical Company, PO Box 355, Milwaukee, WI 53201 (800-558-9160). Sigma-Aldrich says that they offer technical support on all of the software products that they sell regardless of whether they are the producer of the software.

Some of the CD and/or diskette products available from Sigma-Aldrich include the following:

Aldrich FT-IR Spectral Libraries on CD-ROM. Condensed-phase FT-IR (Fourier transform–infrared) library with 10,607 spectra and vapor-phase FT-IR library with 5010 spectra. These are available in both CD-ROM and diskette form and correspond with the printed volumes. Peak search capability is available separately from Sigma-Aldrich. Exclusive.

Bretherick's Reactive Chemical Hazards Database, 5th ed. Diskettes or CD-ROM.

CD-CHROM Chromatography Database. Includes over 106,000 abstracts covering literature from early 1958 to early 1994—updated every 12–18 months.

ChIM. Chemical Inventory Management System. A "cradle to grave" system that includes more than 60,000 Sigma and Aldrich chemical records with "unlimited" capacity for additional records. Another related file includes safety and regulatory information on 20,000 products. Also available on diskettes. Updates not provided at this time.

Gardner's Electronic Chemical Synonyms and Trade Names, 10th ed.

Industrial Surfactants Electronic Handbook. Over 16,000 surfactants are described. Diskettes or CD-ROM.

Sax's Dangerous Properties of Industrial Materials, 9th ed. and *Hawley's Condensed Chemical Dictionary,* 12th ed. Diskettes or CD-ROM. Now also available as part of a suite of products called *Comprehensive Chemical Contaminants Series on CD-ROM.*

Sigma-Aldrich Chemical Directory. Lists over 140,000 chemical products that are available from Sigma Chemical Co., Aldrich Chemical Co., Fluka Chemie, and Supelco. Most of the chemicals can be searched by structure or substructure. Searching by physical property, synonym, molecular formula, and other entry points is also possible. CD-ROM. Exclusive.

Sigma-Aldrich Material Safety Data Sheets. Over 77,000 data sheets for Sigma, Aldrich, and Fluka. CD-ROM or magnetic tape. Exclusive.

WIMP95. Structure drawing program. Diskettes. Exclusive.

In addition to the more general criteria for evaluating online databases, the chemist or engineer should consider the following for CDs and other portable databases:

1. Understand fully the type of hardware and software required and compatibility requirements relative to one's local computer. Full compatibility is essential.
2. Read the reviews in such sources as the *Journal of Chemical Information and Computer Science.* If there are no reviews, evaluate the source and consider how the product was constructed.
3. Ask for a free demonstration and/or trial disk that can be tried on a local computer. It is also a good idea to examine the user manual before purchase.
4. Inquire about the availability of technical support, and the hours and costs, if

any, of such support. Is there a toll-free help phone number? Free technical support is highly desirable.

5. Inquire about the return and/or warranty policy. A no-obligation 30-day return period should be available.
6. Consider the cost relative to budget.
7. Consider the availability and frequency of updates of the data.
8. Consider the availability of features that facilitate multidisk searching.

10.17. MDL INFORMATION SYSTEMS, INC.

A number of firms offer powerful in-house (within a company) systems that can supplement and augment online and Internet use. These systems permit and facilitate input and searching of in-house data such as proprietary chemical structures and internal test and screening results. Some systems also offer collections of databases from external sources for internal use on a company's own computers; in this case, the chemist can have access to both internal (company) data and external data in an integrated package.

Clearly one of the worldwide leaders in the field, and the best known of the various companies that are active today, is MDL Information Systems, Inc., 14600 Catalina Street, San Leandro, CA 94577 (800-326-3002). The Internet address is http://www.mdli.com. MDL (originally Molecular Design Limited) was founded in 1978 by Stuart Marson and W. Todd Wipke, who was then a professor of chemistry at the University of California at Santa Cruz. The company made an almost immediate mark—both Marson and Wipke were subsequently honored with the American Chemical Society's prestigious Skolnik Award, and the company's products have received outstanding acceptance.

MDL became part of Maxwell Communications in 1987, and in 1993, it became an independent public company. Plans to merge with Reed Elsevier were announced in 1997. Revenues for fiscal 1996 were in the range of $61 million, and there are now over 350 employees. Most of the interest in MDL products appears to come from companies in the pharmaceutical, agrochemical, and biotechnology industries, especially the larger companies, but a number of other well-known chemical companies are also customers. Some MDL products are priced to be purchased by companies, and some are explicitly priced for use by individuals. MDL points out that the company serves many lone users in academia as well as many users in smaller companies.

MDL's principal products consist of software for management of chemical information within an organization; databases of published syntheses and reactions, and of chemical suppliers, all aimed at in-house use; and consulting and other services. Steven D. Goldby is chairman and chief executive officer.

One product offered by MDL Information Systems, Inc. is *The Available Chemicals Directory* (*ACD*), a database covering some 500,000 products and 200,000 substances from over 240 suppliers worldwide. Most of the vendors included are suppliers of research or fine chemicals. Pricing information is included. Updating is twice a year. This database can be searched by such entry points as chemical name or synonym, structure, substructure, molecular formula, molecular weight, or CAS Registry Number. Price information is included. It is configured for installation on in-house server computers.

(Note: ISIS, MDL, OHS MSDS ON DISC, OHS SAFETY SERIES, MDL SCREEN, and PROJECT LIBRARY are trademarks of MDL Information Systems, Inc.)

MDL offers an outstanding collection of synthetic chemistry databases for in-house client/server installation. These include

1. *The ChemInform Reaction Library.* A database of reactions from 1946 to date. Updating is annual through FIZ (Fachinformationszentrum) Chemie (located in Berlin, Federal Republic of Germany) based on survey of 100 key journals. The emphasis is on new reactions and syntheses. Approximately 50,000 new reactions are covered each year. This "corresponds" to the printed weekly *CHEMINFORM.*
2. *REACCS-JSM.* An electronic version of the *Derwent Journal of Synthetic methods* from 1980 to date. Approximately 3000 new reactions per year. Annual updates.
3. *Reference Library of Synthetic Methodology.* Novel methods of organic synthesis from 1946 to 1991 as based on a combination of five MDL databases: *Theilheimer's Synthetic Methods of Organic Chemistry* 1946–1980; Current Literature File, primary literature 1983–1991; *Core,* primary literature through 1991; *CHIRAS,* asymmetric synthesis literature 1975–1991; and *METALYSIS,* metal-mediated transformation literature, 1974–1991).
4. *ORGSYN.* Electronic version of the classic book series *Organic Synthesis* from 1921 as published by John Wiley & Sons, New York, NY. Updating is annual and includes approximately 100 new reactions per year.
5. *Comprehensive Heterocyclic Chemistry.* Compendium of heterocyclic synthesis and reactions through 1983 as based on the Pergamon Press book *Comprehensive Heterocyclic Chemistry,* 1984. Includes 43,000 reactions. No updates.

MDL's reaction databases are structure-searchable in the following ways:

Reaction substructure search.

Substructure search on reactants and products.

Exact match on reactants and product.

Reaction similarity search.

Molecule similarity search.

Data searching is also offered.

In addition to these, MDL offers important products that help companies manage in-house data and information.

ISIS (*Integrated Scientific Information System*) is a client/server software system that provides extensive chemical information management for companies. Its components consist of the following:

1. *ISIS/Draw.* This permits drawing of chemical structures. Results can be utilized to build and search databases. This software can be implemented on a personal computer (PC) basis (client application only). A version (ISIS/Draw, version 2.1.1) is available free on the Internet under the conditions described at the MDL site.
2. *ISIS/Base.* This is a database management system focussing on molecules, reactions, and the related data. It can be operated on a personal computer or, using ISIS/Host, can connect to multiple servers. Thus, the user can store, search, and retrieve research and other data by chemical structure or access databases in the organization.
3. *ISIS for Microsoft Excel.* This is an *Excel* add-in that makes it possible to search *ISIS* databases (the same ones that *ISIS/Base* can access) within *Excel.* Search results, including structure or reactions, are automatically pulled into *Excel,* and utilities provided by MDL make it possible to sort (keeping the structures with the appropriate data) and to rearrange or "pivot" data (e.g., reorganize test data into separate columns by test).
4. *ISIS SAR Table.* This is an add-on to ISIS/Base. It allows the user to create a spreadsheet of structures and data. Data can be extracted from either local or server-based databases.
5. *ISIS/Host.* This is the server portion of the *ISIS* client/server database management system and can be installed on one or more server computers serving multiple users through a network. It offers the ability to integrate data from chemical databases and relational databases. Molecule and reaction gateways of ISIS/Host have capabilities not found on local database systems, including 3D searching, similarity search, and R-group searching. Similarity searching is available on molecules and reactions. Three-dimensional and R-group searching are available only for molecule gateways. ("R-group searching" means that the user specifies a set of fragments that are allowable in that molecular position.)

ISIS/Draw and *ISIS/Base* (items 1 and 2, above) can be utilized independently by individual scientists on their PCs to create databases of reactions, structures, associated numerical data and text, and graphics.

Other *ISIS* products include *ISIS Procedure Language,* a tool for modifying and customizing ISIS features, building applications, and communicating between *ISIS* and other programs; *ISIS/Object Library,* a tool that allows access to ISIS functionality through other application development environments; *ISIS/Host Open Gateway,* which allows one to create new points of entry to data sources not already supported by MDL; and *ISIS/Host API* (Application Program Interface), which is a collection of routines that can be called directly by other programs to provide ISIS functionality such as structure searching, polymer structure searching, and data integration. *ISIS/Host Open Gateway* can link onto a structure identified as of interest, but about which little is known, and can then search CAS's *Registry File* (assuming that the user has appropriate access and password) in order to identify information about that structure. The product can also be utilized as a general search tool for *Registry,* again assuming proper access and password.

The *ISIS* software programs noted above are among the newest and most rapidly growing MDL products. *MACCS-II* and *REACCS* (described below) are older products, and are still in very active use.

1. *MACCS-II* (*Molecular Access System*). This is a mainframe application used to construct and manage a company's two- and three-dimensional chemical structures and related research data.

2. *REACCS* (*REaction ACCess System*). This is a mainframe application used to construct and manage a company's databases of chemical reactions and associated information. *REACCS* can search any MDL reaction database, and *MACCS-II* can search any MDL molecular database.

The new *Chemscape* products are aimed at delivering molecule (2D and 3D) and reaction searching capabilities in both the World Wide Web and intranet environments. *Chime Pro* is a plug-in for *Netscape Navigator* that can be used to build chemical applications in the HTML world. This product, when combined with *Chemscape Server* and *Netscape Enterprise Server,* can be used to search chemical databases via *ISIS/Host. Chime Pro* can display chemical structures and reactions in the HTML page (rather than externally in a helper application), and structures can be manipulated in the HTML page. *Chime,* a free version of *Chime Pro,* and accessible to all through MDL's web site, http://www.mdli.com, can also be used to view structures in the HTML pages, but it cannot be utilized for searching. Both *ISIS for Microsoft Excel* and *Chime Pro* work smoothly with *ISIS/Draw.* In summary, *Chime Pro* makes it possible to (a) search, retrieve, and manipulate structures and related data; and (b) take search results (structures) and sort and otherwise manipulate these in *Excel. Chime* can not only enable the display of structures right in HTML pages but also change the appearance (ball and stick, ribbons, strands, etc.) of those structures.

Several recent MDL products are aimed at helping companies in the pharmaceutical industry deal with combinatorial chemistry and high throughput screening.

1. MDL's *Project Library,* intended for use with a personal computer, is a software program designed to help manage the chemical and biological data developed in combinatorial chemistry efforts. It permits full enumeration of compound libraries. A related product, *Central Library,* is a client/server system aimed at controlling data and tracking the results of large-scale combinatorial experiments throughout a company.

2. *MDL SCREEN,* intended for client/server systems, is software technology designed to help companies perform and manage the data generated by high-throughput screening, beginning with experiment design and extending through every stage

of the screening process. This product is utilized to manage data generated during the screening process. The compounds screened may, or may not, be from a combinatorial synthetic activity. The product tracks the ID numbers of the compounds, the assays being tested, and the results of that testing. The Project Library and the Central Library, on the other hand, are used to manage the actual structural identity of compounds created in combinatorial libraries. These compounds are typically then submitted for screening, and the screening efforts can be managed by MDL SCREEN.

3. *MDDR* (*MDL Drug Data Report*) is a database of drug candidates in various stages of testing. It is a rich source of biological activity information as well as competitive information. This product is based on the hard-copy publication *Drug Data Report* by J. R. Prous.

MDL's molecule databases are also available with 3D structures, a feature that can be quite valuable to research workers interested in examining structure–property relationships via 3D studies. This benefit applies particularly to the *Available Chemicals Directory* and to the *MDL Drug Data Report.*

One of MDL's latest projects is a joint venture with Current Science Group, London. This is a new firm, ChemWeb, which has built and is marketing a World Wide Web site (http://www.chemweb.com) to provide access to certain scientific journals, databases, references, discussion groups, electronic meeting rooms, news services, a job exchange, and a shopping mall (software, equipment, services, and publications). At this point, most of the journals and databases have what can be called a biomedical orientation. Subscriptions are available, but access to discussion groups, meeting rooms, and job exchange is planned to be free and unlimited. Structure searching on the Web is planned. None of the American Chemical Society journals or databases are currently available through this source.

Virtually all MDL products are intended for use in conjunction with a client/server system. However, in addition, *ISIS/Draw, ISIS/Base, ISIS for Microsoft Excel, Chime Pro,* and *Project Library* can be utilized as standalone products with workstations, or with desktop IBM or Macintosh computers.

In the Fall of 1997, it was reported that ChemWeb Holdings is to be acquired from the Current Science Group by Elsevier Science Publishers, Amsterdam. Other firms or products involved in the reported "definitive agreement" include BioMedNet Limited, Current Biology Limited, Current Chemistry Limited, Electronic Press Limited, and Current Science Limited. MDL Information Systems, one of the co-founders of ChemWeb, is now a United States-based subsidiary of Elsevier.

10.18. INFORMATION FIRMS AND CHEMICAL INFORMATION CONSULTANTS

There are a number of independent information firms (sometimes improperly called "brokers") who can plan and conduct both on- and offline searches of the literature. These firms can offer a wide variety of information services and can be of particular value to smaller organizations or to larger organizations whose staff is overloaded.

One good source for obtaining the names and locations of some of the most capable of the firms is the Information Industry Association. This organization includes as its members most of the profit-making organizations that produce databases and systems. Two of the larger, better known information firms are listed here by way of example and illustration:

Research on Demand, Inc.
P.O. Box 479
Santa Barbara, CA 93102
(800-200-4095)
www.researchondemand.com
(A complete range of information services is offered.)

Wendy Warr & Associates
6 Berwick Court
Holmes Chapel
wendy@warr.com
Cheshire, CW47 HZ, England
(44-1477-533837)

Wendy Warr & Associates specializes in chemical information, combinatorial chemistry, molecular diversity, electronic publishing, and computer structure handling. Consulting services are offered. Firms of all sizes are served.

There are also chemical industry consultants who specialize both in new chemical technologies and in chemical information. For example

Technology Information Consultants
163 Hearn Lane
Hamden, CT 06514
(203-281-3693)
5 Science Park, Box 39
New Haven, CT 06511
(203-786-5030)
tic@snet.net

(This firm was founded in 1985 by Dr. R. E. Maizell, previously manager of Business and Scientific Intelligence Centers and Consulting Scientist for New Technologies at Olin Corporation.)

REFERENCES

1. *Online* and *Database,* both published by Online, Inc., Wilton, CT; *Information Today,* Learned Information, Oxford, United Kingdom and Medford, NJ. There are other such journals as well.
2. M. Alampi, Ed., *Gale Directory of Databases,* Detroit, Gale Research, 1996. Frequently updated.
3. P. Benichou, C. Klimczak, and P. Bourne, "Handling Genericity in Chemical Structures Using the Markush DARE Software," *J. Chem. Inf. Comput. Sci.* **37,** 43–53 (1997).
4. S. R. Heller and G. W. Milne, *Online Chemical Information Workshop Notes,* 1982 (mimeographed).
5. J. Witiak, "Chemical Substructure Searching—Ready or Not, Here It Comes," *Proceedings of the Online Conference,* Dallas, TX, 1981, pp. 143–147; "Substructure Searching Using *CAS ONLINE,*" *Proceedings of the Second National Online Meeting,* New York, 1981, pp. 537–538.

6. D. D. Ridley, *Online Searching: A Scientist's Perspective,* Wiley, New York, 1996.
7. H. Schulz and U. Georgy, *From CA to CAS Online,* 2nd ed., Springer-Verlag, Berlin, 1994.
8. D. Weininger, "SMILES, A Chemical Language and Information System," *J. Chem. Inf. Comput. Sci.* **28,** 31–36 (1988).
9. S. Ash et al., "SYBYL Line Notation (SLN): A Versatile Language for Chemical Structure Representation," *J. Chem. Inf. Computer Sci.,* **25,** 71–79 (1997).
10. S. M. Bachrach, Ed., *The Internet—a Guide for Chemists,* American Chemical Society, Washington, DC, 1996.
11. T. H. Pierce and T. J. Cozzolino, *Chemical Indexing and the Internet,* in S. M. Bachrach, Ed., *The Internet—a Guide for Chemists,* American Chemical Society, Washington, DC, 1996, pp. 277–308.
12. P. Murray-Rust, H. S. Rzepa, and B. J. Whitaker, "The World-Wide Web as a Chemical Information Tool," *Chemical Society Reviews* (in press). (Web site: http://www.rsc.org/csr398.htm.)
13. S. M. Bachrach, "Electronic Conferencing on the Internet: The First Electronic Computational Chemistry Conference," *J. Chem. Inf. Comput. Sci.,* **35,** 431–441 (1995).
14. Manfred E. Wolff, Ed., *Burger's Medical Chemistry and Drug Discovery,* 5th ed., Wiley, New York, 1995. See especially Volume I, Chapters 2, 4, 10, 13, and 14, on information and methods in the pharmaceutical industry.

11 Reviews

Comprehensive and evaluative reviews can minimize or obviate the need for the chemist to do further literature searching on the subject in question. Reviews save time and help provide perspective. Along with encyclopedia articles (see Chapter 12), they can be excellent starting points.

Some questions that can help the user evaluate the quality of a review include the following:

1. Is the author of the review a known expert in the field?
2. Is there a clear statement of scope, methodology, and purpose of the review?
3. What is the time frame covered? How up to date are the most recent sources included, and how far back in time does the author go? The more comprehensive, the better.
4. Are all types of literature covered, including patents?
5. Is the author critical or evaluative? Are the most important and valid data identified?
6. Is the material well organized and presented?
7. Is there a complete bibliography with all sources clearly identified?
8. Is there any reason to expect significant lack of objectivity?

The best known review journal is *Chemical Reviews* (*CR*). Initiated in 1925, and now published eight times a year by the American Chemical Society, *CR* contains critical, comprehensive reviews by experts in the field, with emphasis on the more theoretical aspects of chemistry. Among the most significant review sources are those published by The Royal Society of Chemistry. These include (names of editors given as available):

- *Annual Reports on the Progress of Chemistry* (annual). Martin J. Sugden. Divided into three sections: inorganic, organic, and physical chemistry. Online through STN. The file name is *CJRSC.*
- *Chemical Society Reviews* (bimonthly). K. J. Wilkinson. Bimonthly now; was quarterly. Online through STN.
- *Contemporary Organic Synthesis* (bimonthly). Sheila Buxton. Initiated in 1994. Critical reviews of reactions reagents, strategies, and designs for specifically defined areas including more specialized areas. Online through STN.

- *Issues in Environmental Science* (two issues per year). Initiated in 1994. Includes review articles. Covers the scientific aspects, and also other aspects such as economic and political.
- *Natural Product Reports* (bimonthly). Critical reviews of the published literature on alkaloids, terpenoids, and steroids, and aliphatic, aromatic, and *O*-heterocyclic compounds. Online through STN.
- *Pesticide Outlook* (bimonthly). Initiated in 1989. Reviews many aspects of the pesticide business.
- *Russian Chemical Reviews* (monthly). Cover-to-cover translation of the *Uspekhi Khimii,* the chemistry review journal of the Russian Academy of Sciences.

RSC's *Specialist Periodical Reports* series has a number of titles, some of which are published annually and others as biennial volumes. Recent titles include the following volumes:

Amino Acids, Peptides and Proteins
Carbohydrate Chemistry
Catalysis
Electron Spin Resonance
General and Synthetic Methods
Organometallic Chemistry
Organophosphorus Chemistry
Photochemistry
Spectroscopic Properties of Inorganic and Organometallic Compounds

Annual Reviews publishes a series of annual review volumes in various sciences. Those of most interest to chemists include volumes on physical chemistry, biochemistry, pharmacology and toxicology, biophysics and biophysical chemistry, and materials sciences.

CRC Press, Boca Raton, FL, publishes a number of review journals. Those believed to be of most interest to chemists include the following (names of editors given as available):

- *Critical Reviews in Analytical Chemistry* (quarterly). Charles H. Lochmüller, Duke University.
- *Critical Reviews in Biochemistry and Molecular Biology* (bimonthly). Gerald D. Fasman, Brandeis University.
- *Critical Reviews in Biotechnology* (quarterly). Graham G. Stewart, Heriot-Watt University, Edinburgh, Scotland, and Inge Russell, Labatt Brewing Co.
- *Critical Reviews in Environmental Science and Technology* (quarterly). Terry J. Logan, Ohio State University.
- *Critical Reviews in Toxicology* (bimonthly). Roger O. McClellan, Chemical Industry Institute of Toxicology, Research Triangle Park, North Carolina.

- *Critical Reviews in Solid State and Materials Science* (quarterly). Joseph E. Greene, University of Illinois, and Paul Holloway, University of Florida.

Profile of CRC Press. CRC Press was, at the outset, a laboratory supply house that was founded in 1900 in Cleveland, OH by Arthur Friedman, a student at what was later to become the Case Institute. The company first sold rubberized chemistry aprons and then rubber stoppers and tubing, and it soon adopted the name *Chemical Rubber Company.* It started the *CRC Handbook of Chemistry and Physics* in 1913, and the volume soon became a basic reference book for chemists and engineers throughout the world (see Chapter 15). In the 1960s, the firm was sold to Florence and Bernard Starkoff, who expanded company activities to include a proprietary line of scientific equipment and the publication of several new handbooks in other scientific subjects. The manufacturing division was subsequently sold, and the publishing company changed its name to CRC Press. It moved to Boca Raton, FL and, in 1986, became a subsidiary of the Times Mirror Company. In 1990, it acquired Lewis Publishers of Ann Arbor, MI, as its environmental science publishing unit, and in 1992 it acquired Food Chemical News, Washington, DC. Currently, CRC Press publishes over 350 volumes and 16 journals annually. It employs over 200 full-time people and over 100 freelance persons for specialized publishing functions.

Chemical Abstracts (*CA*) editors have recognized the importance of reviews by designating these with the symbol "R" online and in the Volume Indexes to *CA* (beginning in 1967). Many chemists examine citations to reviews before they look at abstracts for other index entries, because a good review can provide all that is immediately needed on a topic.

ISI publishes *Index to Scientific Reviews.* This covers over 200 review serials and provides selective coverage of reviews in 3000 primary research journals. In includes indexes by subject, source bibliographic information, and author organization. Publication is semiannual, and volumes are available back through 1974. About 50,000 items per year are included.

A number of publishers make available "Advances in" or "Progress in" series. For example, Technomic Publishing, Lancaster, PA publishes several advances series of interest to chemists and engineers in the plastics industry. These include *Advances in Polymer Blends and Alloys Technology, Advances in Urethane Science and Technology, International Progress in Urethanes,* and *Advances in Interpenetrating Polymer Networks.*

Not all reviews appear in journals and books specifically devoted to reviews. A good review can appear in almost any chemical journal or other information source. Reviews are frequently labeled as such or in similar terms, for example, *Recent Advances* or *Recent Trends.* However, any article or book with an extensive list of references, and a good discussion of these, is potentially a review, regardless of whether it is so labeled. Patent sources can also provide an excellent review of prior art (see Chapter 13).

Another way of reviewing technologies is through citations in articles and patents. Once the user has located a key reference, additional pertinent literature may be located, based on citations to or from this reference. There are those who are highly enthusiastic about the merits of this approach, while others believe that there are limitations. The *Institute for Scientific Information* (see Section 8.3) has developed the use of citations in articles to its highest form. Sources of patent citations are discussed in Section 13.20.

List of Addresses

Annual Reviews, Inc.	**CRC Press**
4139 El Camino Way	2000 NW 24th Street
Palo Alto, CA 94306-9981	Boca Raton, FL 33431

12 Encyclopedias and Other Major Reference Books; Journals

I. ENCYCLOPEDIAS AND OTHER MAJOR REFERENCE BOOKS

12.1. INTRODUCTION

Journals and patents almost always contain the latest information, and that is frequently what chemists need. The chemist, however, cannot afford to ignore books. Books are usually the best starting point for searching the chemical literature. They can often provide quick, straightforward answers that could otherwise take many hours to locate in widely scattered journal articles and patents on the subject. Books provide the foundation from which to launch any further search of the literature required. In 1996, *CA* covered 5472 books, of which 4529 (about 83%) were in English.

An increasing number of reference works in chemistry will be found in whole or in part in electronic forms, in addition to the print versions. The full texts of several important chemical reference sources can be found online, and may also be found in CD-ROM form. The advantages and disadvantages of full-text searching were discussed in Chapter 10. The trend toward electronic availability of these sources is expected to increase and strengthen. The Internet is already an important source for descriptive information about books (every major publisher provides useful descriptions of reference and other books), and it is expected that the Internet will become increasingly important as to reference information.

When the literature on a chemical or reaction is voluminous, use of an appropriate book is highly recommended to help sort out what is important and vital from what is not. When judiciously utilized, reference books can provide a reliable anchor and offer an initial mechanism for beginning to deal with what otherwise might seem confusing or even chaotic. Some *advanced* books are evaluative or critical and can help minimize further searching. Review journals and other review sources (see Chapter 11) perform a similar function.

The chemist will find some older books, seemingly obsolete, of value. As noted later in this chapter, older editions may contain some information not included in more recent editions, and this may be precisely what the chemist needs.

If a search is being made to see if an idea being considered for a patent is novel, consultation of pertinent books is mandatory.

Various kinds of books are referred to throughout this volume. This chapter focuses on some of the major reference works found in most good chemistry libraries.

First, there is a discussion of significant encyclopedias in chemistry and chemical engineering. This is followed by brief remarks on the German *Handbuch* concept and a review of some important reference works in organic and inorganic chemistry.

12.2. ENCYCLOPEDIAS: INTRODUCTORY REMARKS

The expert or specialist in a field will usually find treatment of this field by encyclopedias too general and elementary. Encycopedias are also relatively quickly dated. By the time the last volume is completed, some of the articles in initial volumes may be obsolete. Nevertheless, most chemists and engineers beginning work on a new project about which they know little or nothing will find that a major chemical encyclopedia is a good starting point. For further details, users can then consult monographs, patents, and articles cited in the encyclopedia, and more recent sources as required. The relative conciseness, which is a feature of encyclopedias, is usually a plus because it gives the chemist a quick overview of a field.

Encyclopedias, however, are by no means all-inclusive. For example, encyclopedia articles seldom cover the recent process technology and economics for important commercial products with the depth of analysis and detail some users require. The special tools that provide this kind of information are described in Chapter 17.

The best-known encyclopedias of interest to chemical engineers and chemists include the following:

Kirk-Othmer Encyclopedia of Chemical Technology (commonly referred to as *Kirk–Othmer,* after the names of the original editors).

Ullmann's Encyclopädie der Technischen Chemie, now called *Ullman's Encyclopedia of Industrial Chemistry.*

Encyclopedia of Polymer Science and Engineering. This was originally edited by Herman F. Mark, the famous polymer chemist.

Each is described in the sections that follow, as are several other works.

A. Specific Encyclopedias

1. Kirk-Othmer. Most chemists and chemical engineers agree that, with the exceptions of *Chemical Abstracts* (*CA*) and the desk handbooks (see Chapters 6 and 7, on *Chemical Abstracts* and Chapter 14, on desk handbooks and other physical properties sources), *Kirk–Othmer* (*KO*) (1) is the most indispensable tool in any chemistry or chemical engineering library.

One of the most important uses of *KO* continues to be its value as a starting point for research workers and others beginning work in a less familiar field of chemistry or related technology. In this regard, *KO* is often unexcelled in its helpfulness.

KO editors place major, but by no means total, emphasis on the more applied aspects—the application of chemistry and chemical engineering to industrially important concepts, products, processes, and uses. Most articles in *KO* provide good coverage of significant aspects of the topics covered. Numerous references to patent, journal, and other literature permit chemists and engineers using *KO* to pursue fields of interest in more depth.

KO is not all-inclusive and is characterized by the uneven treatment that is inevitable in an encyclopedia in any field. Also, if an article on a chemical is written by an employee of a manufacturer of that product, the article may provide strong coverage for properties and uses but weak coverage of manufacturing details. This approach helps users of the product but is otherwise incomplete.

The founders and original editors of *KO* were the late Raymond E. Kirk and Donald F. Othmer, both of whom were on the faculty of what is now known as *Polytechnic University* (Brooklyn, NY). Othmer was also known as an inventor and philanthropist of considerable note.

Publication of the first edition began in 1947, and it was completed in 1960. The second edition began to appear in 1960, and it was completed in 1972. The third edition began in 1978 and was completed in 1984. The fourth edition began publication in 1991, and is to be complete in 1998.

The executive editor of the fourth edition is Dr. Jacqueline I. Kroschwitz, who is also publisher of the *Encyclopedia of Polymer Science and Engineering* and who was formerly professor of chemistry at Kean College. The editor of the fourth edition is Dr. Mary Howe-Grant, who was formerly on the faculty at Fordham University.

The full texts of all volumes of the third edition are currently online through the DIALOG and DataStar systems. In addition, the text of some of the volumes of the fourth edition are online. This means that the online searcher has the enviable choice of being able to select whichever edition is desired. In addition, the volumes of the fourth edition are becoming available on CD-ROM as published. The electronic files can permit more detailed, specific, and sophisiticated searching than is possible with the conventional printed index to *KO*.

It is important to note that older editions of *KO,* like that of other encyclopedias and of other reference works and books, can still be found useful. One reason is that the newer editions may, for one good reason or another, omit information that was in the previous edition. Some of this information may still be valid and useful. In addition, older editions can provide a useful historical perspective and may shed insight that can be useful for patent or other reasons. For example, earlier technology once thought to be obsolete may become unexpectedly pertinent in the future if, for example, there is a shortage of certain materials or if new discoveries suddenly make something feasible or desirable that had not been before.

Table 12.1, which lists the topics covered by *Kirk–Othmer,* will help give the reader some idea of the comprehensiveness and scope of this set.

When completed, *KO* will contain 26 volumes with over 26,000 pages and an index. A one-volume "concise" version of the *Encyclopedia* is planned for 1998, and this should prove extremely popular as did the corresponding volume for the third edition. Two "spinoff" volumes have already appeared: *Chemotherapeutics and Dis-*

TABLE 12.1. Contents of *Kirk-Othmer Encyclopedia of Chemical Technology*

Volume 1
Ablative Materials
Abrasives
Absorption
Acetaldehyde
Acetal Resins
Acetic Acid and Derivatives
Acetone
Acetylene-Derived Chemicals
Acrolein and Derivatives
Acrylamide
Acrylamide Polymers
Acrylic Acid and Derivatives
Acrylic Ester Polymers
Acrylonitrile
Acrylonitrile Polymers
Actinides and Transactinides
Adhesives
Adipic Acid
Adsorption
Adsorption, Gas Separation
Adsorption, Liquid Separation
Advanced Ceramics
Aeration
Aerosols
Air Conditioning
Air Pollution
Air Pollution Control Methods
Alcohol Fuels
Alcohols, Higher Aliphatic
Alcohols, Polyhydric
Aldehydes
Alkali and Chlorine Products
Alkaloids

Volume 2
Alkanolamines
Alkoxides, Metal
Alkyd Resins
Alkylation
Alkylphenols
Allyl Alcohol and Derivatives
Allyl Monomers and Polymers
Aluminum and Aluminum Alloys
Alluminun and Aluminun Compounds
Amides, Fatty Acid
Amine Oxides
Amines
Amines by Reduction
Amino Acids
Aminophenols
Amino Resins and Plastics
Ammonia
Ammonium Compounds
Amyl Alcohols
Analgesics, Antipyretics, and Anti-inflammatory Agents
Analytical Methods
Anesthetics
Anthraquinone
Antiaging Agents
Antiasthmatic Agents
Antibacterial Agents
Antibiotics

Volume 3
Antibiotics
Antifreezes and Deicing Fluids
Antimony and Antimony Alloys
Antiobesity Drugs
Antioxidants
Antiozonants
Antiparasitic Agents
Antistatic Agents
Antiviral Agents
Aquaculture Chemicals
Arsenic and Arsenic Alloys
Arsenic Compounds
Asbestos
Asphalt
Atmospheric Modeling
Automated Instrumentation
Aviation and Other Gas Turbine Fuels
Azine Dyes
Azo Dyes
Bakery Processes and Leavening Agents
Barium
Barium Compounds
Barrier Polymers
Batteries

Volume 4
Bearing Materials
Beer
Benzaldehyde
Benzene

TABLE 12.1. (*Continued*)

Benzoic Acid
Benzyl Alcohol and β-Phenethyl Alcohol
Beryllium and Beryllium Alloys
Beryllium Compounds
Beverage Spirits, Distilled
Biopolymers
Biosensors
Biotechnology
Biphenyl and Terphenyls
Bismuth and Bismuth Alloys
Bismuth Compounds
Bleaching Agents
Blood, Artificial
Blood, Coagulants and Anticoagulants
Boron, Elemental
Boron Compounds
Brake Linings and Clutch Facings
Bromine
Bromide Compounds
BTX Processing
Building Materials
Butadiene
Butyl Alcohols
Butylenes
Butyraldehydes
Cadmium and Cadmium Alloys
Cadmium Compounds
Calcium and Calcium Alloys
Calcium Compounds
Caprolactam
Carbamic Acid
Carbides
Carbohydrates
Carbon

Volume 5
Carbon and Graphite Fibers
Carbonated Beverages
Carbon Dioxide
Carbon Disulfide
Carbonic and Carbonochloridic Esters
Carbon Monoxide
Carbonyls
Carboxylic Acids
Cardiovascular Agents
Castor Oil
Catalysis
Catalysis, Phase-transfer
Catalysts
Cell Culture Technology
Cellulose
Cellulose Esters
Cellulose Ethers
Cement
Ceramics
Ceramics as Electrical Materials
Cerium and Cerium Compounds
Cesium and Cesium Compounds
Chelating Agents
Chemicals in War
Chemicals from Brine
Chemometrics
Chemotherapeutics, Anticancer
Chemurgy
Chloramines and Bromamines
Chlorine Oxygen Acids and Salts
Chlorocarbons and Chlorohydrocarbons

Volume 6
Chlorocarbons and Chlorohydrocarbons
Chlorohydrins
Chlorophenols
Chlorosulfuric Acid
Chocolate and Cocoa
Choline
Chromatography
Chromium and Chromium Alloys
Chromium Compounds
Chromogenic Materials
Cinnamic Acid, Cinnamaldehyde, and Cinnamyl Alcohol
Citric Acid
Clays
Coal
Coal Conversion Processes
Coated Fabrics
Coating Processes
Coatings
Coatings, Marine
Cobalt and Cobalt Alloys
Cobalt Compounds
Coffee
Colloids
Color

continued

TABLE 12.1. *(Continued)*

Colorants for Ceramics
Colorants for Food, Drugs, Cosmetics, and Medical Devices
Colorants for Plastics
Color Photography
Color Photography, Instant
Combustion Science and Technology

Volume 7
Composite Materials
Computer-Aided Design and Manufacturing (CAD/CAM)
Computer-Aided Engineering (CAE)
Computer Technology
Contact Lenses
Contraceptives
Controlled Release Technology
Conveying
Coordination Compounds
Copolymers
Copper
Copper Alloys
Copper Compounds
Copyrights and Trademarks
Corrosion and Corrosion Control
Cosmetics
Cotton
Coumarin
Cryogenics
Crystallization
Cumene
Cyanamides
Cyanides
Cyanine Dyes
Cyanocarbons
Cyanohydrins
Cyanuric and Isocyanuric Acids
Cyclohexanol and Cyclohexanone
Cyclopentadiene and Dicyclopentadiene
Dairy Substitutes
Databases
Defoamers
Dental Materials
Dentrifrices
Desiccants
Design of Experiments
Detergency

Volume 8
Deuterium and Tritium
Dewatering
Dialysis
Diamines and Higher Amines, Aliphatic
Diatomite
Dicarboxylic Acids
Dietary Fiber
Diffusion Separation Methods
Dimensional Analysis
Dimer Acids
Disinfectants and Antiseptics
Dispersants
Distillation
Distillation, Azeotropic and Extractive
Diuretic Agents
Driers and Metallic Soaps
Drug Delivery Systems
Drying
Drying Oils
Dye Carriers
Dyes and Dyes Intermediates
Dyes. Anthraquinone
Dyes. Application and Evaluation
Dyes. Environmental Chemistry
Dyes. Natural
Dyes. Reactive
Dyes. Sensitizing
Economic Evaluation
Eggs
Elastomers, Synthetic

Volume 9
Elastomers, Synthetic
Electrical Connectors
Electrically Conductive Polymers
Electroanalytical Techniques
Electrochemical Processing
Electroless Plating
Electronic Materials
Electronics, Coatings
Electrophotography
Electroplating
Electroseparations
Embedding
Emulsions
Enamels, Porcelain or Vitreous

TABLE 12.1. (*Continued*)

Energy Management
Engineering, Chemical Data Correlation
Engineering Plastics
Environmental Impact
Enzyme Applications
Enzyme Inhibitors
Enzymes in Organic Synthesis
Epinephrine and Norepinephrine
Epoxy Resins
Esterification
Esters, Organic
Ethanol
Ethers
Ethylene
Ethylene Oxide
Evaporation
Exhaust Control, Automotive
Exhaust Control, Industrial
Expectorants, Antitussives
and Related Agents
Expert Systems

Volume 10
Explosives and Propellants
Extraction
Extraterrestrial Materials
Fans and Blowers
Fat Replacers
Fats and Fatty Oils
Feeds and Feed Aditives
Feedstocks
Fermentation
Ferrites
Ferroelectrics
Fertilizers
Ferroelectrics
Fertilizers
Fiber Optics
Fibers
Fillers
Film and Sheeting Materials
Filtration
Fine Arts Examination and Conservation
Fine Chemicals
Flame Retardants
Flame Retardants for Textiles

Volume 11
Flavor Characterization
Flavor and Spices
Flocculating Agents
Flotation
Flow Measurement
Fluidization
Fluid Mechanics
Fluorescent Whitening Agents
Fluorine
Fluorine Compounds, Inorganic
Fluorine Compounds, Organic
Foamed Plastics
Foams
Food Additives
Food Packaging
Food Processing
Food, Nonconventional
Food Toxicants, Naturally Occuring
Forensic Chemistry
Formaldehyde
Formic Acid and Derivatives
Fractionation, Flood
Fracture Mechanics
Friedel–Crafts Reactions
Fruit Juices
Fuel Cells

Volume 12
Fuel Resources
Fuels from Biomass
Fuels from Waste
Fuels, Synthetic
Fungicides, Agricultural
Furnaces, Electric
Furnaces, Fuel-Fired
Fusion Energy
Gallium and Gallium Compounds
Gas, Natural
Gasoline and Other Motor Fuels
Gastrointestinal Agents
Gelatin
Gemstones
Genetic Engineering
Geotextiles
Geothermal Energy
Germanium and Germanium Compounds

continued

TABLE 12.1. (*Continued*)

Glass
Glass–Ceramics
Glasses, Organic–Inorganic Hybrids
Glassy Metals
Glycerol
Glycols
Gold and Gold Compounds
Grignard Reactions
Groundwater Monitoring
Growth Regulators
Gums
Hafnium and Hafnium Compounds
Hair Preparations
Hardness
Hazard Analysis and Risk Assessment
Heat-Exchange Technology
Heat-Resistant Polymers
Heat Stabilizers

Volume 13
Helium Group
Hemicellulose
Herbicides
High Performance Fibers
High Pressure Technology
High Temperature Alloys
Histamine and Histamine Antagonists
Hollow-Fiber Membranes
Holography
Hormones
Hydantoin and Its Derivatives
Hydraulic Fluids
Hydrazine and Its Derivatives
Hydrides
Hydroboration
Hydrocarbon Oxidation
Hydrocarbon Resins
Hydrocarbons
Hydrogen
Hydrogen Chloride
Hydrogen Energy
Hydrogen-Ion Activity
Hydrogen Peroxide
Hydroquinone, Resorcinol, and Catechol
Hydrothermal Processing
Hydroxybenzaldehydes
Hydroxycarboxylic Acids
Hydroxy Dicarboxylic Acids
Hypnotics, Sedatives, Anticonvulsants, and Anxiolytics

Volume 14
Imaging Technology
Imines, Cyclic
Immunoassay
Immunotherapeutic Agents
Incinerators
Inclusion Compounds
Indium and Indium Compounds
Indole
Industrial Antimicrobial Agents
Industrial Hygiene
Information Retrieval
Information Storage Materials
Infrared Technology and Raman Spectroscopy
Initiators
Inks
Inorganic High Polymers
Insect Control Technology
Insulation, Acoustic
Insulation, Electric
Insulation, Thermal
Insulin and Other Antidiabetic Agents
Integrated Circuits
Iodine and Iodine Compounds
Ion Exchange
Ion Implantation
Ionomers
Iron
Iron by Direct Reduction
Iron Compounds
Isocyanates, Organic
Isoprene
Itaconic Acid
Ketenes, Ketene Dimers, and Related Substances
Ketones
Kinetic Measurements
Laboratory Information Management Systems
Laminated Materials, Glass
Laminated Materials, Plastic
Lanthanides

TABLE 12.1. *(Continued)*

Volume 15
Lasers
Latex Technology
Lead
Lead Alloys
Lead Compounds
Leather
Leather-Like Materials
Lecithin
Licensing
Light Generation
Lignin
Lignite and Brown Coal
Lime and Limestone
Liquefied Petroleum Gas
Liquid Crystalline Materials
Liquid Level Measurement
Lithium and Lithium Compounds
Lubrication and Lubricants
Luminescent Materials
Machining Methods, Electrochemical
Magnesium and Magnesium Alloys
Magnesium Compounds
Magnetic Materials
Magnetic Spin Resonance
Magnetohydrodynamics
Maintenance
Maleic Anhydride, Maleic Acid, and Fumaric Acid
Malonic Acid and Derivatives
Manganese and Manganese Alloys
Manganese Compounds
Market and Marketing Research
Mass Spectrometry

Volume 16
Matches
Materials Reliability
Materials Standards and Specifications
Meat Products
Medical Diagnostic Reagents
Medical Imaging Technology
Membrane Technology
Memory-Enhancing Drugs
Mercury
Mercury Compounds
Metal Anodes
Metallic Coatings
Metallurgy
Metal-Matrix Composites
Metal Surface Treatments
Metal Treatments
Methacrylic Acid and Derivatives
Methacrylic Polymers
Methanol
Mica
Microbial Polysaccharides
Microbial Transformations
Microencapsulation
Microscopy
Microwave Technology
Milk and Milk Products
Mineral Nutrients
Minerals Recovery and Processing
Mixing and Blending
Molecular Sieves
Molybdenum and Molybdenum Alloys
Molybdenum Compounds
Naphthalene
Naphthalene Derivatives
Naphthenic Acids
Neuroregulators

Volume 17
Nickel and Nickel Alloys
Nickel Compounds
Niobium and Niobium Compounds
Nitration
Nitric Acid
Nitrides
Nitro Alcohols
Nitrobenzene and Nitrotoluenes
Nitrogen
Nitrogen Fixation
Nitroparaffins
N-Nitrosamines
Nomenclature
Nondestructive Evaluation
Nonlinear Optical Materials
Nonwoven Fabrics
Nuclear Reactors
Nucleic Acids
Nuts
Ocean Raw Materials
Odor Modification
Oils, Essential

continued

TABLE 12.1. (*Continued*)

Oil Shale
Olefin Polymers
Olefins, Higher
Opioids, Endogenous
Oxalic Acid
Oxo Process
Oxygen
Oxygen-Generation Systems
Ozone
Packaging
Paint

Volume 18
Paper
Papermaking Additives
Patents and Trade Secrets
Patents, Literature
Perchloric Acid and Perchlorates
Perfumes
Peroxides and Peroxide Compounds
Pesticides
Petroleum
Pharmaceuticals
Pharmaceuticals, Chiral
Pharmacodynamics
Phenol
Phenolic Resins
Phosgene
Phosphine and Its Derivatives
Phosphoric Acid and Phosphates
Phosphorus
Phosphorus Compounds
Photochemical Technology
Photoconductive Polymers
Photodetectors
Photography
Photovoltaic Cells
Phthalic Acids and Other Benzenecarboxylix Acids
Phthalocyanine Compounds
Pigment Dispersions

Volume 19
Pigments
Pilot Plants
Pipelines
Piping Systems
Plant Layout
Plant Location
Plant Safety
Plasma Technology
Plasticizers
Plastics Processing
Plastics Testing
Platinum-Group Metals
Platinum-Group Metals, Compounds
Plutonium and Plutonium Compounds
Polishes
Polyamides
Polycarbonates
Polyesters, Thermoplastic
Polyesters, Unsaturated
Polyethers
(Polyhydroxy)benzenes
Polyimides
Polymer Blends
Polymers
Polymers Containing Sulfur
Polymers, Environmentally Degradable
Polymethine Dyes
Polymethylbenzenes
Potassium
Potassium Compounds
Powders, Handling

Volume 20
Power Generation
Pressure Measurement
Printing Processes
Process Control
Process Energy Conservation
Product Liability
Propyl Alcohols
Propylene
Propylene Oxide
Prostaglandins
Prosthetic and Biomedical Devices
Protein Engineering
Proteins
Psychopharmacological Agents
Pulp
Pumps
Pyrazoles, Pyrazolines, and Pyrazolones
Pyridine and Pyridine Derivatives
Pyrotechnics
Pyrrole and Pyrrole Derivatives

TABLE 12.1. (*Continued*)

Quality Assurance
Quaternary Ammonium Compounds
Quinolines and Isoquinolines
Quinones
Radiation Curing
Radioactive Tracers
Radioisotopes
Radiopaques
Radiopharmaceuticals
Radioprotective Agents
Reactor Technology
Recreational Surfaces
Recycling

Volume 21
Recycling
Refractories
Refractory Coatings
Refractory Fibers
Refrigeration
Regulatory Agencies
Reinforced Plastics
Release Agents
Renewable Energy Resources
Repellents
Research/Technology Management
Resins, Natural
Reverse Osmosis
Rhenium and Rhenium Compounds
Rheological Measurements
Roofing Materials
Rubber Chemicals
Rubber Compounding
Rubber, Natural
Rubidium and Rubidium Compounds
Salicylic Acid and Related Compounds
Sampling
Sealants
Sedimentation
Selenium and Selenium Compounds
Semiconductors
Sensors
Separation
Separations Process Synthesis
Shape-Memory Alloys
Silica
Silicon and Silicon Alloys

The chapters below are to be published in suceeding volumes.

Volume 22
Soap
Sodium and Sodium Alloys
Sodium Compounds
Soil Chemistry of Pesticides
Soil Stabilization
Solar Energy
Solders and Brazing Alloys
Sol-Gel Technology
Solvent Recovery
Solvents, Industrial
Sorbic Acid
Soybeans and Other Oilseeds
Space Processing
Speciality Polymers
Spectroscopy
Sprays
Stains, Industrial
Starch
Steam
Steel
Sterilization Techniques
Steroids
Stilbene Dyes
Stimulants
Strontium and Strontium Compounds
Styrene
Styrene Plastics
Succinic Acid and Succinic Anhydride
Sugar
Sulfamic Acid and Sulfamates
Sulfolanes and Sulfones
Sulfonation and Sulfation
Sulfonic Acids
Sulfoxides
Sulfur

Volume 23
Sulfur Compounds
Sulfur Dyes
Sulfuric Acid and Sulfur Trioxide
Sulfuric and Sulfurous Esters
Sulfurization and Sulfurchlorination
Sulfur Removal and Recovery

continued

TABLE 12.1. (*Continued*)

Superconducting Materials
Supercritical Fluids
Surface and Interface Analysis
Surfactants
Sutures
Sweeteners
Syrups
Talc
Tall Oil
Tanks and Pressure Vessels
Tantalum and Tantalum Compounds
Tar and Pitch
Tar Sands
Tea
Technical Service
Tellurium and Tellurium Compounds
Temperature Measurement
Terpenoids
Textiles
Thallium and Thallium Compounds
Thermal, Gravimetric, and Volumetric Analysis
Thermal Pollution
Thermodynamic Properties
Thermoelectric Energy Conversion
Thiazole Dyes
Thin Films
Thioglycolic Acid
Thiols
Thiophene
Thiosulfates
Thorium and Thorium Compounds
Thyroid and Antithyroid Preparations
Tin and Tin Alloys

Volume 24
Tin Compounds
Tire Cords
Titanium and Titanium Alloys
Titanium Compounds
Toluene
Tool Materials
Toxicology
Trace and Residue Analysis
Transportation
Triphenylmethane and Related Dyes
Tungsten and Tungsten Alloys
Tungsten Compounds
Ultrafiltration
Ultrapure Materials
Ultrasonics
Units and Conversion factors
Uranium and Uranium Compounds
Urea
Urethane Polymers
Uric Acid
UV Stabilizers
Vaccine Technology
Vacuum Technology
Vanadium and Vanadium Alloys
Vanadium Compounds
Vanillin
Vegetable Oils
Veterinary Drugs
Vinegar
Vinylidene Chloride and Poly(vinylidene chloride)
Vinyl Polymers
Viscometry

Volume 25
Vitamins
Waste Reduction
Wastes, Industrial
Water
Waterproofing and Water/Oil Repellency
Water-Soluble Polymers
Waxes
Weighing and Proportioning
Welding
Wheat and Other Cereal Grains
Wine
Wood
Wood-Based Composites and Laminates
Wool
Xanthates
Xanthene Dyes
X-ray Technology
Xylenes and Ethylbenzene
Xylylene Polymers
Yeasts
Yttrium and Yttrium Compounds
Zinc and Zinc Alloys
Zinc Compounds
Zirconium and Zirconium Compounds
Zone Refining

ease Control and *Fluorine Chemistry.* In addition, a two-volume "enhanced" set is planned on separation technology.

2. Ullmann's Encyclopedia of Industrial Chemistry. The fifth edition of this important work (2, 3) is entirely in English, a distinct advantage over earlier German editions.

This edition, which began publication in 1985 and was completed in 1996, consists of 36 volumes and one cumulative index volume with more than 80,000 carefully selected keywords and a German–English dictionary of technical terms. Part A (28 volumes) is broadly similar in scope to *KO.* Part B (8 volumes) contains detailed articles regarding general aspects such as chemical engineering topics, industrial safety, and analytical methods. There are a total of 27,000 printed pages. In addition to the printed volumes, a CD-ROM version of the fifth edition was issued in 1997. It is believed that the current edition will be the last printed version. The sixth edition, which is to begin publication in 1998 and is to be complete in about 8–10 years, is to consist of annual updates that are to appear in CD form containing the entire database. The managing editor is Thomas Kellersohn.

There is full and accurate treatment of important topics of interest. The literature and patent citations accompanying each article are of special value, with appropriate emphasis given to pertinent sources, especially patents.

In most searches for chemical information, both *Kirk* and *Ullmann's* should be consulted since the two are complementary. Each includes some information that the other does not in many cases. In addition, the two encyclopedias are written from different perspectives; thus, *KO* reflects a largely North American view, whereas *Ullmann's* reflects a European and Japanese view. This difference is exemplified in the references, especially the citation to patents. Furthermore, as mentioned above, in its Part B, *Ullmann's* contains volumes on more general topics, for instance, four volumes dealing with the fundamentals of chemical engineering, unit operations, and plant design and construction.

3. Encyclopedia of Polymer Science and Engineering. This well-known compilation (5) is intended for the many chemists and engineers concerned with polymer chemistry and technology.

In addition to the full printed encyclopedia of 19 volumes, the second edition is available online through DIALOG, and a one-volume "concise" version was printed. Jacqueline Kroschwitz is the editor.

Publication of a third edition is tentatively planned to begin by about 1999.

4. McKetta. Volumes of this comprehensive encyclopedia, (4) targeted at engineers, were first published in 1977 and are still being issued. As of early 1997, some 58 volumes had appeared. The full title is *Encyclopedia of Chemical Processing and Design,* and the editor is John J. McKetta. The completion of this tome is in sight since a total of 62 volumes are planned. The purpose of this publication is to detail current developments in the field of chemical technology and related industries.

As the title of this work indicates, emphasis is placed on chemical processing, equipment, operations, and design aspects.

5. Polymeric Materials Encyclopedia. This important encyclopedia (6), of interest to polymer chemists and engineers, was published in 1996 in both print and CD-ROM formats. The editor is Joseph C. Salamone, formerly at the University of Massachusetts. This 12-volume work, published by CRC Press, Boca Raton, FL, in 1996, focuses on the chemistry and synthesis of polymeric materials, with information on properties and applications as well. The organization is primarily by type of polymer. Contributors are from organizations around the world.

6. Materials Science and Technology. R. W. Kahn, P. Haasen, and E. J. Kramer are the editors of this comprehensive treatise. The 22-volume series began publication in 1993 (VCH Publishing Group).

B. Which Encyclopedia to Use

In their use of major chemical encyclopedias, chemists and engineers should consider *all* those mentioned as appropriate, because each can provide a different perspective, may have different or additional information, and because topics not adequately covered in one encyclopedia may be more fully and adequately covered in another. The staggered and overlapping time frames within which the encyclopedias are issued is another reason for looking at all those listed. However, *KO* is the most important chemical encyclopedia at this time and should be consulted first.

C. Other General Reference Sources: Choices for the Desktop

To proceed beyond encyclopedias to much less comprehensive works, and especially those that the individual chemist may wish to keep handy for desktop use, one thinks first of all about the desk handbooks such as described in Chapter 15. Thus, especially worthwhile first choices include the *Merck Index* (particularly recommended because of its broad content and low cost), the classic essential *Handbook of Chemistry and Physics,* and the complete product catalogs of the Aldrich Chemical Company or of Lancaster Synthesis, Inc., which contain data about thousands of chemicals. One of the general chemical buyer's guides, such as mentioned in Chapter 16, is also good to keep close at hand. Also very handy for desktop use is a good chemical dictionary; one of the best known examples is the *Condensed Chemical Dictionary* (7). In addition, the field of plastics technology is so specialized, with a great deal of "lingo," that a specialized dictionary, directory, or other such work often need to be consulted. Examples include the dictionary edited by Rosato (8) and the *Modern Plastics Encyclopedia,* which is published annually by *Modern Plastics* magazine, New York, NY. Moreover, one or two of the books dealing with the specific chemicals, processes, or chemical applications of interest to the chemist or chemical engineer may be good desktop choices. A guide to sources of chemical information, such as this volume, is also recommended. A forthcoming publication, *The Laboratory Companion* (9a), which provides some basics on the materials, equipment, and techniques required in a laboratory and particularly emphasizes how to work with glassware and associated equipment, appears to be another good choice.

For chemists interested in the pharmaceutical sciences, another desktop choice in addition to the *Merck Index* is *Remington: The Science and Practice of Pharmacy* (9b), edited by Alfonso R. Gennaro, Philadelphia College of Pharmacy and Science. This text–reference volume includes an extensive listing of pharmaceuticals, their preparations, action, dosages, and the like. There is also extensive textual material for students in pharmacy.

See also Chapter 18 for a discussion of two process economics subscription services. These are not desktop choices, but they are valuable reference sources to keep in mind.

12.3. THE *HANDBUCH* CONCEPT

The *Handbuch* is a valuable type of publication, significantly different from most handbooks published in the United States. It is a multivolume work, far more extensive in scope, and provides more in-depth coverage, than the typical U.S. handbook.

Handbuch coverage usually goes back to the "beginnings" of chemistry—long before *CA* began in 1907—and brings that coverage up to a reasonably current time, as is discussed later in this chapter. Ongoing updating is another feature. For the chemist, this can mean that any needed search of later literature can begin where *Handbuch* coverage ceases.

Although consultation of the *Handbuch* can in many cases obviate the need for further literature searching (assuming the *Handbuch* has the desired information), it would be a mistake for the chemist to conclude that *Handbuch* coverage is 100% complete and without some errors or omissions.

The two most important examples of the German *Handbuch* in chemistry are *Beilstein* and *Gmelin.* Both are edited at the Carl Bosch Haus in Frankfurt-am-Main, Germany.

The *Handbuch* is an old tradition in German chemistry. In 1817, Leopold Gmelin published his first *Handbuch,* which was a quick success. At the start of the fifth edition (1852), the field of organic chemistry was split off and has been carried on separately as *Beilstein.*

Gmelin is now on its eighth edition, and *Beilstein* is in its fourth. It is, unfortunately, all too easy (and incorrect) to regard *Handbuch* volumes as (a) obsolete; (b) sometimes incomprehensible, because large parts are written in languages other than English; or (c) otherwise difficult to use because of large size (many volumes) and "complex" organization. This view, although prevalent among many chemists, can be counterproductive and should be carefully reconsidered, especially because of recent developments in *Handbuch* publication.

A. The Question of Obsolescence

It is true that some volumes in major *Handbuch* series go back many years. But chemists can ignore or downplay older literature only at the risk of unnecessary repetition of older work that may still be valid. Moreover, some newer *Handbuch* volumes bring literature coverage up to much more recent years.

B. On Potential Language Problems

Although most of the older volumes are in German, and will stay that way, this is not an insurmountable hurdle, as indicated later in this chapter in the discussion of *Beilstein*. In addition, all new volumes are now being published in English for at least two of the major *Handbuch* series, that is, *Gmelin* and *Beilstein*.

C. On Potential Complexity of Use

Handbuch volumes occupy many linear feet of shelf space, and this alone may intimidate some at first. There are, however, helpful independent-study-type aids available without charge from *Handbuch* publishers. Trained librarians or information professionals can assist by introducing the use of the volumes to those less familiar. In any event, it is often helpful to motivate oneself to take any extra effort required to use a *Handbuch* on the basis that a *Handbuch* has the potential to save much needless duplication of work. The complexity is often more imaginary than real. Even more to the point, the major handbooks are now widely available in easily searchable electronic format.

Prices of *Handbuch* sets, like that of almost all scientific literature, have been escalating sharply. This difficult situation is more severe with *Handbuch* volumes because their prices have always been relatively high. Some organizations have been forced by budget constraints to curtail or suspend purchase of one or more of the *Handbuch* sets. In such cases, volumes already on the shelves should, of course, be retained because much older data continues to be valid as previously noted. It is a well-accepted principle of chemical literature use that older volumes and editions frequently have data omitted in later volumes or editions.

12.4. ORGANIC CHEMISTRY: SOME IMPORTANT REFERENCE WORKS

A. Beilstein Handbook of Organic Chemistry

Beilstein (10) is the basic reference work on organic compounds. It takes its name from Friedrich Konrad Beilstein, who published the first edition in 1881–1882. Many readers of this book will already have used, seen, or heard about this essential tool for organic chemists. It is available (or should be) in every first-class chemistry library in either electronic or printed form, or both. Furthermore, *Beilstein* makes available a 35-page brochure, *How to Use Beilstein* (11) free of charge, and there is a forthcoming book (Stephen R. Heller, Ed., *The Beilstein Database and Search System—System for the 21st Century,* American Chemical Society, Washington, DC, 1997) that describes *Beilstein* and how to use it in considerable detail. Accordingly, it would be redundant to present an in-depth description of *Beilstein* here and more appropriate to concentrate on key features. Sample pages from *Beilstein* are shown in Figure 12.1.

Beilstein celebrated its 100th anniversary in May 1981. Ever since its origins, its value to chemists throughout the world has continued to increase.

83–85° [from EtOH]; $[\alpha]_D$: –36.5° [EtOH; c = 1]; ^{1}H-NMR; ^{13}C-NMR; IR (*Wi. et al.*).

(3*R*)-6*t*-Methyl-5-oxo-2*t*-((1*R*,2*R*)-1,2,3-trihydroxy-propyl)-morpholine-3*r*-carbaldehyde, Muramic acid lactam $C_9H_{15}NO_6$, formula XI and cycl. taut.

Prep. From O^1-benzyl-α-(or β)-muramic acid lactam [H_2; Pd-C] (*P. Sinay et al.*, Carbohydr. Res. **21** [1972] 339–346). – $[\alpha]_D^{20}$: +35° [MeOH; c = 0.32]; IR.

Hydroxy-Oxo-Compounds $C_nH_{2n-5}NO_6$ and $C_nH_{2n-7}NO_6$

6,8*c*-Dimethoxy-5,7-bis-methylsulfanyl-(7*rH*)-4-thia-1-aza-bicyclo[5.2.0]non-5-en-9-one $C_{11}H_{17}NO_3S_3$, formula XII + mirror image.

Configuration: see *J.L. Fahey et al.*, J. Chem. Soc. Perkin Trans. 1 **1977** 1117–1122.

Prep. From 6-methoxy-5,7-bis-methylsulfanyl-2,3-dihydro-[1,4]thiazepine and methoxyacetyl chloride [Et_3N; CH_2Cl_2] (*A.L. Bose et al.*, J. Heterocycl. Chem. **10** [1973] 791–794). – mp: 54–56° [from Et_2O + hexane]; ^{1}H-NMR; IR (*Bose et al.*). ^{1}H-NMR [lanthanide-induced shifts] (*Fa. et al.*).

6-Methoxy-5,7-bis-methylsulfanyl-8*c*-phenoxy-(7*rH*)-4-thia-1-aza-bicyclo[5.2.0]non-5-en-9-one $C_{16}H_{19}NO_3S_3$. Configuration: inferred after reference to *Fa. et al.* – mp: 100–101°; ^{1}H-NMR; IR (*Bose et al.*).

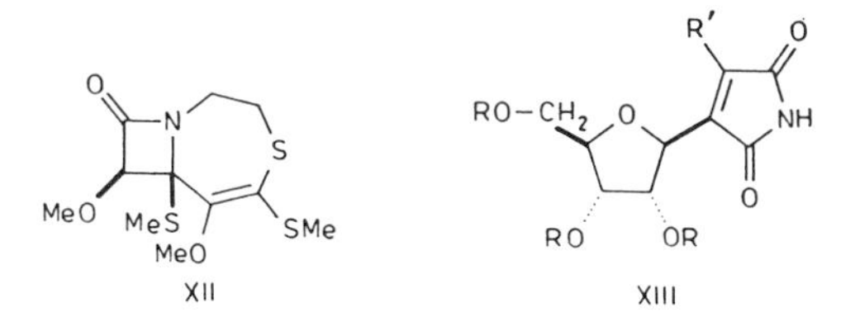

3-β-D-Ribofuranosyl-pyrrole-2,5-dione, Showdomycin $C_9H_{11}NO_6$, formula XIII (R = R' = H).

Isolation from cultures of Streptomyces showdoensis (*H. Nishimura et al.*, J. Antibiot. Ser. A **17** [1964] 148–155).

Synthesis (*L. Kalvoda et al.*, Tetrahedron Lett. **1970** 2297–2300; *G. Trummlitz, J.G. Moffatt*, J. Org. Chem. **38** [1973] 1841–1845; *L. Kalvoda*, J. Carbohydr. Nucleosides Nucleotides **3** [1976] 47–55; *R. Noyori et al.*, J. Am. Chem. Soc. **100** [1978] 2561–2563; *T. Sato et al.*, Tetrahedron Lett. **1978** 1829–1832; *J.G. Buchanan et al.*, J. Chem. Soc. Perkin Trans. 1 **1979** 225–227). – Preparation of ^{14}C-labelled showdomycin (*H. Minato et al.*, J. Labelled Compd. Radiopharm. **14** [1978] 455–460). – Conformation [^{1}H-NMR] (*M.P. Schweizer et al.*, J. Am. Chem. Soc. **95** [1973] 3770–3778). Interatomic distances and angles [X-ray diffraction]; mp: 160–161° [from acetone]; density of the crystals: 1.527 (*Y. Tsukuda, H. Koyama*, J. Chem. Soc. B **1970** 1709–1712). mp: 154.5–156° [from acetone + benzene] (*Tr., Mo.*). mp: 153–154° [from acetone + benzene]; $[\alpha]_D^{22.5}$: +49.9° [H_2O; c = 1]; IR and UV/VIS spectra (*Ni. et al.*). mp: 152–154° [from acetone + Et_2O]; $[\alpha]_D^{20}$: +47.1° [H_2O; c = 0.425]; MS (*Bu. et al.*). ^{1}H-NMR; IR and UV/VIS spectra (*Y. Nakagawa et al.*, Tetrahedron Lett. **1967** 4105–4109). ^{13}C-NMR (*M.-T. Chenon et al.*, J. Heterocycl. Chem. **10** [1973] 427–429). MS (*L.B. Townsend, R.K. Robins*, J. Heterocycl. Chem. **6** [1969] 459–464). Distribution between H_2O [pH 7] and $CHCl_3$ (*T. Yamashita et al.*, Experientia **35** [1979] 1054–1056). – Reaction with glutathione [H_2O (pH 7); 25°]: kinetics (*T. Miyadera, E.M. Kosower*, J. Med. Chem. **15** [1972] 534–537).

3-D-Ribofuranosylidene-pyrrolidine-2,5-dione, Isoshowdomycin $C_9H_{11}NO_6$, formula XIV.

Prep. From showdomycin above, by microbiological transformation (*M. Ozaki et al.*, Agric. Biol. Chem. **36** [1972] 451–456). – mp: 215–216° [from H_2O]; $[\alpha]_D^{22.5}$: –34.0° [H_2O; c = 1]; IR and UV/VIS spectra.

Figure 12.1. Representative pages from *Beilstein Handbook of Organic Chemistry,* Volume 17/1, Fifth Supplementary Series. (Reprinted with permission of Beilstein Institut fur Literatur d. Org. Chemie.)

3-(Tri-*O*-acetyl-D-ribofuranosylidene)-pyrrolidine-2,5-dione, Tri-*O*-acetyl-isoshowdomycin $C_{15}H_{17}NO_9$. ^{1}H-NMR spectrum (*Oz. et al.*).

3-(Tri-*O*-benzyl-*β*-D-ribofuranosyl)-pyrrole-2,5-dione, Tri-*O*-benzyl-showdomycin $C_{30}H_{29}NO_6$, formula XIII (R = CH_2-C_6H_5, R′ = H).

Synthesis starting from tri-*O*-benzyl-D-2,5-anhydro-allose (*G. Trummlitz, J.G. Moffatt*, J. Org. Chem. **38** [1973] 1841–1845). – *Prep.* From (tri-*O*-benzyl-*β*-D-ribofuranosyl)-maleic acid anhydride, by treatment with (i) NH_3 [Et_2O], and (ii) AcCl [DMF] (*J.G. Buchanan et al.*, J. Chem. Soc. Perkin Trans. 1 **1979** 225–227). – Preparation of 3-(tri-*O*-benzyl-*β*-D-ribofuranosyl)-[5-^{14}C]pyrrole-2,5-dione (*H. Minato et al.*, J. Labelled Compd. Radiopharm. **14** [1978] 455–460). – mp: 64–65° [from Et_2O+hexane]; $[\alpha]_D^{23}$: +96° [$CHCl_3$; c = 0.55]; ORD; ^{1}H-NMR; IR; MS (*Tr., Mo.*). mp: 63–64° [from Et_2O+PE]; $[\alpha]_D^{20}$: +93.4° [$CHCl_3$; c = 0.91]; ^{1}H-NMR; IR; MS (*Bu. et al.*).

3-(Tri-*O*-acetyl-*β*-D-ribofuranosyl)-pyrrole-2,5-dione, Tri-*O*-acetyl-showdomycin $C_{15}H_{17}NO_9$, formula XIII (R = CO-CH_3, R′ = H).

Prep. From showdomycin (see above), by treatment with Ac_2O/H_3PO_4 (*Y. Nakagawa et al.*, Tetrahedron Lett. **1967** 4105–4109). – mp: 115–116°; ^{1}H-NMR.

3-(Tri-*O*-benzoyl-*β*-D-ribofuranosyl)-pyrrole-2,5-dione, Tri-*O*-benzoyl-showdomycin $C_{30}H_{23}NO_9$, formula XIII (R = CO-C_6H_5, R′ = H).

Prep. From (*E*)-tri-*O*-benzoyl-3-cyano-D-*allo*-4,7-anhydro-2,3-dideoxy-oct-2-enonic acid *tert*-butyl ester [H_2SO_4; AcOH; Ac_2O] (*L. Kalvoda*, Collect. Czech. Chem. Commun. **43** [1978] 1431–1437). – $[\alpha]_D^{25}$: −28.5° [$CHCl_3$; c = 0.5]; IR.

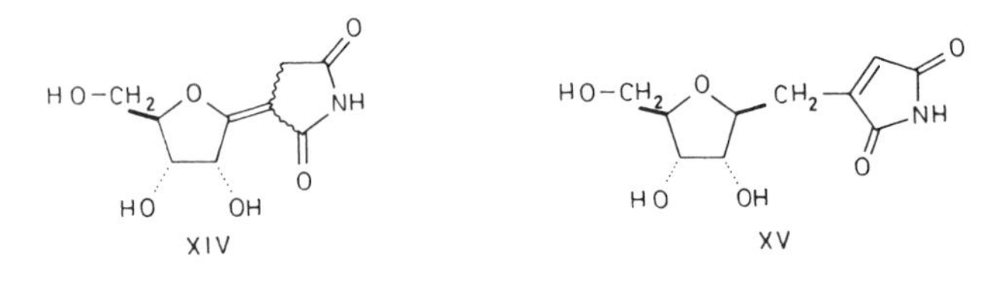

3-(D-2,5-Anhydro-allitol-1-yl)-pyrrole-2,5-dione, Homoshowdomycin $C_{10}H_{13}NO_6$, formula XV.

Prep. From 3-(O^3,O^4-isopropylidene-O^6-trityl-D-2,5-anhydro-allitol-1-yl)-pyrrole-2,5-dione [aq. CF_3CO_2H] (*T. Sato, R. Noyori*, Heterocycles **13** [1979] 141–144). – mp: 150–154°; $[\alpha]_D^{20}$: −24° [MeOH; c = 0.2]; ^{1}H-NMR; ^{13}C-NMR; UV/VIS.

3-Methyl-4-*β*-D-ribofuranosyl-pyrrole-2,5-dione, 4-Methyl-showdomycin $C_{10}H_{13}NO_6$, formula XIII (R = H, R′ = CH_3).

Prep. From 3-methyl-4-(tri-*O*-benzyl-*β*-D-ribofuranosyl)-pyrrole-2,5-dione [BCl_3; CH_2Cl_2] (*G. Trummlitz et al.*, J. Org. Chem. **40** [1975] 3352–3356). – mp: 165–166° [from acetone+benzene]; $[\alpha]_D^{23}$: −39.5° [MeOH; c = 1]; ORD; ^{1}H-NMR; UV/VIS; MS.

3-Methyl-4-(tri-*O*-benzyl-*β*-D-ribofuranosyl)-pyrrole-2,5-dione, Tri-*O*-benzyl-4-methyl-showdomycin $C_{31}H_{31}NO_6$. *Prep.* From tri-*O*-benzyl-D-*allo*-3,6-anhydro-[2]heptulosonic acid methyl ester and (1-carbamoyl-ethyl)-dimethyl-phenyl-phosphonium bromide [NaH/DMSO; benzene] (*Tr. et al.*). – mp: 86–87° [from Et_2O+hexane]; $[\alpha]_D^{23}$: −6.0° [$CHCl_3$; c = 0.14]; ^{1}H-NMR; UV/VIS.

Hydroxy-Oxo-Compounds $C_nH_{2n-13}NO_6$

2-(3,4,5-Trimethoxy-phenyl)-[1,3]thiazine-4,6-dione $C_{13}H_{13}NO_5S$, formula I.

Prep. From 3,4,5-trimethoxy-thiobenzamide and C_3O_2 [THF/Et_2O] (*E. Ziegler, R. Wolf*, Monatsh. Chem. **95** [1964] 1061–1067). – mp: 168° [from PhCl].

Figure 12.1. *(Continued)*

TABLE 12.2. Editors of the *Beilstein Handbook*

Years	Editor	Years	Editor
1881–1896	F. K. Beilstein	1933–1961	F. Richter
1896–1918	P. Jacobson	1961–1978	H. G. Boit
1918–1923	P. Jacobson and B. Prager	1978–1996	R. Luckenbach
1923–1933	B. Prager and F. Richter	1996–	W. Brich and R. Michaelis

Beilstein is a major undertaking, with over 515 individual volumes printed to date, and with several types of electronic files available. A staff of over 100 chemists and other scientists are engaged in evaluation and compilation of data for *Beilstein.* The executive editors are Drs. Werner Brich and Ralf Michaelis, who assumed these functions in 1996. Previous *Beilstein* editors are listed in Table 12.2.

The Beilstein Institute is located in the Carl Bosch Haus in Frankfurt-am-Main, several hundred yards from the Frankfurt fairgrounds and exhibition halls and near the University of Frankfurt. Also located in the same large and modern building is the Gmelin Institute (see Section 12.5.A).

One reason why *Beilstein* is so highly regarded by chemists is that it covers organic chemistry from 1771 up to near the present; the most recent Supplementary Series covers 1960–1979, and the electronic files bring the coverage up to more recent years, with about a 6-month lag period. The printed volumes include in one place (actually several adjacent volumes including supplements) much of the most pertinent data (primarily physical and chemical properties) about a very large number of organic compounds. *Beilstein* editors critically review the data to be included.

For the individual laboratory chemist to attempt to locate all the data included in *Beilstein* about specific compounds of interest is often prohibitively time-consuming. All the journals and other sources that would need to be reviewed to accomplish this task are seldom readily available.

Industrial chemists will frequently need to augment their use of *Beilstein* with additional examination of the literature. *Beilstein* is not a complete resource for commercial applications or use data, and patent documents published after 1979 are not included. It is, however, an extremely important source to *all* chemists, industrial and other. Table 12.3 is a convenient guide to the several series in which *Beilstein* appears.

As noted earlier, *Beilstein's* coverage goes back to the beginnings of organic chemistry. Ongoing supplementing brings the data up to more recent years, as shown in the compilation of data about the *Beilstein* series in Table 12.3.

For each compound included, selected information is presented on

Constitution and configuration
Natural occurrence and isolation from natural products
Preparation, formation, and purification
Structural and energy parameters of the molecule
Physical properties

Chemical properties
Characterization and analysis
Salts and addition compounds

Input is based on a critical review of the published literature, including journals, and, selectively, important theses and conference reports (12). Each piece of data is accompanied by reference to the publication that is the source of the data.

Beilstein editors present material in a highly condensed, telegraphic form and make extensive use of abbreviations as shown in Table 12.4. A list of abbreviations and symbols is provided in the front of each volume.

The printed *Handbook* covers about 3.5 million compounds.

As noted in Table 12.3, *Beilstein* is currently publishing its *Supplementary Series V* (*E, V*), which covers the literature from 1960 to 1979. As of this writing, new volumes are being published at the extremely rapid rate of one volume approximately every 4 weeks.

On completion of the printed volumes on heterocyclic compounds (Vols. 17–27), due for April 1998, the volumes on acyclic and isocyclic compounds will begin to be published. These are to cover literature from 1960 to date, and there will be a CD-ROM index for fast searching. According to plans, the new printed *Handbook* volumes will be current, that is, within approximately one year of the most recent literature. The entire Series V is scheduled to be complete by the year 2002.

TABLE 12.3. *Beilstein* Series Coverage

Series	Abbreviations	Literature Year Covered	Spine Label Color	Status
Basic Series	H	Up to 1909	Green	Complete
Supplementary Series I	EI	1910–1919	Dark red (on brown cover)	Complete
Supplementary Series II	EII	1920–1929	White	Complete
Supplementary Series III	EIII	1930–1949	Blue	Complete
Supplementary Series III/IV[a]	EIII/IV	1930–1959	Blue/black	Complete
Supplementary Series IV	EIV	1950–1959	Black	Complete
Supplementary Series V[b] (in English)	EV	1960–1979	Red (on blue cover)	Began in 1984

[a]Volumes 17–27 of Supplementary Series III and IV, covering heterocyclic compounds, are combined in a joint issue.

[b]The first volumes published in this series relate to heterocyclic compounds (Vol. 17–27), based on a survey of user requirements. Other volumes follow thereafter.

Note: Some of the electronic products of *Beilstein* bring coverage up toward more recent vintage.

TABLE 12.4. Some Abbreviations Frequently Used in *Beilstein*

Abbreviation	Meaning	Abbreviation	Meaning
A.	Ethanol	opt.-inakt.	Optically inactive
Acn.	Acetone	PAe.	Petroleum ether
Ae.	Diethyl ether	Py	Pyridine
alkal.	Alkaline	RRI	*Ring Index* (2nd ed., 1960)
Anm.	Annotation, footnote		
B.	Formation, synthesis	RIS	*Ring Index* Supplement
Bd.	Volume	S.	Page
Bzl.	Benzene	s.	See
Bzn.	Petroleum ether	s	Second
bzw.	Or, respectively	s.a.	See also
Diss.	Dissertation	s.o.	See above
E	Supplemental volume in *Beilstein* series	sog.	So-called
		Spl.	Supplement
E	Freezing point	. . . stdg.	For . . . hours
E.	Ethyl acetate	s.u.	See below
Eg.	Acetic acid	Syst. Nr.	System Number in *Beilstein*
engl. Ausgabe	English edition		
		Tl.	Part
Gew-%	Percentage by weight	U.S.P.	U.S. Patent
H	Main volume in *Beilstein* series	unkorr.	Uncorrected
		unverd.	Undiluted
h	Hour(s)	verd.	Diluted
konz.	Concentrated	vgl.	Compare
korr.	Corrected	W.	Water
Kp	Boiling point	wss.	Aqueous
Me.	Methanol	z.B.	For example
min	Minute	Zers.	Decomposition

Note: Applies to Series H to E, IV only. Series E, V is totally in English.

Heterocyclic Volumes 17–26 and 27/1-30 are now available; Volume 27, to be completed in 1998, deals with heterocycles whose parent registry compounds contain both oxygen (or sulfur, selenium or tellurium) and nitrogen atoms. This volume, when completed, is to contain data on over 300,000 compounds and thus will be the largest volume ever published by *Beilstein.*

Series V, which began in 1984, is totally in English, so that any translation problems for English-reading chemists that may have existed with earlier Series have been completely eliminated. Table 12.4 shows the translations of German abbreviations used in the earlier volumes.

All compounds in *Beilstein* are given a *Beilstein* Registry Number (BRN) and some have a *Beilstein* Preferred Registry Number (BPRN). The latter is required because a single compound may have more than one CAS Registry Number due to the matching process; the BPR denotes the preferred CAS Number. Only approximately 65% of all compounds in *Beilstein* currently have CAS Registry Numbers. Additional Numbers are added on a current basis if these numbers appear in the original literature.

1. Electronic Versions of Beilstein. Beginning in 1988, *Beilstein* began to add a number of major new electronic delivery formats. In addition to offering very convenient, desktop access, these provide significant increases in searching power.

a. Online File. The STN International system was the first to offer an online version of *Beilstein* in 1988. This was followed by DIALOG in 1989.

There are three different data sources for the file:

1. *The Beilstein Handbook* from the Basic Series to Supplement IV covering the literature from 1779 to 1959. For more than 1.1 million compounds, the complete *Handbook* information is available.
2. Primary literature data from 1960 to 1979. This source contains several million additional compounds, bringing the total to about 6.5 million compounds. Specific data are available for melting point, boiling point, density, refractive index, optical rotatory power, isolation from natural products, and chemical derivatives. All other physical and chemical properties are available as keywords together with corresponding references to the original literature. This data source provides the basis for the production of Supplement V. This part of the file is being continuously updated to provide data for all data fields covered.
3. Primary literature from 1980 forward from approximately 160 key journals. In contrast to item 2 above, detailed information for all physical and chemical properties are abstracted from the literature. All data fields contain references as well as data. The category "Physiological Behavior and Application" is also included. Annual growth rate is about 300,000 new compounds, with about one million updates for existing compounds per year.

The following broad categories are present for chemical data:

Reaction
Nongraphical reaction
Isolation from natural product
Derivative
Constitutional data

The following broad properties are present for physical properties:

Structure and energy parameters
Physical state
Mechanical and physical properties
Transport phenomena

Thermochemical data
Optical properties
Spectra
Magnetic properties
Electrical properties
Electrochemical behavior

The following broad categories are present for multicomponent compounds:

Solution behavior
Liquid–liquid systems
Liquid–solid systems
Liquid–vapor systems
Mechanical and physical properties
Transport phenomena
Energy data
Boundary surface phenomena
Adsorption
Association

The following broad categories are present for physiological data:

Use
Toxicity
Biological function
Ecological data

(*Note:* The current data structure for *Beilstein* is to be updated and reorganized toward the end of 1997. This will include additional fields for the classification of new data, particularly in the area of physiological behavior.)

Online searching of *Beilstein* is highly convenient for those many chemists who do not have ready fingertip access to the printed volumes. As compared to the *CrossFire* and *NetFire* products described below, there are no up-front fees. In addition, there are a number of powerful special features that permit types of search that are not easily possible in the printed books. For example, one type of search that is quite feasible in the STN online file is structure searching. In addition, looking for compounds by their CAS Registry Numbers is facilitated in the online version, although as noted, only approximately 65% have CAS Registry Numbers. If the chemist wishes to search by physical property, such as an exact property, a range, or merely the presence of a property regardless of value, that can be done as well.

With regard to any of the properties included, the searcher can use any one of a number of operators. These include

=	equal to
<	less than
>	greater than
<= or =<	less than or equal to
>= or =>	greater than or equal to
–	range (e.g. 10–12)

The range and equal operators may be the most useful in many cases. Otherwise, there may be so many entries that the computer may stop processing.

As compared to *CrossFire* (see Section 12.4.A.1.b), *Beilstein* Information Systems staff point out that the online file has deficiencies. These include the inability to conduct reactions searching, use hyperlinks, or employ context switches. Also they believe that the online interface is not as user-friendly as *Beilstein Commander,* the *CrossFire* user interface.

Plans call for release in 1997 of a new client to be called *CrossFire for Excel.* This will allow for downloading of data to a local hard drive, as well as the sorting capabilities of the *Excel* software.

Browsing can be done online, but some users still prefer the printed books for this.

Some of the information taken from the literature for 1960–1979 has not been fully evaluated, but rather is still being checked. This can be identified in the online file as follows. If a preparation includes experimental details plus a reference and the words *Handbook Data,* evaluation is complete; if only a reference is given, evaluation is not complete. Similarly, as to physical properties, if, for example, in melting-point data, if *solvent* is given and the words *Handbook Data* appear, then the data have been evaluated. For the time period 1960–1979, only five physical properties are given: melting and boiling point, density, optical rotary power, and refractive index.

Beilstein staff do not officially recommend any book on the use of the online files because none are considered up-to-date. Accordingly, pending publication of the book by Stephen R. Heller (see Chapter 10, page 266), the best available written source is probably the user guide available from STN International.

b. CrossFire. This exciting new product, introduced in 1995, has definitively catapulted the image of *Beilstein* as a major chemical information provider from perhaps a bit stodgy in the minds of some to a position of significant leadership and innovation.

The basic *CrossFire* product contains the entire *Beilstein* structure file, including some 350 chemical and physical property fields available for each compound, loaded onto hard disks for use in the customer's inhouse client/server-based system. The server is defined as IBM RISC (Reduced Instruction Set Computer) System/6000, with certain other specific requirements, and the client can be PC equipment that meets certain straightforward requirements.

A consortium of 15 universities in the midwestern United States, known as the *Virtual Electronic Library of the Committee on Institutional Cooperation,* announced in

1995 that the *CrossFire* system is being made available to students and faculty at the member universities over the Internet. The U.S. universities that utilize this system include Chicago, Illinois, Indiana, Iowa, Michigan, Michigan State, Minnesota, Northwestern, Ohio State, Pennsylvania State, Purdue, and the University of Wisconsin, Madison. Three partner institutions include Wayne State, Iowa State, and the University of Cincinnati. The server is located at the University of Wisconsin, Madison. This successful implementation of *CrossFire* at these academic institutions is now being extended nationwide in a project known as *Minerva.* A somewhat similar project is underway in the United Kingdom under the auspices of the Combined Higher Education Software Team (CHEST); this makes the system available to all higher education institutions in the United Kingdom. In addition, the system is utilized by chemists at chemical and pharmaceutical firms on a virtually worldwide basis.

Updating is done on a quarterly basis. It is estimated that some 300,000 new structures, and the associated data, are added annually.

A unique and powerful hypertext feature permits the user who is searching for ways to synthesize an organic molecule to concurrently also quickly determine how the starting materials are synthesized in a retrosynthetic fashion.

c. CrossFireplusReactions and CrossFire Abstracts. Reactions are also now included in the database in a newer and enhanced product known as *CrossFireplusReactions.* The benefit of this new feature is that, in addition to all the data in the standard database, the approximately 5 million graphical and 5 million nongraphical reactions already in the system are to be combined and merged in 1997 so that all 10 million reactions are to be fully structure-searchable. Beilstein officials say that *CrossFire* is the largest and most comprehensive database of chemical reactions available. Reactions may be searched using either full structures or substructures, and reaction attributes may be used as well. These attributes include reaction centers, bond fate, and atom-to-atom mapping. In addition, *CrossFire Abstracts* is to be introduced in 1997; coverage will permit searching of titles and abstracts starting with 1980. The total system thus permits searching by substance, reaction, citation, title, abstract, and factual (property) data.

Reaction details provided in *CrossFireplusReactions* include reagents, solvents, times, yields, temperatures, and citations.

The inclusion of reaction data as described permits chemists to specify a starting material and/or product as a substructure; on the basis of these data, the system will suggest routes to achieve the desired synthesis.

Hyperlinks are provided to: substance, based on the BRN (*Beilstein* Registry Number); reaction, based on the RX.ID (Reaction Structure Identification Number); and citation, using the Citation Number. The citation linkage not only permits quick location of references, but also permits one to quickly scan the work of a given investigator. The citation mode lists the reference and all compounds and reactions in the citation (and soon will list title and abstract as well).

d. Internet File: NetFire. This Internet file provides access to titles, abstracts, and authors from selected journal literature from 1980 to date. Approximately 160 organic and medicinal chemistry journals are planned to be included in the coverage; patents are not included. After an initial free trial, an annual subscription is required.

The Internet address for *Beilstein* is www.beilstein.com. It includes a large number of excellent links to other key chemistry locations on the Internet.

e. Potential of CrossFire. The *CrossFire* product line is probably aimed more at direct use by the individual laboratory chemist than by an intermediary such as an information professional. The product offers the potential of significant advantages in enhancing productivity, creativity, and cost-effectiveness. It has been, however, a relatively expensive product as compared to the usual cost of journals, books, patents, and other literature. The ongoing development of new pricing plans and accessibility is much welcomed. In the future, it is believed that purchase and use of a *CrossFire* system will probably become virtually a requirement for any major firm in pharmaceutical industry and related organic chemicals industry that wants to remain fully competitive.

f. Gmelin Database on CrossFire. Another new addition to the *CrossFire* family is the *Gmelin Database* (see also Section 12.5.A) with over 1 million compounds (470,000 coordination compounds, 55,000 alloys, 14,000 glasses and ceramics, and 3200 minerals).

g. Other Electronic Products from Beilstein

1. *Current Facts in Chemistry.* This quarterly CD-ROM produce covers papers in approximately 160 key journals in organic and medicinal chemistry. It is intended as a handy tool for the bench chemist and is priced accordingly. The data provided include structures, reactions, preparations, physical properties, physiological behavior, keywords, citations, and author names. The series began with coverage of the literature for 1990. The file is said to cover some 300,000 new structures, and associated data, with each new disk. Substructure searching capabilities are available for all 300,000 structures. The search files do not need to be copied to the hard disk.

2. *AUTONOM.* Another interesting *Beilstein* product is software that provides automatic naming of compounds based on a drawn structure. This is known as *AUTONOM* (*Automatic Nomenclature*). The name is provided is an IUPAC-compatible name. System requirements are an IBM, Macintosh, or VAX (Digital Equipment's Virtual Access Extension) computer with graphics capability and a mouse. It is claimed that success is achievable 85% of the time; if the system cannot provide a name, this is indicated in the response of the software. All compounds in *Beilstein* have been processed with *AUTONOM.*

3. *MOLTERM.* Beilstein's *MOLTERM* software is a terminal emulation program that runs under Microsoft Windows. It can be used in structure searching of *Beilstein* in DIALOG, STN International, and Questel•Orbit's DARC. This is now designated as freeware.

4. *SANDRA.* *SANDRA* (*Structure and Reference Analyzer*) is a diskette product that, on the basis of drawn chemical structures, identifies the correct volumes in *Beil-*

stein. Copies of this software are available without charge to subscribers to the printed volumes of *Beilstein.*

2. Finding Compounds in Beilstein. The electronic versions of *Beilstein* offer a variety of approaches for locating compounds, and the printed books also offer several approaches.

When looking for specific compounds in the electronic files, the best approach is to use a structure search. Alternatively, if chemist does not know the structure, combinations of any of the other searchable fields, perhaps with substructure searches, can be considered. The CAS Registry Number, *if included in Beilstein,* can be a helpful access point.

As an alternative to the electronic tools, the user can search the printed books manually in one of several ways: (1) molecular formula index (the quickest and most reliable approach for the printed books), (2) subject (or compound name) index (second choice, because of the potential vagaries of chemical nomenclature, although *Beilstein* editors are aware of these vagaries, and have attempted a compromise by providing more than one name for some chemical compounds in the subject index), and (c) *System Number,* which is based on a kind of hierarchy or classification unique to *Beilstein.* Each volume comes complete with a formula and subject index. Once a compound has been located, the chemist need merely use the same System Number for further searching, and the search procedure is further aided by the *coordinating references* found at the tops of each odd-numbered page.

Another option is to utilize *Beilstein* System rules, details about which are available from the publisher. The first principle is that the *Beilstein* volumes (actually groups of volumes) are divided as follows: acyclic compounds, Volumes 1–4; isocyclics, Volumes 5–16; and heterocyclics, Volumes 17–27. Further determination on placement of compounds is then based primarily on types and numbers of functional groups according to a series of rules. For chemists who want to learn how to use the *Beilstein* System efficiently, a few hours of study and practice are strongly recommended.

In addition to the tables of contents and indexes in individual volumes, cumulative (general) indices are published periodically. The 23-volume *Centennial Index* to *Beilstein* is a major reference source. It covers all the compounds included from 1779–1959. Indexes are provided by compound name and molecular formula. This index replaces the older General Indices H–E, II, which covered the Basic Series and the first two Supplementary Series.

For Series V, only collective indexes are planned. The following is the current schedule:

Vol.	Type of Index	Availability
17	Compound name and formula index	Available
20–22	Compound name and formula index	Available
23–25	Compound name and formula index	Available
26	Compound name and formula index	Available

3. Costs, Availability, and Other Topics. Sets of the printed *Beilstein* are available to chemists in hundreds of major university and industrial libraries throughout the world. The cost of maintaining an up-to-date set is believed to be approximately $30,000 per year at this writing. This expenditure permits purchase of all newly published volumes each year. Separate price schedules apply for universities and for industry. As mentioned, *Beilstein* is also available online (STN International and DIALOG), and beginning in late 1996, Internet accessibility became possible through *NetFire.* For the online file, there is no search or connect-hour charge, but there are charges if data are displayed, "typed," or printed. In the case of the *CrossFire* and *NetFire* products, an up-front annual fee provides for unlimited use of the products.

The issue of cost is mentioned here because, in many organizations, especially those with modest budgets, availability of funds, or lack of such, cost is a major factor in the hoped-for continuing availability to the printed *Beilstein.* Another factor is the sheer size: the volumes take up considerable shelf space. The online, Internet, and *CrossFire* versions are significant advances in making the contents more widely available to working chemists, and have a number of unique features as compared to the printed product, but lack the convenience of physical browsing that some users prefer. Initial *CrossFire* pricing put it within the range of major university and industrial libraries, but essentially out of range for smaller colleges and universities and smaller companies. However, a new program, *CrossFire Direct,* has made the product be more attainable for smaller organizations. This is a client/server configuration; the server is located at the firm's computer facility in Engelwood, CO. Further, a program named *CrossFire-Minerva* has been announced jointly with the University of Wisconsin—Madison with the goal of making the product accessible at PhD-degree-granting institutions throughout North America.

Another welcome move is the establishment of a North America help desk (800-275-6094) as an addition to the main help desk in Frankfurt.

Since 1994, the Beilstein Institute has been offering printed and electronic products through a "daughter" organization Beilstein Informationssysteme GmbH, known as Beilstein Information Systems, Inc., in the United States (However, the future relationship between these two organizations became unclear in 1997.) The managing director of Beilstein Informationssysteme is Professor Dr. Clemens Jochum. The electronic products operations are a joint venture with Information Handling Services (IHS). IHS headquarters are in Englewood, CO (800-275-6094). IHS is, in turn, is wholly owned by Thyssen-Bornemizsa Group (TBG), a firm located in Monaco.*

*According to reports in December, 1997, Elsevier Science Publishers, Amsterdam, the Netherlands, announced plans to acquire shares of Beilstein Informationssyteme GmbH, and to obtain an exclusive commercial license to the Beilstein Database and Handbook.

It is difficult to attempt a comparison of *Beilstein* with *Chemical Abstracts;* this is almost like comparing apples to oranges. Both are historic and monumental sources of information that are important to most chemists today. Both offer sophisticated and powerful online searching capabilities. However, each serves somewhat different functions. The two are competing sources of information for chemists, but they are also complementary sources, and the chemist may often need to consult both, as well as other important chemical information sources, depending on the objectives of the chemist in each specific case.

CA is widely believed to be the most current overall chemical information source, and it is also the most comprehensive overall source. Its sources include over 8000 journals as well as patents from 28 patent-issuing offices. Coverage, which includes all branches of chemistry, not just organic chemistry, begins with 1907. *Beilstein* meticulously organizes and pulls together every piece of factual information for every characterized organic molecule based on the sources that it covers. *Beilstein* coverage begins with 1771. The CAS Registry System includes over 17 million structures approximately over 12 million of which are organic, including large biomolecules and organics, and nearly 2 million are inorganic. *Beilstein* includes some 8 million organic molecules, and it offers what it says is the world's largest reaction database. *Beilstein* includes patents, but it does not include those published after 1979. Both sources require that physical data be documented to added a substance to the database, although *CA* will index compounds hypothesized in patent claims, for which *CA* staff say that the total is not significant. *CA* is, of course, totally in English, and the *Beilstein* databse is also in English (except for uncategorized free-text fields from the printed *Handbook* time period); the printed *Beilstein* began to publish totally in English in 1984. Both *CA* and *Beilstein* have well-deserved distinguished reputations for top quality.

B. Dictionary of Organic Compounds

The *Dictionary of Organic Compounds* (13) is a nine-volume set (plus annual supplements) that has concise data on many common compounds. The current sixth edition was published in November 1995. For a typical entry, see Figure 12.2.

Some older chemists may still prefer to identify these volumes as *Heilbron,* after the late Ian Heilbron, Professor of Chemistry at the University of Manchester, who edited the first three editions beginning in 1934. His name remains associated with the publication, but the more correct designation is now *DOC,* an acronym taken from the words in the title. The executive editor of *DOC* is John Buckingham.

A comparison of *DOC* with *Beilstein* is helpful in providing perspective. *DOC* covers many chemicals, but far fewer than *Beilstein.* In addition, data provided are less comprehensive than those in *Beilstein.*

Although *DOC* is not intended, nor should it be used, as a direct replacement for *Beilstein,* it does have a number of attractive features. One is its relatively low original cost ($5200), which may make it more readily available in chemistry libraries than the much more expensive *Beilstein.* Chemists should find *DOC* handy for quick lookup of basic data such as chemical constitution, some selected physical and chemical properties, references to some preparative methods, types of uses, and hazards. Figure 12.2 shows a typical entry in *DOC.*

Finally, many chemists find *DOC* easy to use. The main volumes have a straightforward alphabetical arrangement, and there are indexes by chemical name (including systematic, trivial, and trade names) molecular formula, heteroatom, and CAS Registry Number.

2-Amino-1-butanol, 9CI **A-0-00018**
[96-20-8]

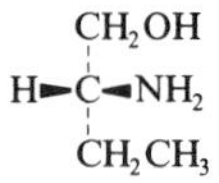

$$H\blacktriangleright \overset{\displaystyle CH_2OH}{\underset{\displaystyle CH_2CH_3}{C}} \blacktriangleleft NH_2 \quad (R)\text{-}form$$

$C_4H_{11}NO$ M 89.1
Cheaply available intermed. which can be readily resolved on an industrial scale. The parent amine has too low a m.w. to give cryst. salts with most acids and the *O*-benzyl deriv. is recommended as a resolving agent (Touet *et al*).

▶ Fl. p. 74° (oc). LD_{50} (mus, orl) 2300 mg/kg. EK9625000.

(R)-form [5856-63-3]

Oxalate: $[\alpha]_D^{20}$ −11.3 (H_2O).
O-*Benzyl:*
$C_{11}H_{17}NO$ M 179.2 Resolving agent.
Viscous oil. $[\alpha]_D$ −18.9 (c, 1.5 in EtOH).
O-*Benzyl; hydrochloride:* Cryst. $[\alpha]_D$ −18 (c, 1.1 in EtOH).

(S)-form [5856-62-2]

Bp_{11} 80°. $[\alpha]_D^{20}$ +9.8 (H_2O).
Oxalate: Mp 190-192°. $[\alpha]_D^{20}$ +11.3 (H_2O).

(±)-*form* [13054-87-0]

Bp 172-174°, Bp 178°.
Picrate: Mp 130.5-132.5°.

Stoll, A. *et al*, *Helv. Chim. Acta*, 1943, **26**, 922 (*synth*)
Johnson, K. *et al*, *J.O.C.*, 1943, **8**, 7 (*synth*)
Kjaer, A. *et al*, *Acta Chem. Scand.*, 1962, **16**, 71 (*abs config*)
Touet, J. *et al*, *Synth. Commun.*, 1994, **24**, 293 (*bibl, use, derivs*)
Lewis, R.J., *Sax's Dangerous Properties of Industrial Materials*, 8th edn., Van Nostrand Reinhold, 1992, AJA250.

Figure 12.2. Sample entry from *Dictionary of Organic Compounds,* 6th ed. (Reprinted with permission of Chapman & Hall. Copyright 1997.)

The total database from which *DOC* is produced has undergone continuous expansion and development since its initial construction in 1979–1980, and it now contains records covering approximately 400,000 compounds. It now covers inorganic and organometallic compounds as well as organic. This part of the database was published as the *Dictionary of Inorganic Compounds* (1992) and the *Dictionary of Organometallic Compounds* (second edition, 1995).

For organic compounds, the compilation covers not only organics but also (extensively) natural products. In 1992, a decision was made to publish a separate *Dictionary of Natural Products,* which is intended to be a comprehensive record. Extensive retrospective searches are conducted by the contributors and editors of *DNP* to provide a resource that is both extensive and well-organized. *DNP* was published in 1992, and it updates and supersedes previously published specialized compilations such as

Omeprazole, BAN, INN, USAN **O-00002**

5-Methoxy-2-[[(4-methoxy-3,5-dimethyl-2-pyridinyl)methyl] sulfinyl]benzimidazole, 9CI. Antra. Audazol. Gastroloc. Losec. Mopral. Omapren. Parizac. Prilosec. H 168/68. Many other names

[73590-58-6]

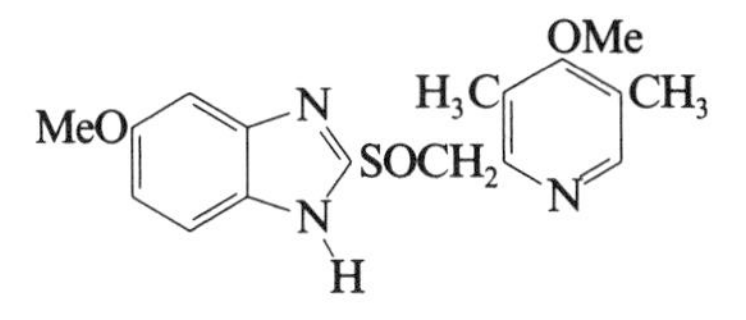

$C_{17}H_{19}N_3O_3S$ M 345.4

Proton pump inhibitor, inhibits gastric acid secretion. Antiulcer agent. Used in combination with Amoxicillin for eradication of *Helicobacter pylori*. Launched 1988. Worldwide 3rd best selling prescription drug ($1.9 bn) (Astra + Merck) (1994, Pharma Business). Mp 156°. Log P 2.23 (pH 6.4) (exp).

▶ LD_{50} (rat, orl) 2210 mg/kg. LD_{50} (rat, ipr) >100 mg/kg. DD9087000.

Eur. Pat., 5 129, (1979) (*Aktiebolag Hassle*); *CA*, **92**, 198396z (*synth, pharmacol*)
Adams, M.H. *et al*, *Clin. Pharm.*, 1988, **7**, 725 (*rev*)
Brändstrom, A. *et al*, *Acta Chem. Scand.*, 1989, **43**, 536, 549, 569, 577, 587, 595 (*props, bibl*)
Ohishi, H. *et al*, *Acta Cryst. C*, 1989, **45**, 1921 (*cryst struct*)
Maton, P.N., *N. Engl. J. Med.*, 1991, **324**, 965 (*rev*)
Hetzel, D.J., *Digestion, Suppl.* 1, 1992, **51**, 35 (*clin trials, rev*)
Ferner, R.E. *et al*, *Human Exp. Toxicol.*, 1993, **12**, 541 (*tox, human*)
Martindale, The Extra Pharmacopoeia, 30th edn., Pharmaceutical Press, London, 1993, 896.
Massoomi, F. *et al*, *Pharmacotherapy*, 1993, **13**, 46 (*rev*)
Yeomans, N.D., *Adverse Drug React. Toxicol. Rev.*, 1994, **13**, 145 (*tox, rev*)
Creutzfeldt, W., *Drug Saf.*, 1994, **10**, 66 (*rev*)
Wilde, M.I. *et al*, *Drugs*, 1994, **48**, 91 (*rev*)
Negwer, M., *Organic-Chemical Drugs and their Synonyms*, 7th edn., Akademie-Verlag, Berlin, 1994, 5818 (*synonyms*)

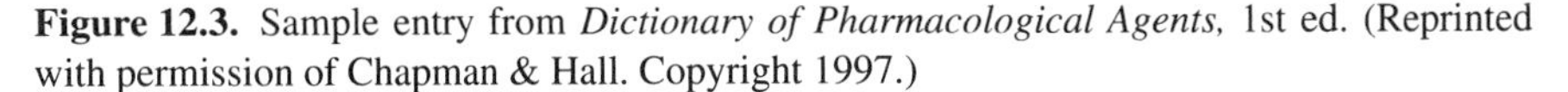

Figure 12.3. Sample entry from *Dictionary of Pharmacological Agents,* 1st ed. (Reprinted with permission of Chapman & Hall. Copyright 1997.)

the *Dictionary of Alkaloids*. *DOC* includes entries for some of the most representative and important natural products (e.g., strychnine, abietic acid), the entries being sometimes abbreviated compared with the same entries in *DNP*.

Each of these two dictionaries is available separately on a CD-ROM that permits searching by both text–numerical values and substructures, with rapid interchange of results between the two modules. The inorganic and organometallic dictionaries are also available in this format as the combined CD product *Dictionary of Inorganic and Organometallic Compounds on CD-ROM*. In addition, there is a specialized pharmaceutical product, the *Dictionary of Pharmacological Agents,* which contains compounds of data of particular interest to medicinal chemists. For a typical entry, see

Figure 12.3. The CD version of this compilation is called *Pharmasource.* It contains an additional data source, *PDR Generics,* which is complementary to that included in the chemical dictionary, and which is coupled to it through extensive hypertext links.

Annual supplements are published for the print versions of *DOC* and *DNP.* Each annual supplement covers the literature to the second half of the previous calendar year. The CD-ROM versions are updated every 6 months, but rather than receiving an additional disk, subscribers to the CD-ROM version are sent a new disk containing the whole dictionary every 6 months, updated not only in terms of the data but also frequently in terms of software and search features. According to the editors, the database is continually edited to ensure that the best and most recent syntheses, references to spectra, and so on are included; that obsolete and less useful data are removed; and that the data are presented in the most user-friendly manner.

Advances in data compression technology have permitted all the dictionaries noted above to become available on a single CD known as *The Combined Chemical Dictionary,* first issued in 1997. This permits searches of the entire database of nearly 400,000 compounds. It is updated every 6 months.

Chapman & Hall dictionaries are also available online through a single database, the *Chapman & Hall Chemical Database* on the DIALOG system, but the online version is not currently substructure-searchable, and is updated only "irregularly."

In addition to sources such as *Beilstein* and the Chapman & Hall dictionaries just noted, Chapter 15, which deals with physical properties, describes other compilations of interest, although most are of somewhat more limited scope. However, CRC Press, Boca Raton, FL, produces a number of products of significant interest, as, for example, their *Properties of Organic Compounds.*

C. Some Other Important Organic Chemistry Reference Works

1. A major English-language reference work in organic chemistry has been *Rodd's Chemistry of Carbon Compounds* (14), the second edition of which began publication in 1964 under the editorship of S. Coffey or M. F. Ansell. This is an outstanding work of broad scope. Although not as comprehensive as *Beilstein,* it provides an excellent starting point.

2. *Houben-Weyl Methods of Organic Chemistry* (15), first published in 1952, is an extensive set of volumes on preparative organic chemistry that is organized according to compound class or functional group. A major change for this ongoing series, now in its fourth edition, is that since 1995, all volumes are only in English (previous volumes were in German).

3. The ongoing series *Organic Syntheses,* first published in 1921, contains "the most convenient laboratory methods for preparing organic chemical reagents in one-half pound to five-pound lots." Every experiment has been conducted in at least two laboratories. Collective volumes every 5 years (previously 10 years) revise and update the annual volumes as necessary, and there are cumulative indexes as well. In addition, the contents of this series, and of other similar publications, is available electronically for use as an in-house database through MDL Information (see Chapter 10). The secretary of the Editorial Board for the collective volumes is Jeremiah P. Freeman, University of Notre Dame.

Recent volumes, beginning with Volume 41 (1961), emphasize

> widely applicable, model procedures that illustrate important types of reactions; . . . many of the procedures selected . . . have major significance in the synthetic method, rather than in the product that results. However, preparations of reagents and products of special interest are also included, as in previous volumes (16).

The change in policy is easily explained. When the series was founded by Roger Adams at the end of World War I, there was an urgent need to prepare useful compounds in reasonable quantities, because supplies of organic chemicals from abroad were no longer available. Today, a wide variety of fine organics is commercially available, and reactions can be studied with decreasing starting material quantities because of improved techniques.

4. *Organic Reactions* (17). Leo A. Paquette, Ohio State University, is editor of the well-known continuing annual series *Organic Reactions,* which began in 1942 and is now in Volume 50 (1997), which deals with the Stille reaction. This series contains "critical discussions of widely used organic reactions or particular phases of a reaction." The material is written from a preparative viewpoint. Most of the authors are in academia, but some are industrial research workers. Each volume contains a cumulative list of chapter titles by volume. A typical volume will cover two to four reactions, although a few volumes, such as Volume 50, deal with a single reaction. New volumes are currently being published at the rate of approximately two per year. Extensive references to the literature are included. Although detailed procedures are included, unlike *Organic Syntheses,* these have not been subjected to careful testing in two or more laboratories. *Organic Reactions* editors have included some of the most outstanding research workers in organic chemistry: Roger Adams, Volumes 1–10; Arthur C. Cope, Volumes 11–16; William G. Dauben, Volumes 17–32; A. S. Kende, Volumes 33–37; and Leo A. Paquette, Volume 38 to date. Paquette is also editor of the *Encyclopedia of Reagents for Organic Synthesis* (18), an eight-volume set that contains descriptions of the use of "all" reagents in organic chemistry. It comprises approximately 3500 articles.

5. In the field of medicinal chemistry, the five-volume *Burger's Medicinal Chemistry* (19) is a well-known classic. The editor of the 5th edition is Manfred W. Wolff, Immuno-Pharmaceutics, Inc. This book includes a discussion of the drug discovery process and follows this with a discussion of individual types of drugs such as gastrointestinal, cardiovascular, and other agents.

6. One of the most complete collections of chemical reactions in a single place is *Synthetic Methods of Organic Chemistry* (20). This is often referred to simply by the last name of the original editor, William Theilheimer, but it is now edited by A. F. Finch, Derwent Information, Ltd. The set began in 1948 (Vol. 1 covers literature for 1942–1944), and Volume 50 was published in 1996.

7. The Theilheimer/Finch volumes noted above been adapted and updated by Derwent Information into electronic format. Thus, the chemist has the choice of either printed or online access (*Derwent Journal of Synthetic Methods* on STN and Orbit). The material is also available from MDL Information (see Chapter 10).

8. *Compendium of Organic Synthetic Methods* (21), originally edited by Harrison and Harrison, and now by Michael B. Smith, University of Connecticut, is a compilation of published organic functional group transformations. Synthetic methods are presented in the form of reactions with references.

9. Another important work is *Feiser and Fieser's Reagents for Organic Synthesis* (22). This well-known multivolume series covers the reagent literature. As of 1997, 17 volumes had appeared covering the literature from 1967. A collective index covers Volumes 1–12.

10. John Wiley & Sons, Inc., New York, have been publishing a series of monographs on the chemistry of heterocyclic compounds (23), originally under the editorship of the late Arnold Weissberger, who was for many years with Eastman Kodak Co., Rochester, NY. This extensive series goes back a number of years and is still in progress.

D. Name Reactions

Over the years, there have been a number of reference sources that compile so-called name reactions. One convenient source is the 12th edition of the *Merck Index*, which includes succinct descriptions and reaction schemes as well as carefully selected references for 425 reactions. An example of a more extensive treatment is *Name Reactions and Reagents in Organic Synthesis* by Mundy and Ellerd (24). This volume contains cross-references to even more complete information about both reactions and reagents. A volume by Hassner and Sturmer (25) describes some 450 reactions, including first reported reference, reference to experimental procedure, a review article if found, and an example reaction that outlines reaction conditions.

E. Some Online and Other Sources of Organic Reactions

Some online and other electronic sources of reaction information were noted in Chapter 10. These included the products of Beilstein Information Systems, Chemical Abstracts Service, Derwent, INPI, ISI, and MDL Information.

For a discussion of Chemical Abstracts Service reaction and other products, see especially Chapters 6 and 7. Derwent's reaction products are noted also in Chapter 13, and ISI's products are discussed in Chapter 8. MDL is discussed in Chapter 10, and INPI is discussed in Chapters 10 and 13.

12.5. INORGANIC CHEMISTRY: SOME IMPORTANT REFERENCE WORKS

A. Gmelin Handbook of Inorganic and Organometallic Chemistry

Chemists working on elements and inorganic chemistry and compounds, including organometallic compounds, should usually consult *Gmelin* (26) as the next step after studying any available chemical encyclopedia article on the topic of interest.

As compared to the other classic sources of information on inorganic chemistry, *Gmelin* is by far the source of first choice. Gmelin is better organized, more complete, much more current, and easier to use, and it is now published totally in English. Most of the older *Gmelin* volumes are, however, in German.

The Gmelin Institute for Inorganic Chemistry is part of the Max Planck Society for the Advancement of Science. It is staffed by about 100 full-time employees, of whom about 65 have doctorates.

Gmelin is located in Frankfurt in the same building as Beilstein. The director (from 1979–1997) was Ekkehard Fluck, also a professor at the University of Stuttgart and an eminent inorganic chemist. The current (1997) Acting Director is H. W. Vollmann. The Institute has two groups: Handbook, group leader Dr. Jörn von Jouanne and Database, group leader Dr. Gottfried Olbrich.

About 450,000 compounds are described in the *Gmelin Handbook,* and the online file includes about 1 million compounds. The CAS Registry files include more compounds (about 1.85 million inorganic compounds) and offer more current coverage than does *Gmelin.* However, *Gmelin* offers other advantages. The functions and objectives of these two great sources are different, and they complement each other. Other important sources of information about inorganic chemistry also need to be consulted if a complete search is desired. In a comparison with another major source, *Gmelin* is organized by chemical compounds, whereas *Landolt-Bornstein* (see Section 15.10) is organized by physical properties.

Whereas all *Beilstein* editorial work is in-house, some *Gmelin* contributions have been by outside experts, including several from the United States, such as Therald Moeller, Earl Muetterties, and Glenn Seaborg. The *Handbook* will continue, and will even increase, the number of manuscripts from outside expert scientists; the editorial work is done in-house.

The *reporting period* for *Gmelin,* like that of the *Landolt-Börnstein Handbook* (see Section 15.10), begins with the literature closing date of the previous volume and extends up to about 6 months before publication date of the new volume.

There are no plans for a ninth edition of *Gmelin.* Rather, the current (eighth) edition, begun in 1922, will continue to be updated by more recent volumes. This edition now contains over 747 volumes and more than 230,000 pages. About 300 of the volumes are in English.

Thus, *Gmelin* consists of:

- Main Series (Hauptband) volumes—began in 1922.
- Supplement (Ergänzungsband) volumes—began in 1954. These are now the principal vehicle for the updating of *Gmelin.*
- New Supplement Series volumes—began in 1970. This was the first of the *Gmelin* series to introduce text in English. These volumes should now be found shelved among the regular Supplement volumes of the corresponding elements.

Supplements thus update the coverage of the Main Series volumes and are frequently much more extensive. For the latest, and possibly the *best,* data, users of *Gmelin* should consult Supplements first, and then, only if needed, Main Series volumes.

The first edition of *Gmelin* was published in 1817–1819 in three volumes under the title *Handbuch der Theoretischen Chemie* under the editorship of Leopold Gmelin, a professor at the University of Heidelberg. At that time, "theoretical" referred to work being done in universities as compared to work done in industry. The first edition covered both inorganic and organic chemistry. The limitation to inorganic chemistry began with the fifth edition, 1852–1853.

The English title *Gmelin Handbook of Inorganic Chemistry* was adopted in 1982. The current title *Gmelin Handbook of Inorganic and Organometallic Chemistry* was adopted in September 1990 to reflect the rapid growth in the importance of organometallic compounds. All organometallic compounds containing a metal–carbon bond are said to be covered in *Gmelin* except for carbides, cyanides, cyanates, and thiocyanates. Organometallic coverage began with the publication of the *Organovanadium* and *Organochromium Compounds* volumes in 1971. There is no overlap in coverage of organometallics between *Gmelin* and *Beilstein.* Those organometallic compounds with central atoms Li, Na, K, Rb, and Cs (not Fr) and Mg, Ca, Sr, and Ba (not Be and Ra) are covered by *Beilstein.* Organometallic compounds with all other central atoms (including Fr, Be, and Ra) are covered by Gmelin.

From 1978 on, the New Supplement Series has been fully incorporated in the eighth edition, and is no longer designated as such.

As an aid to using *Gmelin,* the chemist will find that the reverse of the title page for each volume contains the latest date through which literature for that system volume is evaluated. This means that the chemist can elect to begin searching of other sources, if needed, issued subsequent to that date and thereby save considerable time.

Gmelin provides key information in concise form. The original source of the data is referred to, as, in many cases, are abstracts of these sources.

One of the most significant features of *Gmelin* is inclusion of many valuable tables of numerical data, curves, and other graphics material, including diagrams of apparatus. Another significant feature is extensive coverage of the more applied or "practical" aspects. Considerable attention is paid to commercial manufacturing practices. The staff covers patents liberally, but selectively. About 20% of patents considered are included.

Gmelin staff has kept pace with major developments affecting chemists by increasing scope of coverage. Thus, since about 1975–1977, more attention is paid to toxicology than in the past because of the great surge of interest in environmental matters, including occupational health and safety. In addition, there is more emphasis on uses and applications than in the past because much research, especially in industry, is more focused in this direction.

Beginning in 1982, *Gmelin* volumes are published totally in English, although some volumes prior to then are also in English, or may have tables of contents, marginal captions, and indexes in English. Most American and other English-speaking chemists find even those older *Gmelin* volumes in German easy to use because of the high degree of organization throughout the treatise. For example, chemists using *Gmelin* will find its indexes invariably helpful. This is especially so for elements that are covered in more than one volume, such as sodium and its compounds. To help in use of volumes on sodium, the editors have provided this series with its own formula and subject indexes. Another example of the strength of *Gmelin* indexes is in helping chemists locate data on multicomponent systems. The English-language *General Formula Index* (consisting of 12 volumes and covering the years 1924–1974) is augmented by the First Supplement 1974–1979 (eight volumes), the Second Supplement 1980–1987 (10 volumes), and the Third Supplement 1988–1992 (six volumes). Other indexes appear in a number of the volumes of the main work and of the supplement volumes.

All compounds in *Gmelin* have a unique *Gmelin* Registry Number (GRN). Unfortunately, only about 20–25% of the compounds in the *Gmelin* online file also have CAS Registry Numbers. For this reason, searching by molecular formula is preferable. To complete the assigning of CAS numbers to all compounds in *Gmelin* would be a very large and difficult task because of the differences in criteria utilized for registration.

As a further aid to helping chemists use *Gmelin* more easily, subscribers should shelve the volumes as recommended by *Gmelin* staff, for example, by atomic symbol, rather than by System Number (see Figures 12.4 and 12.5). This outmodes previous distinctions among the several types of Supplemental volumes. Special dividers are supplied by *Gmelin* to facilitate this arrangement. Finally, helpful descriptive booklets about *Gmelin* are available without charge from the publisher, Springer Verlag. Sample pages from *Gmelin* are shown in Figure 12.6.

In addition to the printed volumes, since 1991, *Gmelin* has been available online through the STN International system. The online file includes *Gmelin Handbook* data from 1924 through 1975, and, in addition, data taken from what *Gmelin* editors believe are to be the 111 most important periodicals since 1986. (Plans call for the gap between 1975 and 1986 to be closed by the end of 1997 by excerpts from the periodicals covered.) The online file is updated four times a year.

Gmelin and *Beilstein* may be searched on STN in similar ways. For example, the same structure–substructure approach may be utilized for searching *Beilstein, Gmelin,* and the *CAS Registry* file on STN.

The *Gmelin* online file contains information on the identification of a compound, a great number of chemical physical properties, and bibliographic data. The physical data, for instance, are represented in about 120 fields, which may also contain searchable parameters such as temperature and pressure.

To assist in online searching of *Gmelin,* a diskette has been developed by the editors to support and facilitate the process. This is known as *GMELIN-HELP.*

The *Gmelin* Database is also available for electronic searching under the *Beilstein CrossFire* system as mentioned earlier in this Chapter. The *Gmelin* Database relies on only 110 preselected journals; the *Handbook,* however, is much more complete.

Like *Beilstein* and other major works of this type, the printed *Gmelin* is expensive. In 1994, for example, 22 volumes were published at a total list price of approximately $25,000. As of 1997, volumes, occupied over 50 linear feet of shelf space. Approximately 20 supplement volumes are published annually.

Gmelin has been published and distributed since 1974 by Springer-Verlag (800-SPRINGER), which has offices in Berlin, New York, and several other major cities worldwide.

System Number	Symbol	Element
1		Noble Gases
2	H	Hydrogen
3	O	Oxygen
4	N	Nitrogen
5	F	Fluorine
6	**Cl**	**Chlorine**
7	Br	Bromine
8	I	Iodine
	At	Astatine
9	S	Sulfur
10	Se	Selenium
11	Te	Tellurium
12	Po	Polonium
13	B	Boron
14	C	Carbon
15	Si	Silicon
16	P	Phosphorus
17	As	Arsenic
18	Sb	Antimony
19	Bi	Bismuth
20	Li	Lithium
21	Na	Sodium
22	K	Potassium
23	NH_4	Ammonium
24	Rb	Rubidium
25	Cs	Caesium
	Fr	Francium
26	Be	Beryllium
27	Mg	Magnesium
28	Ca	Calcium
29	Sr	Strontium
30	Ba	Barium
31	Ra	Radium
32	**Zn**	**Zinc**
33	Cd	Cadmium
34	Hg	Mercury
35	Al	Aluminium
36	Ga	Gallium

System Number	Symbol	Element
37	In	Indium
38	Tl	Thallium
39	Sc, Y La–Lu	Rare Earth Elements
40	Ac	Actinium
41	Ti	Titanium
42	Zr	Zirconium
43	Hf	Hafnium
44	Th	Thorium
45	Ge	Germanium
46	Sn	Tin
47	Pb	Lead
48	V	Vanadium
49	Nb	Niobium
50	Ta	Tantalum
51	Pa	Protactinium
52	**Cr**	**Chromium**
53	Mo	Molybdenum
54	W	Tungsten
55	U	Uranium
56	Mn	Manganese
57	Ni	Nickel
58	Co	Cobalt
59	Fe	Iron
60	Cu	Copper
61	Ag	Silver
62	Au	Gold
63	Ru	Ruthenium
64	Rh	Rhodium
65	Pd	Palladium
66	Os	Osmium
67	Ir	Iridium
68	Pt	Platinum
69	Tc	Technetium[1]
70	Re	Rhenium
71	Np, Pu . . .	Transuranium Elements

HCl → 6; $ZnCl_2$ → 32; $CrCl_2$ → 52; $ZnCrO_4$ → 52

Material presented under each Gmelin System Number includes all information concerning the element(s) listed for that number plus the compounds with elements of lower System Number.

For example, zinc (System Number 32) as well as all zinc compounds with elements numbered from 1 to 31 are classified under number 32.

[1] A Gmelin volume titled "Masurium" was published with this System Number in 1941.

Figure 12.4. Key to the *Gmelin* System of Elements and Compounds. (Reprinted with permission of Springer-Verlag GmbH & Co.)

Another *Gmelin* publication is *TYPIX* (four volumes), which is a critical compilation of crystallographic data for over 3200 compounds representing structure types found among inorganic compounds. Structure types found exclusively among halides or oxides are included only in some specific cases. Also included is condensed chemical information about individual structure types, and there is a chapter on the crystal chemistry of particular structure families. These volumes were published in 1993–1994, and are all totally in English.

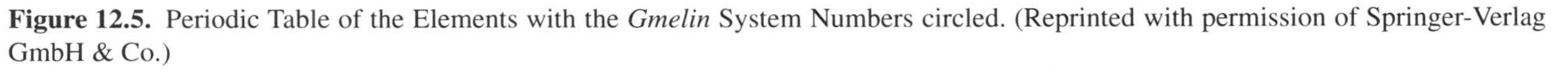

1 H (2)																1 H (2)	2 He (1)
3 Li (20)	4 Be (26)											5 B (13)	6 C (14)	7 N (4)	8 O (3)	9 F (5)	10 Ne (1)
11 Na (21)	12 Mg (27)											13 Al (35)	14 Si (15)	15 P (16)	16 S (9)	17 Cl (6)	18 Ar (1)
19 * K (22)	20 Ca (28)	21 Sc (39)	22 Ti (41)	23 V (48)	24 Cr (52)	25 Mn (56)	26 Fe (59)	27 Co (58)	28 Ni (57)	29 Cu (60)	30 Zn (32)	31 Ga (36)	32 Ge (45)	33 As (17)	34 Se (10)	35 Br (7)	36 Kr (1)
37 Rb (24)	38 Sr (29)	39 Y (39)	40 Zr (42)	41 Nb (49)	42 Mo (53)	43 Tc (69)	44 Ru (63)	45 Rh (64)	46 Pd (65)	47 Ag (61)	48 Cd (33)	49 In (37)	50 Sn (46)	51 Sb (18)	52 Te (11)	53 I (8)	54 Xe (1)
55 Cs (25)	56 Ba (30)	57** La (39)	72 Hf (43)	73 Ta (50)	74 W (54)	75 Re (70)	76 Os (66)	77 Ir (67)	78 Pt (68)	79 Au (62)	80 Hg (34)	81 Tl (38)	82 Pb (47)	83 Bi (19)	84 Po (12)	85 At	86 Rn (1)
87 Fr	88 Ra (31)	89*** Ac (40)	104 (71)	105 (71)													* NH_4 (23)

**Lanthanides (39)	58 Ce	59 Pr	60 Nd	61 Pm	62 Sm	63 Eu	64 Gd	65 Tb	66 Dy	67 Ho	68 Er	69 Tm	70 Yb	71 Lu
***Actinides	90 Th (44)	91 Pa (51)	92 U (55)	93 Np (71)	94 Pu (71)	95 Am (71)	96 Cm (71)	97 Bk (71)	98 Cf (71)	99 Es (71)	100 Fm (71)	101 Md (71)	102 No (71)	103 Lr (71)

Figure 12.5. Periodic Table of the Elements with the *Gmelin* System Numbers circled. (Reprinted with permission of Springer-Verlag GmbH & Co.)

[2-CH_3O(5-$C_3H_4N_2$)C_6H_6Fe(CO)$_3$]BF_4 (Table **9**, No. **21**) is quantitatively prepared in situ by mixing equimolar amounts of [2-$CH_3OC_6H_6$Fe(CO)$_3$]BF_4 and imidazole in acetone-d_6 [81].

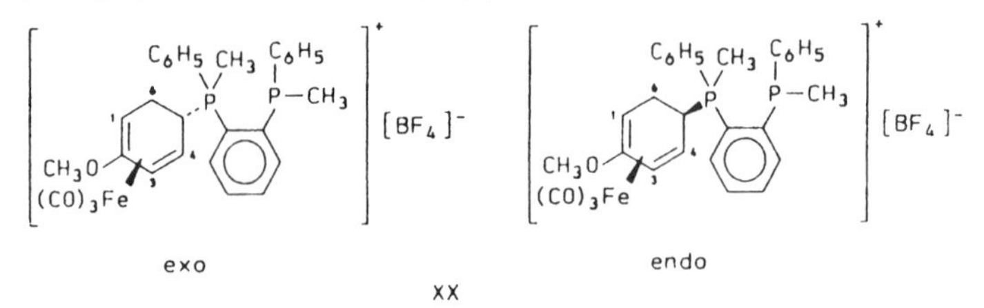

[2-CH_3O(5-(2-CH_3(C_6H_5)PC_6H_4P(C_6H_5)CH_3)C_6H_6Fe(CO)$_3$]BF_4 (Table **9**, No **26**) is prepared by attack of racemic [2-$CH_3OC_6H_6$Fe(CO)$_3$]BF_4 by $(-)_{589}$(S,S)-CH_3(C_6H_5)PC_6H_4P(C_6H_5)CH_3-2. This reaction exhibits considerable kinetic diastereoselectivity in CH_3CN and OC(CH_3)$_2$. The corresponding reaction in CH_2Cl_2 is of lower selectivity [78]. The phosphine preferentially selects the $(-)_{589}$(R)-enantiomer of the salt [72]. Thus, recovery of unreacted dienyl salt from reactions between racemic [2-$CH_3OC_6H_6$Fe(CO)$_3$]BF_4 and the $(-)_{589}$(S,S)-phosphine (1:0.5 molar ratio) provides a convenient method for preparing optically active [2-$CH_3OC_6H_6$-Fe(CO)$_3$]BF_4 [78] (6 to 11% enantiomeric excess [78, 79]). Unexpectedly, the 50/50 mixture of diastereoisomers XX, obtained immediately upon reacting both compounds in a 1:1 molar ratio, equilibrates over 3 d in CH_3CN to a 60:40 mixture [78]. However, this equilibration is not confirmed by [79] for no change in the CD or ^1H NMR spectra is observed over a period of a week at room temperature [79]. Treatment at room temperature in acetone gives, upon immediate evaporation, a high yield (>80%) of the yellow salt XX. The ^1H NMR spectrum (acetone-d_6) confirms a 50:50 mixture of the two diastereoisomers XX by the presence of $\delta = 3.48$ and 3.68 (OCH_3), and 5.40 and 5.65 (H-3) ppm for the endo and exo isomer. The IR spectrum (acetone) shows bands at 1990 and 2060 cm^{-1}. Preliminary attempts to separate the diastereoisomers XX by crystallization from various solvents are unsuccessful [78], cf. Nos. 27 and 28.

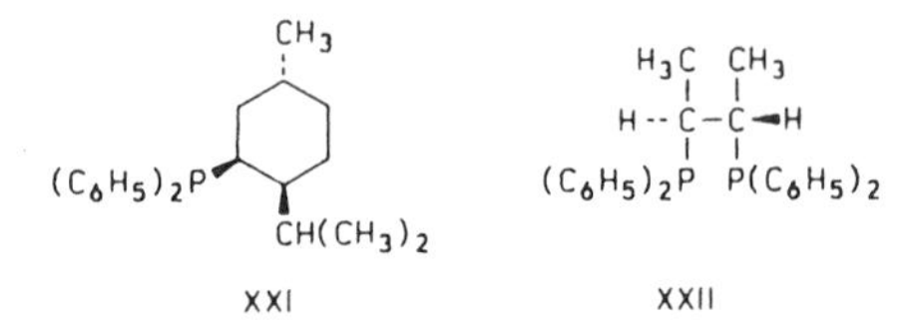

[2-CH_3O(5-R″(C_6H_5)$_2$P)C_6H_6Fe(CO)$_3$]BF_4 (Table **9**, Nos. **27** and **28** with R″ = cyclo-C_6H_9(C(CH_3)$_2$H-6)CH_3-3 and CH(CH_3)CH(CH_3)P(C_6H_5)$_2$) are prepared by the reaction of racemic [2-$CH_3OC_6H_6$Fe(CO)$_3$]BF_4 with equimolar amounts of $(-)_{589}$-XXI or $(-)_{589}$(S,S)-XXII in high yield. In contrast, a 1:0.5 molar ratio mixture of racemic [2-$CH_3OC_6H_6$Fe(CO)$_3$]BF_4 and $(-)_{589}$(S,S)-XXII gives a quantitative yield of XXIII. Each compound, No. 27 and 28, is expected to be a 50:50 mixture of diastereoisomers (see No. 26). However, in the ^1H NMR spectra in CH_3CN, distinct OCH_3 or H-3 resonances are not observed for the two diastereoisomers in each case even in the presence of Eu(fod) (for fod see No. 6). This coincidence of diastereoisomer proton resonances prevents quantitative in situ determination of the chiral discrimination. However, the unreacted dienyl salt recovered from the

References on pp. 228/30

Gmelin Handbook
Fe-Org Comp B8

Figure 12.6. (*a,b*) Representative pages from *Gmelin Handbook of Inorganic and Organometallic Chemistry,* 8th ed., Fe: Organoiron Compounds, Part B8; Mononuclear Compounds 8. (*c*) Representative page from *Gmelin Handbook of Inorganic Chemistry,* 8th ed., S: Sulfur-Nitrogen Compounds, Part 2: Compounds with Sulfur of Oxidation Number IV. (*d*) Gmelin presentation of organometallic compounds in tabular form. (Reprinted with permission of Springer-Verlag GmbH & Co.)

reaction of $[2\text{-}CH_3OC_6H_6Fe(CO)_3]BF_4$ with XXI (1:0.5 molar ratio) in acetone is found to contain an 11% enantiomeric excess of $[(+)_{589}(S)\text{-}2\text{-}CH_3OC_6H_6Fe(CO)_3]BF_4$. Thus $(-)_{589}$-XXI preferentially selects the $[(-)_{589}(R)\text{-}2\text{-}CH_3OC_6H_6Fe(CO)_3]BF_4$. Similarly a 5% enantiomeric excess of $[(-)_{589}(R)\text{-}2\text{-}CH_3OC_6H_6Fe(CO)_3]BF_4$ is found for the reaction with $(-)_{589}(S,S)$-XXII (molar ratio 1.5:0.5). However, the relative low enantiomeric excess of recovered $[(-)_{589}(R)\text{-}2\text{-}CH_3OC_6H_6Fe(CO)_3]BF_4$ from such a reaction in acetone suggests that $(-)_{589}(S,S)$-XXII is less diastereoselective than either $(-)_{589}(S,S)\text{-}CH_3(C_6H_5)PC_6H_4P(C_6H_5)CH_3\text{-}2$ (see No. 26) or $(-)_{589}$-XXI [79].

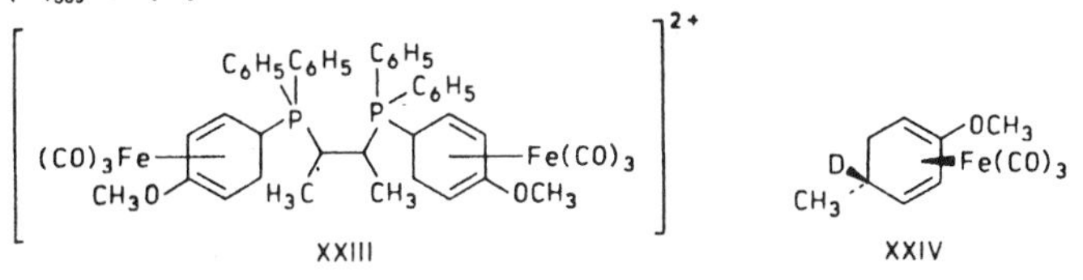

$2\text{-}CH_3O(5\text{-}CH_3)C_6H_6Fe(CO)_3$ (Table 9, No. 30) can also be prepared by reduction of $[2\text{-}CH_3O(5\text{-}CH_3)C_6H_5Fe(CO)_3]BF_4$ with $NaBH_4$ in H_2O. The reaction mixture is extracted with ether and the organic layer is washed with H_2O, dried ($MgSO_4$), and evaporated. The residue is dissolved in pentane, and filtered through silica. Evaporation gives a mixture of $1\text{-}CH_3O\text{-}(4\text{-}CH_3)C_6H_6Fe(CO)_3$ and No. 30 (ratio 2:3) in 90% yield. Chromatography has shown compound No. 30 to be a 1:1 mixture of 5-exo and 5-endo isomers. This mixture of the exo and endo isomers of No. 30 reacts with $[C(C_6H_5)_3]BF_4$ in CH_2Cl_2 to give $[2\text{-}CH_3O\text{-}(5\text{-}CH_3)C_6H_5Fe(CO)_3]BF_4$ in 50% yield. Chromatography of the neutral washings gives pure $2\text{-}CH_3O(5\text{-endo-}CH_3)C_6H_6Fe(CO)_3$ [11]. Similar reduction of $[2\text{-}CH_3O(5\text{-}CH_3)C_6H_5Fe(CO)_3]BF_4$ with $NaBD_4$ gives XXIV [19]. Optical isomers of compound No. 30 are prepared as follows: a cooled solution of $[(-)(2R)\text{-}2\text{-}CH_3OC_6H_6Fe(CO)_3]PF_6$ ($[\alpha]^{20}_D = -107°$, c = 9 mg/100 mL, CH_3CN) in CH_3CN is added to a solution of $LiCu(CH_3)_2$ at −45 °C. After 2 min the reaction is quenched by iced 5% aqueous HCl and extracted with ether. The extract is washed with H_2O, brine, dried ($MgSO_4$), and filtered through silica. Evaporation gives a yellow oil of $(+)(2R,5R)\text{-}2\text{-}CH_3O(5\text{-}CH_3)C_6H_6Fe(CO)_3$ ($[\alpha]^{20}_D = +11.7°$, c = 4 mg/100 mL, $CHCl_3$) in 75% yield [62]. 36 mg of fully resolved $[(-)(2R,5S)\text{-}2\text{-}CH_3O(5\text{-}CH_3)C_6H_5Fe(CO)_3]PF_6$ (see Formula XXVI, $[\alpha]^{20}_D = -138°$, c = 4 mg/100 mL, CH_3CN) is mixed with 324 mg of racemic $[2\text{-}CH_3O\text{-}(5\text{-}CH_3)C_6H_5Fe(CO)_3]PF_6$ to obtain a sample with $[\alpha]^{20}_D = -13.2°$. $NaBH_4$ is added to this sample in CH_3CN at 5 °C. After 10 min the mixture is filtered and evaporated. The residue is taken up in pentane and filtered through basic alumina. Evaporation affords a pale brown gum (84% crude yield). Chromatography on silica with C_6H_6/hexane (1:1) gives the more polar $(-)(1R,4S)\text{-}1\text{-}CH_3O(4\text{-}CH_3)C_6H_6Fe(CO)_3$ (see Formula XXVII, $[\alpha]^{20}_D = -12°$, c = 1 mg/100 mL, $CHCl_3$) in 7% and $(+)(2R,5S)\text{-}2\text{-}CH_3O(5\text{-}CH_3)C_6H_6Fe(CO)_3$ (see Formula XVIII, $[\alpha]^{20}_D = +13.1°$, c = 2 mg/100 mL, $CHCl_3$) in 31% yield and in an enantiomeric excess of 9.5% [65].

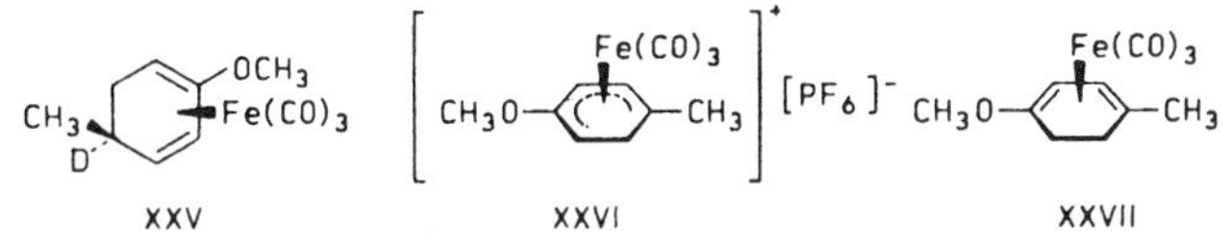

The (+)(2R,5S) isomer (see Formula XXVIII) shows in the ^1H NMR in $CDCl_3$ chemical shifts at δ = 0.86 (d, CH_3; J = 7 Hz), 1.20 (ddd, H-6 exo; J = 14 Hz, J = 5 Hz), 1.63 (m, H-5 exo),

Gmelin Handbook
Fe-Org Comp B8

References on pp. 228/30

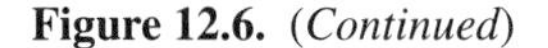
Figure 12.6. (*Continued*)

B. Some Other Inorganic Chemistry Reference Works

A newer source in inorganic chemistry is edited by Professor A. Bruce King, University of Georgia. This is the *Encyclopedia of Inorganic Chemistry* (27), which is an extensive treatment as is indicated by the dimensions: eight volumes, 6000 pages. There are some 260 main articles.

$S_3N_3Cl_3 + 9H_2O \rightarrow 3NH_4Cl + 3H_2SO_3$ [7]. When $S_3N_3Cl_3$ is treated with 10 to 30% alkali sulfite solutions, the following reactions occur simultaneously:

$$S_3N_3Cl_3 + 9OH^- \rightarrow 3SO_3^{2-} + 3Cl^- + 3NH_3 \quad (1)$$

$$S_3N_3Cl_3 + 3SO_3^{2-} + 9H_2O \rightarrow 3S(OH)_2 + 3SO_4^{2-} + 3Cl^- + 3NH_4^+ \quad (2)$$

With decreasing pH-value the importance of reaction (2) increases. At the same time the rate of decomposition continuously decreases. The intermediate sulfoxylic acid, $S(OH)_2$, decomposes to give $S_2O_3^{2-}$ and $S_3O_6^{2-}$ [8]. When $S_3N_3Cl_3$ is treated with ice-cold water in a moderate acidic medium, SO_2, NH_4^+, SO_4^{2-}, and a dark blue compound of composition $H_4N_2S_2O$ form, probably via $4S_3N_3Cl_3 + 27H_2O \rightarrow 3OS_2(NH_2)_2 + 6(NH_4)HSO_4 + 12HCl$ [6, 9], see also "Schwefel" B 3, 1963, p. 1850.

Reactions with Nitrogen Compounds

NH_3. The ammonolysis of $S_3N_3Cl_3$ in liquid NH_3 (−78°C) is said to produce the unstable compounds $(NSNH_2)_3$ and $S(NH)_2$. The metal derivative of $S(NH)_2$, $Hg[N_2S] \cdot NH_3$, can be isolated if a mixture of $S_3N_3Cl_3$ and NH_3 is treated with HgI_2 [10]. According to recent investigations the reaction of $S_3N_3Cl_3$ with NH_3 at −78°C produces an orange-red solid, $NH_4^+S_4N_5^-$, probably via the intermediate compounds $(NSNH_2)_3$ or $S(NH)_2$ [11]:

$$S_3N_3Cl_3 \xrightarrow[-3\,NH_4Cl]{+6\,NH_3} \{(NSNH_2)_3 \text{ or } 3S(NH)_2\} \rightarrow \tfrac{3}{4}NH_4^+S_4N_5^- + \tfrac{1}{4}N_2 + NH_3$$

C_6H_5S–NH–C_6H_5. C_6H_5S–NH–SC_6F_5. C_6F_5S–NH–SC_6F_5. C_6F_5S–NH–$Si(CH_3)_3$. The reaction of $S_3N_3Cl_3$ with sulfenyl amines produces sulfur diimides in the presence of an HCl trapping base B ($B = C_5H_5N$ or $(C_2H_5)_3N$) per $R\text{–}NH\text{–}R' + \tfrac{1}{3}S_3N_3Cl_3 \xrightarrow{B} R\text{–}N{=}S{=}N\text{–}R' + B \cdot HCl$, with $R = C_6H_5S$, $R' = C_6H_5$ (in CCl_4/petroleum ether, at 0°C), $R = C_6H_5S$, $R' = C_6F_5S$ (CCl_4/ether, 50°C, 1 h reflux) [12], $R = R' = C_6F_5S$ (CCl_4, 60°C) [13], or $R = C_6F_5S$, $R' = Si(CH_3)_3$ (CCl_4, 25°C, 2 h) [14].*

$[(CH_3)_3Si]_2NH$. $S_3N_3Cl_3$ reacts with $[(CH_3)_3Si]_2NH$ and C_5H_5N (mole ratio ~1:4:4) in CCl_4 at 25 to 60°C in 12 h to produce $(CH_3)_3SiN{=}S{=}NSi(CH_3)_3$ along with S_4N_4 and $(CH_3)_3SiN{=}S{=}N\text{–}S\text{–}N{=}S{=}NSi(CH_3)_3$ as side products [14, 15].

$(CH_3)_3SiN_3$. The reaction of $S_3N_3Cl_3$ with trimethylsilylazide, $(CH_3)_3SiN_3$, leads to a variety of products depending on the reaction conditions: temperatures −15 to +40°C, reaction time 0.5 to 48 h, mole ratios ~1:4 to ~1:15, solvents CH_3CN, tetrahydrofuran (THF), and CH_2Cl_2. Main products have been found to be up to 67% S_4N_4 and up to 72% $(SN)_x$ in CH_3CN. No $(SN)_x$ is obtained in THF and CH_2Cl_2; instead $S_4N_3^+Cl^-$ and $[S_3N_2Cl]_2$, respectively, form in addition to S_4N_4. Other products are $(CH_3)_3SiCl$ and NH_4Cl [16]. In an earlier paper the reaction of $S_3N_3Cl_3$ with $(CH_3)_3SiN_3$ (mole ratio ~1:4) in CH_3CN was found to produce 40 to 50% $(SN)_x$, along with $(CH_3)_3SiCl$, NH_4Cl, S_4N_4, N_2 and sulfur [17].

NO. NO_2. Treating a suspension of $S_3N_3Cl_3$ in CH_3NO_2 with gaseous NO gives the green five-membered ring compound $[S_3N_2Cl]_2$: $8S_3N_3Cl_3 + 24NO \rightarrow 3[S_3N_2Cl]_2 + 12NOCl + 3S_2Cl_2 + 12N_2O$ [6]. $S_3N_3Cl_3$ is oxidized by NO_2 and the ring completely destroyed. The reaction proceeds similarly to that of NO_2 with S_4N_4 (see "Schwefel" B 3, 1963, p. 1548), forming the compound $(NO)_2S_2O_7$ [6, 9].

$(CF_3)_2NO$. $S_3N_3Cl_3$ reacts with excess $(CF_3)_2NO$ to form $S_4N_4[ON(CF_3)_2]_4$ [18, 19]. Reaction of $S_3N_3Cl_3$ with iodine and a large excess of $(CF_3)_2NO$ at room temperature for 12 h also yields $S_4N_4[ON(CF_3)_2]_4$ [19].

$(CF_3)_2CN_2$. When $S_3N_3Cl_3$ is reacted with bis(trifluoromethyl)diazomethane, $(CF_3)_2CN_2$, in a 1:3 mole ratio in a glass bomb at 40°C, ring cleavage occurs: $S_3N_3Cl_3 + 3(CF_3)_2CN_2 \rightarrow 3(CF_3)_2C{=}N\text{–}SCl + 3N_2$ [20].

Gmelin Handbook
S-N Comp. 2

Figure 12.6. (*Continued*)

The *Dictionary of Inorganic and Organometallic Compounds on CD-ROM,* published by Chapman & Hall, was referred to earlier in this chapter.

Inorganic Syntheses, published by John Wiley & Sons, New York, is an ongoing series that provides a compilation of current techniques and ideas in inorganic synthetic chemistry. Volume 31 appeared in 1996. Another ongoing series from the same publisher is *Progress in Inorganic Chemistry;* some 46 volumes are available as of 1997.

Compounds of the Type 2-R(5-R')$C_6H_6Fe(CO)_3$ and [2-R(5-R'_3M)$C_6H_6Fe(CO)_3$]X (M = N or P).
Further information on numbers preceded by an asterisk is given at the end of the table, pp. 216/28.
For abbreviations and dimensions see p. 484.

No.	5-R'	con-forma-tion	method of preparation (yield in %), properties and remarks	Ref.
24	-P(OC_4H_9-n)$_3$]BF_4	–	IVs IR (Nujol): 1030 (BF_4^-) IR (acetone): 1980, 2055 mass spectrum: $[M - C_4H_9]^+$	[28, 41]
25	-P(C_4H_9-n)$_3$]BF_4	–	IVs	[41, 71]
*26	-P(C_6H_5)(CH_3)C_6H_4P(C_6H_5)CH_3-2]BF_4	exo	see "Further information"	[72, 78]
		endo	see "Further information"	[72, 78]
*27	-P(C_6H_5)$_2$$C_6H_9$(C($CH_3$)$_2$H-6)$CH_3$-3]$BF_4$	–	see "Further information"	[79]
*28	-P(C_6H_5)$_2$CH(CH_3)CH(CH_3)P(C_6H_5)$_2$]BF_4	–	see "Further information"	[79]
29	-P(C_6H_5)$_3$]BF_4	–	IVo	[71]
*30	CH_3	exo	Ia, IVk (94), IVl (50) b.p. 78 to 80°/2.5 Torr (as mixture from Method Ia), oil ^{1}H NMR (CCl_4): 0.96 (d, CH_3; J = 7), 1.28 (m, H-6 exo; J = 10), 2.10 (m, H-5, 6 endo), 2.70 (dd, H-4; J = 3), 3.34 (m, H-1), 3.73 (s, OCH_3), 5.15 (dd, H-3; J = 1.5)	[9, 11, 26, 45]
		endo	see "Further information"	[11, 83]

Figure 12.6. *(Continued)*

12.6. KEEPING CURRENT WITH AND IDENTIFYING BOOKS

There are a number of ways to search for and keep up-to-date with books. *Chemical Abstracts* is a very good source for books on topics of interest to chemists and chemical engineers, even though its coverage of books is not as thorough as its coverage of journal papers and patents. Nevertheless, *CA* is one of the most complete sources for listings of new books of chemical interest. In the printed *CA,* books are listed in a special place in each section, and they may be searched for online by use of the convention "B/DT" (document type = book). In addition, one of the *CA SELECTS* series, *New Books in Chemistry,* offers a very convenient form of current alerting.

The Library of Congress catalog, available on the Internet (one address is http://lcweb.loc.gov/z3950/gateway.html) and through commercial hosts, offers outstanding coverage for books of all types on a worldwide basis. In addition, the catalogs of the libraries of The Royal Society of Chemistry and of the Canada Institute for Scientific and Technical Information, also available on the Internet as noted in Chapter 10, offer excellent coverage of books related to chemistry, as do the online catalogs of many university libraries.

An example of a newer source for book information is UnCover, Engelwood, CO (800-547-7707), which provides an electronic alerting service about new books on subjects as specified by the user. Delivery of these alerts is by e-mail; this is a companion to the comparable service from the same firm that offers current alerting information about new journal papers.

ISI, discussed more fully in Chapter 8, publishes the *Index to Scientific Book Contents,* which contains complete bibliographic data on the individual chapters of the world's most recently published multiauthored scientific and technical books as well as single-authored books that are part of a multiauthored series. Available in print, it is published quarterly. Approximately 36,000 chapters from 2400 books and book series are covered annually.

There are many other ways to keep current with new books. For example, *Chemical and Engineering News* provides extensive listings at regular intervals. The *Journal of Chemical Education* has, for years, offered a book buyer's guide in its September issues. The book exhibits at national American Chemical Society meetings offer an excellent way to inspect first hand the latest offerings from the leading publishers. Many of the leading research journals in chemistry will offer book reviews.

Some of the most important chemistry book publishers over the years have included the following examples: American Chemical Society, CRC Press (now including Lewis Publishers), Elsevier, Georg Thieme Verlag, John Wiley & Sons (now including Verlag Chemie), Marcel Dekker, McGraw-Hill, Prentice-Hall (primarily college textbooks), The Royal Society of Chemistry, Springer-Verlag, and others. The preceding list is not intended to be complete; there are other important publishers in chemistry as well. In addition, mergers, acquisitions, and publisher policy changes will affect this list. Thus, ACS is in the process of repositioning its book publishing efforts so that it will focus on publishing of symposia and proceedings. On the other hand, CRC Press is believed to be one of the fastest-growing chemistry book publishers, with special emphasis on volumes for reference use.

II. JOURNALS

12.7. INTRODUCTION

To provide an initial perspective, *Chemical Abstracts* coverage includes some 8000 chemistry journals from all over the world. By way of further categorization, *CA* designates about 1350 journals as "core"; these are covered by the online file *CAplus* cover-to-cover and on an especially timely basis. As previously noted (Chapter 10), only a few of the journals covered by *Chemical Abstracts* are exclusively Internet journals.

Almost all reputable journals in chemistry are refereed; that is, before manuscripts are accepted and published, they must go through a time-honored peer review process, usually involving two or more other chemists whose identity is usually kept confidential. The reviewers decide whether the science is sound and whether the paper belongs in the journal to which it has been submitted.

Most chemists agree that the referee system works well, but others complain that refereeing may introduce unnecessary delays and needless attention to details. Also, it is claimed that "unknown" authors from less well known locations may have trouble getting published. A new journal, *Materials Research Innovations,* edited by Professor Rustum Roy, Pennsylvania State University, and published by Springer-Verlag, will accept papers based to a great extent on the record of the author as evidenced by past publications in reputable journals. Prospective authors must also explain to the editors why their material is appropriate and new. In the case of new authors, "sponsors" will be required.

To circumvent the peer review system and/or to get the results of research published more quickly, some investigators post their work on the Internet; this is not common in chemistry, but it is more common in other sciences such as physics. Most chemistry journal editors will not accept for publication any work that has been previously posted on the Internet. Furthermore, such postings can cause problems with any patents that the authors of the postings may wish to file. In general, unrefereed research posted on the Internet is not recommended as a reliable source.

On the other hand, electronic submission of papers to journal editors is expected to considerably speed up the entire publications process. It is significant that over 90% of all accepted manuscripts for American Chemical Society primary journals are now submitted electronically. A mechanism for transmittal of such manuscripts through the Internet is being considered. In a landmark development, the American Chemical Society has announced that, beginning in 1998, journal articles will be posted on the Internet as soon as they have been peer reviewed, copy edited, and checked by authors for accuracy. This will make these articles accessible to subscribers weeks before the print edition.* (Similarly, it is reasonable to expect that books will begin appearing on the Internet once security methodology is further improved.)

The most important and widely read journals in chemistry are the primary research journals, such as the *Journal of the American Chemical Society.*

In addition, some journals strive for especially rapid publication of results. Rapid publication journals often contain the word "Letters" in their titles. They may alternatively use the word "Communications" in their titles. According to Chemical Abstracts Service staff, the latest *CASSI* compilation of journals contains the rather astonishing total of approximately 100 "Letters" journals. These figures alone show how challenging it can be to keep up with all that is new on a current basis.

Other types of journals in addition to the primary or basic research journals, include, for example, review journals, technoeconomic publications, and general chemical news publications. The latter are, in fact, magazines rather than journals, and they are usually unrefereed, but they often provide clues as to important papers appearing

*This service has the name *"ASAP";* see pubs.acs.org/journals/asap. ASAP = As Soon As Publishable.

in the primary journals. In addition, there are a number of important magazines dedicated to specific branches of the chemical industry. These so-called trade publications typically contain some technical papers, frequently unrefereed, and in addition, news and analysis about what is going on in the particular industry covered from the standpoint of business, markets, and technology. One good example is *Modern Plastics,* published monthly in New York, NY. Such publications are essential for those working or interested in the specific industry covered.

The major journal publishers in chemistry include both professional societies and commercial publishers. The leading professional societies that publish research journals of interest to chemists include the American Chemical Society and The Royal Society of Chemistry. Commercial publishers of research journals of interest to chemists include Elsevier Science Publishers B. V. and John Wiley & Sons, Inc., to name just two of the leaders. Some specific publishers and journals are noted below. This list is meant to be representative, but it is by no means all-inclusive.

12.8. AMERICAN CHEMICAL SOCIETY JOURNALS

The Publications Division of ACS publishes primarily basic research journals. The flagship publication is *Journal of the American Chemical Society,* more commonly known as *JACS,* which began publication in 1879. The quality and quantity of information published by the ACS journals is impressive. High quality is achieved because there is a group of distinguished editors, and because all manuscripts submitted to the basic journals are subject to a stringent referee system before being accepted and published. In 1995, ACS journals published a total of 120,250 pages that contained 16,541 articles. Frequency of publication is on a weekly, biweekly, monthly, or bimonthly basis. In addition, ACS acts as a U.S. agent for a few of the publications of its sister societies in other countries. Virtually all of the editors are affiliated with universities. Dr. Mary E. Scanlan is Director of Journal Publishing Operations at the ACS offices in Washington, DC.

A few of the ACS publications are of a technoeconomic nature. An example is *Chemical and Engineering News,* which is oriented mostly to current technical, business, and public policy news. Another such publication is *Today's Chemist at Work,* and *CHEMTECH* also contains technical articles as well as articles that are of commercial or nonscientific interest.

All the ACS journals are available in conventional printed form. In addition, a number of alternative delivery methods are available. All ACS journals are available in microfilm, and some are also in microfiche form. In addition, the text of *Chemical and Engineering News* is searchable and displayable on STN from 1991. This is an important weekly news publication that often provides early announcement of significant technical developments, as well as product, business, and market news.

Primary ACS journals have been available online through STN International for some years. Full-text display and searching is available from 1982 to date, with page images, including half-tones, available from 1992 to date. ACS journals in full-text format first became available on the Internet, in 1996, with the *Journal of Physical Chemistry, Environmental Science and Technology,* and *Biochemistry;* all primary research journals from ACS became available on the Internet in 1997 through such addresses as www.pubsacs.org. Availability will be on a subscription basis, with deep discounts anticipated for those who also subscribe to the printed journal.

Beginning in 1995, for all ACS journals, the so-called "Supporting Information" (formerly known as "Supplemental Material"), began to be available to subscribers only on the Internet under the location http://pubs.acs.org/. Readers can continue to obtain photocopies of this material for all the journals. Subscriptions to the microfilm versions for each journal include "Supporting Information." For some journals, microfiche editions are available, and these include "Supporting Information." "Supporting Information" is usually voluminous material (such as tables) of relatively secondary interest that is not published with the original journal paper.

ACS primary journals are also available through the CAS *SciFinder* client/server product as described in Chapters 6 and 7 of this book. In addition, ACS offers on CD-ROM the *Journal of the American Chemical Society* and *Biochemistry,* with all illustrations, tables, and photographs.

The titles of journals published by ACS, frequency of issue, names of their editors at this writing, content (if not obvious from journal titles), and years of first publication are listed below.

Accounts of Chemical Research (monthly). Joan S. Valentine, UCLA. Features short critical articles to keep readers current on major advances in chemical research. 1968.

Advance ACS Abstracts (semimonthly). Includes author abstracts for papers accepted for publication in 23 journals. Also includes tentative publication date. Covers over 14,000 abstracts each year. Bibliographic content appears online in STN (*CAplus File*). An excellent source of advance information. (Discontinued at the end of 1997.)

Analytical Chemistry (semimonthly). Royce W. Murray, University of North Carolina. Subscription includes the buyer's guide (*LabGuide*); this guide is publicly available on the Internet. 1929.

Biochemistry (weekly). Gordon G. Hammes, Duke University Medical Center. 1962.

Bioconjugate Chemistry (bimonthly). Claude F. Meares, University of California, Davis. Multidisciplinary research in conjugation chemistry. 1990.

Biotechnology Progress (bimonthly). Joint with AIChE since 1990. Jerome G. Schultz, University of Pittsburgh. Covers research reports, reviews, and news. 1985.

Chem Matters. A chemistry magazine for high-school students. David A. Robson, Towson State University. 1983.

Chemical and Engineering News (weekly). Madeleine Jacobs, ACS staff. 1923. Spun off as News Edition of *Industrial and Engineering Chemistry.*

Chemical Health and Safety (bimonthly). Dr. Carl Goettschall. A news magazine about handling chemicals in the workplace. 1994.

Chemical Research in Toxicology (eight issues per year). Lawrence A. Marnett, Vanderbilt University. 1988.

Chemical Reviews (eight issues per year). Josef Michel, University of Colorado. Comprehensive expert reviews in major branches of chemistry. Some issues focus on a single theme. 1924.

Chemistry of Materials (monthly). Leonard V. Interrante, Rensselaer Polytechnic Institute. Includes research on preparation, processing and analysis of organic and inorganic solids, ceramics, polymers, liquid crystals, and composites. 1989.

CHEMTECH (monthly). Abraham P. Gelbein, Washington, DC. Primarily technical, but also management and business articles on all phases of the chemical industry. 1971.

Energy & Fuels (bimonthly). John W. Larson, Lehigh University. Nonnuclear energy sources. 1987.

Environmental Science & Technology (monthly). William H. Glaze, University of North Carolina. Research and news reports. A condensed version entitled *Environmental Science & Technology: News and Research Notes* is also available; this contains a full magazine, except for the research papers that are presented in abstract form only. 1967.

Industrial and Engineering Chemistry Research (monthly). Donald R. Paul, University of Texas. 1962.

Inorganic Chemistry (biweekly) M. Frederick Hawthorne, UCLA. 1962.

Journal of Agricultural and Food Chemistry (monthly). Irvin E. Liener, University of Minnesota. 1953.

Journal of the American Chemical Society (weekly). Allen J. Bard, University of Texas. 1879.

Journal of Chemical and Engineering Data (bimonthly). Kenneth N. Marsh, Texas A & M University. High-quality experimental data on organic and inorganic compounds and their mixtures. Also includes reviews, evaluations, and predictive schemes. 1959.

Journal of Chemical Information and Computer Sciences (bimonthly). George W. A. Milne, National Institutes of Health. 1961.

Journal of Medicinal Chemistry (biweekly). Philip S. Portoghese, University of Minnesota. 1963.

Journal of Natural Products (monthly). Published jointly with the American Society of Pharmacognosy beginning in 1996. 1938.

The Journal of Organic Chemistry (biweekly). Clayton H. Heathcock. University of California, Berkeley. 1936.

Journal of Pharmaceutical Sciences (monthly). Published jointly with the American Pharmaceutical Association since 1994. William A. Higuchi, University of Utah. 1912.

Journal of Physical and Chemical Reference Data (bimonthly). Published jointly with the American Institute of Physics, and the National Institute of Standards and Technology. Jean W. Gallagher, National Institute of Standards and Technology. Covers, for example, data compilations and reviews. 1972.

Journal of Physical Chemistry, The (weekly). Mostafa A. El-Sayed, Georgia Institute of Technology. 1896.

Langmuir (biweekly). William A. Steele, Pennsylvania State University. Covers surface and colloid chemistry. 1985.

Macromolecules (biweekly). Robert W. Lenz, University of Massachusetts. 1968.

Organic Process Research & Development (bimonthly). Trevor Laird, Scientific Update, Mayfield, East Sussex, UK; John F. Arnett, Recombinant BioCatalysis, Sharon Hill, PA; and Richard Pariza, C & P Associates, North Zion, IL. Copublished with The Royal Society of Chemistry beginning in 1997. This new journal is to report original work-in-process chemistry in the fine organics and specialty chemical industries, including pharmaceuticals and a number of other areas.

Organometallics (biweekly). Dietmar Seyferth, Massachusetts Institute of Technology. 1982.

Today's Chemist at Work (monthly). Patrick P. McCurdy, American Chemical Society. Features of broad general interest to chemists working in industry. 1988.

Some of the sister society journals that are not published by ACS but that are available from ACS are as follows:

Angewandte Chemie (semimonthly). International edition in English. A publication of the German Chemical Society. Includes results of research in organic and inorganic chemistry. Available online through STN International under the name CJVCH and produced by VCH Verlagsgesellschaft.

Bulletin of the Chemical Society of Japan (monthly). Tadamasa Shida, Kyoto University, Japan. A publication of the Chemical Society of Japan. The most cited international chemical journal of Japan.

Chemistry & Industry (semimonthly). Andrew Miller, Society of Chemical Industry, UK. A publication of the Society of Chemical Industry, UK. News, reviews, and comments on research, business, and public policy from around the world, with emphasis on Europe.

Chemistry Letters (monthly). A publication of the Chemical Society of Japan. Distinguished by rapid distribution of results.

Beginning in 1998, individual American Chemical Society journal papers are to be available on the Internet after they are peer reviewed, copyedited, and checked again by the authors. This will make them available several weeks before the printed versions of the journals in which they will appear.

12.9. JOURNALS AND RELATED PUBLICATIONS OF THE ROYAL SOCIETY OF CHEMISTRY

A toll-free phone number is conveniently available in the United States (800-473-9234), and the Internet address is http://chemistry.rsc.org/rsc/.

The Royal Society of Chemistry (RSC) publishes a number of significant journals and related publications. These include basic research journals, technoeconomic publications, review journals, and current alerting services. Some of these journals represent some highly innovative publishing methods as described below. Robert Parker is responsible for the principal journal and review publications.

Origins of some RSC publications go back many years. The Chemical Society of London first published its proceedings in 1841, and that is the basis of today's *Journal of the Chemical Society. The Analyst* was first published by the Society of Public Analysts in 1876. The *Faraday Journals, Transactions, and Discussions* had their origins in 1903 when the Faraday Society was formed. The Faraday Society was subsequently amalgamated with The Chemical Society, The Royal Institute of Chemistry, and the Society for Analytical Chemistry (previously Society of Public Analysts). RSC was granted its Royal Charter on June 1, 1980, as the sole heir and successor to the previously mentioned groups.

The RSC copublishes a few journals with its sister societies in other countries, and it distributes American Chemical Society books in the United Kingdom. ACS distributes RSC books in the United States. The RSC and ACS are to copublish a journal from January 1997: *Organic Process Research and Development.*

In addition to the printed versions, the following RSC journals are online in full text through STN International: *The Analyst, Analytical Communications, Annual Reports on the Progress of Chemistry, Chemical Communications, Chemical Society Reviews, Contemporary Organic Synthesis, Dalton Transactions, Faraday Discussions, Faraday Transactions, Journal of Analytical Atomic Spectrometry* (also available in diskette form for the PC), *Journal of Chemical Research, Journal of Materials Chemistry, Mendeleev Communications, Natural Product Reports, Perkin Transactions 1,* and *Perkin Transactions 2.* They are also available through Chemical Abstracts Service's *SciFinder* client/server electronic product. In 1997, all 12 primary RSC journals became available, in fully searchable form, on the Internet, and plans call for other versions of specific journals on additional electronic hosts. Electronic access to the contents pages is free; full journal access requires that the user (institution) be a subscriber and pay a small surcharge.

The flagship publication of the RSC is *Chemical Communications* (24 issues annually), Robert Parker, Editor. From January, 1996, the prefix *Journal of the Chemical Society* was dropped. This covers all branches of chemistry with emphasis on rapid publication of short communications that may be followed up by a full paper in the corresponding section of *Journal of The Chemical Society* when fuller results are known. Since January 1996, it has included a "Feature Article" in each issue.

Another very important publication of the RSC is the *Journal of The Chemical Society.* This is subdivided into the following separate journals (unless otherwise shown, managing editors are located at The Royal Society of Chemistry offices, Cambridge; when there are several editors, only one name is given below):

1. *Dalton Transactions,* including *Dalton Communications* and *Dalton Perspectives.* Janet L. Dean. Inorganic, bioinorganic, and organometallic chemistry. 24 issues annually.
2. *Faraday Transactions,* including *Faraday Research Articles* and *Faraday*

Communications (24 issues yearly). Rosemary A. Whitelock. Physical chemistry and chemical physics.

3. *Perkin Transactions 1* (24 issues yearly). Sheila R. Buxton. Organic and bioorganic chemistry. Includes full papers and a "Communications" section.
4. *Perkin Transactions 2* (monthly). Sheila R. Buxton. Physical organic chemistry. Includes full papers, "Keynote Articles," and a "Communications" section.

Other Royal Society of Chemistry journals and related publications include the following:

Analyst, The (monthly). Harpal S. Minhas. A leading international journal in analytical chemistry. Contains full articles, reviews, tutorial reviews, perspectives, and news items.

Analytical Communications (monthly). Harpal S. Minhas. Communications are refereed. Formerly titled *Analytical Proceedings.*

Chemistry in Britain (monthly). Includes scientific articles of general chemical interest, professional and industrial matters affecting science, news, announcements, and book reviews.

Education in Chemistry (bimonthly).

Faraday Discussions (three volumes yearly). Forum for exchange of views and new results in physical chemistry and chemical physics. Contains original research and discussion comments. Rosemary Whitelock.

Faraday Symposia. Between January 1986 and July 1994, this material was included in *Faraday Transactions.* It is now published as part of *Faraday Discussions.*

Journal of Analytical Atomic Spectrometry (monthly). Brenda Holliday. Includes primary papers, Atomic Spectrometry Updates (bimonthly), and rolling list of references to journal articles and conference papers (10 times a year). Also available are diskettes containing full bibliographic references to journal articles and conference papers since 1985 fully indexed by analyte, matrix, technique, or mode of sample preparation. Atomic Spectrometry Updates tables and references are available in diskette form for the PC.

Journal of Chemical Research (monthly). Marcus Ennis. Sponsored jointly by The Royal Society of Chemistry, Gesellschaft Deutscher Chemiker, and Societe Francaise de Chimie. Covers all areas of chemistry. Part S is a conventionally printed journal that consists of synopses of research papers. Each synopsis is in English and is one or two pages long. Part S also contain short papers that are brief but give a complete description of work. Part M contains reproductions of full texts of author manuscripts, typed in English, French, or German, and corresponding to the synopses; it is available in microfiche or miniprint form. It is the only pure chemistry journal to be published in both synopsis and microform formats concurrently.

Journal of Materials Chemistry (monthly). Janet L. Dean. Interdisciplinary coverage of materials. Cotains full papers, communications, and feature articles.

Mendeleev Communications (bimonthly). Andrew Wilkinson. Sponsored jointly by The Royal Society of Chemistry and the Russian Academy of Sciences and published in English. Provides rapid access to preliminary accounts of work from the Russian Federation.

12.10. PUBLICATIONS OF AIChE

Journals published by AIChE (American Institute of Chemical Engineers, New York, NY) include the following:

AIChE Journal, edited by M. V. Tirrell, University of Minnesota. Includes both theoretical and applied research papers.

Chemical Engineering Progress (*CEP*), edited by M. R. Rosenzweig, AIChE staff. There is also a separate Student Edition. Includes technical articles, book reviews, and news stories. CEP subscribers receive a valuable software directory once a year with the December issue. This directory includes well over 1500 programs of potential interest to chemical engineers and others in the chemical industry. The compilation embraces offerings from 500 vendors worldwide, and the titles are grouped in 32 categories such as environmental control, waste management, materials, and process design–simulation. Available for separate purchase by others.

Biotechnology Progress, edited by J. S. Shultz, University of Pittsburgh. Joint with the American Chemical Society. Bimonthly joint with the American Chemical Society.

Environmental Progress, edited by G. F. Bennett. Quarterly publication on all aspects of pollution control.

Process Safety Progress, edited by T. A. Ventrone. Quarterly publication on design, operation, and maintenance of safe installations in the chemical process industries.

12.11. ELSEVIER SCIENCE PUBLISHERS, B.V.

Elsevier (Amsterdam), a commercial firm that is believed to be the world's largest science publisher, offers over 1100 journals, many of which are of interest to chemists. In addition to the conventional printed versions, a number of interesting electronic options are available. For example, almost all of these journals are available through what is called *EES* (*Elsevier Electronic Subscriptions*). Under this plan, libraries at companies or universities can subscribe to complete electronic editions either in addition to or in lieu of print editions. For the libraries who get the electronic subscriptions, the principal benefit is, of course, very substantial savings in space. For the chemist utilizing the electronic subscriptions, the benefits can include access to the material at one's desktop, and fully searchable article titles and abstracts.

In addition to *EES,* Elsevier offers several electronic current awareness options for its science journals. Most of these options are free, although some are available to subscribing institutions only. These can be accessed through, or are described on, the Elsevier Internet page, www.elsevier.nl. These services include *ESTOC* (*Elsevier Science Tables of Contents*), *ContentsDirect,* and *Contents Alert ESTOC. ESTOC* coverage begins in January 1995 and includes the tables of contents for almost all 1100 Elsevier Science journals within about 6 weeks after original publication.

ContentsDirect provides tables of contents through e-mail about 2–4 weeks before publication for over 100 Elsevier journals.

Contents Alert provides prepublication e-mail tables of contents for certain subjects such as fullerenes and "surfaces, interfaces, and thin films," as covered in Elsevier journals.

A good example of innovative journal publishing is Elsevier's journal on molecular biology, *GENE-COMBIS.* This is a peer-reviewed online product available to those who already subscribe to the print journal *Gene.* Manuscripts are submitted electronically and then are submitted to the editorial board for what is said to be quick peer review. The subject matter is computing problems in the field of molecular biology. The user is provided with other features such as online access to major databases of nucleotide and protein sequences as well as databases of bibliographic references and a subscriber's e-mail directory.

Elsevier's *Tetrahedron* family of journals has an enviable reputation for both good quality and prompt publication. The family includes *Tetrahedron Letters, Tetrahedron, Tetrahedron: Asymmetry, Bioorganic & Medicinal Chemistry,* and *Biorganic and Medicinal Chemistry Letters.* The Internet-based system *Tetrahedron Alert* provides chemists with access to graphical abstracts for all accepted papers in the *Tetrahedron* family 4 weeks prior to publication. Abstracts may be searched and viewed for approximately the most recent 3 months.

The following Elsevier journals are available for online full-text searching on STN International: *Analytica Chimica Acta, Applied Catalysis, Carbohydrate Research, Journal of Organometallic Chemistry,* and *Vibrational Spectroscopy.*

Elsevier ScienceDirect (www.sciencedirect.com) is an emerging Internet site that is to include all of Elsevier Science's academic journals and those of participating publishers. As mentioned, hundreds of Elsevier journals are already available through subscriptions to the electronic formats in many libraries.

12.12. JOHN WILEY & SONS, INC.

John Wiley & Sons, New York, NY, publishes a number of journals of interest to chemists, in addition to the Wiley line of reference and textbooks in chemistry. A number of these journals are online in full text through STN International. The Wiley home page on the Internet, www.wiley.com, includes descriptions of the Wiley journals and a number of valuable links to other chemistry-related information.

A listing of Wiley journals of interest to chemists includes the following:

Advanced Materials for Optics and Electronics (combines *Chemtronics* and *The Journal of Molecular Electronics*). Professor D. J. Cole-Halmilton, Editor-in-

Chief, University of St. Andrews, Fife, Scotland. Provides a forum for the emerging knowledge of advanced materials whose focus is information technology.

Advances in Polymer Technology. Dr. Marino Xanthos, Editor-in-Chief, Stevens Institute of Technology, Hoboken, NJ. Important developments in polymeric materials, production, and processing methods, and equipment and product design.

Analytical Methods and Instrumentation. H. Michael Widmer, Editor-in-Chief, Ciba, Ltd., Basel, Switzerland. New developments in methods and applications; comparisons between and limitations of particular methods of general analysis; novel instrumental developments, including analytical systems; significant applications from industry.

Applied Organometallic Chemistry. P. J. Craig, Editor-in-Chief, DeMontfort University, Leicester, UK. Reviews, original papers, short communications, and reports of relevant conferences.

Biomedical Chromatography. C. K. Lim, Editor-in-Chief, University of Leicester, Leicester, UK. Original papers on the application of chromatography and allied techniques in the biological and medical sciences.

Biopolymers. Murray Goodman, Editor, University of California at San Diego, La Jolla, CA. Original, innovative research papers on the structure, properties, interactions, and assemblies of biomolecules. Searchable online through STN.

Biospectroscopy. Laurence A. Nafie, Editor, Syracuse University, Syracuse, NY. Interdisciplinary journal concerned with the application of spectroscopic techniques to problems of biological significance.

Color Research and Application. Ellen C. Carter, Editor, Arlington, VA. Research reports and applications-oriented articles from the world's leading color authorities.

Fire and Materials. John P. Redfern, Editor-in-Chief, Surrey, UK. International journal for communications directed at fire properties of materials and the products into which they are made.

Flavour and Fragrance Journal. Dr. Roger Stevens, Editor, Threlkeld, Keswick, Cumbria, UK. International journal devoted to the rapid publication of scientific and technical papers on essential oils and related products.

Heterogeneous Chemistry Reviews. Professor David Avnir, Editor-in-Chief, The Hebrew University, Jerusalem, Israel. Broad-spectrum interdisciplinary review journal, aimed at facilitating information flow between diverse scientific fields, which commonly deal with chemistry in multiphase environments.

International Journal of Chemical Kinetics. David M. Golden, Editor, SRI International, Menlo Park, CA. Quantitative relationships between molecular structure and chemical activity.

International Journal of Quantum Chemistry. Per Olov Löwdin, Editor-in-Chief, Uppsala College, Uppsala, Sweden. Advanced information on quantum mechanics, fundamental concepts, mathematical structure, and applications to atoms, molecules, crystals, and molecular biology.

Journal of Applied Polymer Science. Eric Baer, Editor, Case Western Reserve University, Cleveland, OH. Progress and significant results in the systematic practical application of polymer science. Searchable online through STN.

Journal of Bioluminescence and Chemiluminescence. L. J. Kricka, Editor-in-Chief, Hospital of the University of Pennsylvania, Philadelphia, PA. Fundamental and applied aspects of chemiluminescence and bioluminescence.

Journal of Chemical Technology and Biotechnology. J. Melling, Chairman, Center for Applied Microbiology and Research, Porton Down, UK. Provides vital information relating scientific discoveries and inventions in biotechnology and chemical technology to their conversion into commercial products and processes.

Journal of Chemometrics. Professor Steven D. Brown, Editor-in-Chief, University of Deleware, Newark, DE. Fundamental and applied aspects of chemometrics.

Journal of Computational Chemistry. Dr. Norman L. Allinger, Editor. University of Georgia, Athens, GA. Concerned with all aspects of computational chemistry.

Journal of Labelled Compounds and Radiopharmaceuticals. J. R. Jones, Editor, University of Surrey, Guildford, UK. Research and development leading to and resulting in labeled compound preparation.

Journal of Mass Spectrometry. Richard M. Caprioli, Editor, University of Texas Medical School, Houston, TX. Organic and biological mass spectrometry and gas-phase ion processes.

Journal of Molecular Recognition. Irwin M. Chaiken, Editor-in-Chief, Smithkline Beecham, Philadelphia, PA. Multidisciplinary journal devoted to communicating original research articles and reviews on the principles, characterization, methods, and biotechnological applications of molecular recognition.

Journal of Peptide Science. Conrad J. Schneider, Editor-in-Chief, Institut fur Klinische Immunologie, Bern, Switzerland. Original articles of significant experimental research in peptide science.

Journal of Physical Organic Chemistry. Professor Joseph B. Lambert, Editor-in-Chief, Northwestern University, Evanston, IL. International forum for the rapid publication of original scientific papers dealing with physical organic chemistry in its broadest sense.

Journal of Polymer Science. Eric J. Amis, Editor, National Institute of Standards and Technology, Gaithersburg, MD. Fundamental research in all areas of polymer chemistry and physics. Available in Parts A, B, and C, all of which are also searchable online through STN.

Journal of Raman Spectroscopy. D. A. Long, Editor-in-Chief, University of Bradford, Bradford, West Yorkshire, UK. Original work and reviews in all aspects of Raman spectroscopy.

Journal of the Science of Food and Agriculture. M. P. Tombs, Chairman, University of Nottingham, UK. Original research and critical reviews in agriculture and food science, with particular emphasis on interdisciplinary studies at the agriculture–food interface.

Journal of Thermal Analysis. J. Simon, Editor, Technical University, Budapest, Hungary. Thermal investigations.

Magnetic Resonance in Chemistry. Professor H. Günther, Editor-in-Chief, Universität Siegen, Siegen, Germany. Application of NMR, ESR, and NQR sectrometry in chemistry.

Mass Spectometry Reviews. Dominic M. Desiderio, Editor, University of Tennessee at Memphis, Memphis, TN. Review articles of current research on mass-spectrometry instrumentation and application in chemistry, biology, environmental science, medicine, agriculture, engineering, and physics.

NMR in Biomedicine. John R. Griffiths, Editor-in-Chief, St. George's Hospital Medical School, London, UK. Original papers in which nuclear magnetic resonance methods are used for investigating basic biochemical and medical problems.

Packaging Technology and Science. F. A. Paine, Editor, Chidding Fold, Surrey, UK. Research papers and review articles about developments in this field.

Pesticide Science. G. T. Brooks, Chairman, Brighton, UK. All aspects of the research and development, application, use, and impact on the environment of products designed for pest control and crop protection.

Polymer International. Professor J. F. Kennedy, Chairman, The North East Wales Institute, Deeside, Clwyd, UK. Papers and critical reviews from macromolecular science and technology.

Polymers for Advanced Technologies. Manachem Lewin, Editor-in-Chief, Polytechnic University, Brooklyn, NY. New areas of polymer research and development related to advanced technologies.

Rapid Communications in Mass Spectrometry. Professor John H. Benyon, FRS, Editor-in-Chief, University of Wales, Swansea, UK. Rapid publication of original research ideas and results on all aspects of the science of gas-phase ions.

Surface and Interface Analysis. David Briggs, Editor-in-Chief, Wilton Research Centre, Wilton, Middlesborough, Cleveland, UK. Development and application of characterization techniques for surfaces, interfaces, and thin films.

X-Ray Spectrometry. John V. Gilfrich, Editor-in-Chief, Bethesda, MD. Rapid publication of papers dealing with the theory and application of X-ray spectrometry.

12.13. CRC PRESS (BOCA RATON, FL)

CRC offers several journals of interest to chemists, with emphasis on review-type publications. These include the following examples:

Critical Reviews in Analytical Chemistry

Critical Reviews in Biochemistry and Molecular Biology

Critical Reviews in Biotechnology

Critical Reviews in Environmental Science and Technology
Critical Reviews in Microbiology
Ozone: Science and Engineering
Critical Reviews in Solid State and Materials Science
Critical Reviews in Toxicology

CRC has begun putting some of its journals online through its Internet home page (www.crcpress.com) on an experimental basis.

12.14. SPE (SOCIETY OF PLASTICS ENGINEERS) (BROOKFIELD, CT)

Journals published by SPE include.

Journal of Injection Molding Technology
Journal of Vinyl and Additive Technology
Polymer Engineering & Science
Plastics Engineering
Polymer Composites

In addition, SPE publishes preprints for RETEC (Regional Technical Conference) and ANTEC (Annual Technical Conference), both of which are important for polymer chemists and engineers.

12.15. SPRINGER-VERLAG (BERLIN AND NEW YORK)

This well-known German publisher announced in 1997 that most of its printed journals are to be available online through the Internet (http://science.springer.de/chemol/chemol.htm).

Accreditation and Quality Assurance
Amino Acids
Applied Magnetic Resonance
Bioprocess Engineering
The Chemical Educator (electronic edition only)
The Chemical Intelligencer
Chinese Journal of Chemical Engineering
Colloid & Polymer Science
Computing and Visualization in Science
Experiments in Fluids
Fresenius' Journal of Analytical Chemistry

Journal of Biological Inorganic Chemistry
Journal of Molecular Modeling (electronic edition only)
Journal of Planar Chromatography—Modern TLC
Materials Research Innovations
Mikrochimica Acta
Molecules (electronic edition only)
Monatshefte für Chemie
Physics and Chemistry of Minerals
Polymer Bulletin
Theoretica Chimica Acta
Zeitschrift für Ernaehrungswissenschaft
Zeitschrift für Lebensmittel-Untersuchung und Forschung
Zeitschrift für Physik D

12.16. TECHNOMIC PUBLISHING COMPANY

Technomic Publishing, Lancaster, PA, publishes several useful journals relating to polymer science and technology, including the following:

Journal of Bioactive and Compatible Polymers (quarterly)
Journal of Cellular Plastics (bimonthly)—new developments in foamed plactics chemistry and technology
Journal of Coated Fabrics (quarterly)—advances in coated fabric technology
Journal of Elastomers and Plastics (quarterly)—chemistry, processing, properties, and applications
Journal of Fire Sciences (bimonthly)—research reports on fire behavior and fire retardance of materials and products
Journal of Plastic Film and Technology (quarterly)
Journal of Reinforced Plastics and Composites (monthly)
Journal of Thermoplastic Composite Materials (quarterly)

As previously noted, the same firm also publishes abstracting bulletins on plastics as well as books in the field. Thus, Technomic is a resource to be checked when looking for pertinent literature regarding polymers and plastics, especially from an industrial perspective.

12.17. VCH (WEINHEIM AND NEW YORK)

This firm has recently been acquired by John Wiley & Sons, Inc., New York, NY. As the selected titles below indicate, it publishes some of the best known and oldest journals in chemistry.

Acta Polymerica
Chemical Engineering & Technology
Chemical Technology Europe
Chemical Vapor Deposition
ChemInform
*Chemische Berichte**
*Chemistry—A European Journal**
Helvetica Chimica Acta
Journal of Chemical Research
*Liebig's Annalen**
Materials and Corrosion
Spectroscopy Europe

12.18. CONSULTANTS BUREAU JOURNALS (PLENUM PUBLISHING)

As noted in Chapter 5, Consultants Bureau (CB), New York, NY, and its parent Plenum Publishing, have specialized in English-language versions of Russian journals for many years.

Other journals by Plenum (all English language) include these examples:

Journal of Chemical Crystallography
Journal of Chemical Ecology
Journal of Environmental Polymer Degradation
Journal of Inorganic and Organometallic Polymers
Journal of Protein Chemistry
Journal of Solution Chemistry
Plasma Chemistry and Plasma Processing
Structural Chemistry

12.19. SELECTING JOURNALS; LISTS OF JOURNALS IN CHEMISTRY

There are so many journals (at least 8000) in chemistry and related fields that selecting the most appropriate ones to read and subscribe to is difficult.

Some factors that can bear on this decision include subscription cost, space required to shelve, language, reputation and quality of the journal, availability online or other electronic form, availability in nearby libraries, and time available to read. Inclusion of a journal in a current awareness profile can also be a factor. If a group of chemists is willing to circulate and share the same journal on a prompt basis, that can be advantageous. In considering subscription costs, note that many journals offer three separate rates: academic, industry, and personal or individual; the individual rate is usually the most favorable.

*See important note on page 313.

If reprints or copies of issues are continuously obtained from a specific journal, that is one good indicator that a subscription may be in order. For journals that seem to be constantly disappearing from the shelves, or that circulate around the laboratory very slowly, more than one subscription may very well be in order.

To evaluate a journal, it is prudent to ask the publisher for a few sample issues on a complimentary basis. In addition, most publishers have pages on the Internet, and descriptions and tables of contents of their journals are frequently available there.

Some clues as to the "most important" journals in chemistry may be gleaned from the carefully selected lists of those covered in *Index Chemicus,* The Royal Society of Chemistry's *Methods in Organic Synthesis,* or in the *Beilstein Current Facts* series. These are limited to organic chemistry. *Gmelin* has a comparable list of what it considers key journals in inorganic chemistry. Another limited set of journals is the group of some 1350 core journals covered by the online CAS service, *CAplus;* the titles of the journals on this list can be viewed on the Internet.

One quantitative way to evaluate the relative importance of chemistry and other scientific journals is the product *Journal Citation Reports* (*JCR*), a product of the Institute for Scientific Information, Philadelphia. Citation analysis provides the basis for this tool, which is published annually in microfiche or CD-ROM formats.

JCR answers such questions as how frequently has a journal been cited, by which journals is it cited, how soon after publication do the citations occur, and which journals are cited by a particular journal. Journals are ranked by impact factor (average number of cites for recent articles) within several subject categories. There are several categories related to chemistry, such as general, analytical, inorganic, organic, clinical, physical, and polymer science.

This tool can not only be helpful in selecting journals but also help authors decide where to publish. Citations are, of course, just one factor in making a judgment as to the potential value of journals; there are other important factors as well.

Other related tools of ISI, as described elsewhere in this book, provide additional citation evaluation capability.

A factor that enhances the practical value of journals is the availability of cumulative indexes. For example, the *Journal of Physical and Chemical Reference Data* makes available valuable cumulative indexes in printed form by authors and by properties and classes of substances. Indexes are also on the NIST Internet location (www.nist.gov./srd/jpcrd.htm).

Several good lists of journals of interest to chemists are available. *CAS Source Index* (*CASSI*), a product of Chemical Abstracts Service, is an outstanding source of information about both current and discontinued journals and related publications. It includes extensive data about 70,000 serial and nonserial sources published worldwide.

Its most unique feature is that, for the publications listed, the holdings of these publications in more than 350 research libraries around the world are included as applicable. In addition, there is a KWOC (Keyword Out of Context) index that permits the user to identify titles even if only one word in the title is known. Other data included is information about variant titles, publication histories, English translations of foreign language titles, a directory of publishers and sales agencies, and a guide to the depositories of unpublished works. There is also identification of titles that can be ordered from the CAS Document Detective Service.

The 1907–1994 edition of *CASSI* includes documents covered by CAS services from 1907 to 1994 and, in addition, materials cited in *Beilstein* and *Chemisches Zentralblatt* as far as 1830. *CASSI* is now available in both printed and CD-ROM form. The print edition is updated quarterly, and the CD-ROM product is updated twice a year. The product is also available online through Questel•Orbit.

Another good list is *Ulrich's International Periodicals Directory,* a product of Reed Reference Publishing, New Providence, NJ. The print edition is updated annually (the latest is the 34th edition with a 1996 cover date).

This major five-volume work is an invaluable tool in helping chemists, chemical librarians, and information professionals identify and select journals and other serials for reading, subscription, and publication.

Ulrich's includes over 147,000 journals and other serials in all disciplines, worldwide, all categorized by subject matter covered. There is a separate section devoted to chemistry that includes breakouts by field, such as analytical chemistry and organic chemistry. For each journal, complete basic information is given as to pricing, location, phone, e-mail address, name of editor, description of content, reprint sources, and other valuable information. Any previous titles of the journal (many journals have changed their titles over the years) are also listed and cross-referenced to the most recent title. There are over 3300 basic entries for journals in the chemistry categories and, in addition, many thousands more in related categories.

Unique features for each journal include information on whether a journal is refereed, data on where a journal is indexed (abstracting and indexing services that cover a journal), and notation as to whether the journal is available online or in CD-ROM form and from which vendors. Over 100 journals in the chemistry categories are available in both printed and online form, and over 60 journals in the chemistry are available in both CD-ROM and printed forms.

In addition to being available in printed form, *Ulrich's* is available online through DIALOG. It is also available as a CD-ROM product and on microfiche.

In deciding on which journals to read or skim, the chemist should also be aware of the various current awareness options as described in Chapter 4.

REFERENCES

1. J. I. Kroschwitz, Ed., *Kirk-Othmer Encyclopedia of Chemical Technology,* 4th ed., Wiley, New York, 1991–1998. Also available online.
2. *Ullmann's Encyclopedia of Industrial Chemistry,* 5th ed. (in English), Verlag Chemie, Weinheim, Germany and Deerfield Beach, FL, 1984 to date.
3. *Ullmann's Encyclopädie der Technischen Chemie,* 4th ed., 1972–1984.
4. J. J. McKetta and W. A. Cunningham, Eds., *Encyclopedia of Chemical Processing and Design,* Marcel Dekker, New York, 1976 to date.
5. J. I. Kroschwitz, Ed., *Encyclopedia of Polymer Science and Engineering,* 2nd ed., Wiley, New York, 1990.

6. J. C. Salamone, Ed., *Polymeric Materials Encyclopedia,* CRC Press, Boca Raton, FL, 1996.
7. Richard J. Lewis, Sr., Ed., *Condensed Chemical Dictionary,* 12th ed. (Van Nostrand Reinhold), New York, VNR, 1992.
8. Dominic V. Rosato, *Rosato's Plastics Encyclopedia and Dictionary,* Hanser, Munich, 1993.
9. (a) G. S. Coyne, *The Laboratory Companion: a Practical Guide to Materials, Equipment, and Technique,* Wiley, New York, 1998; (b) A. R. Gennaro, Ed., *Remington: The Science and Practice of Pharmacy,* 19th ed., Williams & Wilkins, Baltimore, 1995. Updated every 5 years. Includes CD-ROM.
10. W. Brich and R. Michaelis, Eds., *Beilstein Handbook of Organic Chemistry,* 4th ed., 4th and 5th Supplements, Springer-Verlag, Berlin, 1972 to date. Now in English.
11. *How to Use Beilstein* (pamphlet); *Stereochemical Conventions* (pamphlet); *What Is Beilstein* (pamphlet).
12. R. Luckenbach, R. Ecker, and J. Sunkel, "The Critical Screening and Assessment of Scientific Results Without Loss of Information—Possible or Not," *Angew. Chem., Int. Ed. Engl.* **20,** 841–849 (1981).
13. J. Buckingham, Ed., *Dictionary of Organic Compounds,* 6th ed., Chapman & Hall, London, 1995, and supplements thereafter, as well as online and CD versions.
14. S. Coffey or M. F. Ansell, Ed., *Rodd's Chemistry of Carbon Compounds,* 2nd ed., Elsevier, Amsterdam, 1964 and subsequent years.
15. *Houben-Weyl Methods of Organic Chemistry,* 4th ed., Georg Thieme Verlag, Stuttgart and New York, ongoing.
16. From *Organic Syntheses,* Vol. 55, p. IX, reprinted with permission of John Wiley, & Sons, Inc. New York.
17. L. A. Paquette, Ed., *Organic Reactions,* New York, Wiley, 1942 to date.
18. L. A. Paquette, Ed., *Encylopedia of Reagents for Organic Synthesis,* Wiley, New York, 1994.
19. M. E. Wolff, Ed., *Burger's Medicinal Chemistry,* 5th ed., Wiley, New York, 1994–1997.
20. A. Finch and W. Theilheimer, Eds., *Synthetic Methods of Organic Chemistry,* S. Karger, A. G. Basel, New York, 1946 to date (approximately annual). Also available online.
21. Michael B. Smith, Ed., *Compendium of Organic Synthetic Methods,* Vol. 8, Wiley, New York, 1995.
22. M. Fieser, Ed., *Fieser and Fieser's Reagents for Organic Synthesis,* Wiley, New York, 1967 to date.
23. Various authors, *The Chemistry of Heterocyclic Compounds—a Series of Monographs,* Wiley, New York, ongoing series.
24. B. P. Mundy and M. G. Ellerd, *Name Reactions and Reagents in Organic Synthesis,* Wiley, New York, 1988.
25. A. Hassner and C. Sturmer. *Organic Syntheses Based on Name Reactions and Unnamed Reactions: Tetrahedron Organic Chemistry Series,* Vol. 11. Pergamon (Elsevier Science, Ltd.) Oxford, UK, 1994.
26. E. Fluck, Ed., *Gmelin Handbuch der Anorganischen Chemie,* 8th ed., Springer-Verlag, Berlin, 1924 to date. New Acting Director is H. W. Vollmann.
27. A. Bruce King, Ed., *Encyclopedia of Inorganic Chemistry,* Wiley, New York, 1994.

References Pertinent to Beilstein

1. W. T. Donner, "Economic Aspects of Chemical Information," *J. Chem. Inf. Comput. Sci.,* **36**(5) 937–941 (1996)
2. L. Goebels, A. J. Lawson, and J. L. Wisniewski, "AUTONOM: System for Computer Translation of Structural Diagrams into IUPAC-Compatible Names. 2. Nomenclature of Chains and Rings," *J. Chem. Inf. Comput. Sci.,* **31**(2), 216–225 (1991).
3. M. G. Hicks, "Surfing the Organic Chemistry Hyperdocument with CrossFire plus Reactions," *J. Chem. Inf. Comput. Sci.,* **37,** 146–147 (1997).
4. M. G. Hicks, "CD-ROM Chemical Databases: The Influence of Data Structure and Graphical User Interfaces on Information Access." *J. Chem. Inf. Comput. Sci.,* **34**(1), 32–38 (1994).
5. R. Lukenbach. "Past Perfect, Present Perfect, Future Perfect—Quality Assessment and Quality Control Mechanisms at Beilstein, *J. Chem. Inf. Comput. Sci.,* **36**(5), 923–929 (1996).
6. J. L. Wisniewski, "AUTONOM: System for Computer Translation of Structural Diagrams into IUPAC-Compatible Names. 1. General Design," *J. Chem. Inf. Comput. Sci.,* **30**(3), 324–332 (1990).
7. L. Wisniewski, "Let An Expert Name That Compound," *CHEMTECH,* pp. 14–16 (1993).
8. T. Stankus, *Making Sense of Journals in the Physical Sciences: From Specialty Origins to Contemporary Assortment,* Binghamton, NY: Haworth Press, 1992.

List of Addresses

Beilstein Information Systems
Varrentrapstrasse 40/42
D-60486 Frankfurt am Main
Germany
[49-69-7917-258 (Germany)]
[800-275-6094 (U.S.)]

Gmelin Institute
Varrentrappstr 40/42
D-60486 Frankfurt am Main
Germany
[49-69-7917-583 (Germany)]

*Note: many publishing changes have occurred in recent years. For example, six of the leading European journals are to merge into two: *Chemische Berichte/Recueil* and *Liebigs Annalen/Recueil.* The new names of these are to be *European Journal of Inorganic Chemistry* and *European Journal of Organic Chemistry,* both to appear in English in 1998. Discontinued journals include the former Belgian, French, and Italian journals that had existed for over 100 years. Details are reported in *C&EN,* November 25, 1997, pages 14–15.

Many chemists, engineers, and information professionals are very concerned about the soaring costs of many important chemical journals. Many libraries, faced with budget reductions, have responded to the cost situation by extensive cancellations of journal subscriptions. The abstracting, current alerting, tables of contents, and document delivery services (described elsewhere in this book) may help bridge the gap for many. It is not clear as yet, but the electronic (Internet and other) availability of full journals, including individual issues, individual articles, and tables of contents, may offer the most viable solution in the long range.

13 Patent Documents (with a Brief Section on Trademarks)

13.1. INTRODUCTION

Patent documents are the most significant forms of literature about chemical technologies. In the United States, granting of a patent indicates that inventors have the right to *exclude others* from making, using, or selling their inventions in the United States, and its territories and possessions, for 20 years after the patent or its parent case is filed at the U.S. Patent and Trademark Office.

A U.S. patent may be granted to the inventor of any *new, unobvious,* and *useful* process, machine, manufacture (includes all manufactured articles), or composition of matter (may include new chemical compounds or mixtures). For a more complete legal definition and explanation, the reader should consult publications (1) of the Patent Office, or a patent attorney or agent registered to practice before that office. See also the Internet site www.uspto.gov. Other countries have similar provisions.

The patent literature is voluminous. According to the World Intellectual Property Organization, Geneva, Switzerland, some 670,000 patents were granted worldwide in 1994, and some 4 million patents were in force. In 1996, *Chemical Abstracts* abstracted over 121,000 patent documents.

13.2. TYPES OF PATENT DOCUMENTS

The two principal types of published patent documents discussed in this chapter are published patent applications and issued or granted patents. *Published patent applications* are documents filed by inventors with a national or multinational patent office and subsequently published (made available to the public or open to inspection) prior to the official examination process that may be necessary to determine whether a patent meets requirements to be granted and issued. *Granted or issued patents* are published only after successful completion of the examination process in countries that require this; only these patents permit the patent owner (inventor or assignee) to exclude others as previously described. For ease of expression, this chapter usually refers to all patent publications as "patents," unless otherwise specified.

13.3. WHY PATENTS ARE IMPORTANT IN INDUSTRY, UNIVERSITIES, AND GOVERNMENT

Chemists and engineers ought to display a keen interest in patents for two principal reasons: (a) many chemists discover new products, processes, or uses that can be patented and can lead to substantial financial and other rewards for inventors or their organizations; and (b) patents are a rich and unique source of chemical information. More specifically, the benefits of the patent as an information tool include these:

1. It permits one to evaluate an idea rapidly and accurately.
2. It tells whether the idea belongs to anyone else.
3. It gives a feel for directions in which others are working, thus making it possible to evaluate one's own competitive situation.
4. It gives a trading position, for several inventors might have parts of an idea and might need each other to make an economical operation.
5. It is a good source of ideas, and especially component parts, to build a total system.

For a company in the chemical and related industries, patents are often the key to success. A strong patent position can give a company an exclusive or proprietary position from which it can exclude its competitors and thereby strengthen its profitability. If a chemist or engineer in industry develops inventions that result in patents, especially commercially significant patents, this can be vital to all aspects of career success and in establishing a reputation among colleagues.

For the university and government researcher, patents have become increasingly important, both to keep up with and understand the state-of-the-art and for other vital reasons as well. Particularly in recent years, both university and government researchers have become more and more interested in patents. Most major universities and government laboratories now have strong technology transfer offices that make known the results of research at the university or government laboratory to industrial firms that may be interested.

These technology transfer offices attempt to negotiate licenses with industry that will permit a firm to utilize the technology on either an exclusive or nonexclusive basis. In many cases, the originating laboratory (university or government) realizes sizable licensing royalties that can be plowed back into the research effort. In addition, inventors in both universities and government laboratories may share in the royalties and thereby significantly increase their personal incomes.

As an example of benefits from technology transfer efforts, Yale University's Office of Cooperative Research announced [*Yale Bulletin and Calendar,* **24,** 3 (July 22, 1996)] that the University earned some $15 million in royalties from licenses and U.S. and foreign patents during 1982–1996. There were some 200 license agreements during this period. Many other universities have earned at least that amount or more; some of the leaders include Stanford University, MIT, and the University of Wisconsin.

University technology transfer officers have formed a group known as Association of University Technology Managers (AUTM), which is headquartered in Norwalk, CT. The group meets periodically to discuss matters of mutual interest and maintains an Internet site that lists many of the inventions that universities wish to license. Technologies for license are also listed in several other databases. For example, government inventions are listed in the online NTIS database.

Despite all the very extensive formal and public technology transfer mechanisms, some believe that the best inventions are not available through any of these mechanisms, but rather through informal one-on-one contacts and negotiations. An allied line of reasoning is that by the time a patent document is published, the invention may be several years old, considering all the time it takes for the paper work to be accomplished.

13.4. PATENT STRUCTURE

The parts of a typical patent include

1. The front page, which contains basic bibliographic information and an abstract.
2. The drawings (if any).
3. The so-called specification, which constitutes the main body of the patent (in terms of size) and contains a description of the invention.
4. The claims, which always appear at the end of the patent and delineate the scope of the monopoly granted under law.

If chemists read the abstract and the first few paragraphs of the specification, they can identify the general purpose and nature of the invention. They should look for sections captioned *Background of the Invention* and/or *Summary of the Invention.*

The specification usually contains in the first few paragraphs indication of the uniqueness (that which is new) and of the advantages of the invention over "prior art" (that which is already known). In most chemical patents, the specification includes *examples,* which are so labeled. These examples are useful to the chemist because they give specific experimental details.

Claims express the legal essence of the patent and are of special interest to chemists in industrial organizations, particularly to patent attorneys, who are trained to interpret what is, and is not, covered by the patent from a legal standpoint. The first claims of most patents tend to be broad (generic); succeeding claims tend to become more and more specific. Almost all claims are, quite properly, highly legalistic in phrasing.

Official patent titles, particularly those of several years ago and many issued today, may be too vague to be a meaningful guide to the technical content of a patent.

Good patent information tools, such as most of those mentioned in this chapter, try to provide the chemist adequate and reliable clues to patent content in several ways:

1. Patent titles are enriched with words that help make the context of the patent more understandable to the chemist. For example, in the title "Compound XYZ, *A Solvent, Manufacture by Process ABC Using Catalyst C,*" the italicized words represent hypothetical enrichment of a hypothetical title.
2. The content of patents is indicated by abstracting and indexing of key features, especially examples and novel subject matter covered by the invention.

13.5. THE ROLE OF PATENTS IN IDEA GENERATION AND CREATIVITY

Patents are invaluable in helping guide chemists and engineers in evaluation and development of ideas. Although the patent system is available to chemists looking for information, it is frequently not used to the extent that it should be. Effective use of patents can lead to more productive research and development in chemistry. Such use often requires the cooperative efforts of the research chemist, the patent attorney, and the literature chemist or information specialist.

Since the beginning of time man has had an urge to create. To *create* means to carry a new idea to the public. This creation may take the form of a book, a painting, a musical arrangement, a floral design, a structure such as a house or bridge, or in the case of chemistry, a new chemical compound, process, or use.

Those who create things usually want to be identified with them. All men and women recognize that progress is made by these creative people. The government has seen the need to develop systems to encourage them to create. The three most important ways used by the government to encourage creativity are (a) the copyright for creative works, as writing and music; (b) trademarks; and (c) patents.

In the patent system: (a) the inventor describes the invention; (b) files the detailed description, including claims, with government patent offices in one or more countries; (c) has it examined by others (officially designated patent examiners in the United States and many other countries) to determine if it is really new, unobvious, and useful; and (d) if it is, to grant the right to exclude others from using the claimed invention, for 20 years from filing in the case of patents granted by the U.S. Patent Office.

The purpose of granting patents is to encourage people to create. It is anticipated by the government that these concepts will be used or that the dissemination of the information contained therein will promote the advance of the technology.

The patent system was developed to tell chemists and others what has been going on and what areas of technology have been staked out by others. So the patent system was, and is, not only a *legal* tool or system, but an *information* tool. A limited monopoly is granted by the government in exchange for disclosure of details of the invention.

Not to use the patent system is a mistake, because if an idea is old, the sooner the chemist knows it, the better. Even if the idea is old and patented, the chemist might still be able to use it. For example, the first person to conceive a specific chemical idea could have been 20 years ahead of his or her time. If this is so, then present-day chemists could have the concept *free*. Also, it may be that the basic concept was conceived, but something was missing to make it satisfactorily useable. This may permit a modern-day chemist to modify an older invention to make it more useful. This is building on the technical foundation of the past and is what the creators of the patent system wanted to happen.

13.6. PATENTS AS INFORMATION TOOLS

How can the chemist or engineer use the patent system as an information tool? Searching the literature, reading, and thinking are cheaper and faster than experimenting. So, a chemist's approach to a problem could be as follows:

1. Think and guess what he or she would like to accomplish.
2. Make a quick survey of patent and other scientific literature.
3. Study.
4. Think and guess again.
5. Make a more detailed study of patent and other scientific literature.
6. Read, study, think.
7. Experiment.

To do the literature searching noted, the laboratory investigator often needs expert help—people who know their way around the patent system and other parts of the scientific literature. The patent system has become too complex for most laboratory chemists to "go it alone." That is why it is important for the laboratory chemist to confer with the information chemist, tell that person what is being thought and guessed about, and ask for help or advice in pulling together the works of others for study.

In this type of interactive relationship, the laboratory person tells the information professional the type of search needed and also suggests how much time is worth spending on the search. Time to be spent is based on such factors as potential profit or other impact and priority considerations.

Next, the laboratory chemist studies the results of the search, does some additional searching as needed, and makes decisions based on these studies.

The decision could be one of these:

1. Dig a little deeper.
2. Modify the original idea—shift to the left or right.
3. It's old stuff—forget it.
4. It's old stuff, but by adding A or B it can be used.

The benefits of this approach include learning, using appropriately the skills of others, and determining as quickly and economically as possible if an idea is good and novel. The next, or perhaps concurrent, step could be to work in close cooperation with a patent attorney or agent toward the ultimate filing of a patent application if that is perceived as a goal.

13.7. PATENTS COMPARED TO OTHER INFORMATION SOURCES

One way in which patents are unique sources of chemical information is that they are primarily legal documents. Some chemists find patents more difficult to read and understand than books and journal articles because patents are usually written in legal phraseology by patent attorneys or agents who act for inventors. But this should not deter the chemist from taking full advantage of the unique and rapid access to the enormously broad range of chemical information that patents provide.

Similarly, patents should not deter laboratory investigators from using journals, books, and other forms of chemical information. All forms are important, and all have certain advantages that vary with the situation and intended use. Although journals and books are more familiar to most investigators, patents are at least on a par with books and journals in value of information provided with regard to chemical technology.

Some people do not look on patents as a source of reliable information. It has, however, been the experience of many that it is as easy to replicate an example in a patent as it is to duplicate an experimental description in a scientific journal. In both cases, the inventors usually report the highest yield obtained, but in both cases this yield was not always obtained in exactly the way the process was described. Nevertheless, with a little ingenuity and a knowledge of the field, the scientist can easily duplicate the majority of examples in patents.

Because investigators do not usually publish anything proprietary in journals and books, at least until there is fully adequate patent coverage, patents are usually good clues to what the competition is thinking. It is believed that much, although by no means all, of what appears in patents cannot be found published in journals or in other forms.

Patent documents frequently contain numerous detailed examples, discussion of variables, and drawings. The amount of highly specific experimental detail given in patent examples can be outstanding. These details can include not only methods of synthesis or manufacture, but also results of physical tests and application studies. In contrast, journals may be constrained as to length, because of budget and other limitations, and they are also very often limited in their coverage of the most modern commercial methods, of synthesis, manufacture, and testing. Patents are not subject to conventional editorial or refereeing processes, so their content can be broad. In addition, because generic scope is frequently sought to provide the broadest possible patent coverage, the number and variety of examples given can be considerable and the detailed discussion of variables can be useful.

Another unique and important advantage of patents is the invaluable perspective and background they can provide on technological history and development. For example, perusal of patents in individual subclasses of the U.S. Patent Office usually provides chemists with an excellent picture of how a technology has unfolded and progressed. Reading of the initial pages of patents, prior to the examples, can provide insights into the state of the art. These pages usually provide background and help identify what is important. Furthermore, references cited by U.S. patent examiners, as noted on the front pages of U.S. patents, can provide access to other relevant literature. All of this can be very helpful in creating and planning new research and development programs.

The exercise just described can be carried out in the Public Search Room of the U.S. Patent Office where chemists can find all U.S. patents conveniently arranged in the appropriate subclasses and can also obtain advice as to which subclasses to examine. This exercise can sometimes be carried out elsewhere, including at one's own desk, particularly if a subclass is small enough such that all patents in that subclass can be economically purchased. As discussed later in this chapter, however, use of the subclass approach requires considerable expertise.

An alternative approach, that is very convenient and that can be done at one's desktop computer, is to do the kind of scanning described above utilizing one or more of several online databases, which are described later in this chapter. However, none of these systems, except for that operated at the U.S. Patent Office (and at some of its associated Patent and Trademark Depository Libraries), go back in time to the beginnings of the U.S. patent system in 1790.

Patent literature has largely replaced conventional trade and scientific journals as the most up-to-date source of information on technological progress in numerous fields. This is partly because most countries have revised their laws so that patent applications are published without examination 18 months after the earliest filing date. Completion of the examination process to determine whether a patent application should be granted usually takes additional time, typically an additional 1–3 years in examination countries.

The publication of patent applications as described above was adopted by the Netherlands in 1964, by West Germany on October 1, 1968, and by Japan on January 1, 1971, for the following reasons:

1. To eliminate backlogs of unexamined patent documents
2. To make the contents of these documents available more quickly
3. To defray legal costs until inventions are more fully developed
4. To examine only the more worthy documents

All important patent issuing authorities have adopted this plan, except for the United States. As mentioned again below, legislation now (1997) pending in the Congress proposes to put U.S. practice in harmony with the rest of the world on this matter; enactment into law is believed to be probable, but timing is uncertain at this time because of strong opposition in some quarters. Thus, for the present at least, unexamined applications from other countries can give important advance clues to technologies for which patents may later issue in the United States.

In the United States and most other countries, before a patent application can be granted and the patent issued, it must be examined by patent examiners on the basis of such criteria as unobviousness, novelty, and usefulness. These are demanding criteria that can vary somewhat depending on the specific patent issuing authority and are more specifically defined by laws and regulations.

13.8. OFFICIAL GOVERNMENT SOURCES OF PATENT DOCUMENT INFORMATION

An important way for chemists to learn about new patents is to read the patent gazettes usually published weekly by patent offices of most major industrialized nations. Here a chemist will find the title, and usually the principal claim, depending on the country, of new patent documents.

A chemist can use the gazettes to keep up to date with patents; this is usually a quick way for any single country. But this can be pose problems for a variety of reasons. One is that key patents may turn up in material issued by unexpected countries and in unfamiliar languages. Furthermore, many chemists find that information given in the gazettes is much less useful than the fully descriptive patent abstracts that appear in such services as *Derwent* and *CA* described elsewhere in this chapter. The gazettes, however, provide some information useful to laboratory chemists, for example, in helping decide whether to try to locate an abstract that would provide more information or to obtain a copy of the full patent.

13.9. SOME RECENT CHANGES AFFECTING U.S. PATENTS

On June 8, 1995, the patent laws of the United States changed significantly. These changes implement the Uruguay Round Agreement of the General Agreement on Tariffs and Trade (GATT) that included also 110 other countries. Congress passed the new laws in November 1994, and President Clinton signed them on December 8, 1994. The changes align the U.S. system more closely with the systems in most other industrialized countries.

Some background on the Uruguay Round is as follows. The General Agreement on Tariffs and Trade (GATT) was signed in Geneva in 1947 and, since then, has played an important role in promoting world trade. Subsequently, there have been several rounds of trade liberalization negotiations. The "Uruguay Round" was so called because the initial meeting took place in Punte del Este, Uruguay, in September 1986. This round of negotiations lasted over 7 years because the problems were difficult. Improved protection of intellectual property, including patents, was among the issues negotiated. The new U.S. patent law is sometimes known as the URAA or Uruguay Round Agreements Act.

These changes are as follows:

1. *Term of Patent.* Since 1861, and continuing until this change, the term or life of a U.S. patent was 17 years from date of patent grant. As of June 8, 1995, for patents issued on applications filed on or after June 8, 1995, the patent rights begin when the patent is granted and end 20 years from the day when the inventor filed the patent application. If the applications are continuations, continuations-in-part, or divisionals, the term ends 20 years after the date of the earliest filed parent nonprovisional application. For example, if an inventor files a continuation application after June 8, 1995 and relies on a filing before that date, the application will be subject to a 20-year term from the earlier filing date. For patents issued on applications filed before June 8, 1995, the patent term will expire in either 20 years from the earliest U.S. filing date or 17 years from the date the patent is granted, whichever is longer. For example, if an inventor files an application on April 1, 1995, which issues as a patent on December 31, 1996, the patent term would end on April 1, 2015 (20 years from the application date). For patents that are in force on June 8, 1995, the new laws specify that the term will be the greater of either 20 years from earliest U.S. filing date or 17 years from date of patent grant. Thus, an inventor filed an application on May 1, 1992 and

the patent was granted on July 1, 1994. The new laws change the expiration date from July 1, 2011 (17 years from date of grant) to May 1, 2012 (20 years from the date of application), thus providing 10 additional months of patent protection. These changes "harmonized" the U.S. patent system so that it is now essentially consistent with the patent systems of most other major nations as to patent duration.

2. *Provisional Patent Application.* Under the new laws that went into effect on June 8, 1995, inventors can file a provisional application that will allow them to obtain an early filing date with a few formalities and at a lower cost than filing a nonprovisional application. A provisional application differs from a nonprovisional application in several ways. The patent term does not start from this filing date, the application is not examined, and it cannot issue as a patent. The nonprovisional application is retained in confidence, and it is automatically abandoned 12 months after filing. The fee is only $150 for larger firms and $75 for a "small entity." In effect, this may extend patent term another year, that is, from 20 years to 21 years. Before the application expires, if the applicant desires to obtain a patent, a nonprovisional application must be filed. Even though the provisional application apparently offers a relatively simple, inexpensive procedure, and can be attractive in certain situations, just as with any matter involving inventions and patents, it is advisable that a patent attorney be consulted first.

As mentioned, yet another proposed change to U.S. patent law is still pending. This legislation would require that all patent applications (other than provisional applications and design applications) be published 18 months after the earliest filing date. This is now the case in most other industrialized nations. The proposal that is now in the Congress called for initiation of this major change in January 1996, but it is still (1997) pending. If approved, a special *Gazette,* abstracting all such published applications, is to be issued, according to Patent Office plans.

13.10. OBTAINING INFORMATION ABOUT UNITED STATES PATENTS: THE BASICS

The *Official Gazette: Patents,* compiled since 1872 by the U.S. Patent and Trademark Office, provides a listing of claims (usually only the main claim) for all U.S. patents issued each week. Copies of the *Gazette* can be found in many university and public libraries and in almost all industrial research organizations.

Patents listed in the *Gazette* are arranged in three major categories: (1) general and mechanical, (2) chemical, and (3) electrical. Note that there may be some overlap among these categories. A patent of chemical interest may be found in one of the other sections. Within each category, patent claims are grouped in numerical sequence according to a classification system established and maintained by the Patent Office.

Other categories in the *Gazette* are Statutory Invention Registrations (SIRs), Reissue Patents, and Reexamined patents. An SIR is not a patent. It has the defensive attributes of a patent, but not the enforceable attributes of a patent. The SIR series replaces the previous Defensive Publication Series. A *reissue* patent is one that is granted when a substantive error is discovered in a previously issued patent.

Patent applications are filed with the U.S. Patent Office at the rate of approximately 212,000 (1995) applications per year. In 1996, a record total of 109,646 utility patents were issued. (Utility patents are mechanical, chemical, and electrical patents.) By definition, this total does not include design or plant patents, which are of little or no interest to most chemists. Of the utility patents granted in 1996, 55.7% were to U.S. inventors and 44.3% were to non-U.S. inventors. The percentage of non-U.S. inventors has remained at approximately the same level for several years, with inventors from Japan as the leaders in terms of numbers of patents granted to non-U.S. inventors. In addition to the utility, design, and plant patents granted in 1996, there were 109 SIRs, and there were 279 reissued patents. Although large companies are granted most patents, over 17,000 patents were granted to independent inventors in 1996—an encouraging sign as to the health of innovation in the United States.

According to data compiled by IFI/Plenum Data Corp. (see Section 13.16), the total number of U.S. chemical and chemically related patents that issued in 1996 was 33,467.

The U.S. Patent and Trademark Office has established a procedure for review of granted patents as a way to reduce the amount of patent related litigation presented to the courts. Under the rules, on payment of a large fee, anyone may apply for reexamination of any U.S. patent. If the reexamination procedure results in changes to the patent, the patent is assigned a reexamination number, and the affected portions of the patent are published in the *Official Gazette.*

To use the *Gazette* effectively, investigators must identify pertinent classes and subclasses that reflect their interests or the interests of their organizations. This identification can best be achieved with the aid of a patent attorney or by direct communication with the U.S. Patent Office, since the classification system is complex and is revised at intervals. This approach to locating patents (scanning of the *Gazette*) is at best very risky and can result in significant information loss unless properly done.

An obvious weakness of the *Gazette* is lack of cross-references to related patents. This means that important subject matter will often appear in subclasses other than that in which a patent of interest appears in the *Gazette.*

The annual indexes to the *Gazette* have been appearing many months or even several years after completion of the calendar year.

Furthermore, the annual index to the *Official Gazette* does not contain a subject index, although all U.S. patents issued that year are classified by class and subclasses. This feature is useful to the patent attorney or agent. It can be a difficult means of access for the working scientist because of the complexities of the classification, as noted previously, but use becomes much easier if the scientist studies the classification and knows the field.

If a specific patentee is of interest, access by this key is relatively easy. It involves scanning the indexes to the *Gazette* that appear weekly and are cumulated annually.

When looking for organizations in patent indexes, the searcher must consider corporate name changes and other variations. Companies merge, acquire partially or totally owned subsidiaries, disinvest or spin off, or change names to reflect new activities and policy changes. For example, a historical search over a period of years for patents issued to one company might include such variations as

Mathieson Alkali Works
Mathieson Chemical Corp.
Olin Corporation (this is the current name)
Olin Industries, Inc.
Olin Mathieson Chemical Corp.

Alternatively, some patents exhibit organization name variations for other reasons, including simple error.

Several organizations that provide online patent databases, as described later, have attempted to facilitate searching for alternative organization names by standardizing these names whenever possible. These include Derwent, IFI, and *INPADOC.* Even so, patents issued to subsidiaries or affiliated companies may be missed, as may patents that are not initially assigned.

Nomenclature is even more highly variable than organization names and can also lead to loss of information, unless the searcher is on the lookout for all possible variations. Use of CAS Registry Numbers in online databases can sometimes eliminate or minimize problems, and another approach is use of one of the systems that permit graphic searching for structures (see Chapter 10).

Another way to search for and identify U.S. patents is to visit the Public Search Room of the Patent Office located at Crystal Plaza, near Washington National Airport. This is the only place where all U.S. patents are arranged by subject matter according to a *Manual of Classification* (2) which helps the user decide which of over 400 main categories or *classes* to search. These classes are further broken down into more than 200,000 subclasses. The *Manual of Classification* contains an alphabetic subject index. A further aid in searching by class and subclass is *Class Definitions,* a Patent Office product that defines the scope of each class and subclass and also refers the searcher to related classes. The searcher removes the selected subclass bundles of patents from their open-stack arrangement, takes them to a desk, and inspects or reads them in their chronological order.

If a patent discloses subject matter classified in two or more subclasses, a copy is placed in each subclass. To distinguish these copies from each other, the copy placed in the subclass selected as the principal basis of classification is called an *original,* and all others are called *cross-references.* Both original and cross-reference patents are contained in classification bundles in the search room.

It is important to be aware that patents are constantly being reclassified by the U.S. Patent Office, especially in the most active technologies. For this and other reasons, the class and subclass groupings need to be used with both care and skill.

The class–subclass approach can, of course, also be utilized for online searching as described later in this chapter. Several online sources are available to search the

Manual of Classification, identify patents by subclass, and help minimize the chances of missing reclassified patents.

Note that almost all U.S. patent applications being examined for possible issue are *not* currently available to the public. (Exceptions to this rule include patent applications from the U.S. government that meet certain criteria; these applications are published by the U.S. National Technical Information Service.

Unfortunately, the public search room at the U.S. Patent Office is geographically inconvenient for regular use by most chemists, except those who live in the Washington, DC area, or whose business brings them to that location. All are, however, always welcome to use this facility.

Although the search room is available to all, it is designed primarily for use by patent attorneys, patent agents, and other trained patent searchers. Laboratory chemists and other inventors desiring to use the search room for the first time should seek assistance from the staff of the room, or ideally be accompanied and guided by a patent attorney or agent from their own organization who knows all Patent Office facilities and how to use them efficiently. Chemists and other search room users can make good use of Patent Office booklets on how to use the search room and associated facilities. In addition to the patents in the public search room, unofficial subclasses and foreign patents and literature can be found in examiner search areas and may be searched with the permission of appropriate officials in these areas.

When using the public facilities, it is important to note that individual patents or groups of patents may sometimes be missing because they are in use or because they may have been misfiled (the public is asked to refile all patents). For this reason, integrity of the files is always a concern. The staff conducts integrity checks to minimize such problems, and use of the automated systems that are in place can help ensure completeness of a search as well. The Public Search Room is open from Monday through Friday from 8 A.M. to 8 P.M., except for holidays. The manager of this room is Edith Wilkniss.

The Public Search Center previously operated by Derwent Information, Limited at the Public Search Room of the U.S. Patent Office, has been discontinued. In its place, the public can use the Patent Office's computerized APS system (described in the next section) on payment of a modest fee, and, of course, the balance of the facilities open to the public continue to be available. In addition to the resources previously mentioned, these facilities include an excellent scientific library that contains official journals of patent offices of other countries and millions of non-U.S. patents, as well as many other reference sources. The scientific library is the best U.S. source of non-U.S. patents.

As a convenience for those who cannot visit its facilities in Arlington, the Patent Office will supply, for a fee, lists (patent numbers) of original or cross-reference patents contained in the subclasses constituting the *field of search.* The same information may be obtained on a visit to the Patent Office or to any of the Patent Depository Libraries (see Table 5.1) through the Patent Office's Classification and Search Support Information System (CASSIS), or it is available through the PTO's Internet site, www.uspto.gov.

The chemist should seek the help of a patent attorney or of other appropriate experts in defining the pertinent subclasses.

13.11. COMPUTERIZED SEARCH TOOLS AVAILABLE THROUGH THE U.S. PATENT AND TRADEMARK OFFICE AND SOME OTHER ASPECTS OF PATENT OFFICE OPERATIONS

The PTO offers to the general public, at a relatively modest fee for use, several computerized search tools at its headquarters in Crystal City, VA. Some of the key tools are also available at the official Patent and Trademark Depository Libraries (PTDLs), and some may, in addition, be purchased by anyone wishing to open a subscription. However, the APS system described below is not available for purchase.

PTO's Automated Patent System (APS) text offers text search capability for the full text of U.S. patents issued since 1971. The system is up to date as of the date of issue of the patents, and at the same time, reclassification information is added automatically; these two features are not available that quickly anywhere else at this writing. Reexaminations and certificates of correction are also included. It also has English-language abstracts of unexamined Japanese patent applications from 1980 to the present. It may be used at the public search facilities of the PTO in Arlington, VA, which has some six terminals for text searching ($40 per hour of use) and 15 terminals for image or text searching ($50 per hour); these terminals are heavily used. It is also available at the PTDLs shown in Table 13.1 at inexpensive rates determined separately by each of the 28 PTDLs participating; in some cases, there may be no charge.

APS-CSIR (Classified Search and Image Retrieval) provides text (since 1971) and image search capabilities for patents issued since 1790 by classification, keyword or patent number. In addition to the public search facilities of the Patent Office, it is also available for use at the PTDLs in Sunnyvale, CA (Sunnyvale Center for Innovation, Invention, and Ideas) and Detroit, MI (Great Lakes Patent and Trademark Center, Detroit Free Public Library). No other database in the world has the complete collection of U.S. patents online.

CASSIS/BIB and CLASS are CD-ROM products of the PTO that permit search of U.S. patents for two time periods. The BIB product covers since 1969. However, abstracts are included and searchable only for the most recent 3 years. Otherwise, the user must search by assignee code, classification, or title. The CLASS product provides classification numbers back to 1790 and permits searching using this feature. The products are updated every 2 months. These products are available for use without charge at all PTDLs. They may also be purchased from the main PTO facility in Crystal Plaza. As compared to the Chemical Abstracts Service, Derwent, and IFI patent products, with their sophisticated indexing and searching features, these particular PTO files can be a useful, no-cost first approach, but are much more rudimentary.

In addition, in November 1995, the PTO began to make patent information available for searching on the Internet at no charge for the use of the file. The information provided is not extensive or sophisticated as compared to commercially sponsored tools. It includes merely the front-page information (except the images) for U.S. patents for the most recent 20 years. The Patent Office decision came in the midst of substantial controversy. Proponents argued that since the Patent Office's electronic system was developed with hundreds of millions of dollars of public funds, it ought

TABLE 13.1. U.S. Patent and Trademark Depository Libraries that Offer Automated Patent System Searching (in Order by State)

Auburn University
ArizonaState University Science and Engineering Library
Arkansas State Library
Los Angeles Public Library
San Francisco Public Library
Sunnyvale Center for Innovation, Invention, and Ideas
Broward County Main Library, Fort Lauderdale
Miami–Dade Public Library
Hawaii State Library
Ablah Library, Wichita State University
Louisville Free Public Library
Boston Public Library
Great Lakes Patent and Trademark Center, Detroit Public Library
Minneapolis Public Library and Information Center
Linda Hall Library, Kansas City
Engineering Library, Nebraska Hall, Lincoln
Buffalo and Erie County Public Library
North Carolina State University
Cleveland Public Library
Toledo/Lucas County Public Library
Paul L. Boley Law Library, Lewis and Clark College, Portland
The Free Library of Philadelphia
Texas A&M University
Dallas Public Library
University of Utah
Virginia Commonwealth University
University of Washington, Engineering Library

to be made freely available to the public. In contrast, others have argued that this dissemination ought to be a function of the private sector.

The basic PTO Internet site is http://www.uspto.gov/, and the search features may be accessed through links on this site. The site includes information about other patent and trademark resources on the Internet to the extent known.

13.12. PATENT AND TRADEMARK DEPOSITORY LIBRARIES (PTDLS)

To facilitate the searching and viewing of patents on a nationwide basis, and relieve the burden on its central facility, the U.S. Patent and Trademark Office has developed a network of Patent and Trademark Patent Depository Libraries (PTDLs). The concept began operation in 1871 and now consists of at least one in every state except Connecticut, a deficiency to be corrected in 1998. The public may visit these without charge although modest charges usually apply to make copies of any of the materials and to use the online search system.

The PTDLs offer an extensive array of paper, CD-ROM and online tools, including, in some cases, the full-text searchable patent database covering the 1.8 million patents issued from 1971, as noted above. Almost 50% of the PTDLs have complete U.S. patent collections dating back to 1790, and all have at least 20 years of backfiles, in addition to currently issuing patents on microfilm. The collections are in numerical order by patent number, rather arranged by class and subclass as at the Patent Office. The complete backfiles of U.S. patents disclose information not commercially available online or on CD-ROM—none of the commercial products goes as far back as 1790 at this writing. In addition, as mentioned, the PTDLs at Sunnyvale, CA and Detroit, MI, have begun to offer electronic access to the scanned images of all 5.8 million patents back to 1790, and they also offer video conferencing capability so that patent applicants may talk with patent examiners at the Patent Office. Some of the larger PTDLs also offer collections of non-U.S. patents.

One major advantage of the PTDLs is that they are staffed by specially trained librarians who are available to provide specific instruction in the ways to use the available tools and who have direct connections with Patent Office resources and personnel. The service is popular; use levels of the PTDLs average 10,000 persons a week. The facilities are especially well suited to independent inventors and to persons connected with small companies. All PTDLs have readily available the basic information about how the patent system works and how to file a patent application.

The PTDLs also have extensive search facilities and information concerning trademarks.

Copies of patents can be inspected at the U.S. Patent and Trademark Office or at any of the PTDLs that are (hopefully) within a reasonable distance. Please see Chapter 5 for a complete list of PTDLs. These facilities will also provide copies of patents on a fee basis. Copies of U.S. and foreign patents are also readily available quickly from a large number of document delivery firms of the type described in Chapter 5. In addition, the full text of many patent documents is available online as will be described later in this chapter.

13.13. OTHER REMARKS ABOUT THE U.S. PATENT AND TRADEMARK OFFICE

The U.S. Patent and Trademark Office is an impressive operation. The total U.S. Patent and Trademark Office staff is some 5200 persons, and the most recent budget (1997) is approximately $650 million. The corps of examiners has been increased from about 1500 persons some 10 years ago to about 2400 as of 1997. The position of Assistant Secretary of Commerce and Commissioner of the U.S. Patent and Trademark Office (PTO) is a presidential appointment, and the occupant usually changes depending on who is in the White House; the present Commissioner is Bruce A. Lehman. Legislation currently pending in the Congress would make the PTO a government-owned corporation, although regarding policy matters, and as to the top position, no change would be made.

With regard to the examination process for U.S. patents, the average pendency period (time from patent filing to patent issue or abandonment) in fiscal year 1994 was 19 months (it had been as high as 30 months in 1969). This improvement was accomplished by an increase in the number of examiners, as noted above, and implementation of a very successful computerization program. Chemical Abstracts Service was one of the organizations that participated in this automation program. However, because of significant budget budget cuts already implemented, the average pendency at this writing (1997) has, unfortunately, increased to about 24 months. Since further budget cuts are proposed, average pendency is expected to increase further and the number of patents granted is expected to drop, if these proposed budget cuts are implemented. Pendency can vary depending on the field of technology. If a field is crowded (many issued patents and patent applications), the examination process can take longer.

13.14. DERWENT INFORMATION LIMITED: WORLDWIDE PATENT AND RELATED PRODUCTS

Derwent has been a well-established leader in worldwide patent information for many years. Its products and services are very highly regarded.

Derwent was founded in 1951 by Montagu Hyams, a British chemist and the principal executive for many years, and its main offices are located in London. It is now part of The Thomson Corporation, the publishing industry giant, and the CEO is Martin J. Nathan. In addition to its strong coverage of patents, Derwent now provides coverage of nonpatent information in certain areas as will be described later.

A major enhancement for chemists in North America is that Derwent has a full-time staff in the Washington, DC area (telephone 800-451-3551). This staff offers expert help in use of the online files and other services that are available. There is also an office in Tokyo. The Internet site, which provides descriptions of Derwent products, is www.derwent.com. or www.derwent.co.uk.

The first Derwent product was an abstracting service covering new British patents. This was followed by a series of similar bulletins covering German, Belgian, French, Dutch, Soviet, United States, and Japanese patents. In the 1960s, Derwent initiated a number of discipline-based current awareness and retrospective services to supplement the previously existing country-based current awareness services. The new subject services included pharmaceutical patents (*FARMDOC,* founded in 1963), agricultural chemical patents (*AGDOC,* 1965), and polymer patents (*PLASDOC,* 1966). These three services were consolidated, and chemical patent coverage was expanded, to produce the *Chemical Patents Index* (CPI), which began in 1970 as the *Central Patents Index.*

The premier Derwent product today is the *Derwent World Patents Index* (*DWPI*) which began in 1974, although coverage for some fields of chemistry begins in 1963 as outlined above. This online and hard-copy product covers all technologies, both chemical and nonchemical.

There are more than 14 million patents in the database, of which some 7.3 million are chemical, that is, have the Derwent classes A–M (see Table 13.4 for a list of Derwent classes). These figures do not reflect the current update rates which were drastically changed with the introduction of complete Japanese unexamined patent document coverage during 1995.

The weekly average input (1996) is 17,400, broken down as follows for the main sections:

Chemical Patents Index (CPI)	6420	37%
Electrical Patents Index (EPI)	6260	36%
Engineering Patents Index (ENGPI)	4720	27%

DWPI includes patent document coverage from 40 patent-issuing authorities and two journal sources as shown in Table 13.2.

TABLE 13.2. Country Coverage and Codes, *Derwent World Patents Index*[a]

Australia	AU	Norway	NO
Austria	AT	PCT	WO
Belgium	BE	Philippines	PH
Brazil	BR	Portugal	PT
Canada	CA	Romania	RO
China	CN	Russian Federation	RU
Czech Republic	CZ	Singapore	SG
Czechoslovakia	CS	Slovakia	SK
Denmark	DK	South Africa	ZA
East Germany	DD	South Korea	KR
Europe	EP	Spain	ES
Finland	FI	Sweden	SE
France	FR	Switzerland	CH
Germany	DE	Taiwan	TW
Hungary	HU	United Kingdom	GB
Ireland	IE	USA	US
Israel	IL	USSR	SU
Italy	IT		
Japan	JP		
Luxembourg	LU	*Plus:*	
Mexico	MX	International Technology	
Netherlands	NL	Disclosures	TP
New Zealand	NZ	Research Disclosure	RD

[a]The 40 patent issuing authorities and two journals covered by Derwent, together with their associated two-character country codes are given.

Derwent is actively working on efforts to significantly improve speed of coverage, but 1996 data show the following average delays (in calendar weeks) between patent document publication date and online appearance:

Great Britain	3.0
Germany	5.0
European Patent Office	5.0
Patent Cooperation Treaty	5.9
United States	6.1
Japan (unexamined)	10.7

Derwent is re-engineering its production systems and, since mid-1997, British documents have appeared online, complete with chemical and/or polymer indexing, some 2.0 weeks after publication. The improvements in timeliness will be extended in 1997 and 1998 to include the other countries covered in DWPI.

1. Abstracting by Derwent. Abstracting for all "basic" patent documents (basic means not previously covered by Derwent) is done intellectually. The other patents covered by Derwent are equivalents. These can be defined simply as further patent documents covering the same invention as the basic ones, but published later in the patent process or by other countries. See page 351 for a further discussion of equivalents and patent families. For European Patent Office and British Patent Office granted patent documents, all of which are equivalents, the main claim is provided as the abstract; for U.S. equivalents, the main claim is provided; and for all other equivalents, the basic abstract is given. Note that, prior to 1996, U.S. patents had been re-abstracted.

The following are not given abstracts unless and until another country equivalent appears: Czechoslovakia, Czech Republic, Finland, Italy, Luxembourg, Norway, and Slovakia. Abstracts for basics of some other countries were not provided from the start of coverage, but were introduced later: Austria (1993), Brazil (1987), China (1995), Denmark (1990), Hungary (1983), Israel (1983), South Korea (1990, Korean nationals only), Portugal (1983), Romania (1983), and Spain (1989).

Like some other major patent information tools, Derwent modifies or expands patent titles as required to make these titles more meaningful.

In addition, names of assignees, which can vary widely in spelling and in other ways, are standardized by Derwent, and, in fact, assignee codes are assigned by Derwent to further simplify searching. Some other patent information tools also provide this service. Nevertheless, the searcher must continually be vigilant for unexpected assignee name variations, for subsidiaries, and for changes caused by spin-offs, mergers, etc.

Abstracts appear first online in *DWPI* (the Alerting version). The first Derwent abstracts to appear in hard-copy print are their *Alerting Abstracts Bulletins,* published in two editions: (1) by section in country order and (2) by section in Derwent class order. Edition (2) appears one week later than (1). A further week later, the *Docu-*

mentation Abstracts Journals appear. Documentation abstracts are much more detailed than the Alerting abstract and also include manual codes; Documentation abstracts do not appear online, but are available in CD-ROM and microfilm/fiche version. Typical Derwent Alerting and Documentation Abstracts appear in Figures 13.1 and 13.2.

Another printed product, *World Patents Abstracts* (WPA), is published at the same time as the first Alerting Abstracts Bulletin (by section in country order, see Table 13.3). These WPA products cover all sections A-M as individual countries.

Reading one or more of the *CPI* Alerting Bulletins on a regular basis is one of the single best ways for laboratory chemists and engineers to keep up with new advances in patented chemical technology. For those who don't have the time to read the Altering Bulletins, a reasonable shortcut is to set up an online SDI (Selective Dissemination of Information) profile using key words, International Patent Classification Numbers, and, possibly best of all, Derwent Manual Codes.

CPI coverage is divided into the 12 separate sections shown in Table 13.4. The user may purchase any or all of the separate printed Alerting Bulletins that are available for each section. Each appears weekly.

The countries covered by Derwent are shown in Table 13.2. Some of these countries (Czechoslovakia, East Germany, Soviet Union) no longer exist, although their documents are still of interest. Mexico is being added in 1997.

2. Finding Information in Derwent. *DWPI* is available online through any of four major hosts: DIALOG, Orbit, Questel, and STN International. There is a gateway to *DWPI* on Westlaw.

On all hosts, *DWPI* is cross-file-searchable with other patent-containing databases. This means, for example, that given a set of results in *DWPI,* the user can cross over to *Chemical Abstracts* on STN International to determine how CAS abstracts the same documents. This crossover can be done without the need for rekeying the *DWPI* results. Similarly, given results from *Chemical Abstracts,* one can readily cross over to the Derwent files to determine such information as equivalents or to read Derwent abstracts that may contain additional information. Cross-file searching also permits the searcher to utilize the unique searching capability of each file separately as described by Simmons and Kaback (3). Note that *DWPI* now provides representative images (drawings, etc.) for chemical patent documents since 1992 and for nonchemical patents since 1988. Images can be displayed online and ordered as offline prints.

Access points available to all oneline searchers include the following:

Derwent Accession Number—a 6-digit serial number preceded by a 2-digit year and a hyphen, assigned uniquely to each basic (e.g., 93-348385).

Related Accession Numbers—these are pointers to related patents.

Title—in English and expanded by Derwent to be informative. Utilizes British rather than American spellings (although, when searching, both should be used).

Derwent Classification—the broad classification includes 138 chemical classes.

Patent Assignee and Company Code—the assignee is the organization to which the inventor assigns the patent. Derwent staff "standardize" company names so

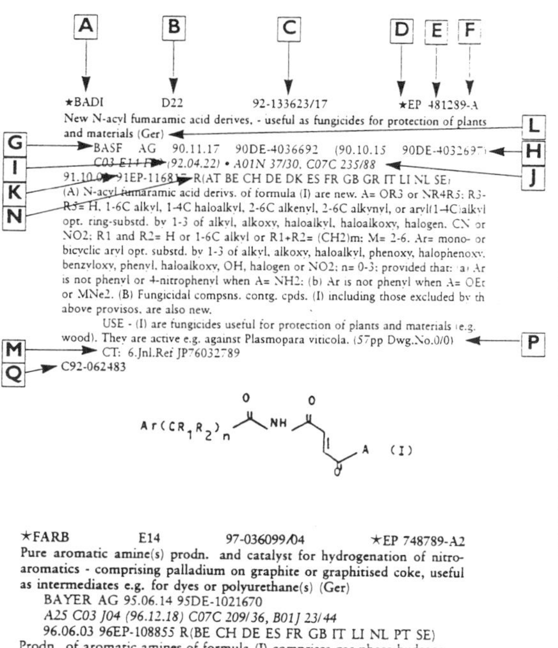

★BADI D22 92-133623/17 ★EP 481289-A
New N-acyl fumaramic acid derives, - useful as fungicides for protection of plants and materials (Ger)
BASF AG 90.11.17 90DE-4036692 (90.10.15 90DE-4032697)
C03 E11 F?? (92.04.22) • A01N 37/30, C07C 235/88
91.10.0? 91EP-11681? R(AT BE CH DE DK ES FR GB GR IT LI NL SE)
(A) N-acyl fumaramic acid derivs. of formula (I) are new. A= OR3 or NR4R5; R3-R5= H, 1-6C alkyl, 1-4C haloalkyl, 2-6C alkenyl, 2-6C alkynyl, or aryl(1-4C)alkyl opt. ring-substd. by 1-3 of alkyl, alkoxy, haloalkyl, haloalkoxy, halogen, CN or NO2; R1 and R2= H or 1-6C alkyl or R1+R2= (CH2)m; M= 2-6. Ar= mono- or bicyclic aryl opt. substd. by 1-3 of alkyl, alkoxy, haloalkyl, phenoxy, halophenoxy, benzyloxy, phenyl, haloalkoxy, OH, halogen or NO2; n= 0-3; provided that: (a) Ar is not phenyl or 4-nitrophenyl when A= NH2; (b) Ar is not phenyl when A= OEt or MNe2. (B) Fungicidal compsns. contg. cpds. (I) including those excluded by th above provisos, are also new.

USE - (I) are fungicides useful for protection of plants and materials (e.g. wood). They are active e.g. against Plasmopara viticola. (57pp Dwg.No.0/0)
CT: 6.Jnl.Ref JP76032789
C92-062483

Ar(CR1R2)n ... NH ... A (I)

★FARB E14 97-036099/04 ★EP 748789-A2
Pure aromatic amine(s) prodn. and catalyst for hydrogenation of nitro-aromatics - comprising palladium on graphite or graphitised coke, useful as intermediates e.g. for dyes or polyurethane(s) (Ger)
BAYER AG 95.06.14 95DE-1021670
A25 C03 J04 (96.12.18) C07C 209/36, B01J 23/44
96.06.03 96EP-108855 R(BE CH DE ES FR GB IT LI NL PT SE)
Prodn. of aromatic amines of formula (I) comprises gas phase hydrogenation of nitroaromatics of formula (II) with hydrogen over a stationary catalyst of palladium on a graphite or graphitised coke support with a BET surface area of 0.2-10 m^2/g. The catalyst contains more than 1.5 - 7 wt.% Pd w.r.t. total catalyst and the H_2 supply to the catalyst is 30-6000 equiv./NO_2 gp. equiv.. In the formulae, R_1 = H, Me, Et or NH_2; R_3 = H, Me, Et or NO_2; R_3, R_4 = H, Me or Et.

Also claimed is the catalyst per se.

USE - The catalyst is used esp. for the hydrogenation of nitroaromatics (claimed). (I) are valuable intermediates, e.g. for dyes, polyurethanes and plant protection prods.

ADVANTAGE - The catalyst allows higher GHSVs and gives higher selectivity than usual. It can operate in simple reactors. The productivity is an order of magnitude higher than usual. The selectivity for (I) is high at over 99.4% initially and over 99.95% (esp. more than 99.999%) after a short time. This eliminates the need for distn. of the condensate. (8pp Dwg.No.0/0)
CT: No-SR.Pub
C97-011221

(II) → (I)

Key to Flagged Terms

A - Derwent Patentee Code
B - Derwent Classification
C - Primary Accession Number *(Year, Serial, Week)*
D - Basic (★) or Equivalent (=) indicator
E - Patent Number
F - Patent Status *(A-1st, B-2nd, C-3rd Publication)*
G - Patentee Name
H - Priorities *(date, country, number)*
I - Publication date of Patent Document
J - International Patent Classification
K - Local filing details
L - Language of Application
M - Citations
N - Designated States *(N-National, R-Regional)*
• - Basic Patent Number, *given on equivalent headings*

For Basic Only
P - Number of pages and Drawing Reference
Q - Secondary Accession Number *(Microfilm auto-indexing number)*

Patent Families

A document relating to an invention not already recorded on the Derwent database is designated as a "**Basic**" and its abstract is identified by an asterisk (★) in the heading.

Further documents relating to the same invention are designated as "**Equivalent**", and abstracts for these are identified by an equals sign (=) in the heading.

The Basic document and its Equivalents form a patent family to which a unique Derwent **Primary Accession Number**, (**PAN**) is assigned.

The PAN is always shown in the printed abstracts and is one of the key reference points for patent records in Derwent's World Patents Index online database. It also forms the basis of the Accession Number Index which allows readers to determine quickly where a patent has been filled and the availability of documents in different languages.

Figure 13.1. Derwent Alerting Abstracts © 1997 Derwent Information.

as to minimize company name variations; a four-letter code identifies major companies and subsidiaries.

Inventors—last names and initials only (not given for Japanese documents).

Number of patents—number of basic and equivalent patents. Patent family members are shown as part of the online record.

Patent number and country.

97-036099/04 A25 C03 E14 J04 **FARB 95.06.14**
BAYER AG *EP 748789-A2
95.06.14 95DE-1021670 *(96.12.18)* C07C 209/36, B01J 23/44
Pure aromatic amine(s) prodn. and catalyst for hydrogenation of nitro-aromatics - comprising palladium on graphite or graphitised coke, useful as intermediate e.g. for dye or polyurethane (Ger)
C97-011221 R(BE CH DE ES FR GB IT LI NL PT SE)
Addnl. Data: LANGER R, BUYSCH H, PENTLING U
96.06.03 96EP-108855

A(1-E2) C(10-B4A) E(10-B4A1) J(4-E) N(2-F1) .1

Prepn. of aromatic amines of formula (I) comprises gas phase hydrogenation of nitroaromatics of formula (II) with hydrogen over a stationary paladium catalyst on a graphite or graphitised coke support with a Brunauer-Emmett-Teller (BET) surface area of 0.2-10 m^2/g. The catalyst contains more than 1.5 to 7 wt.% Pd w.r.t. total catalyst and the H_2 supply to the catalyst is 30-6000 equiv./nitro gp. equiv.:

R_1 = H, Me, Et or NH_2;
R_2, R_4 = H, Me or Et; and
R_3 = H, Me, Et or NO_2.

R3 R4 NO2 (II) → R1 R2 NH2 (I)

Also claimed is the catalyst per se.

USE

The catalyst is used esp. for hydrogenation of nitroaromatics. (I) are valuable intermediates, e.g. for dyes, polyurethanes and plant protection prods.

ADVANTAGE

The catalyst allows higher gas hourly space velocities (GHSV's), gives higher selectivity than usual and can operate in simple reactors.

The productivity is an order of magnitude higher than usual and selectivity for (I) is high (over 99.4% initially and over 99.95% (esp.

EP 748789-A+

more than 99.999%) after a short time). This eliminates the need for distn. of the condensate. The process gives quantitative conversion of (II).

PREFERRED CATALYST

The catalyst is prepd. by applying the Pd to the support in 1-50 impregnation stages and drying in a hot gas stream after each stage. After impregnation and before use, it is activated in a H_2 stream at 1-10 bar and 250-450°C.

PREFERRED MATERIALS

(II) is nitrobenzene or nitrotoluene.

PREFERRED CONDITIONS

The educt gas mixt. contg. (II) and H_2 is at a temp. of 200-400°C before the catalyst bed and the max. catalyst temp. is 600°C.

Reaction is carried out in an uncooled reactor and at a pressure of 1-30 bar.

A catalyst loading of 0.5-40 kg (II)/l catalyst and hr. is used.

EXAMPLE

A catalyst had a support of graphite granulate with a BET surface area of 0.4-0.8 m^2/g, particle size of 1-3 mm, bulk density of 650-1000 g/l and acetonitrile absorption of 7 ml/100 g.

200 g support were impregnated with 1.6% paladium by impregnating 7 times with a soln. of 0.95 g paladium acetate in 14 g acetonitrile. Each impregnation step was followed by drying for 5 min in an air stream at 40 °C. The catalyst was activated for 20 hr. in a hot H_2 stream at 370 °C and normal pressure.

A 220 ml (219.0 g) catalyst layer was placed in a well-insulated reactor and treated with H_2 at 200 °C and normal pressure for 10 hr. After adjusting the H_2 stream to 1620 l/hr., 110 g/hr. nitrobenzene were supplied at a starting temp. of 201 °C, corresp. to a H_2/nitrobenzene molar ratio of 81:1. Quantitative conversion under adiabatic conditions gave a temp. difference between the educt and prod. gas streams of ca. 200 °C.

After a few hrs., the temp. profile in the catalyst bed corresponded to a 10% heat loss through the reactor walls. There was no sign of catalyst deactivation after 1000 hrs. The GHSV was 7460/hr. The selectivity was 99.49% after 4 hrs., 99.54% after 40 hrs., 99.63% after 214 hrs. and 99.73% after 1004 hrs. and the prod. was free from nitrobenzene in all cases.

If a catalyst of 9 g Pd, 9 g V on α-Al_2O_3 was used at a GHSV of

EP 748789-A+/1

97-036099/04

7460/hr., the selectivity was 98.0% after 2 hrs., 98.5% after 60 hrs. and 98.9% after 301 hrs. and the nitrobenzene content of the prod. was 0 ppm after 2 and 60 hrs. but 100 ppm after 301 hrs. This catalyst was not suitable for use with large H_2 excess and nitrobenzene throughputs. (AF)
(8pp0016DwgNo.0/0)

EP 748789-A/2

Figure 13.2. Derwent Documentation Abstract © 1997 Derwent Information.

TABLE 13.3. ***World Patents Abstracts***

British Patents Abstracts (unexamined plus granted by section or unexamined only by numerical order
European Patents Abstracts (unexamined plus granted by section or unexamined by numerical order
German Patents Abstracts
Japanese Patents Abstracts Examined
Japanese Patents Abstracts Unexamined
PCT Patents Abstracts
Russian Patents Abstracts
United States Patents Abstracts

Status letter—WIPO code for stage of publication (A, B, etc.).

Publication date.

Derwent week—week added to database; it is derived from the Accession number, e.g., 93-348385/44 signifies the Derwent week 9344.

Designated States—countries designated for European or PCT patent documents.

Language of the original patent document—the patent family information included can, in many cases, help identify an equivalent in a language that the searcher can read. Language is included for all EP and PCT documents, and also for other documents published in a language other than the major language of the country.

TABLE 13.4. ***CPI Sections***[a]

A	Plastics and Polymers (*Plasdoc*)—covers polymers, fabrication, additives, uses and specified monomers; began in 1966.
B	Pharnaceuticals (*Farmdoc*)—patents of pharmaceutical, veterinary, and related interest; began in 1963.
C	Agricultural and Veterinary Chemicals (*Agdoc*)—compounds of agricultural and specified veterinary interest; began in 1965.
D	Food, Detergents, Water Treatment, and Biotechnology—includes disinfectants; began in 1970.
E	General Chemicals (*Chemdoc*)—general organic and inorganic compounds and dyestuffs; began in 1970.
F	Textiles, Paper Making—began in 1970.
G	Printing, Coating, Photographic—began in 1970.
H	Petroleum—began in 1970.
J	Chemical Engineering—began in 1970.
K	Nucleonics, Explosives, Protection—began in 1970.
L	Refractories, Ceramics, Cement, Electro(in)organics—began in 1970.
M	Metallurgy—began in 1970.

[a]Above list corresponds to main Derwent classes.

Priority number, country, and date—information that can be useful in tracking related patents.

Application date.

Cited patents—for European and PCT patents only. Derwent has two additional files that can identify cited patents. These are described on page 355.

International Patent Classification (IPC)—these are much more extensive and detailed than the Derwent classes and provide an additional means for subject searching. IPCs can be searched at any level of generality (e.g., IC = A, IC = A61, IC = A61K, IC = A61K-012, IC = A61K-012/34). IPCs are added to the patent family from new equivalents if they differ in the group level (e.g., A61K-012, or broader).

Abstract—in English. Full-text searchable. Utilizes British rather than American spellings. Note that British terminology is also employed (e.g., "nappies" for diapers).

For those searchers who meet certain minimum subscription levels or who pay a special access fee, there is online access to additional indexes. These include manual codes, chemical indexing, enhanced polymer indexing, Derwent Registry Numbers, and indexing updates. Markush searching and API indexing are also available. Some of these additional access modes are described elsewhere here.

In the online files, not all index fields will be in all patents. For example, only patents in Derwent's Section A (PLASDOC which covers polymer patents) will have Derwent's special polymer indexing. Only patents in Derwent's sections B, C, and E will have special chemical indexing, and, of course, only those patents with suitable rings will have Derwent Ring Index Numbers (RINs).

In the printed Alerting Abstracts Bulletin, the patent documents are arranged by Derwent class which can be utilized as a very broad search mechanism. Each Alerting Bulletin issue contains indexes by patentee/assignee, accession number, and patent number. There are no printed cumulative indexes; the online files serve that purpose.

Some of the special indexing features noted above are described below.

3. Chemical Fragmentation Codes; TOPFRAG. Fragmentation codes permit searching for chemical structures (e.g., rings and functional groups) and nonstructural concepts (e.g., uses and properties of chemicals). These codes are applied on much the same basis as Markush DARC (see discussion below). Codes are applied to natural products, processes, dyes, formulations, and activities and fields of use. They are applied to specific chemicals (e.g., benzene or penicillin) and generic disclosures (also known as *Markush structures,* e.g., CH_3–X, where X may be H, a halogen, or any pharmaceutically acceptable moiety). Each patent can have several subrecords, each subrecord for a different structure, and each subrecord can have an average of 40–50 code descriptors. Derwent's *TOPFRAG* (*Topological Fragmentation*) software, available on a diskette, permits the chemist to draw chemical structures offline and then automatically generates structure search strategies for use in searching *Derwent World Patents Index* and its Markush counterpart.

4. Manual Codes. Each patent can have up to about 70 manual codes in rare instances (complex pharmaceutical or chemical patents), but the average is about 2–3.

5. Polymer Indexing and PILOT. Like the Chemical Fragmentation Codes mentioned above, Polymer Indexing is deep, covering the full patent, especially the claims and examples, not just the documentation abstract and the IPCs.

Derwent's *PILOT* (*Polymer Indexing Language Online Translation* software), also on disk, permits the chemist to create search strategies offline for patent documents relating to polymers and plastics. It can thus be utilized in the searching of *Plasdoc/Polymer Indexing* in *DWPI* from 1966 to date.

6. Derwent World Patents Index Markush. For this online file, available on Questel, with gateways available on DIALOG, and Orbit, indexing covers new structures, end products of reactions, ingredients in compositions, compounds detected, and detecting agents. Both specific and generic structures are included. The file begins coverage in 1987, and includes 25 patent issuing offices plus *Research Disclosure.*

7. Derwent World Patents Index with API Indexing. On the Orbit online system only, the user can search the *Derwent World Patents Index with API* (*American Petroleum Institute*) *Indexing.* This gives the searcher the further advantage of in-depth API indexing for patent documents of petroleum and petrochemical interest and is in addition to the previously described Derwent indexing. However, this combined file is available only to those who are subscribers to both Derwent and API. Meanwhile, on DIALOG and STN International, the API database *APIPAT* continues to be available by itself. Coverage dates from 1964.

8. Examples of Some Other Derwent Products.

1. *Chemical Innovations on CD-ROM* (*Weekly*). Covers patents in certain key areas. Titles available are: Adhesives, Cosmetics, Food and Food Technology, Household & General Cleaning, and Packaging & Containers.
2. *Standard Patents Profiles.* Printed current awareness books available in over 100 chemical titles. Issued every 1–4 weeks, depending on title.
3. *Biotechnology Abstracts.* Online, print, and CD. Covers journals, conferences, and patents from 1982.
4. *Derwent Crop Protection File.* Online, print, and CD. Covers journals, conferences and patents from 1968. Accompanied by the Crop Protection Registry which contains chemical structures and activities re crop protection.
5. *Derwent Geneseq.* Online. Patented nucleic acid and protein sequence information. From 1981.

6. *Derwent Reaction Service.* Online (STN, Orbit), in-house via MDL Information Systems, Inc., and print (*Derwent Journal of Synthetic Methods*). Begins in 1975. Continuation of the classic Theilheimer series of volumes (the data provided by Derwent). Covers details of reactions reported in journals and patents.
7. *Derwent United States Classification.* Online. Classification of U.S. patents since 1978. On Orbit only.
8. *Derwent United States Patents.* One of a Derwent series of CD products (also online on Orbit) that can provide weekly updates of information ranging from front-page information to full text. From 1971.
9. *Derwent Drug File/Derwent Drug Registry.* Online, print, and CD. Technical journals and conference papers from an industrial pharmaceutical perspective. The Registry contains chemical structures and activities. From 1964.
10. *LitAlert.* An online file, available on Orbit and DIALOG, that enables the searcher to monitor patent and trademark infringment suits filed in 94 U.S. District Courts. From 1973.
11. *Patent Status File.* An online file (Orbit) that provides coverage of 25 types of postissue actions that may affect U.S. patents. Specifically, this product covers changes to U.S. patents after issue as published weekly from 1973 in the *Official Gazette of the U.S. Patent and Trademark Office.* These include

 Adverse Decision
 Adjudicated Patent
 Certificate of Correction
 Disclaimer and Dedication
 Dedication
 Disclaimer
 Delayed Payment of Maintenance Fees
 Previous Error Corrected
 Expiration of Patent Due to Failure to Pay Required Maintenance Fees
 Term of Patent Extended
 Lapsed Patent
 Patent Suit
 Reissued Patent (these patents are assigned a new number prefixed by "RE")
 Reissue Application Filed
 Request for Reexamination
 Reexamination Certificate
 Commissioner Ordered Reexamination
 Patent Substituted for Previously Withdrawn Patent

Withdrawn
Withdrawal Notice

9. User Aids. Examples of user aids for Derwent include

1. The summary sheets for Derwent produced by the four online hosts.
2. The database chapters produced by the four online hosts.
3. *Derwent User Manual 3* for each of the online hosts.

A list of Derwent User Guides is shown in Table 13.5.

In addition to the above, Derwent offers frequent seminars on the use of its products.

Derwent help desks are available in the United States, United Kingdom and Japan.

(*Note:* Derwent's new patent citations product and its product for statistical analysis of patent search results are described separately.)

10. Advantages and Disadvantages of Derwent. The principal advantages of Derwent are these:

1. Coverage is comprehensive for all technologies and is virtually worldwide.
2. Abstracts or claims are included for the vast majority of both basics and equivalents.
3. Abstracts are excellent and contain extensive filing details.
4. The Documentation Abstracts are especially useful because they contain so many details.
5. Online searching is quick, flexible, and highly efficient.
6. The coding systems available to subscribers are very powerful.
7. Derwent offers a search service that will conduct searches of its files for a fee.

Some disadvantages of Derwent are these:

1. There are no conventional substance or subject indexes with standardized and systematic nomenclature such as those of *Chemical Abstracts.* Instead, Derwent coding systems should be utilized. Derwent is currently addressing this issue.
2. A number of key Derwent products, notably the coding systems, are available only to subscribers at certain levels or to those who pay access fees.
3. Time coverage is relatively limited, although better than most. Truly comprehensive chemical coverage begins with about 1970.
4. American chemists can miss information unless they are careful to utilize British spellings as well when searching the texts of Derwent patent titles and abstracts.

TABLE 13.5. Derwent Online Users Guides

- *Dewent World Patents Index*
  - Online User Guide (Dialog)
  - Online User Guide (Orbit)
  - Online User Guide (Questel)
  - Online User Guide (STN)
  - Online User Guide (STN German)
  - Online User Guide (STN Japanese)
  - Title Terms
  - Patentee Codes
  - CPI Chemical Indexing Guidelines
  - CPI Manual Codes
  - CPI Registry Compounds
  - CPI Chemical Retrieval (2 parts)
  - CPI Chemical Code Dictionary
  - CPI *Plasdoc* Retrieval Guide (2 parts)
  - CPI *Plasdoc* Code Dictionary
  - Polymer Indexing Reference Manual
  - Polymer Indexing System Description
  - Polymer Indexing Thesaurus
  - Polymer Indexing Hierarchy
  - EPI Manual Codes (English) (2 parts)
  - EPI Manual Codes (Japanese) (6 parts)
  - Guide to Derwent Engineering Products
- *Derwent Patents Citation Index*
  - Online User Guide (Dialog)
- *Derwent Biotechnology Abstracts*
  - Introduction to *Derwent Biotechnology Abstracts*
  - Online User Manual
  - Thesaurus
- *Derwent Crop Protection File*
  - Introduction to the *Derwent Crop Protection File*
  - Thesaurus
- *Derwent Drug File*
  - Introduction to the *Derwent Drug File*
  - Journal List and Selection Guidelines
  - Online User Guide and Sample Searches
  - Thesaurus
- *Derwent Journal of Synthetic Methods*
  - Introduction to the *Derwent Journal of Synthetic Methods*
- *Derwent Veterinary Drug File*
  - Introduction to the *Derwent Veterinary Drug File*
  - Thesaurus
- *Derwent GENESEQ*
  - Introduction to *Derwent GENESEQ*
- *General User Guides*
  - Guide to Reading Japanese Patents
  - Global Patent Sources
  - Guide to Patent Expiries

5. The Derwent families of products are complex, because of their power, and this can be somewhat confusing to the newcomer. This difficulty can be overcome by careful ongoing study by the person who intends to become a regular user of Derwent.
6. If a chemist's organization desires to subscribe to all or most of the Derwent products of chemical interest, a healthy budget is required. Subscribers must pay for certain minimum number of subscription units to have complete access to all the online features.

13.15. COVERAGE OF PATENTS BY CHEMICAL ABSTRACTS SERVICE

Chemical Abstracts (*CA*) is a major source of chemical patent information. Some aspects of the treatment of patents by CAS are touched on in Chapters 6 and 7, and elsewhere in this chapter. In 1996, patents constituted over 17% of all documents abstracted by *CA*. Specifically, 121,682 patent documents were abstracted. The largest percentage was unexamined Japanese patent applications, which began an explosive growth trend in 1972. In addition, *CA* cited, but did not abstract, a total of 141,374 equivalent patent documents.

When *CA* receives a patent document equivalent to one already abstracted (as in the case of the members of a patent family representing several different countries), the equivalent patent is not reabstracted, but rather it is listed in a printed concordance (the *CA Patent Index*) with a reference to the previously abstracted patent along with the *CA* abstract number. The examined version of a previously abstracted unexamined patent document is treated similarly. *CA* concordance data are printed but, unfortunately, are not available online through *CA*.

Published German patent applications (Offenlegungschriften) are designated by *CA* as *Ger. Offen.* Granted patents (Patentschriften) are designated simply as *Ger.* The previous Auslegeschriften series was discontinued by the German Patent Office on January 1, 1981.

CA abbreviates the unexamined Japanese patent application as *Jpn. Kokai Tokkyo Koho.* (Some chemists refer to a Japanese patent application simply as a *Kokai.*) The examined Japanese patent application is designated as *Jpn. Tokkyo Koho.*

CA handling of German and Japanese patent information is especially significant. For both countries CAS provides extensive coverage and thorough indexing of unexamined patent applications. The significance of these two countries is based on their technological strength and on their quick-issue policies noted earlier in this chapter.

Compared to some other patent information sources and tools, *CA* is characterized by such features as the following:

1. *CA* is probably more widely available and used in all types of organizations and libraries than any other comparable tool.
2. Indexing and abstracting of patents is excellent.

3. For *CA* to cover a chemical substance reported in a patent *specification,* there must be sufficient scientific or technical evidence; so-called *paper chemistry* is not enough. This evidence can consist of such data as examples describing synthesis or manufacture; yield; boiling or melting point; or end use test results. (Note that Derwent does not require this kind of evidence but rather merely that a chemical substance be disclosed or exemplified.) However, *CA* indexes chemical substances specifically mentioned in patent *claims,* even when unsupported by any examples or other evidence; this is the result of a gradually implemented policy change first started in 1979, and that was in place for all *CA* sections by 1982.

4. More emphasis is placed on the chemical aspects than the legal aspects, with emphasis on details found in examples.

5. For U.S. chemical patents since the 1970s, CAS offers a unique combination of *CA* indexing and full-text searching capability in the file *USPATFULL.*

As compared to the abstracts in the *Derwent World Patents Index,* especially the Documentation Abstracts, *CA* abstracts are not as lengthy. However, the two sources often complement one another. Other pros and cons of CAS products are discussed in Chapters 6 and 7.

13.16. IFI/PLENUM DATA CORPORATION PATENT SERVICES

IFI (IFI/Plenum Data Corporation, 3202 Kirkwood Highway, Wilmington, DE 19808, 800-331-4955) has long been an important source of information about U.S. patents. The emphasis is on indexing of chemical patents, but all other types of patents are covered as well.

At the end of 1996, this important indexing service had covered over 925,000 U.S. chemical and chemically related patents, dating back from 1950 to within a short time of the most recently issued patents. IFI's *CLAIMS* databases are thought to be the largest computerized collection of fully indexed information about U.S. patents in the world. IFI *CLAIMS* files have been steady, consistent, and reliable workhorses for many years and, furthermore, have benefited recently from significant enhancements and improvements in 1992–1995.

The time period of coverage is longer than for any other online patent file. Chemical patents and chemically related patents are covered since 1950, and all U.S. utility patents, including chemical, mechanical, and electrical, are covered since 1963. Design and plant patents are covered from December 1976.

Two of the *CLAIMS* databases, described below, are distinguished by exceptionally deep indexing of the full text of chemical patents. IFI says that there is an average of 55 indexing entries per chemical patent in its most comprehensive service (CDB; see description below). Indexing is based on the full text of the patents included.

Reissue patents, defensive publications and Statutory Invention Registrations, also known as SIRs are included in all IFI *CLAIMS* files.

All records in the databases include title, patent number, patent issue date, assignee names, inventor names, U.S. Patent Office classification, International Patent Classification, and main claim from the *Official Gazette* (except for Design patents).

Because official titles of many patents are often not sufficiently meaningful, official titles are enhanced (expanded) by IFI staff for about 90% of chemical patents since 1972. In addition, some mechanical and electrical patent titles are enhanced.

For patents issued since 1971, the database records include all data from the front page of the patent, specifically, application data (filing number and date), priority application data, addresses of assignees, addresses of inventors, names of examiners, names of attorney (or agent or firm), U.S. patent examiner classifications utilized in examiner's field of search, references cited by the examiner, and abstract from the front page of the patent. All patent claims are included for most patents during 1971–1974, and for all patents since 1975.

For all chemical patents from 1950 to date, IFI includes patent application numbers and dates, including related filing data, such as continuation and divisional filing data, as well as priority filing information for non-U.S. priority applications.

U.S. classifications, with main and cross-references, are updated annually on all *CLAIMS* records to reflect any changes in the classification system made by the Patent Office during the year. This permits the searcher to utilize current U.S. classifications to search for patents during any included time period.

As mentioned, IPC (International Patent Classification) codes are now present for patents included since 1950, so that these codes can be employed as search tools for all time periods included. Over 99.9% of the utility patents in the *CLAIMS* database are said to include IPC codes. The codes for 1950–1970 were added by creating a concordance using an algorithm that analyzed the frequency of classes assigned to patents over the past 5 years. In this connection, the *IPC Manual on CD-ROM,* also marketed by IFI, allows the searcher to create comprehensive search strategies using IPCs. The CD-ROM product features the sixth edition of the IPC classification, but includes a concordance to all previous editions.

Many chemical patent records also include equivalent patent numbers (Belgium, France, Great Britain, the Netherlands, and West Germany) for 1950–1979, although this information is not necessarily complete. Also included for many chemical patents are CAS Registry Numbers for patents issued between 1967 and 1979, and references to *CA* abstracts when available (added annually).

IFI standardizes assignee names for all major companies, regardless of the form in which the name appears in the *Official Gazette,* thereby permitting consistent searching. For example, International Business Machines Corp. is the standard name utilized to represent such potential variations as IBM, I B M, or International Business Machines.

Full patent text is not offered in *CLAIMS,* but the *CLAIMS* databases are linked to the DIALOG full-text U.S. patent database, that is, the "*PATFULL*" (*U.S. Patents Fulltext*) series of databases so as to permit quick viewing of a full-text patent record while in *CLAIMS,* Alternative electronic or postal delivery options are also available. In this connection, IFI provides the data for assignee names, reassignment data, reexamination data, expirations, extensions, and annual reclassification of U.S. classes in the DIALOG full-text file.

Chemical patent structures that are two dimensional are converted by IFI staff to easily understood linear notations that can be searched and displayed online. *CLAIMS* does not provide figures, drawings or graphs in its files, although their presence is noted. Descriptions of figures are included for design patents (1976 onward).

The basic service is the *CLAIMS U.S. Patents* database. This provides ready access to all the above (full bibliographic, front page, and *CLAIMS* data as noted above), but it does not include keywords or any other indexing effort by IFI staff.

The next level of the service is the *CLAIMS Uniterm* file, which, in addition to basic service, also includes controlled keyword descriptors assigned by IFI for all chemical and chemically related U.S. patents. IFI says that there are about 25 index entries per patent.

The *CLAIMS Comprehensive Database* (*CDB*) is IFI's most powerful and significant product. In addition to keyword indexing, and the other features previously noted for the *CLAIMS* files, *CDB* has other noteworthy attributes. For example, *CDB* features use of "roles" (since 1964) to indicate the function of a chemical substance. Thus, the role of a chemical may be as a reactant or a product, or it may merely be "present." IFI is believed to be the first commercial indexing firm to utilize roles for indexing of chemical patents, and this general approach has since been adopted by others.

Other important features of *CDB* include specialized procedures for indexing polymers, especially since 1972, and the use of a "fragmentation" system for substructure searching. In the fragmentation approach, controlled vocabulary terms are utilized to describe chemical compounds in terms of the substructural segments which characterize them. Thus, IFI says that Markush or generic searching is facilitated from 1950 on. (Some limited fragmentation searching is also possible in the *Uniterm* file for the time period since 1972.) *CDB* depth of indexing is about twice that of the *Uniterm* file notes above. This is probably the deepest indexing of U.S. chemical patents that is available anywhere on a commercial basis.

The *CDB* file is available on the DIALOG, Orbit, and STN International systems, and in addition, IFI staff can search the file for anyone on a fee basis. Unfortunately, direct use of the online database is accessible only to those organizations who also maintain a separate subscription to *CDB* in magnetic tape form for in-house use. This subscription is expensive, and only a relatively few organizations subscribe.

The *CLAIMS Reassignment and Reexamination* database provides information on changes in assignment after issue. This file also includes information on expirations of patents due to nonpayment of maintenance fees, reexamination results, extensions of term for pharmaceutical patents, adverse decisions in interference actions, disclaimers and dedications, and requests for reexamination.) There are weekly updates for expiration, extension, and reexamination data. The time period covered is as follows: reassignments registered, 1980 onward; reexaminations issued, 1981 onward; expirations, September 1985 onward; reinstatements, October 1986; extensions, April 1986 onward; adverse decisions and disclaimers, 1980 onward; and requests for reexamination from the beginning, August 1981. (In 1994, IFI registered 30,000 reassignments recorded in 1994 due to changes in patent ownership or in company name, 275 reexaminations, 42,300 expirations for failure to pay maintenance fees,

630 reinstatements, and 16 extensions.) The file is available on DIALOG, Orbit, and STN.

The *CLAIMS Reference* database includes the classification codes and text from the U.S. Patent Office *Manual of Classification,* and, in addition, the IFI controlled terms and codes that can be employed in the searching of the *Uniterm* and *CDB* files.

A new product available from IFI is *CLAIMS PC Reference,* available on floppy disk. It contains all index terms from the *IFI General Term Thesaurus, Compound Term List, Fragment Term List,* and *Assignee Term List.* Thus, this product facilitates development of search strategy and permits formatting of search strategies and subsequent uploading and searching.

The *CLAIMS Compound Registry* database is a dictionary file for IFI chemical substance terms that permits identification of term numbers, main compound names, synonyms, molecular formulas, fragment codes, and terms. Results can be utilized in searching of the Uniterm or CDB databases. It can be searched online in DIALOG, STN International, and Questel•Orbit.

Another important IFI file is *CLAIMS Citation* (see Section 13.22).

Another IFI product is *Patent Intelligence and Technology Report.* This annual report contains the following sections:

1. Alphabetical list of companies, showing total number of patents granted during the year, rank number, and patent activity profile page. Data are provided on all assignees who receive 10 or more patents in the year.
2. Listing of company by rank.
3. Distribution of patents by company (ranked) within U.S. Classification.
4. Patent activity profile by company for 6 years. Number of patents received within each patent classification.
5. Concordance of U.S. Classification and International Patent Classification, showing the number of patents granted in each class (in both U.S. and IPC order 2 sections).

1. Some Advantages and Features of IFI Services

1. Coverage extends back through 1950 for U.S. chemical and chemically related patents. No other indexed database offers such a broad time spread.
2. The subject indexing is very deep, and structure searching capability of the *Comprehensive Data Base* is very good.
3. The IFI help desk is staffed by chemists, experienced with patents, who provide excellent technical support and advice.
4. Data about electrical and mechanical patents are online beginning in 1963. Many of these patents can be important to chemists.
5. IFI will search any of their files for a fee.
6. Bibliographic and abstract/claims data is now online about one week after the issue date.

7. IFI products are subjected to an editing and error-correction process that IFI says offers more complete and reliable information than that available in the unedited tapes from the U.S. Patent Office. Standardizing of assignee names is one significant feature of this editing process. Another is the inclusion of data for over 126,000 patent documents issued during 1971–1974 and that are missing from the original USPTO full-text tapes.

2. *Some Disadvantages of IFI Services*

1. Content does not include foreign patents, although there are some foreign equivalents listed for older patents (1950–1979).
2. Turnaround for indexing has improved, with indexing available online about 4 months after issue. However, this is still slower than some users would like to see for a service of this importance.
3. Drawings (images) are not included.
4. The unique, highly controlled indexing vocabulary makes for consistency of indexing, but requires that searches be very disciplined. This is good for the information professional, but may be more difficult for some end users.
5. IFI's most powerful tool, *CDB,* is available online only to subscribers to the magnetic tapes.

3. Brief History. IFI began operations in 1955 under the title *Information for Industry* after conversations with leading industrial firms indicated an interest in its product. The first indexing for IFI was done by Documentation, Inc. under contract although IFI now has its own full-time technical staff of more than 30 chemists in Wilmington, DE. (Documentation, Inc was founded by the late Mortimer Taube, formerly of the U.S. Atomic Energy Commission, to commercialize his discovery of the Uniterm indexing concept.) In 1959, coverage was expanded back to 1950 at the request of the DuPont Company. The products were originally available in printed form, and, in 1961, became available in computerized form on a batch basis at first.

In 1971, IFI purchased DuPont's indexing technology (involving use of roles and fragmentation codes) and merged the two files back to 1964 (subsequent back-indexing has extended this coding back to 1950). IFI became a subsidiary of Plenum Publishing in 1967. The technical director of IFI is Darlene Slaughter, and the long-time president/CEO is Harry Allock (not to be confused with the chemist of the same name, a well-known professor of chemistry at Pennsylvania State University).

13.17. FULL-TEXT PATENT DATABASES AND SOME INTERNET SOURCES

A. Full-Text

CAS's *USPATFULL,* available on *STN International* (and its Internet twin *Chemical Patents Plus*), is a unique file. For chemical patents that are covered in the *CA* and

CAplus files, in addition to full-text searching capability, these files also contain full CAS indexing, including CAS Registry Numbers. This provides search capability that is very powerful. In *USPATFULL,* for chemical patent answers that contain Registry Numbers, the structures may be displayed when the proper software (such as STN EXPRESS) is used. Complete data for the images of patents is available for patents since January 1, 1994 (images are to be available on a rolling 3-year basis). This permits the following. Pages containing the drawings of U.S. patents may be downloaded, and structures often embedded in claims and other text of patents may be displayed and printed for chemical patents. In addition, in the Internet version, 2D and 3D chemical structures may be displayed and 3D structures may be rotated, when a Java browser is utilized.

Patents are published by the Patent Office each Tuesday. STN's *USPATFULL* is available on Thursday of the same week. The same schedule applies to the Internet version.

Several other databases offer full-text searching of U.S. patents. One of these is *U.S. Patents Fulltext* (*PATFULL*) on the reliable DIALOG system; however this file does not include CAS indexing.

Another patent database that has attracted much attention is *QPAT-US,* offered by Questel•Orbit on the Internet (www.qpat.com). Coverage is the full text of U.S. patents since 1974. An annual subscription fee ($1995) offers unlimited access, although searching of abstracts is free to registered users at this time. In addition to the usual keyword and boolean search features, there are several special features, most notably natural language (fuzzy logic) capabilities. Fuzzy logic is intended to help the user achieve relevant results based on queries that may be imprecise, a particularly useful feature for end users. Thus "or" logic is utilized unless otherwise specified. In addition, "search term stemming" is offered; this means that the system automatically moves forward and backward within a search word so as to expand the possibilities. The system is good at suggesting alternative spellings and word forms, and statistically related words. Results are presented in relevance-ranked order so that the user can make a selection. Left-hand truncation is also available, but it is slow at this time. As of this writing, images are not available.

The LEXIS-NEXIS search service (Dayton, OH) was the first to offer full-text searching of U.S. patents, and this continues to be available through their *LEXPAT* file, for which coverage begins with 1975.

B. Some Other Internet Sources

1. International Business Machines Corporation. In 1997, International Business Machines Corp. began to offer searching of U.S. patents on the Internet (www.ibm.com/patents). The effective start date of coverage is 1974, but some patents go back as far as 1971. All front-page information and all claims are searchable. The service may ultimately offer full-text searching. Images can be viewed for a number of years, and earlier years are to be added. Patent Cooperation Treaty patent documents and European Patent Office patent documents may be added. One of the most useful features is that most of the patents are "hyperlinked"; this permits the user to navigate through the U.S. patents cited by the examiners as well as citations in later patents. Also included is information on the maintenance fee payment status for all patents. The system does not offer any indexing of the type available from CAS or Derwent. At present, the service is available without charge. Copies of patents can be ordered on a fee basis through Optipat, Inc., Arlington, VA.

In addition to the above, there are a number of other sources that offer patent searching on the Internet. These include, for example, the U.S. Patent and Trademark Office and MicroPatent (see next section). The sites of the U.S. Patent Office (www.uspto.gov) and of the European Patent Office (www.epo.com.at) are among those that provide good links to other sources of patent information both on and off the Internet.

2. MicroPatent. Founded in 1989 by Peter H. Tracy, this company (located at 250 Dodge Ave., East Haven, CT 06512 800-648-6787) offers a broad variety of patent-related products and services, including patent and trademark searching tools, patent copying, and searching. These are available through the Internet, in CD-ROM form, by telephone, through the mail, including e-mail, and via fax.

The core of MicroPatent's products and the additional value they provide are based on creative and high-speed repackaging and delivery of electronic input received from the U.S., European, WIPO (PCT), and Japanese patent offices. MicroPatent does not do any intellectual abstracting or indexing of the patents included in their World Wide Web products, but offers patent abstracts (as reported in the original patent) and claims searching in its CD-ROMs.

Probably its principal competitor in the CD arena is the family of CD-ROM products offered by Derwent.

MicroPatent's products are on the World Wide Web and can be accessed by www.micropat.com. The MicroPatent patent file on the Internet is designated as *PatentWEB,* and the trademark file as *Trademark Checker.*

PatentWEB permits several levels of access:

1. Front-page display for all U.S. patents back to 1994, EP applications back to 1992, and PCT documents back to 1978. There is a small charge for viewing an image if the full patent is not ordered.
2. View and download complete patent documents (US, EP, PCT) beginning 1974 for US, 1992 for EP, 1978 for PCT. The charge for this is $3.00 per document. (This material may be viewed or downloaded but not searched.)
3. Use their *Online Gazette* product to search and display USPTO *Official Gazette* data for all new U.S. patents, beginning in 1997, including front-page drawings and all chemical structures and equations from exemplary claim. This service is free, but registration is required. Weekly updates. Comparable EP and PCT coverage is planned.
4. Search of the most current patents by U.S. Patent Office classification. This is currently limited to a 4-week window but it is planned to go back earlier as well. Output includes a list of patent numbers and titles. There is no charge for this.

5. Full text of U.S. patents for the most recent 2 weeks fully searchable. However, drawings are not displayed in this part of the service. There is no charge for this.
6. Full text [ASCII (American Standard Code for Information Interchange)] of any U.S. patent from 1974 for $1.50 each delivered by e.mail. Drawings not included.

The MicroPatent Internet file permits the user to order full copies of U.S., EPO, or PCT patent documents while online. Delivery options include: bulk or individual patent downloading, e-mail, fax, and U.S. mail.

Future plans for this Internet product include providing access equivalent to the "shoebox" method of searching that has long been a popular feature of the Public Search Room at the U.S. Patent Office. This will permit Internet users of MicroPatent to "flip" through images of the complete text of U.S. patents back to 1830.

Trademark Checker on their *TrademarkWeb* permits searching of the full U.S. federal trademark database back to 1884.

For chemists, one of the most interesting and useful CD-ROM products from MicroPatent is *RetroChem.* This disk contains all chemical patents since 1978 on a single disk, and it is updated annually. For each patent on the disk, the following information is included: patent number, status (withdrawn or expired for failure to pay fees), issue date, application number and date, inventor(s), assignee, U.S. Classification Number, IPC (International Patent Classification), U.S. references, foreign references, other references, priority country, number and date, PCT (Patent Cooperation Treaty) data [publication date, number, filing date and 102(e) date], related data (statements about continuation and divisional applications), Primary Examiner, Agent, and Title/Abstract. Each of these information categories is fully searchable. In addition, the user can combine fields in a search; thus, one can look for a combination of a U.S. class and a keyword. Browsing and complete boolean logic are possible. Results can be sorted in ascending or descending order either alphabetically or numerically.

Other MicroPatent products include

1. Backfiles of U.S. patents from 1975 to date with abstracts or claims. On CD-ROM (*USPS- U.S. PatentSearch*). Monthly.
2. *U.S. PatentImages,* full text, from 1976. On CD-ROM. Weekly. Also available U.S. Chemical PatentImages, subset limited to chemical patents.
3. "World" Patent (PCT) applications—images from 1989 and searching of abstracts from 1992. On CD-ROM.
4. All European patent (EP) applications, including abstracts (1978 to the present). On CD-ROM. Weekly.
5. *MicroPatent Alert.* Current alerting service available on a floppy disk or e-mail for a number of major industries. Topics include polymers, pharmaceuticals, and biotechnology.*
6. *PatentBible.* Reference tools including *Manual of Patent Office Examining*

*A new Micropatent is *World PatentSearch.* This permits searching front pages of U.S. patents and of EP and PCT applications.

Practice with *U.S. Code* and *Code of Federal Regulations, Manual of Classification, Index to Classification,* Classification Definitions, US/IPC Concordance, and current classification indexes for all U.S. patents. Semiannual.

7. *Patent Abstracts of Japan.* English abstracts of Japanese Kokai. Monthly.
8. *Trademark Checker.* CD counterpart of the online file. Monthly.

The *QPAT-US* file on the Internet was previously mentioned.

13.18. ONLINE DATABASES THAT INPUT MARKUSH STRUCTURES IN PATENTS AND FACILITATE SEARCHING OF THESE STRUCTURES

There are three especially noteworthy online databases that input Markush structures from patents and facilitate searching of these.

One of these, *Pharmsearch,* is an excellent abstracting and indexing service for pharmaceutical patents, that covers not only any Markush structures in the patents, but also indexes deeply other details in the patents, including other chemical structures, excipients, and uses, from a pharmaceutical viewpoint. The file is available on Questel and Orbit, and is a product of INPI (Institut National de la Propriété Industrielle or the French Patent and Trademark Office). The U.S. contact is O'Hara Consulting, Washington, DC (800-949-5120). Coverage begins in the mid-1980s for most of the patent document issuing authorities included (U.S., U.K., France, Germany, European Patent Office, Germany, and World Intellectual Property Organization). The entire database is in English. Significantly, all indexing is complete and available within a relatively few weeks after patent issue. For this database only, the Markush DARC features of Questel are significantly enhanced. For example, structure input is speeded up, and can best be done by text input (keyboard) in this case. To cite just two examples of other enhancements, displays can include both structures and bibliographic references, and results are automatically saved for a week.

In connection with Markush DARC, please see the paper by Pierre Benichou and Christine Klimczak (4).

A second online database that inputs Markush structures is the *Derwent World Patents Index Markush* (*WPIM*). With coverage since 1987, this online product permits graphical searching of both specific and generic structures as reported in chemical patent documents worldwide. This capability exists for eligible subscribers only on Questel, although subscribers may also access it through the DIALOG and Orbit gateways to Questel.

MARPAT is a Chemical Abstracts Service product that is graphically searchable on STN and was discussed earlier. Coverage includes patents from all countries covered by *Chemical Abstracts* from 1988 and that include Markush structures for this time period.

In addition, other general patent databases report some Markush *query* capability; these include *Derwent World Patents Index* and IFI's *Comprehensive Data Base.*

13.19. RESEARCH DISCLOSURE

Research Disclosure includes descriptions of inventions that organizations have decided not to pursue, for one reason or another, through the costs and other rigors of the patenting process. Publication of the description in this source is intended to block anyone from getting a patent in the area described. The purpose is thus defensive. This publication, which is issued each month by Kenneth Mason Publications, Ltd., is indexed and/or abstracted by *CA*, Derwent, and IFI/Plenum, among other sources. There had been at least one other similar publication, *International Technology Disclosure*, but this was discontinued in 1993.

13.20. PATENT EQUIVALENTS AND FAMILIES

Many organizations file patent applications in more than one country, especially for inventions believed to have widespread commercial or other importance that transcends national boundaries. This kind of filing cannot be done lightly, because extensive costs can be involved.

Accordingly, if a patent application has been published or issued in a number of countries, one can be reasonably sure that the organization that filed the patent application may be practicing, or at least strongly considering practicing, what is taught in the patent on a commercial basis.

From the point of view of the chemist interested in finding and using chemical information that represents actual practice, what has just been said is important. It is equally important to note that identification of equivalents (or so-called patent families) can be useful in other ways. For example, Section 5.7 on translations notes the importance of finding English-language equivalents of foreign-language patents, thereby possibly obviating the need for a translation. Also, identification of a patent family indicates that the chemist probably does not need to obtain copies of all patents in that family; they are probably, although not necessarily, similar in content. Members of a patent family are subject to the laws and practices of several different countries, and these vary as to what is patentable. Accordingly, the content of the members of a family may vary somewhat.

> The firm ISTA (Rosemont, PA, 800-430-5727) offers relatively inexpensive computer-based translations of Kokai (published Japanese patent documents that are unexamined). This may often be found useful when the usual abstracting and indexing services give incomplete information. Although computer-based translations are not intended to be perfect, they can give useful information at low cost, and they can help indicate whether a full-blown "human" translation is warranted. ISTA services have received a favorable response from industry.

How can the chemist most effectively achieve identification of patent equivalents or patent families? This is best done through use of one of several patent concordances.

The most complete source of patent family (equivalent) information is the INPADOC database, a product of EPIDOS (European Patent Information and Documentation Systems) which is a important agency of the European Patent Office and is located in Vienna, Austria. (The original meaning of *INPADOC* was International Patent Documentation Center.)

The INPADOC database is readily searchable through STN International, DIALOG, and several other sources. INPADOC estimates that it covers some 95% of all patents published worldwide since 1973, although the official start date of the online files is 1968.

Some 65 patent issuing authorities (12 are countries that are no longer active, such as the Soviet Union) and over 25 million documents are included. Approximately 1.5 million new documents are added each year. In addition to patent family coverage, INPADOC provides legal status information for 20 patent-issuing authorities. This permits the user to identify and track the progress of inventions through the various stages of the patent process including the payment of maintenance fees required to continue the full life of a granted patent. The file is updated weekly and is said to be current to within approximately 3 weeks for the major patent offices.

One of several special features permits the searcher to identify only those patent families for which at least one member is in English; the overwhelming number of patent document titles are in English. Another permits the display of U.S. patents only.

The INPADOC database online does not contain abstracts, and the data are not indexed; hence this is not a good source for searching by subject matter. However, search by national and international patent classifications is possible. Searching of INPADOC files by words in the titles of patents is feasible, but this is, of course, not usually a desirable approach since patent titles are typically incomplete and can be misleading. Other types of searches that are possible in INPADOC, in addition to patent number, include, for example, those by assignee, and by inventor. Further, publication and application data are provided. References to abstracts in the *Derwent World Patents Index* and *Chemical Abstracts* are included when available. References to abstracts from JAPIO (the Japanese Patent Information Organization) had also been included, but this was discontinued in 1994.

Another good patent concordance system is that of Derwent. Updates are done weekly. Coverage is comprehensive, and online access is very efficient. Equivalent patents are fully abstracted. A disadvantage of the Derwent patent concordance system is that is does not go back far enough in time—only until about 1970 in many cases.

The patent index to *CA* is issued weekly, but it is a cumulated only semiannually. The concordance features goes back to 1963 and is probably more readily available to most chemists than any other concordance. Until about 1985, however, the number of countries covered was small.

To put the history of CAS reporting of equivalent patents into perspective, it is helpful to review the chronology:

1. Since 1981 (Vol. 94) all patents (abstracted and equivalent) are listed in a single index, *CA Patent Index.*
2. This replaced the *CA Numerical Patent Index* and *CA Patent Concordance.* The latter was introduced in 1963 (Vol. 58) and last appeared in 1980 (Vol. 93).
3. Prior to that, between 1961 (Vol. 55) and 1962 (Vol. 57), cross-references from equivalent patents to patents first abstracted in *CA* were inserted in the *CA Numerical Patent Index.*

4. Prior to that, that is, in 1960 and earlier, title-only patent-abstract cross-references were inserted in *CA* Issues such as:

   CA 54: 25941c (1960):
   Aqueous dispersions of elastomers. B. F. Goodrich CO. Brit. 840,093, July 6, 1960. See U.S. 2,905,649 (*CA* 54:15990d).

   In the *CA Numerical Patent Indexes* for 1960 and earlier, each such equivalent patent was listed with the reference to the title-only abstract without any indication that the *abstract* was in fact merely a cross-reference to another abstract.
5. There is no readily available record of when this title-only practice started, but there are examples as far back as 1930 (Vol. 24).

Several additional comments are appropriate:

a. The original purpose of inserting title-only patent abstract cross-references was simply to eliminate duplicate abstracting of essentially the same document, and not to keep a record of equivalent patents.
b. In those days, manual card files on patents were kept by one or two individuals who often depended on memory in trying to screen equivalent patents from duplicate abstracting.
c. Later efforts in building patent files based on priority records (cards, microfilm) were described by Platau (5).
d. Some aspects of the basically manual operation of determining patent families at CAS until 1980 were described by Pollick (6).
e. Computerized operation based on *INPADOC* tapes was described by Pollick (7).
f. *CA* patent equivalency data are not searchable in *CA* online files. This is a major shortcoming of these files.

IFI's capabilities in searching for equivalents were mentioned earlier.

13.21. FILE WRAPPERS

In addition to reading issued patents themselves, it is often helpful to read the complete files of those U.S. patents that have previously been identified as of interest. These are available once patents have been issued and can be obtained in forms known as "file wrappers." These include all correspondence between the applicants and the Patent Office during prosecution of the application and the original specifications and claims. The complete files can often have information not in issued patents and can include original data, including laboratory notebook pages, and explanations of the rationale behind the invention. File wrappers may be purchased from the U.S. Patent Office.

13.22. PATENT CITATIONS

Patent citations are references cited by patent examiners in the course of their examining patent applications to determine whether these applications should be issued. Citations may be to other literature forms, but most frequently they are to other patents. Literature cited by examiners is to prior art related to the applications in hand.

Patent citations can be useful for a number of reasons. One of these is that the citations provide another way (in addition to subjects, inventors, assignees, and classes) to track what is going on in a given technology and to hopefully better understand the prior art. They can provide clues as to organizations and inventors who may be performing related work of interest. Citations can show who may be building on the foundation of a technology that has been previously patented. To an industrial company, citation of that company's patents by others can help indicate potential markets, customers, or competitors.

If a patent is cited frequently in other patents, this may or may not indicate its importance in the opinion of some patent attorneys. But others believe that frequent citations indicate landmark patents and that citation patterns help indicate the strengths or weaknesses of a company as well as provide other useful competitive information.

However, Kaback, Lambert, and Simmons (8) have pointed out that citations in patents should not be considered as equivalent to citations in journal literature, and, therefore, must be used with a considerable measure of caution. The argument is that citations in the journal literature are generated by the authors of the papers and hence usually indicate work very closely related to that of the author; inclusion of the citation is based on the judgment of the author who did the work. On the other hand, a patent citation, as noted above, is the material cited by a patent office examiner, a person who has no connection with the inventor or the material cited other than the function of examiner of the application submitted by the inventor. A patent citation is made by a patent examiner to determine whether patent claims should be granted, and, accordingly, is not necessarily a clear indicator of significance.

In the case of patent applications examined at the European Patent Office, search examiners follow the helpful practice of assigning relevance ratings to their citations, and these ratings appear in the search reports. The ratings are as follows:

- A Document defining the general state of the art, which is not considered to be of particular relevance.
- D Document mentioned in specification.
- E Earlier document but published on or after the international filing date.
- L Document that may throw doubts on priority claim(s) or that is cited to establish the publication date of another citation or other special reason (as specified).
- O Document referring to an oral disclosure, use, exhibition or other means.
- P Document published prior to the international filing date but later than the priority date claimed.
- T Later document published after the international filing date or priority date and

not in conflict with the application, but cited to understand the principle or theory underlying the invention.

X Document of particular relevance; the claimed invention cannot be considered novel or cannot be considered to involve an inventive step.

Y Document of particular relevance; the claimed invention cannot be considered to involve an inventive step when the document is combined with one or more other such documents, such combination being obvious to a person skilled in the art.

Codes X and Y above are the most crucial ones.

In contrast, examiners in the U.S. Patent Office utilize no rating systems, so that the relevance of their citations is essentially unknown to "outsiders."

Searching of patent citations is ideally adaptable to the use of online computer systems. Online tools available to facilitate patent citation searching include the following:

1. Derwent *Patents Citation Index.* This important new file was first made available in 1995 as File 342 on DIALOG. It is also available on STN International, and it is to become available on both Orbit and Questel. The database now (1997) provides citation data on basic patents from eight patent-issuing authorities: Belgium, European Patent Office, France, Germany, Japan, Patent Cooperation Treaty, United Kingdom, and the United States. This is, unfortunately, a scale-down from the original scope of the file that had embraced a total of 16 countries: Austria, Australia, Belgium, Canada, European Patent Office, France, Germany, Japan, Netherlands, New Zealand, Patent Cooperation Treaty, South Africa, Sweden, Switzerland, United Kingdom, and the United States. For some of these countries, there are a few limitations as to types of patents covered. For all U.S. patent documents since 1973, examiner citations to both patents and literature are included. For European (EP) and Patent Cooperation Treaty (WO) documents since 1978, examiner citations to patents only are given. Examiner citations to nonpatent documents are not included. For all other patent authorities covered, the period of coverage begins in 1994, specifically with Derwent Week 9418. For all authorities, beginning with Derwent Week 9418, coverage includes
   a. Examiner citations to both patents and literature reference and relevance indicators (such indicators are given only with European patent documents).
   b. Bibliographic information including field of search.
   c. Full patent family, as given in *Derwent World Patents Index.*

   Citations by inventors had been included in the original scope of the file, but these are now no longer covered.
2. IFI *Claims/Citation* files, available on DIALOG. Covers U.S. patent citations beginning with U.S. patents issued in 1947, thus going back in time farther than any other service of this type.
3. *Derwent World Patents Index* (online database), searchable on four systems as described earlier in this chapter. This has made possible the searching of citations in European and Patent Cooperation Treaty documents for the time period since 1979.

4. *Derwent United States Patents,* searchable on Questel. Covers essentially front-page data from the 1970s.
5. Full-text patent databases. The various full-text patent databases, such as described in this chapter, also make possible searching of patent citations. Given a patent number, one can easily determine which later patents cite that patent number.
6. The IBM Internet patent database facilitates citation searching, as does the U.S. Patent Office Internet file.

CHI Research, Inc., Haddon Heights, NJ (609-546-0600) has been active since 1968 in the use of patent citations as an information and intelligence tool, and the firm has an extensive database of these citations, which it utilizes to offer a tracking and evaluation service with respect to individual companies as well as overall technological trends. Francis Narin is the company's president. An example of another firm active in the analysis of patent citations is Mogee Research and Analysis Associates, Great Falls, VA. Mary Ellen Mogee is president.

13.23. INTERNATIONAL TREATIES AND OTHER DEVELOPMENTS

International patent cooperation dates back to the Paris Convention for the Protection of Individual Property to which the United States has been a party since 1887. This Convention, since revised, has made it easier to protect inventions across national borders. The most important new international patent agreements since then are the European Patent Convention of 1977 and the Patent Cooperation Treaty of 1978.

The European Patent Convention (EPC) came into effect in 1977 and full operation started in 1978. The first patents were granted on January 9, 1980. Ultramodern headquarters offices of the European Patent Office are located on the banks of the Isar River near the famous Deutsches Museum in Munich. By filing a single patent application in one of the three official languages (English, French, or German), one can obtain patent protection in any or all of the 18 member states of the European Patent Organization. A granted European patent provides the same rights in the designated states (designated by the applicant) as a national patent granted in any of these states. Thus, this is a very efficient way of applying for patent protection in several countries at once.

European patent applications are published 18 months after the date on which the European or national first application was filed (priority date). Because of this, and because European patent applications are seldom filed "frivolously," they are an especially valuable source of the most recent chemical information. European patent documents are identified by the letters "EP," followed by the number. The letter following the number indicates the stage (ranging from first application to grant). Thus, the earliest stage is designated by the letter "A." Copies of European patent documents are readily publicly available from many different sources. Very useful copies of the official search reports by the European Patent Office staff are published with the application or later. The applicant then has 6 months to decide whether to pursue the application through substantive examination. It takes an average of 44 months to obtain a European patent, and the life of the granted patent is 20 years from the date of filing of the application.

The Internet site of the European Patent Office is www.epo.co.at/epo/. In addition to basic descriptions of the EPO patent process, numerous links are provided to other patent sites on the Internet.

Another major source of patent information is WIPO (World Intellectual Property Organization), a branch of the United Nations located in Geneva, Switzerland. The governing treaty is the Patent Cooperation Treaty (PCT). A total of 88 nations were parties to this treaty as of November, 1996. Most major industrialized nations participate, including the United States, Canada, United Kingdom, Germany, Japan, and France. The applicant indicates which of these countries are "designated states"; the effect of the international application in each designated state is as if a national patent application had been filed with the national patent office of the state. The filing is followed by an international search report, from one of the major participating patent offices, which cites published documents, mostly other patent documents, that might bear on the application. The applicant may then withdraw; if not, the application and the search report are published and are readily publicly available. PCT patent applications begin with the letters "WO" (World), followed by the applicable numbers.

PCT filing gives the U.S. patent applicant an opportunity to get an International Patent Search Report, with effect in all designated states, from either the U.S. Patent Office or the European Patent Office. In addition, the applicant can obtain the benefit of an international preliminary examination report by the U.S. or European Patent Offices. A decision can then be made as to whether to proceed with all the other expenses associated with national or regional filing such as the very expensive translation costs. The applicant can delay the national or regional filing phase about 30 months pending results of the international preliminary examination, and much of the work can be done in the offices of a U.S. attorney with minimal use of expensive foreign representation. The net result can be substantial cost savings. PCT filing has become increasingly popular, especially in the United States.

A new regional patent office was established on January 1, 1996, in Moscow, Russia: the Eurasian Patent Office. Some nine countries are members, mostly countries from the former Soviet Union, and the patent documents are designated as Eurasian patents.

Good coverage of European and PCT patents is provided by both Derwent and *CA*. For European Patent Office patents, Derwent offers, in addition to its classified alerting bulletins, the *European Patents Report,* which is in numerical (patent number) order.

An expectation held by some is that a single European patent may ultimately replace all individual country patents in Europe. This simplification, if it happens, would help chemists and other scientists by streamlining document access and eliminating most remaining language barriers.

13.24. STATISTICAL ANALYSIS OF PATENT SEARCH RESULTS

Statistical analysis of patents is now readily possible. This analysis can answer such questions as the following:

Which companies have the most patents in a certain technology?

Who are the leading inventors in a certain technology?

What are the principal technologies in which a certain company tends to patent?

What are the most significant U.S. Patent Office classes and subclasses to search for patents to a specific assignee or on a particular technology?

The answers to these and other questions of this type can be provided in ranked order, with percentages supplied if desired.

Statistical analysis of patents can be done in several ways. Results can be obtained while online with the RANK command on DIALOG, GET on Orbit, MEM and MEMSORT on Questel, and SMARTSELECT on STN.

Alternatively, software packages may be utilized. For example, Derwent Infoview, formerly called PATSTAT, is a program sold by Derwent for statistical analysis of downloaded answer sets. It will work on all files and hosts, not just Derwent's *World Patents Index* (*WPI*).

Another provider of patent analysis capabilities is SmartPatents, Inc., Mountain View, CA. CD-ROMs are provided that include also the complete text of the U.S. patents of interest as well as the capability to analyze in a variety of powerful and sophisticated ways.

13.25. OTHER SOURCES OF PATENT INFORMATION

Databases may specialize in certain types of patents by subject matter. Thus, for example, *Pharmsearch* has already been discussed. Another database specializing in pharmaceutical patents is the IMS Patents International file which is readily available online and in CD format. If covers some 1800 molecules that are in stage 3 or higher. One of the features is the emphasis on patent term extensions and related data so that users can more easily determine when a patent will expire, a matter of special interest in the pharmaceutical industry. Another feature is the inclusion in the indexing of drug names and trade names in addition to chemical indexing. IMS Global Services is headquartered in London. In addition to those patent files that specialize in certain fields, other types of patent files specialize in specific countries. IFI has already been discussed. JAPIO (Japan Patent Information Organization), Tokyo, produces an English-language online file, with abstracts, that is believed to be the most complete online source of information on most unexamined Japanese patent applications (Kokai) from approximately October, 1976 to date. The JAPIO database is available worldwide and covers certain patents that may not be covered elsewhere. Even when abstracts are available from other reliable sources such as the *Derwent World Patents Index* or *Chemical Abstracts,* JAPIO provides the user with another abstract that may well contain additional information of value. All technologies are included. JAPIO also produces a comparable Japanese-language file that appears earlier for those interested in maximum speed of coverage.

Other sources of patent information include the following:

- Numerous other indexing and abstracting services, such as the *Abstract Bulletin* of the Institute of Paper Science and Technology (see Chapter 8), useful for pinpointing patents in specific fields.

- News magazines, such as *Chemical Week,* which in identifying new technology frequently note pertinent new patents associated with that technology.
- Articles, especially review articles.
- Encyclopedias and treatises, such as those mentioned in Chapter 12—good especially for the older patent literature, which can still be useful.
- Programs such as those mentioned in Chapter 17, such as the SRI Process Economics Program, which does an excellent job of technoeconomic interpretation of important patents on manufacture of many major commercial chemicals.

Questel•Orbit, one of the leading online vendors of patent and trademark (and other) information, has provided a summary of the databases that it offers as shown in Table 13.6.

13.26. OTHER REMARKS ON PATENTS*

- Any chemist who undertakes industrial or academic research today and has not first made a study of the state of the art (i.e., a patent search), in addition to looking at the other literature, is wasting time. That has been a theme of this chapter and is worth repeating.
- Many chemists complain that there is too much *secrecy* in information and especially too much secrecy in chemical and other technical processes. That may be because they are not willing to study the patents and determine from them the current state of the art. In addition, patents are an excellent stimulus for new ideas.
- One major goal of research, especially in industry, is to invent new products and patents that can withstand keen competition. Because competition does not stand still, patents and patent systems are a spur to ongoing research.
- Issuance of a patent does not necessarily mean that the invention will be commercialized; the patent may be primarily defensive (to exclude competition), or it could represent a potential interest that may never be exploited because of changes in policy, technology, or economics. Many patents are not commercialized immediately, but rather at a later date, perhaps a few years after issue of the patent, when conditions are more favorable for the technology or use.
- It is worth emphasizing that patent documents can have a noteworthy life cycle, even after granting. For example, granted patents may expire prior to their normal 20-year life if the owner does not pay the periodic maintenance fees required in the United States and some other countries. Other changes may include, for example, change of ownership of a patent (patents can be bought and sold like any other piece of property) or licensing of a patent (licenses may be either exclusive or nonexclusive). An official change or transfer of ownership is known as *reassignment.* Patents can be corrected, in which a Certificate of Correction is issued. They may also be reissued or reexamined, and, in the latter case, scope

*Some of the material in this section reinforces points made earlier.

TABLE 13.6. Some Representative Online Databases that Have Patent Information[a,b]

International Files

Dewent World Patents Index — Questel & Orbit
A comprehensive and authoritative file of data relating to patent specifications issued by the patent offices of major issuing authorities, including European Patent Office and Patent Cooperation Treaty published applications, plus Research Disclosures; produced by Derwent Information, Ltd. *in English*

EDOC — Questel
Contains patent documents published in 18 major industrialized countries as well as European patents and Patent Convention Treaty (PCT) applications; all documents are classified by EPO using ECLA system; produced by Institut National de la Propriété Industrielle (INPI) *in French*

INPADOC/INPANEW — Orbit
Covers patent documents issued by more than 50 national and international patent offices; produced by the European Patent Office *in English and other languages*

National or Regional Files

Claims/IFIPAT — Questel & Orbit
Provides access to the front page and claims information for U.S. patents issued by the U.S. Patent and Trademark Office; produced by IFI/Plenum Data Corporation mostly *in English* (sic)

CHINAPATS — Orbit
Covers all patent applications published under the patent law of the People's Republic of China; produced by the European Patent Office *in English*

EPAT — Questel (new on Orbit)
Contains European patents, published applications and EURO-PCT (European Patents Convention) patent documents; produced by Institut National de la Propriété Industrielle (INPI) *in English, French, and German*

FPAT — Questel
Contains patents applied for and published in France; produced by Insitut National de la Propriété Industrielle (INPI) *in French with English descriptors*

ITALPAT — Questel
Contains Italian patents applications and utility and design model applications; produced by Justinfo Ltd. *in Italian*

JAPIO — Questel & Orbit
Represents the most comprehensive source of unexamined Japanese patent applications, providing abstracts and covering all technologies; produced by Japanese Patent Information Organization *in English*

PCTPAT — Questel (new on Orbit)
PCTPAT covers PCT applications in all disciplines; produced by Institut National de la Propriété Industrielle *in English and French*

U.S. Patents — Orbit
Provides complete patent information, including complete front-page and claims information, for U.S. patents; produced by Derwent, *in English*

TABLE 13.6. (*Continued*)

Legal or Jurisprudence Files	
CLAIMS Reassignments and Reexamination Provides information about U.S. patents that have been reassigned, reexamined, expired, extended or reinstated; produced by IFI/Plenum Data Corporation *in English*	Orbit
JUREP Provides the entire European jurisprudence, published or not, in which a patent is involved; produced by IuK Information Service GmbH *in English and German*	Questel
JURGE Provides all German jurisprudence in which a *trademark or a patent* is involved; produced by IuK Information Service GmbH *in German*	Questel
JURINPI Contains published and unpublished French and European (EPO) jurisprudence in which a *patent or trademark* is involved; produced by Institut National de la Propriété Industrielle (INPI) *in French*	Questel
Legal Status Records actions that can affect the legal status of a patent document after it is published and after the patent is granted; produced by the European Patent Office *in English*	Orbit
LitAlert Included notices of filings and subsequent actions for patent and trademark infringement suits filed in U.S. District Courts and reported to the Commissioner of Patents and Trademarks; produced by Derwent, *in English*	Orbit
Patent Status File Provides a comprehensive alert to more than 20 types of postissue actions affecting U.S. patents; produced by Derwent, *in English*	Orbit
Specialty Files	
Derwent World Patents Index Markush DARC Allows access through structure searching to patents issued in the areas of pharmaceuticals, agricultural chemicals, and general chemicals; produced by Derwent Information, Ltd. *in English*	Questel
Derwent World Patents Index/API Merged The Derwent World Patents Index merged with APIPAT indexing allows selective searching using the American Petroleum Institute's renowned indexing capabilities as well as the unique Derwent classification and coding system; produced by Derwent Information, Ltd. and the American Petroleum Institute *in English*	Orbit
Drug Patents International Provides evaluated product or equivalent process patent coverage for commercially significant pharmaceutical compounds, either marketed or in active R&D; produced by IMSWorld Publications *in English*	Orbit
Patent Fast-Alert (Formerly Current Patents Fast Alert) provides rapid, weekly access to pharmaceutical, biotechnology and agrochemical patent information; produced by Current Drugs Ltd. *in English*	Orbit
Pharm Contains U.S., EPO, French, English, and German patents and PCT applications in the pharmaceutical field; produced by Institut National de la Propriété Industrielle (INPI) *in English and French*	Questel (new on Orbit)

continued

TABLE 13.6. (*Continued*)

	Patent Classifications
CIB	Questel
Represents the full text of the 5th edition of the International Patent Classification, effective January 1990; produced by Institut National de la Propriété Industrielle (INPI) *in French*	
CLAIMS Classification	Orbit
Provides a dictionary index to the U.S. Patent Office's classification code system containing more than 115,000 subclasses; produced by IFI/Plenum Data Corporation *in English*	
ECLATX	Questel
Represents the full text of the internal classification scheme of the European Patent Office (EPO); produced by Institut National de la Propriété Industrielle (INPI) *in English*	
U.S. Classification	Orbit
Contains all U.S. classifications, cross-reference classifications, and unofficial classifications for all patents issued from 1790 to date; produced by Derwent, Inc. *in English*	

[a]This list of online patent databases is offered by Questel•Orbit, which has a very extensive array of such files.

[b]There are a number of other pertinent intellectual property databases and a number of the above databases are available on other systems as well.

Source: Reprinted with permission of Questel•Orbit.

of the patent may be either increased or decreased. Patent life can be extended in the cases of some products (notably pharmaceuticals) that are subjected to a lengthy review process. In addition, patents may be the subject of litigation that can affect a patent. There are computer-based and other files that attempt to track some of the changes during the life cycle of a patent, as for example, those of Derwent, INPADOC, and IFI. But some of the changes are difficult to track. For example, when patents are licensed, this is rarely publicized or recorded in any publicly available source. Change of patent ownership may or may not be recorded in the national or international patent offices. When businesses change names, are spun off, or merge, their patent portfolios may become more difficult to understand except by those who are exceptionally well informed.

- Images of certain drawings or of entire pages of patent documents have become increasingly available in such online sources as the *Derwent World Patents Index* (representative drawings for chemical patents worldwide) and the CAS product *USPATFULL* (full-page images for U.S. patents), for patent documents published during recent years. In the case of *USPATFULL*, and its Internet counterpart, the searcher can view the chemical structures often found in the specifications and claims of patents, as well as other drawings and diagrams. Images have also become available in certain online files that cover trademarks. Display of images can be a relatively slow process as might be expected.
- In 1969 the U.S. Patent Office instituted a defensive publication program whereby an individual or corporation may elect to publish an abstract of a patent application in the *Official Gazette* in lieu of examination by the U.S. Patent Office.

On publication of the abstract, the applicant also agrees to open the complete application to inspection by the general public. On May 8, 1985, this program was replaced by the Statutory Invention Registration series.

- For details of the legal aspects, the reader should consult a patent attorney or agent—this book is written from neither of those perspectives, but rather from the viewpoint of chemical (not legal) information. Other good sources of information include, as previously mentioned, the U.S. Patent Office and its publications; Derwent user manuals; the book by Maynard and Peters (9), *Understanding Chemical Patents;* the guide by Lechter, (10) and the article by Simmons and Kaback in Volume 18 of the fourth edition of *Kirk-Othmer Encyclopedia of Chemical Technology* (3). For keeping up on a current basis, the writings of Nancy Lambert (Chevron Research and Technology Company, Richmond, CA) and Edlyn S. Simmons (Hoechst Marion Roussel, Cincinnati, Ohio), are very important; some of these are cited at the end of this chapter. In addition, those interested in patent information may wish to join the Patent Information Users Group- PIUG (www.piug.org), which focuses on patent searching. This is a small (about 150 members) group that is concerned with all types of patents, not just chemical patents. There is a newsletter, there are annual meetings, and there is a discussion group on the Internet, PIUG-L. An archive of these discussions may be found on the Questel•Orbit Internet location, www.questel.orbit.com. An affiliated group is the Patent and Trade Mark Group (PATMG). Over the years, the magazine *World Patent Information* has also proved of interest to many, although, unfortunately, the insightful column by Stuart M. Kaback (Exxon) is no longer being written.

13.27. TRADEMARKS

Trademarks are an important means for manufacturers to market their products in a distinctive way and for chemists and engineers to identify and purchase chemicals of interest. For the chemist and engineer who wishes to conduct patent and literature searches, trademarks can also be important. Another issue with chemical trademarks is to determine the chemical structures represented by the names. CAS Registry Numbers are now widely used by virtually all chemical manufacturers, and these numbers are usually by far the most definitive and unambiguous way to identify any chemical in terms of both chemical name and structures. However, Registry Numbers may or may not distinguish between grades of the same chemical or between cases in which the same chemical may differ in physical properties such as viscosity. In these cases, trademarks may offer the best additional recourse in achieving these kinds of distinctions.

A trademark can be defined as a word, phrase, symbol, or design, or combination of these, that identifies and distinguishes the source of the goods or services of one party from those of others. The duration of a trademark can be indefinite if the owner continues to use the mark to identify goods or services and follows the proper renewal procedures.

There are many sources of chemical and other trademarks that will identify those already in use. A trademark attorney should always be consulted. Some of these sources also contain other information as well. One of the best sources for identifying trademarks and their owners are the official records maintained by the U.S. and other national Patent and Trademark Offices. Thus, the U.S. Patent and Trademark Office (PTO) has available CD-ROM disks listing the ownership of the trademarks it has registered, as well as those pending. This file can be consulted without charge at any U.S. Patent and Trademark Depository Trademark Library. It is also available for purchase. The U.S. Patent and Trademark Office trademark files at Crystal City, VA, are very useful for searching trademarks, and these may be searched by the general public. PTO publishes a weekly *Official Gazette: Trademarks,* which contains an illustration of each trademark published for opposition, a list of trademarks registered, a classified list of registered trademarks, and official PTO notices.

Trademarks registered in most major nations, as well as in individual states in the United States, are readily searchable through the use of online databases through such hosts as DIALOG. Important providers of online trademark information are the firms Thomson and Thomson, North Quincy, MA, and Compu-Mark (Belgium) whose files cover the United States and major European countries, respectively. Depending on country, these can be found on the DIALOG system and on Questel•Orbit and are also available as a CD-ROM (U.S. Federal and state trademarks) from DIALOG. In addition, MicroPatent, East Haven, CT offers a U.S. (federal) trademark search capability on the Internet (www.micropat.com/trademarkmoreinfo.html).

Another source of all types of trademarks is *Thomas Register of American Manufacturers* (see page 448, Chapter 16).

Another excellent source for alternative names, including some that may be trademarks, is the online *Registry File* of Chemical Abstracts Service. In the case of trade names, this comprehensive source (over 17 million compounds included as of 1997) will not identify the owner directly, but it does provide the Registry Number, and it usually gives full chemical structure information. It leads to pertinent patents and articles (as abstracted in *CA*), and these often do help identify the manufacturer. The companion *CAS Registry Handbook—Common Name* product is available from Chemical Abstracts Service only on microfiche; this is a powerful tool.

The *CA Index Guide* (see Chapters 6 and 7) is also quite extensive and includes many alternatives names, including some that may have been utilized as trade names.

Another excellent source, especially for products of pharmaceutical interest, is the *Merck Index* (see Chapter 15).

Examples of other sources include

Michael Ash and Irene Ash, Eds., *Industrial Chemical Thesaurus,* 2nd ed., VCH Publishers, New York, 1992. This includes not only trademarks, but also CAS numbers, definitions, toxicity data, properties, applications, and names of manufacturers.

D. R. Lide and G. W. A. Milne, Eds. *Names, Structures, Synonyms of Organic Compounds: A CRC Reference Handbook,* 3 Vols. CRC Press, Boca Raton, FL, 1995.

Michael Ash and Irene Ash, Eds, *Gardner's Chemical Synonyms and Trade Names,* 10th ed., Gower, UK, 1994.

Some of the buyer's guides mentioned in Chapter 16 contain indexes of trademarks. The *Chemical Week Buyer's Guide* is a good example.

The Internet is a very useful source for trademark attorneys and others.

REFERENCES

1. One pertinent publication is *General Information Concerning Patents—A brief introduction to patent matters,* which is for sale by the U.S. Superintendent of Documents, U.S. Government Printing Office, Washington, DC 20402. This is revised at frequent intervals.
2. U.S. Patent Office, *Manual of Classification of Patents,* U.S. Government Printing Office, Washington, DC. Revised periodically; includes *Alphabetical Index of Subject Matter.*
3. E. S. Simmons and S. M. Kaback. "Patents (Literature)," *Kirk-Othmer Encyclopedia of Chemical Technology,* 4th ed. (18), Wiley, New York, 1996, pp. 102–156.
4. P. Benichou and C. Klimczak, "Handling Genericity in Chemical Structure Using the Markush DARC Software," *J. Chem. Inf. Comput. Sci.,* **37,** 43–53 (1997).
5. G. O. Platau, "Documentation of the Chemical Patent Literature," *J. Chem. Doc.,* **7**(4), 250–255 (1967).
6. P. J. Pollick, "Patents and Chemical Abstracts Service," *Sci. Technol. Libr.,* **2,** 3–22 (1981).
7. P. J. Pollick, "Processing of Patent Bibliographic Data at CAS," *World Patent Inf.,* **3**(3), 128–131 (1981).
8. E. Simmons and N. Lambert, "Patent Statistics: Comparing Grapes and Watermelons," *Proceedings of the Third Montreux International Chemical Information Conference & Exhibition,* H. Collier, Ed., Springer-Verlag, Heidelberg, 1991; "Comparing Grapes and Watermelons," *CHEMTECH,* **23** (6), 51–59 (1993); S. Kaback, N. Lambert, and E. Simmons, "Patent Citation Data," paper presented at American Chemical Society Meeting, Washington, DC, August 24, 1994.
9. J. T. Maynard and H. M. Peters, *Understanding Chemical Patents,* American Chemical Society, Washington DC, 1991.
10. M. A. Lechter et al., Eds., *Successful Patenting for Engineers and Scientists,* IEEE Press, New York, 1995.

Other Selected Writings of Lambert and Simmons

N. Lambert, "How to Search the IFI Comprehensive Database Online—Tips and Techniques," *Database* **10**(6), 46–59 (1987).

N. Lambert, "After the Grant: Online Searching of Legal Status Information for U.S. Patents," *Database,* 42–48 (1981).

N. Lambert, "Online Searching of Polymer Patents: Precision and Recall," *J. Chem. Inf. Comput. Sci.* **31**(4), 443–446 (1991).

N. Lambert, "The Idiot's Guide to Patent Resources on the Internet," *Searcher,* **3**(5), 34–39 (1995).

N. Lambert, "More Patents—Lots More—on the Internet," *Searcher,* **3**(10), 24–27 (1995).

N. Lambert, "Online Statistical Techniques as Patent Search Tools. Part 1: Patent Indexing, Patent Citations," *Database,* **19**(1), 74–78 (1996).

N. Lambert, "Online Statistical Techniques as Patent Search Tools. Part 2: Patent Classifications," *Database,* **19**(2), 67–73 (1996).

N. Lambert, "Drilling for Gold: Database for Petroleum Information," *Searcher,* **4**(4), 8–12 (1996).

N. Lambert, "QPAT-US: A New Patent Search Tool for the Internet," *Database,* **19**(4), 56 (1996).

N. Lambert, "Patent Searching: What, Why, When, Where?" *Online User,* **2**(6), 45–51 (1996).

E. S. Simmons, "The Central Patents Index Chemical Code, A User's Viewpoint," *J. Chem. Info. Comput. Sci.* **24**(1), 10–14 (1984).

E. S. Simmons, "The Paradox of Patentability Searching," *J. Chem. Info. Comput. Sci.* **25**(4), 379–386 (1985).

E. S. Simmons, "JAPIO—Japanese Patent Applications Online," *Tokyo Kanri* (*Patent Management*), **37**(1), 51–61 (1987); Japanese translation by Kikunoshin Nakamura.

E. S. Simmons, "Patenting Pitfalls . . . and Their Avoidance," *CHEMTECH,* **16**(13), 144–149 (1987).

E. S. Simmons, and S. M. Kaback, "Online Patent Information—Doubleheader," *World Patent Inf.* **10**(3), 204–206 (1988).

E. S. Simmons, "What's In a Claim?" *Proceedings of the Montreux International Chemical Information Conference & Exhibition,* H. Collier, Ed., Springer-Verlag, Heidelberg, 1991, pp. 93–104.

E. S. Simmons, "The Grammar of Markush Structure Searching: Vocabulary vs. Syntax," *J. Chem. Info. Comput. Sci.* **31**(1), 45–53 (1991).

E. S. Simmons, "A Markush Story," *Proceedings of the 1992 Montreux International Chemical Information Conference & Exhibition,* H. Collier, Ed., (Royal Society of Chemistry, Cambridge, UK), CRC Press, Boca Raton, FL, 1992, pp. 33–78.

E. S. Simmons, "Competitive Intelligence from Patents—Volatile: Handle With Care," *J. Assoc. Global Strategic Inf.* **4**(3), 124–133 (1995).

E. S. Simmons, "Intellectual Property and the Internet: 'You Can't Sell It If You Give It Away,'" *Searcher,* **3**(1), 38–41 (1995).

E. S. Simmons, "Patent Family Databases 10 Years Later," *Database,* **18**(3), 28–27 (1995).

List of Addresses

Derwent Information, Ltd.
1420 Spring Hill Road, Suite 525
McLean, VA 22102

Derwent Information, Ltd.
Derwent House
14 Great Queen Street
London WC2 5DF

IFI/Plenum Data Corp.
3202 Kirkwood Highway
Wilmington, DE 19808
(800-331-4955)

LEXIS-NEXIS
9443 Springboro Pike
Miamisburg, OH 45342
(800-227-4908)

Public Search Room of the Patent Office, Crystal Plaza
314 South Clark Place
Arlington, VA

U.S. National Technical Information Service
Springfield, VA 22161

14 Environment, Safety, and Related Topics

14.1. INTRODUCTION

The amount of information on safety, occupational hygiene, toxicity, environmental matters, and related aspects of chemistry and chemical engineering has continued to grow rapidly. This kind of information is especially important before beginning work on any new project or on new or less well known chemicals, but it also has longer-range significance. Pertinent information benefits not only persons directly connected with the investigation. It is equally important for colleagues in the same laboratory, pilot plant, or plant area and for all who may later use or otherwise come into contact with the chemicals.

There are a large number of outstanding literature tools in this field. In a personal communication relating to reactive chemical hazards, David J. Frurip, Dow Chemical Company, Midland, MI, writes that "there is a saying in our laboratory that one hour in the library can save you four days in the laboratory."

Review of literature findings by a qualified environmental hygienist, toxicologist, safety specialist, and/or regulatory specialist, depending on what is at issue, is desirable and is recommended because so much is at stake.

Another important caveat is that it has been estimated that fewer than 5% of the chemicals in use have been studied in detail as to their environmental effects, and hence predictive techniques may need to be utilized.

Although the field is large and of worldwide interest and action, this chapter emphasizes primarily U.S. information sources.

The kinds of environmental safety information with which the chemist or chemical engineer must be familiar includes one or more of the following examples:

1. Effects (toxicity, carcinogenicity, etc.) of specific chemicals on human beings.
2. Effects of chemicals on animals, especially if data on humans are lacking.
3. Effects on aquatic life and vegetation if discharged as liquid effluents or into the air.
4. Proper methods for pollution abatement, transportation, safe handling, and storage.
5. Biodegradability.
6. Hazardous chemical reactions and reactive chemicals information.
7. Flash point and any other flammability or explosive characteristics. Also smoke generation, if any.

8. Safe laboratory, pilot plant, plant, and office practices in general (applies to all chemists and chemical engineers).
9. Pertinent local, state, and federal standards and regulations and "lists." Equivalent information for countries outside the United States as needed. Some of the agencies that may be involved include the U.S. Environmental Protection Agency, Food and Drug Administration, Occupational Safety and Health Administration, Department of Transportation, Coast Guard, Consumer Product Safety Commission, and Federal Trade Commission.

The field is volatile. Standards, regulations, technology, and toxicity/safety/environmental ramifications change rapidly. Because of this, chemists can expect presently available sources of environmental and safety information, as described in this chapter, to be supplemented or replaced continuously by new and improved sources. Organizations such as Chemical Abstracts Service (CAS) and BIOSIS, federal agencies such as the National Library of Medicine, and for-profit information industry organizations such as Cambridge Scientific Abstracts can be expected to take the lead in developing new information tools in this field. Chemists should do their best to maintain appropriate contact with these organizations on a regular basis and to keep up with the newer sources. Most organizations have designated individuals for this function. Table 14.1 lists some key environmental dates that may be of interest.

14.2. LOCATING PERTINENT SAFETY DATA: INITIAL APPROACHES

In industry, and to an increasing degree in universities, much of the needed data may already be available from departments in one's own organization. These departments are typically designated by such names as *Loss Prevention* or *Environmental Hygiene and Toxicology.* Assistance of experts in these departments is essential in evaluating any potential hazards. Toxicological data, because of their biomedical nature, need to be evaluated by a professional toxicologist or a physician.

In smaller, or less well organized laboratories, the bench chemist may need to make a personal investigation of pertinent safety data and should consider the following suggestions:

A. Material Safety Data Sheets

The manufacturer of a chemical, if the chemical is commercially available, is a prime source of environment and safety data and should be contacted as a first step. If trade literature is already on hand, the chemist must check with the manufacturer to ensure that this literature contains the latest information.

Material Safety Data Sheet (MSDSs) are an important source of environment and safety information. Under U.S. government regulations effective as of November 25, 1985, these are required to be available (and updated regularly) for commercial chemicals. Some states also have regulations regarding the MSDS. MSDS forms are prepared by manufacturers for their products and are made readily available to customers

TABLE 14.1. Some Key Environmental Events and Dates

Date	Event
1962	Rachel Carson's *Silent Spring*
1965	Solid Waste Disposal Act
1969	National Environmental Policy Act (NEPA)
1970	Clean Air Act Amendments
1970	Resource Recovery Act
1970	EPA formed
1972	Federal Water Pollution Control Act
1972	Safe Drinking Water Act
1975	Issue of chlorinated organics becomes priority topic to EPA
1976	Resource Conservation and Recovery Act (RCRA)
1976	Toxic Substances Control Act (TSCA)
1977	Clean Water Act (CWA)
1977	Love Canal news stories
1977–1979	National Interim Primary Drinking Water Regulations
1980	Dioxin spills news stories begin to make headlines
1980	Comprehensive Environmental Response, Liability and Compensation Act (CERCLA)
1980	Solid Waste Disposal Act Amendments (to RCRA)
1982	Regulations on volatile organics in drinking water
1984	RCRA Reauthorization
1984	Bhopal disaster
1986	Emergency Planning and Community Right-to-Know Act
1987	International agreement on Montreal on CFC limits
1988	CMA Responsible Care program begins for member companies
1989	Valdez oil spill stirs environmental concerns across the board
1990	Pollution Prevention Act
1992	CMA Product Stewardship Code approved for member companies
1996	Food Quality Protection Act; Safe Drinking Water Act (reauthorization)

and frequently to potential customers. These forms can also be a helpful source of physical property information, as mentioned in Chapter 15.

The standard components of an MSDS normally include chemical product and company identification, composition and information on ingredients (may include information on permissible or threshold exposure limits, etc.), hazards identification; first aid measures, firefighting measures, accidental release measures; handling and storage, exposure controls and personal protection, physical and chemical properties, stability and reactivity, toxicity data; ecological data, disposal, transportation requirements, regulatory information (specific U.S., state, and international regulations); and other information such as label text. Although sometimes difficult, it is always important, to determine whether a MSDS can be counted on to provide complete, current and reliable information for the chemist or engineer who wants to use and handle a chemical safely and properly. At a minimum, the MSDS should be dated and signed by the responsible persons, the originating organization should be clearly identified (including address and phone number), pertinent references should be clearly identified as appropriate, and all 16 categories of the ANSI standard for MSDSs should be addressed. A typical MSDS may be some 13–15 pages in length. Note that MSDS content requirements and appearance may vary from country to country.

Another caveat about MSDSs is that manufacturers sometimes choose to keep the exact chemical identity of some of their products as trade secrets on the MSDS, even though they may provide all of the required environment, health, and safety information.

In addition to manufacturers, a number of other sources of MSDSs now exist. MSDSs can vary widely in quality. Some are more reliable, more complete, and more current than others. A MSDS should be based on a thorough review of the literature (in some cases, laboratory and toxicology tests are conducted) up to the present and should have also been reviewed by an industrial hygienist and/or toxicologist. It may sometimes pay to obtain data sheets from several reliable sources to ensure that all available MSDS information on a specific chemical is at hand.

One well-known producer of MSDSs is Genium Publishing Corp., Schenectady, N.Y. This firm produces MSDSs for some 1000 (1997 count) chemicals with emphasis on those commonly found in the workplace. The firm was originally founded as a "spinoff" from the General Electric Company, and it is now a separate entity. Production of each sheet is said to be based on a careful review of the literature, and these results are, in turn, reviewed by an industrial hygienist. The content is written by humans, not by computer, and the results are intended to be as user-friendly (understandable) as possible. Some 120 new or revised sheets are added each year on a semiannual basis. Indexes by chemical name and by CAS Registry Number are provided. Genium also offers labels for many of the chemicals that it covers, and the company sells a software program that provides a template whereby companies can generate MSDSs for their own products. Another product from Genium is the book *Internet User's Guide for Safety and Health Professionals,* which lists sources of MSDSs on the Internet as well as other useful information. Michael Cinquanti is president of Genium (800-243-6486). Internet address is www.genium.com.

MSDSs are extensively available online for thousands of chemicals. The operations of MDL's OHS Safety Series, a supplier of chemical health and safety information, including MSDSs, are an example. OHS products, beyond full MSDSs, include brief MSDs summaries, and data for such applications as regulatory reporting and transportation requirements. The data are provided in forms compatible with relational databases or other environmental health and safety systems.

MDL's OHS was originally started in 1978 in Nashville, TN, where its Research Operations Center continues to be located. MDL Information Systems (see Chapter 10) purchased OHS in 1995. MDL headquarters are in San Leandro, CA, and the MDL OHS Web site address is www.mdli.com/ohshome.html. MDL OHS writes and regularly updates MSDSs on a custom basis for chemical manufacturers and suppliers; thousands of complete MSDSs are currently available, in addition to other related products.

OHS Safety Series MSDSs are now sold through distributors worldwide, including

DIALOG (online and Internet)
STN International (online and Internet-accessible)
Chem Web (www.ChemWeb.com)
By MDL in the following forms:
  Standalone CD-ROM
  Standard files for integration into customer software systems
  In combination with Intranet search software to form "a complete solution"

MSDs are also available on a custom basis as needed by chemical manufacturers. In addition, 24-hour delivery fax service for urgent MSDS needs is available directly from MDL.

OHS says that the principal distinguishing feature of its MSDSs is the data content and quality. Their data sheets are not manufacturer data sheets; rather, they are originally researched as based on the scientific literature. In addition, these data sheets are timely; these sheets are compiled and updated every 90 days by MDL staff, based on the literature and on keeping up with the regulations. The output is reportedly easy to work with because of consistent format, readability, and consistency with U.S., Canadian, and European Community requirements. Translations of these MSDSs are available in a number of European and Asian languages. In addition, customers can use the same data sets for labels, training materials, and populating their systems which produce transport manifests and the like. Thus, MDL MSDSs are a supplement to manufacturer MSDSs. Customers are said to include many industrial firms, universities, hospitals, and governmental agencies such as the U.S. Environmental Protection Agency and the Korean Industrial Safety Corporation.

As examples of content of MDL OHS MSDSs, percentages of components are listed, regardless of whether they are considered hazardous. Extensive physical and chemical property data are given. Toxicology data are provided not only from *RTECS* (Registry of Toxic Effects of Chemical Substances; see Section 14.5, discussion in text preceding Fig. 14.2) but also from other sources such as NTP (National Toxicology Program), and all of these data are said to be evaluated by OHS staff. Exposure limits given include not only OSHA (U.S. Occupational Health and Safety Administration) and ACGIH (American Conference of Governmental and Industrial Hygienists) but also NIOSH (and National Institute of Occupational Safety and Health), German MAK (maximum allowable concentrations). Regulatory information given embraces TSCA Inventory Status, TSCA Section 12(b) export notification, all five SARA Hazard Categories, OSHA Process Safety, and California Proposition 65. The format utilized is ANSI Z400.1. Peer review is said to be employed.

As mentioned above, options include one- to two-page summary sheets that distill, in simplified language, the essence of hazard information for quick reference or training. MDL OHS will also work with companies to (a) help them track their chemicals through so-called cradle-to-grave systems and (b) establish and maintain right-to-know and hazard communications systems.

Also available on STN, as well as the Internet, is the "MSDS-CCOHS" file of the Canadian Center for Occupational Health and Safety (CCOHS), Hamilton, Ontario.

The file contains nearly 100,000 trade-named materials. The data sheets are as prepared by the manufacturers or distributors. This file is also available on CD-ROM. It is also on CCINFOline, the host system of CCOHS. CCINFOline, in addition to material safety data sheets, contains a number of other files of environmental and safety interest, all together in one convenient place. This network is accessible through an annual subscription fee and is available on the Internet. The Web site of CCOHS is www.ccohs.ca, and this includes links to over 1000 other health and safety sites on the net. CCOHS publishes the book *Using the Internet to Access Health and Safety Resources,* edited by Chris Moore.

Another online source of MSDS data is CIS (Chemical Information System, Division of Oxford Molecular Group, Towson, MD, 800-CISUSER). Three of their databases contain MSDSs. *MALLIN, BAKER,* and *CHEMHAZIS.* In addition, three of their files contain some of the data needed to complete an MSDS; these files are *OHMTADS, CESARS,* and *CHRIS.* Further MSDS data are available from CIS on an offline service bureau basis through the *HMIS* file. CIS is discussed further in Section 14.4.

MSDSs may be found on the Internet in several places in addition to the above. For example, the University of Utah is a well-known provider, and another is Vermont SIRI (Safety Information Resources on the Internet) whose address is http://hazard.com.SIRI is located in Burlington, VT, and, in addition to the MSDSs, offers access to other safety information and links to other environmental and safety and health web sites. One of the products is the book *User's Guide to Safety Information on the Internet* (1996).

Several major suppliers offer collections of MSDSs; these include Aldrich Chemical Co. (available in six languages on CD or magnetic tape) and Fisher Scientific, available on the Internet.

IHS Environmental Information, Exton, PA also has an extensive collection of MSDSs on the Internet (www.ihsenv.com). Some 140,000 data sheets are included. In December, 1994, CMA (Chemical Manufacturers Association, Arlington, VA) began to offer its huge MSDS collection online to the public through its MSDS Central system, but this major effort (which promised to be the largest collection of MSDSs available anywhere) was, unfortunately, discontinued in March 1997. However, CMA continues to maintain its library of 1.5 million data sheets for emergency responders, and many chemical manufacturers want to be contacted directly by persons interested in their data sheets.

B. Some Other Initial Sources

Since published literature in the field for some chemicals can be voluminous, consultation of basic reference tools is often a good next step after the MSDS. (As mentioned the MSDS may or may not be fully complete and current; it is usually too brief to include all pertinent information.

MSDSs in countries other than the United States may be called by other similar names such as SDSs (Safety Data Sheets). International "harmonization" of safety data sheet format and content is still a long sought-after goal.

Federal, state, and local environmental officials may be able to provide advice and offer suggestions on proper handling.

The chemist can find the most recently published safety information, particularly on hazardous chemical reactions or laboratory procedures, in the *letters to the editor* and other sections of chemical news as well as in other chemical magazines and journals. The chemist should scan these sections regularly and carefully, on a timely basis, if possible in cooperation with other chemists in the same laboratory, plant, or R&D team. *Chemical and Engineering News* is a particularly good example for this type of information. In addition, the *Journal of Chemical Education* features a series of columns on safety entitled *Safety Tips,* currently edited by Timothy Champion. Major chemical accidents, such as may occur on a plant scale or during transportation, are reported quickly by mass news media and then in more detail in the technical press. Some chemical organizations keep running tabs of these incidents, wherever they occur, to improve overall safety of their *own* operations.

If safety data on the specific compound being worked with is not found, data on analogous or homologous compounds and other related structures can sometimes, but not always, provide useful guides to what can be expected. Such extrapolations are best made by a toxicologist or other trained safety expert as appropriate.

In watching trends and developments in areas of environment, safety, and toxicity concerns, the chemist needs to pay special attention to information and news from countries such as Japan, Sweden, Canada, and Germany. These countries have shown special concern for the environment and are regarded as bellwether nations in this field. For example, potential hazards of inappropriate use of mercury were first widely publicized in Sweden and Japan. Within the United States, California is often a bellwether state.

Close to the leading edge of environmental developments are well-organized groups of informed citizens. These groups have been among the first in the United States to publicize potential hazards. They can provide clues that may prove to be valid. Examples include

Center for Science in the Public Interest
1501 16th Street, NW
Washington, DC 20009

Environmental Defense Fund
1875 Connecticut Ave., NW
Washington, DC 20009

Natural Resources Defense Council
1200 New York Avenue, Suite 400
Washington, DC 20001

Physicians for Social Responsibility
1101 14th Street, NW
Washington, DC 20005

Sierra Club
85 Second Street
San Francisco, CA 94105

Trade associations such as the Chemical Manufacturers Association, Soap and Detergent Association, Chemical Specialties Manufacturers Association, and Synthetic Organic Chemical Manufacturers Association are important sources of safety and environmental information. See also Section 14.7.

If no published information on hazardous properties or environmental risks of a chemical is found, this does not mean that none exists. The chemist must assume that the chemical may be suspect unless positive and reliable information is found showing that the chemical has been proven to be safe under specific conditions.

Questions that may help the chemist, assisted and advised by toxicologists and environmental specialists, in evaluating the risk (or lack of risk) that a chemical poses include these examples:

1. Will there be substantial or significant human exposure in (a) production, raw materials, intermediates, product, and waste products; (b) distribution; (c) end use; and (d) disposal?
2. How does the chemical compare to existing products already in use that have similar properties and exposure characteristics?
3. How does it compare with such other existing products in physical properties that affect human exposure (vapor pressure, corrosiveness, etc.) and in toxicological properties?
4. Does it contain functional groups that are known carcinogens?
5. Does a toxicologist or environmental specialist have any reason to suspect that the product may pose an unusual risk in manufacture, use, or disposal that could not be minimized by adequate warnings or reasonable standard procedure? Experience with products in similar distribution patterns, end use, and disposal characteristics will also be valuable.
6. Will there be substantial exposure to the environment?
   a. Will it be produced in large quantities? What quantities (if any) could be released to the environment? In what forms?
   b. How does it compare to existing products being distributed that have similar structure or toxicological effects and similar environmental exposure characteristics?
   c. What are the comparative physical properties, molecular structure, persistence, solubility, toxicity, and degradation characteristics?

Because no single published source of environmental and safety data can be all-inclusive, and for other reasons, it is a mistake to use only one such source. The source used could be unevaluated or biased, or data could be taken out of context to mean something entirely different. Multiple *original* sources should be obtained and evaluated by an appropriately qualified expert such as a toxicologist.

14.3. BOOKS AND RELATED INFORMATION SOURCES

Basic reference books on environmental effects, safety, toxic effects, and related aspects of chemicals are often the quickest and most efficient way to gather needed information. Several reference tools attempt to bring together in one place data on these aspects of a number of chemicals. Examples are cited in References 1–8, and a few are described in the following sections.

There are caveats. Since any book can quickly become obsolete, particularly in environmental and safety-related matters, it is imperative that the chemist look at the most recent knowledge available and obtain as much pertinent detail as possible. In addition, no published compilation can cover the gamut of these matters; none is complete. Another concern is peer review or some other form of verification of the data included. Unless otherwise stated in the compilation itself, the data can be assumed to be unverified.

1. A well-known handbook on potentially hazardous chemicals is *Sax's Dangerous Properties of Industrial Materials* (7), which now contains over 3000 pages. Entries for about 20,000 materials are in the eighth edition. This book is helpful but like other books of its type, it is only a starting point. Not all materials or hazards are covered and alternative information sources are sometimes more current, more accurate, or more complete. This book and other similar published works do, however, perform the important function of helping alert the chemist to many hazards—alerts that are then best followed up in detail in more specialized and current sources, preferably in cooperation with safety or toxicity experts.

2. Another well-known reference book is *Clinical Toxicology of Commercial Products* by Gosselin et al., in its fifth edition (2) but out of print. This includes toxicity data for 1642 substances, with extensive references to the literature. In addition, composition data for consumer products are included. Data from this book, frequently known simply as either *Gosselin* or *CTCP*, are available online through *CIS* (see Section 14.4).

3. *Bretherick's Handbook of Reactive Chemical Hazards* (8), now edited by P. G. Urben, is a very highly regarded and essential compendium that is now in its fifth edition (1995). In addition to the print edition, both floppy-disk and CD-ROM counterparts are available. It covers more than 4800 elements and compounds, as well as other entries that cover more than one compound. Data about stability, reactivity, and other hazards are included.

4. The seven-volume set, *Dictionary of Substances and Their Effects* (also known as *DOSE*) (9), is a significant reference source from the Royal Society of Chemistry. This compilation, edited by Mervyn L. Richardson, is arranged alphabetically by chemical. For each chemical included, information given, when possible, is as follows: basic identification, uses and occurrence, physical properties, occupational exposure, ecotoxicity, environmental fate, mammalian and avian toxicity, legislation, and other comments and references. The compilation is also available as a CD-ROM. It covers some 4000 chemicals.

5. An important work in this field is the *Encyclopedia of Occupational Health and Safety*. A fourth edition in four volumes (also on CD-ROM) is currently being readied for planned 1997 publication under the editorship of Jeanne M. Stellman, Columbia University. The publisher is the International Labor Office, with headquarters in Geneva, Switzerland, and an office in Washington, DC.

6. The classic U.S. reference source in industrial hygiene and toxicology is *Patty's Industrial Hygiene and Toxicology* (1), now in its fourth edition and edited by G. D. Clayton and F. E. Clayton. Some volumes edited by Lester V. Crally, et al.

7. Another important book is *Hazardous and Toxic Materials* by H. W. Fawcett (10). It is mentioned here because of Fawcett's record of leadership in safety.

8. A work of key importance to the environmental professional is *Standard Methods for the Examination of Water and Wastewater* (11). Testing procedures and standardized methods help provide the consistent and reproducible results needed to comply with current requirements. Continued revision can be expected as standards and technology change. This volume is available from the American Public Health Association.

9. The proceedings of the Purdue University Industrial Waste Conferences (12) are an invaluable collection of case histories on handling a wide variety of pollution abatement problems. These proceedings have been published annually for 50 years.

14.4. SELECTED ELECTRONIC RESOURCES: CHEMICAL INFORMATION SYSTEM (CIS)

The CIS concept was originally developed in 1973 to aid U.S. government agencies, notably the Environmental Protection Agency (EPA), concerned with chemicals having an environmental or other related regulatory impact. That has been the key thrust of its development, although what was then known as the National Bureau of Standards (now the National Institute for Standards and Technology) also played a key role. CIS has been fully available to the public since 1979.

CIS offers a "valet service" under which a professional searcher will conduct a search of CIS databases on a fee basis.

Chemical Information System was formerly operated by Fein-Marquart Associates, Baltimore, MD, which had worked on initial development of the system. It was later part of the Chemical Informatics Division of PSI International, Inc. and is now owned by Oxford Molecular. CIS is located at 810 Gleneagles Court, Suite 300, Towson, MD 21286 (800-CIS-USER).

Following is a list of the databases available through CIS:

ENVIRONMENT

CERCLA Information System (CERCLIS)—contains over 41,300 records for *all* sites and incidents, including active CERCLA and NFRAP (No Further Remedial Action Planned) sites, proposed or chosen for cleanup under the Superfund program.

Emergency Response Notification System (ERNS)—contains more than 231,312 release notification reports from 1987 to present covering hazardous substances, petroleum and non-petroleum-based oil products, and identified substances.

EPA Civil Enforcement Docket (DOCKET)—civil judicial cases filed by the Department of Justice on behalf of EPA from 1971 to present.

Facilities Index Data Systems (FINDS)—contains more than 710,000 records for sites or facilities regulated by the EPA.

Resource Conservation and Recovery Information System (RCRIS)—contains records describing hazardous-materials handlers of waste, including their operations, classification, permitting, and violations.

VISTA Environmental Information, Inc. (VISTA)—indexes 370 different local, state, and federal databases for more than 3.1 million records dealing with sites where hazardous substances are used, disposed of, and stored.

HAZARDOUS-MATERIALS HANDLING and CHEMICAL–PHYSICAL PROPERTIES

Chemical Hazard Response Information System (CHRIS)—contains information on labeling, properties, hazards, chemical reactivity, water pollution, and hazard classifications for 1218 chemical substances. The database was created by the U.S. Coast Guard.

Environmental Fate (ENVIROFATE)—deals with environmental fate or behavior (transport and degradation) of chemicals released in the environment.

Biodegradation of Substances in the Environment (BIODEG)—contains information concerning the biodegradation of chemical substances in several types of experiments.

Biodegradation Literature References (BIOLOG)—BIOLOG was developed through the collaborative efforts of the EPA's Office of Toxic Substances and the Syracuse Research Corporation (SRC). It contains information concerning published literature on the biodegradation and microbial toxicity of chemical substances.

Hazardous Chemicals Information and Disposal Guide (HAZINF)—detailed instructions on the safe handling and disposal of some 220 chemicals or classes of chemicals.

Oil and Hazardous Materials/Technical Assistance Data System (OHM/TADS)—provides up to 126 different fields of information for 1402 materials.

MATERIAL SAFETY DATA SHEETS (MSDSs)

BAKER and MALLIN—contains more than 1600 MSDSs (BAKER) and more than 1200 MSDSs (MALLIN) from Mallinckrodt-Baker, Inc. To become a single file.

Chemical Hazards Information System (CHEMHAZIS)—over 2280 chemical records drawn from literature and experimentally determined under the sponsorship of the National Toxicology Program (NTP).

TOXICOLOGY AND CARCINOGENICITY

Aquatic Information Retrieval (AQUIRE)—information on acute, chronic, bioaccumulative, and sublethal effects data from experiments performed on freshwater and saltwater organisms.

Chemical Evaluation Search & Retrieval System (CESARS)—evaluated chemical profiles, including toxicity data and environmental fate, taken from primary literature for over 190 compounds.

Chemical Carcinogenesis Research Information System (CCRIS)—assay results and test conditions for more than 1400 chemicals in the areas of carcinogenicity, mutagenicity, tumor promotion, and cocarcinogenicity.

Clinical Toxicology of Commercial Products (CTCP)—toxicity data, including symptom and treatment data, for approximately 230,000 common commercial products derived from some 3000 chemicals. Based on fifth edition of the same name by Gleason, Hodge, Gosselin, and Smith.

Dermal Absorption (DERMAL)—data on toxic effects, absorption, distribution, metabolism, and excretion related to dermal application of chemicals.

Genetic Toxicity (GENETOX)—mutagenicity information on more than 4300 chemical substances as tested against 38 biological systems.

Gastrointestinal Absorption (GIABS)—references to articles in the literature on absorption, distribution, metabolism, or excretion of chemical substances in test animals or humans.

Integrated Risk Information System (IRIS)—680 chemical substances records developed by U.S. EPA workgroups to aid in the process of risk assessment.

Plant Toxicity (PHYTOTOX)—records relating to the biological effects of the application of natural and synthetic organic chemicals to terrestrial plants.

Registry of Toxic Effects of Chemical Substances (RTECS)—toxicity information on over 132,000 substances. The compilers do not evaluate the data in this source.

Toxic Substances Control Act Test Submissions (TSCATS)—information on unpublished health and safety studies submitted to EPA under the provisions of the Toxic Substances Control Act.

Toxic Substances Control Act Inventory (TSCAINV)—a list of chemical substances that have been manufactured, imported, or processed in the United States for commercial purposes since January 1, 1975.

PHARMACEUTICALS

Drug Information Fulltext (DIF)—complete contents of two reference volumes published by the American Society of Hospital Pharmacists: *American Hospital Formulary Service Drug Information* and *Handbook of Injectable Drugs.*

***The Merck Index* Online**—more than 10,360 monographs on chemical substances from the 11th edition of *The Merck Index* plus additions and revisions made to the online version since publication of the printed edition. Updated semiannually. Covers preparation, chemical and physical properties, principal pharmacological action, and toxicity of substances. Full-text is searchable online.

SPECTROSCOPY

Wiley Mass Spectral Search System (WMSSS)—140,000 mass spectra (for 120,000 substances) from the fifth edition of the John Wiley & Sons Registry of Mass Spectral Data (including *all* records from the NBS/NIST mass-spectral database.

STRUCTURE AND NOMENCLATURE SEARCHING

Structure and Nomenclature Search System (SANSS)—information from over 100 different sources on nearly a million chemical names representing over 500,000 substances, including CAS Registry Numbers, structural diagrams, molecular formulas, systematic names, and synonyms. Has structure, substructure, and full or partial name search capabilities.

14.5. SOME OTHER ELECTRONIC SOURCES

The computerized literature retrieval services of the National Library of Medicine (NLM) are available online through a system known as *MEDLARS.* This includes such components as *TOXLINE* and *HSDB.* See also Section 9.6.

Other online sources of environment and safety data include the following examples:

1. HAZARDLINE is a product of MDL Information Systems, San Leandro, CA. It contains MSDS information for some 20,000 chemicals. It is available in CD-ROM, and plans call for it to be available on the CHEMWEB site on the Internet in the near future. (There are many other online sources of MSDS data, as mentioned earlier in this chapter).
2. The databank of the Health and Safety Executive (HSE), which is the organization responsible for safety and health at work in the United Kingdom. This is searchable online through Orbit•Questel and DataStar, for example, and includes references to recent literature.
3. *NIOSHTIC,* the database of the National Institute of Occupational Safety and Health (NIOSH), also searchable through such vendors as DIALOG. This covers about 150 journals, as well as proceedings and books.

One of the most outstanding sources on the Internet that is fully dedicated to the environment is *Environmental RouteNet,* a product of Cambridge Scientific Abstracts (CSA), Bethesda, MD. It was introduced in 1995 and won an Information Industry Association award as "Best Web Site/Business" that same year. Available on a subscription basis, it provides extensive access to a broad spectrum of environmental information sources on the Internet. This important new product is managed by Jill Granader. The current URL address is http://www.csa.com/routenet.

Some of the content includes the following (in almost all cases scope is limited to environmentally related information only):

1. Approximately 18 months of information about federal legislation and federal and state regulatory news as reported by LEGI-SLATE (Washington, DC) and IHS (Engelwood, CO), respectively. Weekly updates to the environmental sections of the *Federal Register* are included.

2. Information about U.S. patents since issued since 1969, updated every 2 months.
3. Funding opportunities for researchers (University of Illinois IRIS Database and from the U.S. Federal Catalog of Domestic Assistance).
4. Environmental journals—access to title pages and, in some cases, abstracts for many journals.
5. Descriptions of, and links to, hundreds of other Internet sites, as annotated by CSA editorial staff.
6. Access to the current year of Cambridge Scientific Abstracts family of environmental and pollution abstracting and indexing databases, including those relating to agricultural and environmental biotechnology, aquatic pollution and environmental quality, bacteriology, environmental impact statements, ecology, health and safety science, industrial and applied microbiology, pollution, risk, toxicology, and water resources. Access to back years of these databases is available as well under separate subscription arrangement.
7. Electronic access to standard reference materials such as dictionaries and glossaries, as well as press releases, newsletters, and many other publications. Electronic access as well to catalogs of worldwide libraries.
8. Pertinent full-text worldwide news stories on a daily basis.

The complete contents are summarized in the "roadmap" in Figure 14.1. Note that the user has the option of searching across all fields simultaneously.

Some of the leading sources of environmental information on the Internet are summarized in a paper by Reichart (13). *Environmental Science and Technology* now includes regular columns about environmental resources on the Internet. As mentioned earlier in this chapter, there are extensive collections of MSDS data on the Internet as well as on other online systems. Other sources of health and safety information on the Internet were discussed earlier in the section on MSDSs.

The most extensive compilation of the toxic effects of chemicals is the *Registry of Toxic Effects of Chemical Substances* (*RTECS*), edited by Doris Sweet, and a product of the National Institute for Occupational Safety and Health (NIOSH), Cincinnati, Ohio. This compilation contains data for over 130,000 chemicals, almost double the number of chemicals covered about 10 years ago.

The *Registry* is available in several formats: in CD-ROM, on magnetic tape, and online in several systems (STN International, DIALOG, the MEDLARS service of the National Library of Medicine, and the Chemical Information System).

The online versions, as compared to the earlier printed volumes, are much more current and permit the user to focus in on data that may otherwise be cumbersome to locate. Figure 14.2 depicts a typical electronic record.

Before using *RTECS,* chemists should read the available user's guide. Absence of a substance from *RTECS* does not imply that it is nontoxic and nonhazardous; the compilation is not complete. Presence of a substance does not necessarily mean that a significant toxic effect has been found, although, of course, all chemicals can be

RouteNet's Roadmap

Global Search of RouteNet	**NEW FEATURE: SEARCH ACROSS ALL FIELDS!**				
Daily News	COMTEX Daily REAL-TIME NewsWire	GREENWIRE's Spotlight Story	Additional News Briefings		
Proprietray Abstract Databases	11 CSA's Environmental & Pollution Databases (Last 12 Months)	NLM'S Toxline Database (Last 12 Months)			
Annotated Internet Directory	Bibliographic Information	Research And Development Programs	Data Sets	Hot Links of the Month	
Patents & Grants	U.S Patent Database	Additional Patent Information	IRIS Grants Database	Government Grants	Additional Grant Sources
Legislation Regulation	IHS' ENFLEX Weekly Update on State Regulations	LEGI-SLATE's Narrative Description of Federal Legislation	International Treaties and Agreements	Federal Information	State Information
Site Specific Resources	VISTA Information Solutions, Inc.	CSA's EIS: Digests of Environmental Impact Statements Database	Additional Site Specific Information		
Reading Room	Reference Desk	Publications	Press Releases	Newsletters	Academic Schools and Programs
Market Place	Associations	Consulting Firms	Directories and Sites with links to lots of companies		
Policy & Compliance Corner	Policy & Compliance Forum: Sponsored by NEPI	18 EPA Sector Notebooks			

Figure 14.1. Environmental RouteNet's Roadmap. (Copyright 1996 Cambridge Scientific Abstracts. All Rights reserved.) There have been changes since this. For latest version, please see Internet site.

```
PC-SPIRS 3.30              RTECS (through July 1996)

RTECS (through July 1996) usage is subject to the terms and conditions of the
Subscription and License Agreement and the applicable Copyright and
intellectual property protection as dictated by the appropriate laws of your
country and/or by International Convention.

                                                                      1 of 1
                                                               Marked Record
AN: LT8524000
PN: Furan-
RN: Current: 110-00-9
UD: 9607
MF: C4-H4-O
MW:   68.08
WL: T5OJ
SY: Axole; Divinylene oxide; 1,4-Epoxy-1,3-butadiene; Furan (DOT); Furfuran;
NCI-C56202; Oxacyclopentadiene; Oxole; RCRA waste number U124; Tetrole; UN2389
(DOT)
CC: Tumorigen (C); Mutagen (M); Reproductive-Effector (T)
ME:
mmo-sat 825 nmol/plate (+/-S9)
   Developments in Food Science. 13,353,86 (DFSCDX);
dnr-bcs 3500 ug/disc
   Developments in Food Science. 13,353,86 (DFSCDX);
msc-mus-lym 1139 mg/L
   Environméntal and Molecular Mutagenesis. 12,85,88 (EMMUEG);
cyt-ham-ovr 184 mmol/L
   Cancer Letters (Shannon, Ireland). 13,89,81 (CALEDQ)
RE:
T02-T12     orl-rat TDLo: 3900 mg/kg (13W male/13W pre)
   National Toxicology Program Technical Report Series.
   NTP-TR-402,93 (NTPTR*)
TE:
V01-L60-L61 orl-rat TDLo: 1040 mg/kg/2Y-I
   National Toxicology Program Technical Report Series.
   NTP-TR-402,93 (NTPTR*);
V01-L60     orl-mus TDLo: 4160 mg/kg/2Y-I
   National Toxicology Program Technical Report Series.
   NTP-TR-402,93 (NTPTR*)
AT:
D28-F07-J22 ihl-rat LC50: 3398 ppm/1H
   Hazelton Laboratories, Reports. HLA468-102,87 (HAZL**);
T/E unlistd ipr-rat LD50: 5200 ug/kg
   American Industrial Hygiene Association Journal. 40,310,79 (AIHAAP);
J15-J30     ihl-mus LC50: 120 mg/m3/1H
   American Industrial Hygiene Association Journal. 40,310,79 (AIHAAP);
T/E unlistd ipr-mus LD50: 7 mg/kg
   American Industrial Hygiene Association Journal. 40,310,79 (AIHAAP);
K01-K13-P01 orl-dog LDLo: 234 mg/kg
   Journal of Pharmacology and Experimental Therapeutics.
   26,281,26 (JPETAB);
F12-P01     ivn-dog LDLo: 140 mg/kg
   Journal of Pharmacology and Experimental Therapeutics.
   26,281,26 (JPETAB);
K01-K13-P01 orl-rbt LDLo: 234 mg/kg
   Journal of Pharmacology and Experimental Therapeutics.
   26,281,26 (JPETAB)
MD:
L30-U01     orl-rat TDLo: 240 mg/kg/16D-I
   National Toxicology Program Technical Report Series.
   NTP-TR-402,93 (NTPTR*);
L02-M03-N74 orl-rat TDLo: 3900 mg/kg/13W-I
   National Toxicology Program Technical Report Series.
   NTP-TR-402,93 (NTPTR*);
J02-L30-Z01 ihl-rat TCLo: 200 mg/m3/4H/60D-I
   Toksikologiya Novykh Promyshlennykh Khimicheskikh Veshchestv.
   9,106,67 (TPKVAL);
L12-U01-Z01 ihl-rat TCLo: 500 mg/m3/2H/9W-I
   Toksikologiya Novykh Promyshlennykh Khimicheskikh Veshchestv.
   10,35,68 (TPKVAL);
A11-H02     ihl-rat TCLo: 5 mg/m3/4H/26W-I
   Toksikologiya Novykh Promyshlennykh Khimicheskikh Veshchestv.
   10,35,68 (TPKVAL);
```

Figure 14.2. Typical online output from *RTECS*. Courtesy of Dr. Doris Sweet.

```
L30-Z01     ihl-rat TCLo: 500 mg/m3/2H/30D-I
   Toksikologiya Novykh Promyshlennykh Khimicheskikh Veshchestv.
   9,106,67 (TPKVAL);
L30-L70-U01 orl-mus TDLo: 1950 mg/kg/13W-I
   National Toxicology Program Technical Report Series.
   NTP-TR-402,93 (NTPTR*);
L02-M03-U01 orl-mus TDLo: 2730 mg/kg/13W-I
   National Toxicology Program Technical Report Series.
   NTP-TR-402,93 (NTPTR*)
TR:
IARC Cancer Review: Animal Sufficient Evidence
   IARC Monographs on the Evaluation of Carcinogenic Risk of
   Chemicals to Man. 63,393,95 (IMEMDT);
IARC Cancer Review: Human Inadequate Evidence
   IARC Monographs on the Evaluation of Carcinogenic Risk of
   Chemicals to Man. 63,393,95 (IMEMDT);
IARC Cancer Review: Group 2B
   IARC Monographs on the Evaluation of Carcinogenic Risk of
   Chemicals to Man. 63,393,95 (IMEMDT)
SR: DOT-HAZARD: 3; LABEL: FLAMMABLE LIQUID
   Code of Federal Regulations. 49,172.101,92 (CFRGBR).
OEL-RUSSIA:STEL 0.5 mg/m3;Skin JAN93
ND: NOHS 1974: HZD 34185; NIS 2; TNF 25; NOS 3; TNE 244;
NOES 1983: HZD 34185; NIS 2; TNF 14; NOS 2; TNE 35; TFE 7
SL: EPA TSCA Section 8(b) CHEMICAL INVENTORY
On EPA IRIS database
EPA TSCA TEST SUBMISSION (TSCATS) DATA BASE, JULY 1996
NTP Carcinogenesis Studies (gavage); clear evidence: mouse, rat
National Toxicology Program Technical Report Series. (NTPTR*) NTP-TR-402,93
```

LEGEND: This version uses 2-letter symbols to label fields as follows:

PN	(Primary) Chemical Name
AN	RTECS Number
RN	CAS Registry Number
UD	Update fields XYMM record last changed
MF	Molecular Formula
MW	Molecular Weight
WL	Wiswesser Line Notation
SY	Synonyms
CC	Compound Descriptor Code
ID	Primary Irritation Data
ME	Mutagenic Effects Data
RE	Reproductive Effects Data
TE	Tumorigenic Effects Data
AT	Acute Toxicity Data
MD	Other Multiple Dose Toxicity Data
TR	Toxicology Reviews
SR	Standards and Regulations
ND	NIOSH Documentation Surveillance Data
SL	Status

Figure 14.2. Continued.

toxic, depending on dose. The lowest published value is given for each route–species combination. Bibliographic references are given. Note that the *RTECS* staff does not go through a peer review process when entering data into this compilation. Accordingly, the full original reference needs be obtained and studied to determine whether or not the data in question can be relied on. Involvement of a toxicologist in the use of *RTECS* is highly recommended.

Donley Technology, Colonial Beach, VA (804-224-9427) offers a number of products that are highly relevant to computer software connected with safety and the environment and that can help the user identify and compare the increasingly large number of products that are available. These include:

1. *Environmental Software Report.* A newsletter that reports on new and upgraded software packages, and online systems relating to environmental software. Eight issues are published per year at a current annual subscription of $125.00. Thus, for example, the November 1995 issue is 24 pages in length. It contains a report on trends in air-quality data management software, a directory of air-quality data management software systems/distributions, descriptions of new software, software industry news, descriptions of software upgrades and forthcoming conferences and courses. The senior editor/publisher is Elizabeth M. Donley.
2. *Special Reports*
   a. *Environmental Management Information Systems Report*
   b. *Air Quality Data Management Software Report*
   c. *Environmental Code Book Software Report*—systems that relate to federal, state and international laws and regulations on CD-ROM, diskettes, or online
   d. *MSDS Software Report*—systems for managing and creating MSDSs

Many software programs and online systems that relate to the environment, pollution control, safety, and/or toxicology are listed and briefly described in the annual compilation of software that is published as part of the AIChE's *Chemical Engineering Progress* December issue.

Other electronic sources are mentioned elsewhere in this Chapter. Please see especially the discussion about *Chemical Abstracts* in Section 14.8.

Examples of other tools containing safety information that can be searched online are mentioned elsewhere in this chapter.

14.6. GOVERNMENT LAWS, REGULATIONS, AND STANDARDS

All chemists and engineers are aware that there are many government regulations and standards relating to chemicals. These government actions have, in many cases, significant impact on the public and on the way chemists and engineers work. Federal, state, and local agencies, legislation, and regulations are all important. The U.S. EPA (Environmental Protection Agency) and U.S. OSHA (Occupational Safety and Health Administration) are probably the best known agencies to scientists in the United States. Regulations of other countries are also crucial for chemical firms and others who do business in those countries.

The interpretation and implementation of regulations have commercial, legal and technical ramifications. Attorneys, toxicologists, engineers, and chemists who specialize in this field are needed to assure appropriate compliance. As in the case of patents, this is an area that is often best handled by specialists, although all chemists should have some basic familiarity with important developments.

General sources of information on government regulations and other actions are discussed next.

The U.S. Environmental Protection Agency and the U.S. Food and Drug Administration both have home pages on the Internet (http://www.epa.gov, and www.fda.gov). These pages can help chemists and engineers with a major interest in regulatory matters keep up to date. Another excellent source for U.S. federal regulations is the *Federal Register.* This is an official daily publication of the U.S. government that provides the complete text of federal regulatory actions and notices, including many regulations that are merely being officially proposed, but that are included for comment. This is, of course, a major primary source, but it is assumed that some of those who are well versed in the field, such as trade association representatives, may know through their contacts of certain actions being considered even before they appear in the *Federal Register.* This publication is available online in several sources (NEXIS, LEXIS, and DIALOG, for example), and it is extracted in a large number of secondary sources which typically cover many other regulatory sources as well.

Current chemical news magazines also have useful discussions of proposed and pending regulations, as do major newspapers, notably the *Wall Street Journal.* For the chemist or engineer to whom regulatory affairs are not a full- or part-time responsibility, these news sources should suffice.

Several major secondary sources are specifically geared to the needs of the chemist or chemical engineer with strong interest in the regulatory aspects of chemicals.

Thus, for example, CHEMLIST is an essential and powerful online database intended to help chemists and others identify current regulatory information associated with chemical substances. It was initiated by the American Petroleum Institute in 1988. Chemical Abstracts Service (CAS) assumed responsibility for building the CHEMLIST database in 1992 and immediately began to broaden the scope to be more international. CHEMLIST is available online exclusively through the STN network. Updating in terms of online availability is said to occur generally within one week of receipt of the sources monitored by CAS.

As of 1996, CHEMLIST included more than 198,000 substances that are regulated or on advisory lists for those who use, manufacture, process, store, or transport chemical substances. More than 80% of the substances have CAS Registry Numbers. In addition to including specific chemicals, CHEMLIST also contains regulatory information pertaining to classes of substances, for example, glycol ethers.

The cornerstones of CHEMLIST coverage are national chemical inventories including the U.S. Toxic Substances Control Act (TSCA) Inventory, the European Inventory of Existing Commercial Chemical Substances (EINECS) as well as new notifications on the European List of Notified Chemical Substances (ELINCS) with both of the latter having chemical names and definitions in English, French, and German. The Canadian Domestic Substances and Non-Domestic Substances Lists (DSL/NDSL) are included with chemical names and definitions in English and

French. (CAS was a partner with the respective government agencies in the compilation of the U.S., European, and Canadian inventories.) In addition, the Korean Existing Chemicals List (ECL) is included with chemical names in English and designations of Toxic Chemicals and/or Specified Toxic Chemicals. CHEMLIST now includes also the contents of the Japan List of Existing and New Chemical Substances (ENCS—Kashin Act of Japan). Over 49,000 chemicals with CAS Registry Numbers are covered in ENCS. Many generic names are included.

CHEMLIST also contains substances subject to regulation under other sections of TSCA, and under Title III of the Superfund Amendments and Reauthorization Act (SARA, Sections 110 and 313). Substances on the following U.S. regulatory lists are included: CERCLA (Comprehensive Environmental Response, Compensation, and Liability Act) List of Hazardous Substances, EPCRA (Emergency Planning and Community Response Act, Title III of SARA) List of Extremely Hazardous Substances, ATSDR (Agency for Toxic Substances and Disease Registry) Toxicological Profiles, Toxic Release Inventory, TSCA Master Testing List, Occupational Safety and Health Administration (OSHA) Highly Hazardous Chemicals List, and Department of Transportation's Marine Pollutants List.

Data from a few states are also included. These include the State of New Jersey Right-to-Know Hazardous Substance List, the California Proposition 65 List, and the State of Pennsylvania Worker and Right-to-Know Hazardous Substance List. Although not all countries and states are included, it is planned that additional inventories and lists are to be added on an ongoing basis to make the CHEMLIST file more comprehensive.

CHEMLIST is a substance-based file so that information from various inventories and lists can be merged with selected CAS data to form a single record for each nonconfidential record for quick access.

The additional value that CAS provides to this database is that chemical substance identification specialists as CAS verify the CAS Registry Numbers and chemical names, make any necessary corrections to the CAS Registry Numbers, and assign new CAS Registry Numbers when appropriate. This is especially important for national inventories, for example, the Korean inventory, which were compiled without CAS participation.

The history of changes in the regulatory status of chemicals in the United States, as announced in the *Federal Register* from 1979 to date in the form of proposed or final U.S. government regulations, are included in a summary format. Other sources monitored include *Chemical Regulation Reporter* (Bureau of National Affairs), EPA's *Chemicals-in-Progress Bulletin,* prepared index entries since 1989 from *Pesticides and Toxic Chemicals News,* and the *Code of Federal Regulations.* In addition, the *Official Journal* of the European Communities and the *Canada Gazette* have been added to the list of sources monitored as CAS continues to expand international coverage of CHEMLIST.

The data content includes citations and summaries to such regulatory activity as premanufacture notifications (PMNs), EPA proposed and final rules, responses to Interagency Testing Committee (ITC) recommendations, and actions on rule violations and citizens' petitions. PMNs in CHEMLIST include company names when these are not confidential, a feature that can be useful in competitive intelligence functions. (Company names appear consistently only since mid-1992, when CAS took over the file, but efforts are being made to provide this information for earlier records as well.) Each substance record also includes the CAS Registry Number, when available, regulated or advisory list names, regulatory list numbers, EPA special flags for TSCA Inventory substances, synonyms from the literature, and generic names for confidential substances. Substance definitions for complex reaction products or process streams from various industries are also available. As an example of a definition, the definition of olive oil is as follows: "Extractives and their physically modified derivatives. It consists primarily of the glycerides of the fatty acids linoleic, oleic, and palmitic. Olea europea." This definition is from TSCA.

TSCATS (Toxic Substances Control Act Test Submissions) data previously on this file have been removed because these are already available as part of the *TOXLINE* file on STN.

Typical displays of information from the CHEMLIST database are shown in Figure 14.3.

Another related product from CAS is their *National Chemical Inventories* CD-ROM product. This includes the following major official national chemical inventories: U.S. (TSCA), Canada (DSL/NDSL), European Commission (ELINCS/EINECS), Korea (ECL), and Japan (ENCS). The letters in parentheses signify the common way that these inventories are referred to. Coverage of Australia is planned for the future. These inventories are very important to any company that does business on an international scale.

One of the most widely utilized sources for helping chemical industry and university personnel keep up with pertinent government laws and regulations is BNA (Bureau of National Affairs), Washington, DC (800-372-1033). One of their foremost products is *Chemical Regulation Reporter,* which is available in both printed form and online (Internet, NEXIS, LEXIS, WESTLAW). The printed product is updated weekly, and there are excellent bimonthly and weekly indices which includes entries by topic, cases, the *Code of Federal Regulations* as affected, and CAS Registry Numbers. The scope includes information on the latest major laws and regulations, and full text of proposed regulations is included. Also included is news and analysis as to new and revised regulations, legislation, research techniques and findings, and enforcement policies and actions. There is an extensive reference file as well.

A comparable *International Environment Reporter,* available in printed form only, covers major countries on a worldwide basis.

Yet another related product is available on CD-ROM. This is BNA's *Environmental Library.* The scope includes pertinent federal laws and regulations as well as similar material for each of the 50 states. The CD is updated on a monthly basis. Certain parts may be obtained separately. Thus, each state may be obtained separately, and federal chemical laws and regulations may be obtained separately if desired.

Examples of other products from BNA are *Daily Environment Report, Air Pollution Control, and Toxics Laws Report,* to name only a few.

Ariel Research Corp., Bethesda, MD (800-982-0064) offers its *Ariel Insight* product (an improvement of over its earlier *International Chemical Regulatory Monitor-*

```
" Ethanol "(abbreviated)

AN     264  CHEMLIST
RN     64-17-5
CN     Ethanol (English, French, German) (TSCA, DSL, EINECS, ECL)
       Alcare Hand Degermer
       Alcohol
       Alcohol anhydrous
       Algrain
       Anhydrol
       Denatured ethyl alcohol
       Desinfektol EL
       Ethicap
       Ethyl alcohol
       Ethyl hydrate
       Ethyl hydroxide
       Jaysol
       Jaysol S
       Methylcarbinol
       SY Fresh M
       Synasol
       Tecsol
       Tecsol C
       UN 1170 (DOT)
FS     CANADA: DSL;  EEC: EINECS;  KOREA: ECL;  USA: DOT, SARA, STATE, TSCA
CBI    Public
RLN    EPA No.: 8EHQ-0288-0722; 8EHQ-0588-0701; 8EHQ-0690-0701;
                8EHQ-0778-0228; 8EHQ-0786-0617; 8EHQ-0886-0617;
                8EHQ-1287-0701
       EINECS No.: 200-578-6
       ECL Serial No.: 2-858
INV    On TSCA Inventory
         January 1995 Inventory Tape.
       On DSL
         Supplement to Canada Gazette, Part I, January 26, 1991.
       On EINECS
         Annex to Official Journal of the European Communities, 15 June 1990.
       On ECL
         Korean Existing Chemicals List, 1992 Ed.
FA     RN; RLN; INV; 4A; 8D; 8E; STY; S110; SNJ; SPA

 ==== U.S. EPA Regulations - TSCA ====
       •
       •
       •
 ==== U.S. EPA Regulations - SARA ====
S110   Superfund Amendment Reauthorization Act Section 110
       Fed. Regist. 56 #201:52166 (17 Oct 1991).
       This previously listed substance is not on the revised priority list of
       hazardous substances, nor will it now be considered for development
       of a toxicological profile. [Also reported in Pesticide Toxic Chem.
       News 19 #51:22 (23 Oct 1991) and in Chem. Regul. Rep. 15 #33:1152
       (15 Nov 1991).]
S110   Superfund Amendment Reauthorization Act Section 110
       Fed. Regist. 54 #206:43615 (26 Oct 1989).
       This and 24 other substances have been selected for the Third Priority
       List of hazardous substances, based on toxicity, potential hazard,
       and other criteria. [A summary of this notice also appears in
       Pesticide Toxic Chem. News 17 #52:35 (01 Nov 1989) and in Chem.
       Regul. Rep. 13 #31:973 (03 Nov 1989). The full text also appears in
       Chem. Regul. Rep. 13 #31:989 (03 Nov 1989).]
S110   Superfund Amendment Reauthorization Act Section 110
       Fed. Regist. 56 #227:59331 (25 Nov 1991).
       Correction to a notice in Fed. Regist. 56 #201:52166 (17 Oct 1991).
 ==== U.S. State Regulations ====
SNJ    New Jersey Right-to-Know
       New Jersey Department of Health Hazard Right-to-Know Program
       Hazardous Substance List, December 1989.
       Special Health Hazard Code(s): F3 (Flammable, Third Degree)
       Common Name(s): ETHYL ALCOHOL; ALCOHOL (ETHYL); DENATURATED ALCOHOL;
       ETHANOL; COLOGNE SPIRITS
SPA    Pennsylvania Right-to-Know
       Pennsylvania Department of Labor and Industry Hazardous Substance
       List 1989.

      Listed Name(s): ETHANOL
```

Figure 14.3. Typical printout from CHEMLIST database. (CAS material is reproduced with the permission of the American Chemical Society. Copyright 1995 American Society.)

ing System), which covers United States and many international regulations or recommendations regarding thousands of chemical substances. In the North American database segment, listings of chemicals in U.S. federal and state, Canadian, and Mexican regulations are included, and the full text is also available. Similar information is available for European Union countries, including nonmember European states as well. Official national inventories are included for the United States, Canada, western Europe, Australia, Korea, Japan, the Philippines, and China. A full scale Pacific Rim database, including India, is planned for 1997. A unique "generics" database tells whether chemicals are included in regulated chemical classes (some 600 classes are included) in North America and western Europe. News briefs about new and proposed regulations are included. The user can license all system capabilities or just those segments of greatest interest. *Ariel Insight* is to be available to licensees on the Internet (using special software) and also in client/server, CD-ROM, and diskette forms. Results may be integrated with inhouse systems. Licenses to the entire package begin at over $30,000 per year, and vary with the number of users.

Access points to the *Insight* product include Chemical Abstracts Service Registry Numbers, molecular formulas, and chemical names and name fragments. Initial output is a summary report that contains immediately available links to full details.

Another Ariel product is *Euro Chemical News,* a monthly report of chemical regulatory compliance news for the European Union and Western European countries.

A related printed publication is *Suspect Chemical Sourcebook,* which is published by the Roytech operation of Ariel. This popular looseleaf reference volume includes over 18,000 chemical listings with respect to U.S. federal, major states, and selected Canadian regulations.

Ariel staff include chemists, chemical engineers, and attorneys, many of whom have international contacts and experience. The total staff consists of more than 30 persons.

IHS Environmental Information, Exton, PA (800-365-2146) produces one of the premier product lines in this field. This includes a CD-ROM that covers environmental, health, and safety regulations in the United States, including both federal and state regulations. Canada, Mexico, and certain other countries are also included. An important part of the environmental scene are regulatory "lists" of chemicals; this product contains some 1000 lists. The product is also available on the Internet on a subscription basis. A related product is an Internet version of the daily *Federal Register,* the official publication that reports on federal regulatory actions and notices, and that is searchable.* Its intended audience is the health and safety professional.

The *Chemical Advisor* is said to contain "fully compiled" regulatory and advisory data on over 11,500 chemicals. The compiler is ChemADVISOR Inc., Pittsburgh, PA. It is available as a CD-ROM product through the Canadian Center for Occupational Health and Safety, Hamilton, Ontario, with quarterly updating available. Data included is not only from the United States and Canada, but other countries as well, including the United Kingdom and Mexico.

The tools mentioned here, especially the compilations of BNA, can also help the chemist make direct contact with appropriate high-level government officials and

*The IHS file is not as up-to-date as the Government Printing Office's site (www.gpo.gov), but it is archived much longer and is said to be more fully searchable.

technical experts by identifying names of these individuals and their functions. Recent government agency organization charts and directories are also important in keeping track of key personnel.

For chemical industry companies, joining one or more of the major trade associations is another very significant way to keep up with the regulatory environment. Some of these associations do an excellent job in keeping their membership apprised of important developments. One excellent example of a trade association that helps its member companies follow and understand the regulatory scene is SOCMA (Synthetic Organic Chemicals Manufacturers Association), Washington, DC. Other associations are discussed in Section 14.7.

On occasion, when a chemist looks through regulatory sources, information that may of interest beyond the environmental aspects may appear.

For instance, environmental filings to obtain permits made in the case of plant expansion may often be part of the public record and may provide competitors with useful information about what a company plans to do.

Other examples are notices in the *Federal Register* relating to PMN (Premanufacture Notices) or "commencement to manufacture." These can also tell competitors something about plans and activities. The fact that a competitor has filed a PMN is meaningful, and announcement of commencement of manufacture is even more meaningful.

In order to properly safeguard their proprietary interests, some companies may not disclose full details or exact chemical composition in their regulatory documents that are available to the public, if the regulations and the various industry codes permit confidentiality while still protecting the public interest.

The *TRI* (EPA's *Toxics Release Inventory*), provides, for approximately 640 chemicals and chemical categories as specified by EPA, facility-by-facility data on the types and quantities of various kinds of discharges and transfers at nearly 30,000 plant sites nationwide. Before a chemical can be listed on the inventory, EPA must show that it is toxic. TRI is based on data supplied annually by the plant sites; about 60% of the material is supplied electronically by the sites to EPA, and the balance, in paper form, must be retranscribed for entry into the database. There are numerous data-quality checks, but errors do occur. TRI data can conveniently be obtained through the *TOXNET* online system (on the MEDLARS system as such, and with Internet availability also due soon) of the National Library of Medicine. The data are also available free on the nonprofit Right-to-Know Network (RTK Net, Washington, DC) along with other related databases from several federal agencies (http://rtk.net). Another source is EPA's *Envirofacts* system described in Chapter 9. CD, diskette, and hard-copy forms are also offered by EPA. The capabilities of some of the various access modes differ somewhat. The 700 Federal Depository Libraries nationwide receive the data in CD format as part of their substantial collections of other government documents in many fields.

14.7. PROFESSIONAL SOCIETIES AND OTHER ASSOCIATIONS

Professional societies and industry associations are active in promoting good environmental practices and safety, and providing information:

1. The American Chemical Society has Committees on Environmental Improvement and on Chemical Safety. There are also Divisions of Environmental Chemistry and of Chemical Health and Safety which have full-scale programs at national meetings. ACS publications in this field include, for example, the journals *Environmental Science and Technology, Chemical Health and Safety Magazine,* and a number of information pamphlets and technical brochures that are available primarily from the ACS Office of Government Affairs.

2. AIChE has long been active in providing information about safety and about the environment. In addition to papers in AIChE journals, other publications include the proceedings of the Loss Prevention Symposia. These meetings have been held annually for many years. Another long-standing series are the volumes of the Ammonia Plant Safety symposia. In 1985, AIChE established its Center for Chemical Process Safety (CCPS). This Center has an active publications program, and also sponsors seminars, training programs and research. The publications program includes a series of guideline volumes such as *Guidelines for Implementing Process Safety Management Systems, Guidelines for Chemical Transport Risk Analysis, Guidelines for Auditing Process Safety Management Systems, Guidelines for Safe Automation of Chemical Processes,* and *Guidelines for Engineering Design for Process Safety,* to name only some of the available volumes. In addition, slides/lectures, problem sets and course outlines are available to assist in safety education. Other safety publications available from AIChE include Dow Chemical Company's *Chemical Exposure Index Guide,* a guide to rating the relative acute health hazard potential of a chemical release, and Dow's *Fire and Explosion Index Hazard Classification Guide* (14), a guide to evaluating the realistic fire, explosion, and reactivity potential of process equipment and its contents. Also available is Dow's "*CHEMPAT*" software, which facilitates assessment of chemical compatibility and reactivity so as to assist in hazard evaluation and management. In addition, AIChE has a Center for Waste Reduction Technologies directorate, and this organization offers a few publications relating to this topic.

3. The Chemical Manufacturers Association (CMA), located in Arlington, VA, provides assistance in safe handling of chemicals. A toll-free telephone number (800-424-9300) provides immediate emergency access to CMA's Chemical Transportation Emergency Center (Chemtrec). This Center was established in 1971 to help emergency responders handle accidents with chemicals in transportation or warehousing. The service (a) determines what chemicals are involved in an accident, (b) supplies preliminary advisory information retrieved from Chemtrec files, and (c) secures involvement of appropriate technical specialists. Another CMA office (Chemical Referral Center) refers inquirers to the names and phone numbers of manufacturers of chemicals as based on the chemical or trade names of these products. The telephone for this CMA office is 800-262-8200.

4. The National Safety Council has a broad range of safety-related activities and publications. Many of these deal with safety in general or with other aspects of safety, but some, such as publications of their Chemical and R&D Sections, have material of specific interest to chemists.

5. The chemist and engineer will find the National Fire Protection Association (NFPA) a valuable source of safety information (15). Of special interest is NFPA's *Manual of Hazardous Chemical Reactions* (15a). This is a compilation of 3550 mixtures of two or more chemicals reported to be potentially dangerous in that they may cause fires, explosions, or detonations as based on a survey of incidents cited in the literature. Before the chemist or engineer embarks on an R&D or other project, this NFPA publication is a key source to be examined. The NFPA publication *Hazardous Chemical Data* (15b), contains basic safety data on approximately 350 chemicals. These data include: description (including odor), fire and explosion hazards (including flash point, flammable limits, and ignition temperature), health hazards, personal protection, fire fighting, usual shipping containers, storage, and other information. These and other NFPA reports are brought together in a single volume, *Fire Protection Guide on Hazardous Materials,* now in its 11th edition published in 1994 (15c). In addition, NFPA has adopted a detailed fire protection standard for laboratories that use chemicals. This 45-page code (16) is one of the most comprehensive of its type. It deals with chemical handling and storage, compressed gases, fire prevention, ventilation, building construction, exit doors, and automatic fire-extinguishing systems. Although NFPA is a nonprofit, private organization, its recommendations have considerable effect on the chemical industry. For example, its reclassification of a product from a Class 2 oxidizer to a Class 1 oxidizer can permit retailers and distributors of a chemical to significantly increase stored quantities of chemical in facilities both with and without sprinkler systems and storage cabinets.

6. The American Conference of Government Industrial Hygienists (ACGIH) publishes threshold limit values (TLVs) for chemical substances and physical agents in workroom air. ACGIH is not an official government agency, but its recommendations are highly regarded. This is a quick reference source for exposure limits of common substances. Many of these TLV data serve as the basis for OSHA (Occupational Safety and Health Administration) permissible limits and are constantly under review. Another excellent source of industrial hygiene data is the American Industrial Hygiene Association, Fairfax, VA.

7. Published information on hazardous reactions and compounds is, of course, very useful, but no matter how thorough, such literature cannot possibly include new reactions and compounds not previously reported. In response to this, ASTM has produced and distributes the computer program CHETAH (Chemical Thermodynamic and Energy Release Program). This program assists in predicting hazards associated with reactions, and in addition, the energy-release (explosive) potential of individual compounds. For a further discussion of CHETAH, please see Chapter 15.

8. The Royal Society of Chemistry, Cambridge, United Kingdom, offers a number of current awareness publications relating to health and safety. These include *Laboratory Hazards Bulletin,* which includes abstracts with a chemical and subject index each issue; *Chemical Hazards in Industry,* which includes abstracts with a chemical and subject index; and *Hazards in the Office,* a bimonthly with abstracts. These three products are put together to form *Chemical Safety NewsBase,* which is available online through STN International, DIALOG, DataStar, and Questel•Orbit.

9. *CHEMSAFE* is a useful safety-data file that is available for searching on STN International. It contains 30–50 safety parameters for each of more than 1650 combustible liquids, gases, and dusts. All of these values are said to have been evaluated, and when possible, marked with a recommendation. Properties covered include, flash point, ignition temperature, explosion and detonation limits, and threshold limit values (MAK). The producer is the German chemical engineering society, DECHEMA, which is located in Frankfurt. A related product of considerable importance is *Environmental Protection and Process Safety,* a current awareness abstracting service (monthly) from DECHEMA that also includes input from The Royal Society of Chemistry. This is to be renamed *Environmental and Safety Technology* for 1997. The coverage includes both environmental and safety aspects as indicated by the title. This is a subset of *Chemical Engineering and Biotechnology Abstracts* which is searchable online through STN.

14.8. *CHEMICAL ABSTRACTS* AND OTHER ABSTRACTING AND INDEXING SERVICES

A vital source in this field of chemistry, as in all others, is *CA*. Some sections that can be scanned selectively on a current awareness basis include the following:

Section 1	Pharmacodynamics
Section 4	Toxicology
Section 50	Propellants and Explosives
Section 59	Air Pollution and Industrial Hygiene
Section 60	Waste Treatment and Disposal
Section 71	Nuclear Technology

All 80 sections, however, may contain some material of interest.

Pertinent safety-related headings in the *CA* General Subject Index include

Accidents
Alarm devices
Combustibles
Disease, occupational
Electric shock
Explosibility
Explosion
Fire, fireproofing, fireproofing agents, fire-resistant materials
Flammability
Health hazard
Health physics
Ignition
Implosion
Injury
Nuclear reactors
Poisoning
Respirators
Safety
Safety devices
Toxicity, toxicology

These entries are helpful when looking for more generic information (such as safety in handling flammable organic liquids or the question of potential toxicity of surfactants to aquatic life). The same entries can also be helpful when looking for specifics (such as explosibility of *o*-nitrophenylsulfonyl-diazomethane), although for specific substances the chemist should also look under the name of that substance in the *CA* chemical substance index.

Beginning in 1976, keyword indexes to weekly issues of *CA* were extended to cover safety more broadly. However, this entry (the keyword safety) by design does not include toxicology. Therefore, to get a complete picture of safe-handling aspects, keyword index entries pertaining to toxicity must also be scanned. Because keyword indexes are not complete for toxicity data, the best places to look in *CA* are online or the 6-month volume and 5-year collective indexes. This is true particularly for toxicity studies reported incidentally to synthesis of a compound.

The *CA SELECTS* series of current awareness bulletins produced by Chemical Abstracts Service (see Chapters 4, 6 and 7) includes several products dedicated to environmental chemistry. These include *Environmental Pollution, Air Pollution, Indoor Air Pollution,* and *Acid Rain and Acid Air.* As to safety-related matters, products in this series include *Occupational Exposure and Hazards, Flammability, Drug and Cosmetic Toxicity, Food Toxicity,* and *Food, Drugs, and Cosmetics—Legislative and Regulatory Aspects.* One or more of the products in these series offer a convenient way to keep up to date on matters relating to environment and safety as defined by the scope of these products.

Similar current alerting products are available from the Institute for Scientific Information (see Chapters 4 and 8). *Index Chemicus* can prove helpful in alerting chemists to explosive reactions. Other products of the same publisher that can be helpful in following safety-related developments are their *ASCATOPICS* products, for example (see Chapter 4).

In addition to *CA,* or the subset groupings noted, the related CAS publication *Chemical Industry Notes* (*CIN*) (see Section 7.16) also needs to be consulted for extracts of reports on specific incidents such as chemical plant explosions, fires, spills, and leakages. If this material is "newsy," and reported in trade and industry publications or a few major national newspapers, it is more likely to be reported in *CIN* than in the full *CA*.

As mentioned in Section 14.7, important abstracting and indexing services on safety and hazards are published by the Royal Society of Chemistry.

BIOSIS Previews (*Biological Abstracts*), available in printed, disk, and online editions, is an indexing and abstracting source that, among its other advantages, sometimes indexes and abstracts documents on biological activity that may not be covered elsewhere. Coverage of this classic, well-known service is from 1926 to date. The publisher is located in Philadelphia, PA. See also Chapter 8.

Environmental Health and Pollution Control, published by Excerpta Medica, subsidiary of Elsevier Science Publishing, Amsterdam, the Netherlands, and Princeton, NJ, is a monthly abstracting service that includes coverage selected from more than 15,000 scientific journals and periodicals. Chemists and engineers will find this particularly useful in covering developments in Europe, although the coverage is reportedly worldwide. Other related abstract bulletins include *Microbiology, Public Health,* and *Occupational Health.* These publications are subsets of the comprehensive *Excerpta Medica* medical abstracting service that is also searchable online from 1974 onward through DIALOG, STN, International and elsewhere in a file known as *Embase*. The online files contain about 25% more abstracts than the printed publications.

Environment Abstracts is another excellent source. It, too, is available in both printed form and online. The online version is *Enviroline* and can be searched by users of several major online systems, including DIALOG.

The scope is broad. Categories covered include air pollution, chemical and biological contamination, energy, environmental education, environmental design and urban ecology, food and drugs, general, international, land use and misuse, noise pollution, nonrenewable resources, oceans and estuaries, population planning and control, radiological contamination, renewable resources—terrestrial, renewable resources—water, solid waste, transportation, water pollution, weather modification and geophysical change, and wildlife. Copies of most of the documents included are available from the publisher, Congressional Information Service, Inc., Bethesda, MD.

Pollution Abstracts published by Cambridge Scientific Abstracts, (see Chapter 10) contains references to worldwide technical literature covering some 3000 journals, conference proceedings and papers, and monographs in the areas of air and water pollution, solid wastes, noise, pesticides, radiation, and general environmental quality. It is available both in conventional printed form and also online via several services.

There are at least two indexing and abstracting services that cover the science of safety defined broadly: *CIS Abstracts* and *Health and Safety Science Abstracts.*

CIS Abstracts is produced by the International Occupational Safety and Health Information Center (*CIS*) of the International Labor Office (ILO), which has headquarters in Geneva, Switzerland, and an office in Washington, DC. Many find this publication especially useful because of its international scope. Coverage includes job safety, worker protection, and disease prevention. Worldwide laws, regulatory documents, and publications of intergovernmental organizations are a particular strength. It is also available online through Questel•Orbit and ESA-IRS (see Section 10.4). A subset of this database is also available through *TOXLINE* (see Section 9.5.A). The card files of *CIS* began in 1960, but the online version begins its coverage in 1972–1974. Despite its name, ILO is not a labor union organization; rather, it is an agency within the United Nations system.

Health and Safety Science Abstracts is a compilation of abstracts that have been selected, classified, and indexed from the world's technical and scientific literature on the science of safety. Each issue contains extensive cross-references and indexes by subject and author. It is published by Cambridge Scientific Abstracts, in association with the Institute of Safety and Systems Management, University of Southern California, Los Angeles.

14.9. QUANTITATIVE STRUCTURE–ACTIVITY RELATIONSHIPS (QSARs)

QSAR (quantitative structure–activity relationships) techniques aim to predict the potency of biological effects of chemicals. Quantification of structure–activity relationships is accomplished by correlating the relative potencies of substances with one or more physicochemical properties into a mathematical equation. Qualitative structure–activity relationship studies utilize correlations and comparisons of chemical structure with biological effects.

Corwin Hansch is widely recognized as the father of QSAR. The most extensive compilation of QSAR results is represented in the *Log P and Parameter Database* assembled by Hansch and colleagues at BioByte Corp., Claremont, CA. More than half of this database is devoted to partition coefficients—the equilibrium ratio of the concentrations of a solute distributed between an organic phase and water. The octanol/water partition coefficient, expressed in logarithmic terms, serves as the "standard" parameter for measuring hydrophobic character. BioByte's *Medchem Database* lists over 40,000 log *P* values in over 300 solvent/water systems, as well as 20,000 electronic and steric parameters for a variety of substituents.

Using a Hammett-type extrathermodynamic approach, Hansch has developed over 9000 regression equations that relate these parameters to toxicity and other types of biological activity. The various fields where this concept has been successfully employed have been summarized by Albert J. Leo (17).

It has been pointed out by Leo (17) that hydrophobicity has a key role in environmental transport and toxicity. An associated "ClogP" software program is available for calculating log *P* octanol/water from structure. This is utilized in the pharmaceutical and pesticide industries and in assessment of environmental hazards; the software is available from BioByte Corp., Claremont, CA.

S. C. deVito, a medicinal chemist with EPA, has written (18) about the design of chemicals from a toxicological perspective. He has commented at length about structural features of chemicals that can affect human health or the environment and how these can be modified so as to minimize adverse effects. EPA utilizes QSAR techniques to predict aquatic toxicity of new chemicals when test data are not present. ECOSAR ("A Computer Program for Estimating the Ecotoxicity of Industrial Chemicals Based on Structure–Activity Relationships") is a free predictive computer program available from EPA that contains more than 100 regression equations of aquatic toxicity for over 40 classes of chemicals. ECOSAR is based on the utilization of log *P* principles.

14.10. TOXICITY TO AQUATIC ORGANISMS; WATER RESOURCES

Potential toxicity of chemicals to fish and other aquatic organisms is of key environmental interest, especially if there is a chance that such organisms might come into contact with a chemical, or its byproducts or waste streams, during manufacture or use. Government agencies (notably the State of California and its War Pollution Control Board) and other sources have compiled much of earlier and more recent published literature. Recent data can also be obtained from some of the online sources and printed abstracting and indexing services mentioned earlier. In addition, there are several more specialized sources:

- Aquatic pollution is one of the fields covered in the *Aquatic Sciences and Fisheries Abstracts Series* published by Cambridge Scientific Abstracts. This is probably the leading abstract series of its type covering virtually all aspects of the field. Several parts of the series are sponsored by four specialized agencies of the United Nations. This material is available online through several sources, as well as in CD-ROM form. A related publication from the same publisher is *Oceanic Abstracts.*

- *Water Resources Abstracts* is a monthly abstracting and indexing service published by the U.S. Department of Interior's Water Resources Scientific Information Center. This covers the subject of water quite broadly, including toxicity to aquatic life. It can be used in conventional (printed) form, and online access is available through DIALOG.
- An important source is *Aqualine Abstracts,* a product of WRc plc, Medmenham, UK. This is available in printed form (published monthly), online, and as a CD-ROM.
- Another online file devoted to water is the *WATERNET* database of the American Waterworks Association, Denver, CO. Coverage begins with 1971.

Despite availability of these and other sources, there is a relative shortage of published data on effects of chemicals on aquatic life. Mammalian toxicity data are more readily available.

Direct mail or phone contact with government experts on the forefront of this work can provide additional leads beyond what is in published sources. Examples of leaders in the field are scientists at the U.S. Environmental Research Laboratory, Duluth, MN.

14.11. BIODEGRADABILITY

Biodegradability of chemical substances is of considerable environmental interest. One of the most extensive known files on this topic is maintained by the National Center for Environmental Toxicology, Medmenham, England.

The online system, Chemical Information System, has two files specifically dedicated to biodegradation, as previously noted.

Other information on biodegradability will be found in more general tools such as *CA* and *BIOSIS Previews* and in specialized environmental tools such as *Pollution Abstracts.*

14.12. WATER-QUALITY DATA

Water-quality data are important in appraisal and management of water sources, pollution surveillance and studies, monitoring of water-quality criteria and standards, and development of energy resources.

One good source of water data is the National Water Data Exchange (NAWDEX). This is part of the U.S. Geological Survey. NAWDEX was established in 1976 to assist users in identifying, locating, and acquiring needed data. It has been described as a confederation of all types of private- and public-sector water-oriented organizations.

To cite just one example, coordination is maintained with the Environmental Protection Agency's STORET file, which has waterways quality data from more than 200,000 collection points. The general public can access NAWDEX through its local assistance centers located throughout the United States.

In addition, NAWDEX may be accessed on the Internet through the address http://h2o.er.usgs.gov/public/nawdex.html>. This location includes access to a directory of Assistance Centers of NAWDEX, a water data sources directory, access to information about STORET (which may be searched online by the public on a fee basis), *Selected Water Resources Abstracts,* national water conditions, and other information. *Selected Water Resources Abstracts* coverage is from the present back to 1939.

14.13. EXAMPLES OF KEY JOURNALS

There are many journals in this field, but it may help the reader to mention a few examples of some of the significant journals and other "serials" related to the environment, toxicology and safety:

Annual Review of Pharmacology and Toxicology
Archives of Environmental Health
British Journal of Industrial Medicine
Bulletin of Environmental Contamination and Toxicology
Chemical Health and Safety
CRC Critical Reviews in Toxicology
Environmental Science and Technology
Environmental Health Perspectives
Food and Chemical Toxicology
Fundamental and Applied Toxicology
Journal of Applied Toxicology
Journal of Hazardous Materials
Journal of Loss Prevention
Journal of Occupational Medicine
Journal of Toxicology and Environmental Health
Mutation Research
Process Safety Progress
Regulatory Toxicology and Pharmacology
Teratology
Toxicology
Toxicology and Applied Pharmacology
Water Environment Research

The last-named journal, *Water Environment Research,* includes a noteworthy annual literature review issue that appears in June. The publisher is the Water Environment Federation, Alexandria, VA.

14.14. NEWSLETTERS

This field is characterized by a number of specialized newsletters. These may contain information that supplements and augments chemical news magazines and sources. Emphasis in the newsletters is on rapid concise reporting. One outstanding example is *Pesticide & Toxic Chemical News,* a product of Food Chemical News Division of CRC Press, Washington, DC. The same publisher also issues *Food Chemical News;* the Internet address, http://www.crcpress.com, contains excellent links to other food chemical sites.

Another good newsletter example is *Air/Water Pollution Report's Environment Week,* published by Business Publishers Inc. (BPI), Silver Spring, MD. Other examples from the same publisher include *Clean Water Report, Environmental Health Letter, Hazardous Waste News, Occupational Health & Safety Letter,* and *Solid Waste Report.* King Publishing, Washington, DC, issues *Bio-Cleanup Report* twice a month; this relates to the remediation industry.

Some newsletters may be found online through such systems as Newsnet and DIALOG, but, interestingly, the print versions may often available more quickly. Timeliness is, of course, crucial in this field.

14.15. OTHER TOPICS

Much of the next chapter has safety and environmental implications.

The Chemical Industry Institute of Toxicology, located in Research Triangle Park, NC, was founded in 1974 by a number of chemical companies to conduct research aimed at better understanding and assessing the potential health effects of chemicals and to develop improved methods. The Institute has a scientific staff of over 30 persons, and, in addition, there is a very active postdoctoral fellowship program that offers important training opportunities. Results form the work of CIIT are published in leading research journals, and there is a newsletter *CIIT Activities.* Roger O. McClellan is president.

A somewhat similar industry-sponsored group is the European Center for Ecotoxicology and Toxicology of Chemicals (ECETOC), Brussels. This group acts is a clearinghouse and reviews toxicology results, but it does not conduct its own laboratory testing.

Another important group is the firm Water Research Center (WRc plc), part of which is the National Center for Environmental Toxicology (NCET), Medmenham, Marlow, Buckinghamshire, UK, (www.wrcplc.co.uk/main.htm) which was established in 1995. A U.S. office is maintained in Trevose, PA. The objective of NCET is to "provide authoritative independent information and advice on the effects of chemical contaminants on the environment." NCET is staffed by toxicologists and environmental chemists who are experienced in their fields in the United Kingdom and continental Europe. One of their unique services is SARTOX, a service that aims to predict the physicochemical properties, biodegradability, and ecotoxicity of chemicals on the basis of input of the chemical structure only. A literature search is employed first, and, then, as needed, predictive models are utilized. These may include octanol–water partition coefficients, water solubility, the organic carbon coefficient, and the gas–water partition coefficient. Results are fed into quantitative structure–activity relationship (QSAR) and environmental distribution models. Underway are approaches to predict the toxic mode of action of compounds, and to predict chemical oxidation processes. NCET maintains, and has available for license, EQualS, a database of Environmental Quality Standards used by U.K. regulators for the protection of aquatic life. WRc is a private company that provides consulting services on a fee basis. Brian Crathorne is director of NCET.

Despite all the toxicological and environmental data and sources described in this book and elsewhere, because little data from testing and experience is available on many thousands of chemicals, assessing the potential risks and hazards for all of these is difficult, if not impossible. Thus, the *Registry of Toxic Effects of Chemical Substances* covers over 130,000 chemicals, but there are more than 17 million chemical substances according to Chemical Abstracts Service. This means that the chances of finding specific risk or hazard data about a chemical of interest, particularly if it is not commercial, are poor. In some cases, cautious estimation or prediction may be quite feasible in the hands of qualified specialists. For the many chemicals not yet studied for risk or hazard potential, similar structures that have been studied may sometimes give useful clues as to potential dangers. In any event, advice of a toxicologist and of other qualified environmental specialists is essential in attempting to plan to extrapolate from what has already been studied to the as yet unknown.

REFERENCES

1. G. F. Clayton and F. E. Clayton, *Patty's Industrial Hygiene and Toxicology,* 4th ed., Wiley, New York, 1991–1994.
2. R. E. Gosselin, H. C. Hodge, and R. P. Smith, Eds., *Clinical Toxicology of Commercial Products,* 5th ed., Williams & Wilkins, Baltimore, 1984.
3. A. Hamilton, H. L. Hardy, and A. J. Finkel, *Industrial Toxicology,* 4th ed., John Wright-PSG, Inc. Littleton, MA, 1982.
4. N. V. Steere, *CRC Handbook of Laboratory Safety,* 2nd ed., CRC Press, Boca Raton, FL, 1971.
5. M. Sittig, *Hazardous and Toxic Effects of Chemicals,* Noyes Data Corp., Park Ridge, NJ, 1979.
6. M. Sittig, *Handbook of Toxic and Hazardous Chemicals,* Noyes Data Corp., Park Ridge, NJ, 1981.

7. *Sax's Dangerous Properties of Industrial Materials,* 8th ed., R. J. Lewis, Ed., Van Nostrand Reinhold, New York, 1992.
8. *Bretherick's Handbook of Reactive Chemical Hazards,* 5th ed., G. Urben, Ed., Butterworths, London, 1995.
9. Mervyn L. Richardson, Ed., *Dictionary of Substances and Their Effects (DOSE),* The Royal Society of Chemistry, London, 1992–1994.
10. H. H. Fawcett, *Hazardous and Toxic Materials,* Wiley, New York, 1988.
11. American Public Health Association, *Standard Methods for the Examination of Water and Wastewater,* 19th ed., Washington, DC, 1995.
12. *Proceedings of the Industrial Waste Conference,* Purdue University, Ann Arbor Press, Ann Arbor, MI, annual.
13. T. Reichart, "Environment Online," *Environmental Science and Technology,* **30**(2), 76A–81A (Feb. 1996).
14. *Dow's Fire and Explosion Index Hazard Classification Guide,* 5th ed., Publication T-71, American Institute of Chemical Engineers, New York, 1981.
15. National Fire Protection Association: (a) *Manual for Hazardous Chemicals Reactions* (NFPA 491M); (b) *Hazardous Chemical Data* (NFPA 49); (c) *Fire Protection Guide on Hazardous Materials,* 11th ed., Quincy, MA, 1994.
16. National Fire Protection Association, *Fire Protection for Laboratories Using Chemicals* (Publication 45), Quincy, MA, 1996.
17. A. J. Leo, "The Octanol–Water Partition Coefficient: *J. Chem. Soc., Perkin Trans. II,* 825–838 (1983); See also A. J. Leo, "Calculating log Poct from Structures," *Chem. Rev.* **93**(2) 1281–1306 (1993); C. Hansch and A. J. Leo, *Exploring QSAR,* American Chemical Society, Washington, DC. Volume 2 includes tables of some 17,000 octanol–water partition coefficients and an extensive listing of electronic and steric parameters utilized in the design and study of bioactive organic compounds. See also A. J. Leo, "Calculation of Partition Coefficients Useful in Evaluation of Relative Hazards in the Environment, in *Structure–Activity Correlations in Studies of Toxicity and Bioconcentration with Aquatic Organisms,* International Joint Commission Symposium Proceedings, 1975 and "The Calculation of Pesticide Hydrophobicity by Computer" in *Classical and Three-Dimensional QSAR in Agrochemistry,* American Chemical Society Symposium Series 606, Chapter 5, Washington DC, 1994.
18. S. C. DeVito, "Designing Safer Chemicals: Toxicological Considerations," *CHEMTECH,* **26**(11), 3447 (1996); S. C. DeVito and R. L. Garrett, Eds., *Designing Safer Chemicals:* Green Chemistry for Polution Prevention, American Chemical Society, Washington, DC, 1996.

List of Addresses

American Conference of Governmental Industrial Hygienists
1330 Kemper Meadow Drive
Cincinnati, OH 45240
(513-742-2020)
(http://www.acgih.org)

American Industrial Hygiene Association
2700 Prosperity Avenue, #250
Fairfax, VA 22031
(703-849-8888)
(http://www.aiha.org/hq.html)

American Public Health Association
1015 15th Street, NW
Washington, DC 20005
(202-789-5600)
(www.apha.org)

ASTM American Society for Testing and Materials
100 Barr Harbor Drive
West Conshohocken, PA 19428
(610-832-9585)
(http://www.astm.org)

BIOSIS
2100 Arch Street
Philadelphia, PA 19103-1399
(610-832-9605)

Bureau of National Affairs
1231 25th Street
Washington, DC 20037
(http://www.bna.com)

Business Publishers, Inc.
951 Pershing Dr.
Silver Spring, MD 20910
(800-274-6737)
(http://www.bpinews.com)

Cambridge Scientific Abstracts
Suite 601
7200 Wisconsin Avenue
Bethesda, MD 20814
(http://www.csa.com)

Genium Publishing Corp.
Genium Plaza
Schenectady, NY 12304-4690
(http://www.genium.com)

Information Handling Services
15 Inverness Way East
Englewood, CO 80150
(303-790-0600; 800-241-7824)
(http://www.ihs.com)

National Fire Protection Association
Batterymarch Park
Quincy, MA 02269

National Safety Council
1121 Spring Lake Drive
Itasca, IL 60143
(630-285-1121)

Technical Database Services
135 West 50th Street
New York, NY 10020
(212-245-0044)

U.S. Environmental Research Laboratory
6201 Congdon Boulevard
Duluth, MN 55804
(218-720-5548)

U.S. Geological Survey
12201 Sunrise Valley Drive
Reston, VA 20192
(703-648-4000)

U.S. National Institute for Occupational Safety and Health (NIOSH)
4676 Columbia Parkway
Cincinnati, OH 45226
(800-35-NIOSH; 513-533-8471)

VCH Publishing Group
Division of John Wiley & Sons
333 7th Avenue
New York, NY 10001
(212-629-6200)

WRC
Henley Road
Medmenham, Marlow
Buckinghamshire SL7 2HD,
England

15 Locating and Using Physical Property and Related Data

15.1. INTRODUCTION

Most chemists and engineers find it difficult to locate physical property and related data for new or little-studied chemicals or chemical systems. The task of locating such data can be tedious, expensive, and often fruitless. The same difficulty applies to location of critically evaluated data for almost any chemical—many chemists and engineers find these among the most difficult-to-locate kinds of chemical information. Location of evaluated, accurate thermodynamic data is especially challenging. If data are inaccurate or not fully understood, this can lead to erroneous conclusions or to syntheses or processes that do not work as expected. Proper use of accurate data is important in the design and engineering of pilot plants and plants that will safely produce chemicals of high quality at the desired yield. Laboratory chemists doing synthesis work find good physical property data equally important in their work. Further, accurate data and correct interpretation of data are vital to major policy decisions at the national level, such as on environment or energy matters.

The problem is complicated by the current policy of some journals not to publish extensive tables of data or full experimental details in an attempt to conserve printing and other costs. The practice has been to make these data and details available in microfilm or other microforms that need to be purchased separately by readers if they are interested. Many scientists consider this a false economy. Fortunately, there is a trend to make such auxiliary data accessible on the Internet.

Another factor is that some scientists are reluctant to evaluate the results of their work critically, or even to provide sufficient information to permit *others* to recognize or evaluate the accuracy of their work. Some estimates are that for about 50% of data reported in scientific literature too little information has been reported for independent evaluation. Even with all the sources described in this chapter, as well as many other sources, it is probably correct to speculate that most properties that scientists need are not available in the literature. This is because there are over 17 million known chemical substances and an uncounted number of mixtures. Nevertheless, informed use of the literature can be extremely helpful in any investigation of properties.

15.2. CATEGORIES OF DATA

The sources and methods reviewed in this chapter can be grouped into several categories:

1. *Original Sources.* These are organizations or systems that generate or critically evaluate data. The data centers of the National Institute of Standards and Technology (NIST) are examples. These and other centers provide a valuable service.
2. *Secondary and Tertiary Sources.* These are compilations of data from original sources and news about projects and activities. Examples include handbooks, journals, and books, including compilations and reports. Abstracting and indexing services, especially *CA,* are valuable for accessing pertinent data, and there are several special bulletins and newsletters in the field.
3. *Evaluation and Estimating.* In this field particularly, qualified chemists and engineers may need to do some of their own evaluation and perform some calculations using appropriate computer programs. Some organizations have designated specialists who concentrate on this activity.

15.3. RECORDING PROPERTY DATA

When any one property of a chemical is being searched for and is located, the chemist should make note of other properties (or at least the source of such properties). Experience has shown that this can save considerable future work, because several properties of a single chemical are often grouped together in one paper or other source.

Many chemists, in addition to recording physical property data in their laboratory notebooks, also make use of a specially designed property data form. Such a form should be kept on file for each chemical and should have separate columns for recording kinds of properties, values obtained, source, and data of source. Recently, software for personal computers has become available for this purpose, which simplifies the task.

If a chemist is part of a team or group working together on the same project, all completed data sheets or computer files should be kept in a central place. This helps avoid repetitive efforts in which different individuals may look for the same values. A central repository for data is especially important in industry. For example, completed data sheets often become part of a process manual used by engineers to design and build pilot plants and plants. For all commercial chemicals, government regulations in the United States and other countries require use of a standard form (Material Safety Data Sheet) with emphasis on properties related to safe handling. This form can be adapted (expanded) to include a complete range of properties about which an organization wishes to maintain a verified, centralized record for all chemicals of interest. Again, rapid progress is being made in computerizing this process.

15.4. NIST STANDARD REFERENCE DATA PROGRAM

The NIST Standard Reference Data Program (SRDP) is one of the key sources available to the chemist and engineer for obtaining reliable physical property and related data. The SRDP was first established in 1963 as a means of coordinating, on a national scale, production and dissemination of critically evaluated reference data in the physical sciences.

Under the Standard Reference Data Act (1968), the National Bureau of Standards (now known as National Institute of Standards and Technology or NIST) of the U.S. Department of Commerce was given the primary responsibility in the federal government for providing reliable scientific and technical reference data. NIST headquarters are located in large campuslike facilities in Gaithersburg, MD, near Washington, DC.

The program coordinates with data evaluation centers located in university and other centers as well as administering data centers within NIST. The data centers compile and critically evaluate numerical data on physical, chemical, material, and engineering properties. The National Standard Reference Data System (NSRDS) comprises the network of data centers, projects, and cooperative programs administered or coordinated by NIST.

The primary aim of the SRDP is to provide critically evaluated numerical data, in convenient and accessible form, to the scientific and technical community of the United States. A second aim is to advance the level of experimental measurements by providing feedback on sources of error in various measurement techniques. Through both these means, the SRDP strives to increase the effectiveness and productivity of research, development, and engineering design.

The technical scope of SRDP has been changed somewhat in recent years, yet still emphasizes physical and chemical properties of substances and systems that are well characterized. Newer activities deal with properties of engineering materials and systems, with critical evaluation the focus.

A. Program Structure

Current activities in the SRDP are conducted in data centers and outside projects located in technical divisions of NIST and in some academic and industrial laboratories. Each activity collects and evaluates available data on a specified set of properties and substances. Activities are grouped into several discipline-oriented program areas.

Principal output of the SRDP consists of computer databases and published compilations of evaluated data and critical reviews of the status of data in particular technical areas. Evaluation of data implies a careful examination, by an experienced specialist, of all published measurements of the quantity in question, leading to the selection of a recommended value and a statement concerning its accuracy or reliability. The techniques of evaluation depend on the data in question but generally include an examination of the method of measurement and the characterization of the materials, a comparison with relevant data on other properties and materials, and a check for consistency with theoretical relationships. Adequate documentation is provided for the selections of recommended values and accuracy estimates.

Data from the SRDP are disseminated in several ways:

National Standard Reference Database Series—databases on diskettes and CD-ROMs. Some NIST databases of interest to chemists have been available for some time through the online STN system (see Chapter 6 and 7). Plans call for the NIST World Wide Web Server to provide many of its Standard Reference databases in chemistry through the Internet address http://www.nist.gov/srd; some of the data will be free, whereas for other databases, a charge is anticipated. A partial list of NIST Database products is shown in Sections 15.4.A.1 and 15.4.A.2 (next two sections); all products are listed at the Internet address given above. Distribution of databases is from: NIST, SRDP, 820/113, Gaithersburg, MD 20899; phone 301-975-2208; e-mail SRDATA@nist.gov. Examples of online NIST databases currently available on the Web include *Physical Reference Data, Electronic Interactions with Plasma Processing Gases,* and the *NIST Chemistry WebBook.* The latter contains thermodynamic data for over 5000 compounds and ion-energetics data for over 10,000 compounds.

Journal of Physical and Chemical Reference Data—journal published jointly with the American Chemical Society and the American Institute of Physics.

Other publications are journal articles and books published by technical society and private publishers.

1. NIST Standard Reference Databases

1. NIST/EPA/NIH Mass Spectral
2. NIST Chemical Thermodynamics
3. NIST Crystal Data Identification File
4. NIST Thermophysical Properties of Hydrocarbon Mixtures
5. NIST Electron and Positron Stopping Powers of Materials—available from the JCPDS, Newton Square Corporate Campus. 12 Campus Blvd., Newtown Square, PA 19073-3273 (610-325-9810)
6. NIST X-Ray and Gamma-Ray Attenuation Coefficients and Cross Sections
7. NIST Thermophysical Properties of Water
8. DIPPR Data Compilation of Pure Compound Properties
9. DIPPR Data Compilation Access Program II-Student DIPPR
10. NIST Thermophysical Properties of Pure Fluids
11. NIST JANAF Thermochemical Tables
12. NIST Mixture Property Program
13. NIST/Scandia/ICDD Electron Diffraction—available from the JCPDS-International Center for Diffraction Data, Newtown Square Corporate Campus, 12 Campus Blvd., Newton Square, PA 19073-3273 (610-325-9810)
14. NACE/NIST Corrosion Performance—available from NACE, PO Box 218340, Houston, TX 77218 (713-492-0535)
15. NIST Chemical Kinetics
16. NIST Estimation of the Thermodynamic Properties for Organic Compounds at 298.15 K

17. (a) NIST Positive Ion Energetics with Structures and Properties Software; (b) NIST Negative Ion Energetics
18. NIST X-Ray Photoelectron Spectroscopy
19. NIST/NASA/CARB Biological Macromolecule Crystallization
20. NIST Tribomaterials Database
21. NIST Thermodynamic Properties of Refrigerants and Refrigerant Mixtures
22. NIST Structures and Properties Database and Estimation Program
23. NIST Vibrational Electronic Energy Levels of Small Polyatomic Transient Molecules
24. NIST Molten Salts
25. NIST Structural Ceramics
26. Phase Diagrams for Ceramists—available from the American Ceramic Society, 735 Ceramic Place, Westerville, OH 43081 (614-890-4700)
27. Lipid Thermotropic Phase Transitions Database (LIPIDAT2)
28. NIST/EPA Gas Phase Infrared
29. NIST/NIH Desktop Spectrum Analyzer Program and X-Ray Database
30. NIST Spectroscopic Properties of Atoms & Atomic Ions
31. NIST Wavenumber Calibration Tables
32. NDR/NIST Solution Kinetics
33. PICT/NIST Heat Capacities of Liquid Hydrocarbons
34. NIST Surface Structure
35. DIPPR/NIST Activity and Osmotic Coefficients in Aqueous Solutions
36. GRI/NIST Orifice Meter Discharge Coefficient
37. NIST Critically Selected Constants of Metal Complexes
38. NIST Tribo-Ceramic Materials
39. CYCLE D:NIST Vapor Compression Cycle Design Program

2. *NIST Special Databases (Selected)*

1. Special Database 5. IVTANTHERMO-PC
2. Special Database 15. COMAR: International Databank on Reference Materials
3. Special Database 16. CHETAH: Chemical Thermodynamic and Energy Release Program
4. Special Database 20. NIST Scientific and Technical Document Database

The most recent product catalog of SRDP, at the Internet address previously mentioned, can be scanned to find out what is available. There is also a printed catalog. Also, SRDP publishes a newsletter, *SRD Data Link,* which has news about data activities, publications, meetings, and other activities related to data evaluation and dissemination. Further information on SRDP activities can be obtained through the office of the Chief, John R. Rumble, Jr.

B. Data Activities

Listed in the following sections are titles and project leaders for continuing data activities that receive at least a part of their support from SRDP. Data activities are listed in alphabetical order by name.

1. NIST Data Centers

Alternative Refrigerants and Applications—David A. Didion
Atomic Energy Levels—William C. Martin
Atomic Transition Probabilities—Wolfgang L. Wiese
Ceramics Machinability—Said Jahanmir
Chemical Kinetics—W. Gary Mallard
Chemical Structure and Properties—Stephen E. Stein
Chemical Thermodynamics—Eugene S. Domalski
Composites Permeability—Donald L. Hunston
Corrosion—E. Neville Pugh
Crystal and Electron Diffraction—Alan D. Mighell
Desktop Spectrum Analyzer—Robert L. Myklebust
Fire Data Management System—Jonathan W. Martin
Fluid Properties—Daniel G. Friend, NIST, Boulder, CO
Fundamental Constants—Barry N. Taylor
High Temperature Superconductors—Ronald G. Munro
Infrared Spectroscopy—Stephen E. Stein
Ion Kinetics and Energetics—Sharon G. Lias
Mass Spectrometry—Stephen E. Stein and Sharon G. Lias
Molecular Energy—Marilyn E. Jacox
Molecular Spectra—Frank J. Lovas
Phase Diagrams for Ceramists—Stephen W. Freiman
Photon and Charged Particles—Stephen M. Seltzer
Protein Crystallization—Gary Gilliland
Radiation Chemistry—Alberta B. Ross
Radiation Laboratory, University of Note Dame, Notre Dame, IN 46556
Surface Structure—Philip R. Watson, PWR Consulting, 1465 NW 15th Street, Corvallis, OR 97330
Structural Ceramics—Ronald G. Munro
X-Ray Photoelectron Spectroscopy—Cedric J. Powell

(*Note:* Except as otherwise indicated, all of these data centers are located at NIST, Gaithersburg, MD.)

Some details on the centers at Texas A&M University and at Purdue University, presented in the next sections, may help illustrate typical data center activity.

15.5. THERMODYNAMICS RESEARCH CENTER

One of the best known data centers in the world is the Thermodynamics Research Center (TRC), Texas A&M University, College Station, (http://www.trcweb.tamu.edu) directed by Kenneth N. Marsh. This center is notably strong in data for organics; inorganics are also covered but not extensively. Some 14 technical and support staff conduct the work.

Two major efforts underway at TRC are described here:

1. TRC Thermodynamic Tables—Hydrocarbons (formerly American Petroleum Institute Research Project 44), which is one of the most complete compilations of physical and thermodynamic data available on these compounds. The project began in 1942 at the National Bureau of Standards with financial support from the American Petroleum Institute (API), under the direction of F. D. Rossini. In 1950 it was moved to the Carnegie Institute of Technology, Pittsburgh. In 1961, Bruno J. Zwolinski assumed the directorship, and the project was moved to Texas A&M University. K. N. Marsh became director in 1985. The material is published in looseleaf form and there are two supplements per year. There are a total of over 5200 tables and more than 2000 reference sheets. These tables are also available on diskettes for IBM-PC or compatible computers.

2. TRC Thermodynamic Tables—Non-Hydrocarbons (*Selected Values of Properties of Chemical Compounds*), which began in 1955 at the Carnegie Institute of Technology, under the auspices of the Manufacturing Chemists Association. Sponsorship changed July 1, 1966 to the Texas A&M TRC. The principal objective is the preparation, publication, and distribution of data on physical and thermodynamic properties of all classes of organic compounds (except hydrocarbons and related compounds covered in the Hydrocarbon Tables), as well as industrially important nonmetallic inorganic compounds. This, too, is one of the most complete compilations of physical thermodynamic data to be found on these compounds. These tables are also available on diskettes for IBM-PC or compatible computers.

The TRC Thermodynamic Tables include data for many chemical compounds of interest and importance to the chemical and petroleum industries, and to science in general. The tables are based on the best experimental data available. This information is critically evaluated and presented in a convenient form for ready use. Furthermore, data are estimated in temperature and pressure ranges that are not easily accessible in the laboratory. Data are often estimated for new compounds not yet synthesized but that may possess desirable properties.

In addition to its work on the Thermodynamic Tables, since the early 1970s, TRC has worked on a project designated as the *International Data Series,* which contains fully documented tabular experimental data for nonelectrolyte mixtures of organic compounds. Properties of mixtures from over 500 pure compounds are included. Data presented include physical and thermodynamic properties. The complete set contains more than 4800 tables published in 64 issues, with approximately 75 tables per issue. The editor is Kenneth N. Marsh.

Further, the TRC Selected Spectral Data Program is a significant reference for identification and formulation of organic chemicals in multicomponent mixtures. Substances included are primarily hydrocarbons, and their oxygen, halogen, sulfur, and nitrogen derivatives. Some silicon substances are included in the NMR portion of the collections.

Examples of other TRC titles include *Spectral Data for Steroids, Thermochemical Data and Structures of Organic Compounds, Thermodynamics of Organic Compounds in the Gas State,* and *Spectral Data for PCBs.*

The *TRC Comprehensive Index* (1) contains information on thermodynamic data and their location in the TRC Thermodynamic Tables (hydrocarbons and nonhydrocarbons), and International Data Series, Selected Data on Mixtures, Series A. Also included are spectral data in TRC Spectral Data publications. This index is available on diskette for IBM-PC-compatible computers and is updated annually.

TRC has other electronic products in addition to those mentioned above. One of these contains software and data to calculate and display tables of vapor pressure for more than 5900 organic, and metal-organic, and nonmetallic inorganic compounds. This product is designated as *TRC Vapor Pressure* (TRCVP). Other electronic databases contain experimental and evaluated data on thermophysical properties of pure fluids and mixtures as developed at the Polish Academy of Science in conjunction with a CODATA Task Group.

The *TRC Thermophysical Property Datafile: Vapor Pressure* is available online through Technical Database Services (TDS), New York, which also offers the information on diskette. The file consists of an experimental database for approximately 5000 organic and inorganic compounds with evaluated data. Also included is a calculated datafile for retrieving boiling points and to generate tables of "smooth" vapor-pressure data for 5500 organic and inorganic chemicals. TRC plans to make additional data available through TDS.

15.6. DESIGN INSTITUTE FOR PHYSICAL PROPERTY DATA (DIPPR)

Extensive physical property data are available through the databases and publications of Design Institute for Physical Property Data (DIPPR) of the American Institute of Chemical Engineers, 345 East 47th Street, New York, NY 10017 (212-705-7338). The first DIPPR projects were initiated in 1980, and there are now about 40 sponsor members, primarily major chemical companies from the United States and other leading industrialized nations. As might be expected, NIST is also extensively involved as a member. DIPPR is one of the largest and most important data evaluation projects in the world.

The purpose of DIPPR is to assemble, determine, and evaluate data on physical properties and predictive methods for selected chemicals and mixtures commonly employed in the chemical industry. An annual budget of approximately $1,000,000 supports 8 research projects, with the work being done in a number of laboratories in the United States and abroad. The technical director for many years (until his retirement at the end of 1995) has been Theodore B. Selover, Jr., formerly of SOHIO. The incoming director is George H. Thomson, formerly with Phillips Petroleum, now located in Santa Fe, NM.

A typical DIPPR project will involve a search and evaluation of existing physical property data, generation and compilation of new data when needed values are not in the existing literature, and creation of databases and publications that contain the results of the work.

Projects are sponsored by the members; funding of each project is year-to-year. Six of these projects are database projects (numerical data are collected, evaluated, and disseminated) and four are experimental projects [vapor–liquid equilibria, vapor pressure of pure liquids, critical properties of pure chemicals, and pure-component enthalpies of combustion (c,l) to derive ideal-gas enthalpies of formation]. Project results are typically made available to members first, and then to the public some 12–18 months later.

The largest and best known DIPPR project is "Data Compilation," with data on over 1600 chemicals released to the sponsors by the end of 1995, and over 1400 to the general public. The principal investigator (contractor) is Professor Thomas E. Daubert, Chemical Engineering Department, The Pennsylvania State University. Another professor at the same university who was active in this work in the past is Ronald P. Danner. In 1998, this work is to shift to Brigham Young University under Professor Richard L. Rowley. Results are available online (STN), in diskette form [from both Technical Database Services and NIST] and in published, hard-copy form (Taylor & Francis, Bristol, PA). The electronic form is the most complete. The five-volume looseleaf compilation contains more than 4000 pages of evaluated physical, thermodynamic, and transport property data for many of the most commonly used chemicals in industry, including 29 property constants and 15 temperature-dependent properties for each chemical.

Other active projects include

Experimental data on mixtures (vapor–liquid equilibria and liquid–liquid equilibria)

Pure-component liquid vapor pressure

Critical properties of pure compounds

Pure component ideal-gas enthalpies of formation

Evaluated data of mixtures

Handbook of diffusion and thermal properties for polymer solutions

Data prediction methods

In addition to these there are two projects related to environment, health, and safety underway at the Chemical Engineering Department, Michigan Technological University. The focus of the first project is a numeric database containing 54 properties of about 500 chemicals from the Clean Air Act and other federal regulatory lists; a second related project relates to estimating these properties. The former project is scheduled to make its first public release in 1998 with data on 238 chemicals in the Clean Air Act and OSHA list.

Other key publications, in addition to the Data Compilation series, include

- *Manual for Predicting Chemical Process Design Data.* (Recommended methods for predicting physical, thermodynamic, and transport properties of pure chemicals and mixtures of defined composition. Emphasis is nonhydrocarbon polar chemicals.)
- *Thermodynamic Analysis of Vapor-Liquid Equilibria—Recommended Models and a Standard Data Base* (with diskettes)
- *Handbook of Polymer Solution Thermodynamics* (with diskettes).
- *Handbook of Aqueous Electrolyte Thermodynamics.*
- *DIPPR/NIST Activity and Osmotic Coefficients in Aqueous Solutions Database* (diskettes).
- *Transport Properties and Related Thermodynamic Data of Binary Mixtures.* Volumes 1 and 2 are already available; Volumes 3 and 4 issued in 1996 and 1997, respectively.
- *Experimental Results for DIPPR 1990–1991 Projects on Phase Equilibria and Pure Component Properties* (DIPPR Data Series No. 2). Publication of this series has been discontinued and has been replaced by publication of these results in the *Journal of Chemical and Engineering Data.*

15.7. AMPTIAC AND CINDAS

AMPTIAC, Rome, NY, began operations on November 1, 1996 as the location for the four materials Information Analysis Centers (metals, ceramics, high-temperature materials, metals matrix composites) previously operated by CINDAS. The AMPTIAC charter includes the functions and services of these analysis centers, and includes additional coverage as well.

AMPTIAC is sponsored by the U.S. Defense Technical Information Center, Fort Belvoir, VA. Operated by the IIT Research Institute, and directed by Steven J. Flint, AMPTIAC consolidates the four preexisting U.S. Department of Defense (DOD) Information Analysis Centers (IACs) mentioned above into a single center. The scope of AMPTIAC includes the following materials components: ceramic and ceramic composites; organic structural and matrix composites; monolithic metals, alloys, and metal-matrix composites; electronic, optical, and photonic materials; and environmental protection, and special-function materials such as advanced coatings and fire-retardant materials.

The mission is to collect, analyze, and distribute information on these technologies for the armed forces, defense contractors, and other authorized users. The public at large may submit inquiries to the facility at no charge, if the question is simple, or on a fee basis, in the case of more complex inquiries. The staff includes materials, metallurgical, and mechanical engineers; physicists; chemists; and information professionals.

All materials properties are included. Applications and process technologies that are covered include material applications; processes of materials; material processing equipment; measurement and testing of materials; quality control; and corrosion–deterioration detection, prevention, and control. Also within AMPTIAC scope are the economic aspects, as are the advanced materials and processes science and technology related to information, documentation, databases, and similar materials.

Some fairly recent publications that have unlimited distribution include the following: *Handbook on Continuous Fiber-Reinforced Ceramic Matrix Composites; Recent Developments in Piezoelectric Ceramic Composites; Availability of Data on the Thermophysical, Thermoradiative, Electronic, Electrical, Magnetic and Optical Properties of Refractory Borides, Carbides, and Silicides; Optical, Thermoradiative, Thermophysical, and Mechanical Properties of Silicon;* and *Properties of Intermetallic Alloys.* Readers who need a complete listing of what is available can consult the Internet Home Page of AMPTIAC at http://rome.iitri.com/amptiac.

CINDAS, located at Purdue University, has a 40-year history of materials property data research, database development, and implementation. The results are published in a variety of books or in the form of electronic media. Examples include forthcoming books on the properties of stainless steels and on methods for thermal expansion measurement, and the already-available, comprehensive 13-volume set *Thermophysical Properties of Matter: The TPRC Data Series.* Other examples include the ongoing CINDAS Data Series on Material Properties, Databooks on Intermetallics, and Databook on Ceramics. Under contract with the Semiconductor Research Corporation, CINDAS develops and upgrades a microelectronic packaging materials database that is available to member companies of SRC.

CINDAS also produces the multivolume *Masters Theses in the Pure and Applied Sciences Accepted by Colleges and Universities of the United States and Canada* (1957 to date). This annual publication covers masters theses in the physical sciences and engineering; theses titles are grouped into 44 categories, one of which relates to chemistry and biochemistry.

CINDAS also manages the updating of publication of handbooks under an agreement with the U.S. Air Force. This includes the five-volume *Aerospace Structural Metals Handbook,* the three-volume *Structural Alloys Handbook,* and the five-volume *Damage Tolerant Design Handbook.*

15.8. COMMITTEE ON DATA FOR SCIENCE AND TECHNOLOGY

At the international level, the Committee on Data for Science and Technology was created in 1966 as a scientific committee of the International Council of Scientific Unions to aid in international coordination of data compilation, evaluation, and dissemination. Material available from CODATA includes key sets of recommended data, guidelines for presentation of data, and articles on methodology of data evaluation. Products of specific chemical interest from CODATA include fundamental physical constants, thermodynamic properties of key substances, and chemical kinetic data.

CODATA has also published directories to sources data on various topics in both hard-copy and electronic forms. Examples include the *CODATA Directory of Data Sources for Science and Technology: Chemical Kinetics* (Chapter 6, 1981) and *Chemical Thermodynamics* (Chapter 11, 1984). CODATA has available on IBM-compatible diskettes its *Referral Database,* which includes 1400 computer-searchable sources of scientific and technical data.

CODATA headquarters are located in Paris, France. The director of the U.S. national committee for CODATA is Paul Uhlir, National Academy of Sciences, Washington, DC (202-334-3061).

15.9. *JOURNAL OF CHEMICAL AND ENGINEERING DATA* AND OTHER JOURNALS

The *Journal of Chemical and Engineering Data* (*JCED*) published since 1956 by the ACS, and edited by Kenneth N. Marsh, is a good example of a current and original source of reliable physical property data for both pure compounds and mixtures. A quick scan of indexes to this publication and of the table of contents of its more recent issues can sometimes produce more useful information than can an equivalent amount of time spent perusing the printed version of online *CA*. No search for the most recent physical property data for specific compounds can be considered complete without consulting *JCED* either directly or through the use of indexing and abstracting services.

In addition to containing experimental or derived data relating to pure compounds or mixtures covering a range of states, *JCED* includes papers based on published experimental information that have made tangible contributions.

JCED sets a high standard. Its guide for authors specifies that experimental methods should be referenced or described in enough detail to permit duplication of the data by others familiar with the field. Published or standardized procedures and their simple modifications need not be described, but a readily available reference should be cited. The data should be presented with such precision that information may be easily obtained from within the paper, within the stated limits of uncertainty of the experimental background. This journal is cited as just one example of several in the field.

Another leading example of other journals in the field is the *Journal of Chemical Thermodynamics.* Others are published by the American Institute of Chemical Engineers, the American Institute of Physics, the American Chemical Society, and other sources. Note that while both scientific journals and trade magazines are rich sources of data, they often offer different kinds of data. Trade magazine data tend to deal more with applications as noted below.

Articles in scientific journals concerned primarily with preparative chemistry are good sources of property data, primarily *basic* properties such as melting point or refractive index. If access to such articles is made through abstracting and indexing services, the chemist must note that these properties are not necessarily indexed or mentioned in the abstract, since their presence is often implicitly assumed. On the other hand, trade-oriented magazines frequently have articles with information on properties affecting end use or applications, such as data relating to handling or processing. Location of these data using magazine indexes is unreliable because properties are frequently not indexed.

15.10. *LANDOLT–BÖRNSTEIN*

One of the most extensive printed sources of physical properties and related data is *Landolt–Börnstein* (3). This series was originated by Hans Landolt and Richard Börnstein in 1883. The sixth edition, consisting of 28 volumes, is almost entirely in German, although some volumes are in German and English. This edition was published from 1950 to 1980 and is now out-of-print. The main divisions of the sixth edition are:

1. *Atomic and Molecular Physics*—basic properties and interactions of nuclei, atoms, ions, molecules, and crystals (5 vols.).
2. *Properties of Matter in its Aggregated States*—mechanical and thermal properties, equilibria, interfaces, transport phenomena, general physicochemical data (8 vols.). Electrical, optical, magnetic properties (5 vols.).
3. *Astronomy and Geophysics*—(1 vol.).
4. *Technology*—nonmetallic materials (1 vol.), metallic materials (3 vols.), heat technology (4 vols.), and electrical engineering, light and X-ray technology (1 vol.).

The New Series volumes, currently with some 210 volumes, have been published totally in English for a number of years now. The editor-in-chief is Werner Martienssen. In 1996, it was planned to publish approximately 15 New Series volumes, and it can be assumed that, in the future, a similar number of volumes will be published.

The *Comprehensive Index 1996* is a printed index that comes with a CD-ROM that includes not only the *Comprehensive Index* but also the update of the *Substance Index 1993* (see paragraph below). The *Comprehensive Index* is an index to the sixth edition and to volumes of the New Series up to the end of 1995.

The *Substance Index 1993* contains the following subvolumes: (a) *Elements and Binary Substances,* (b) *Ternary Substances,* and (c) *Polynary Substances.* The scope is the sixth edition and the New Series to the end of 1992. Indication is given of which property is referred to.

The content of *Landolt–Börnstein* is not online, but there are plans to attach to all new volumes a CD (compact disk) that will contain the contents of the corresponding book. In addition, there is a help desk at the Internet site.

Landolt–Börnstein management apparently sees its compilation as complementary to the online databases rather than competitive. Thus, the purpose of the New Series volumes is said to be to provide the overall picture with exhaustive, critically edited data reviews. Original literature data are critically checked by specialists from all over the world. Data are presented along with the experimental conditions and methods of measurement and evaluation. Graphical presentations are utilized increasingly.

The main groups and topics covered in the New Series, reflecting recent changes, are

General Scientific Methods, Tools and Data
Elementary Particles, Nuclei and Atoms (formerly Nuclear and Particle Physics)—Group I
Molecules and Radicals (formerly Atomic and Molecular Physics)—Group II
Condensed Matter (formerly Solid State Physics)—Group III
Physical Chemistry (formerly Macroscopic Properties of Matter)—Group IV
Geophysics—Group V
Astronomy and Astrophysics—Groups VI
Biophysics—Group VII

15.11. HANDBOOKS

The most heavily used sources of physical property data are the *CRC Handbook of Chemistry and Physics* (4), originally published in 1913 and updated annually; *Lange's Handbook of Chemistry* (5), and *Perry's Chemical Engineer's Handbook* (6). These provide rapid, convenient, desktop access to many different kinds of data and can be purchased for prices within the budgets of most practicing chemists and engineers.

These and other desk handbooks usually contain more than just physical properties—a feature that enhances the value of most handbooks. For example, the essential reference volume *CRC Handbook of Chemistry and Physics* (4) includes such other materials as an extensive section on mathematical tables; information on sources of critical data; rules for chemical nomenclature; and recommendations on symbols, units, and terminology. A new edition of the *CRC Handbook of Chemistry and Physics* appears annually in midyear; thus, the 77th edition appeared in June 1996. The major changes that are made each year appear in the Preface. In recent years, the *Handbook* has undergone especially significant improvements under the outstanding editorship of David R. Lide, Jr. (formerly Chief, National Standard Reference Data Program), and now comprises approximately 2500 pages. Chemistry libraries in universities and colleges and corporate information centers in chemical companies will probably want to purchase each new annual edition as it appears. The purchase cost, now slightly over \$100, seems reasonable when the size and content of the volume are taken into account. The *Chemical Engineers Handbook* includes a section on mathematical tables, data on materials of construction and corrosion, and information on cost and profitability.

Chemists will find the *The Merck Index* (7) exceptionally useful. This source was first published in 1889, and recent editions have been published every 7–8 years. The 12th edition, published in 1996, contains descriptive information on more than 10,000 industrial and laboratory chemicals, drugs, and biological arranged alphabetically by generic, nonproprietary, trivial, or simple chemical name. Although the publisher is a major pharmaceutical company, coverage is across the board for many chemicals important in both commerce and research.

For most of the chemicals included, a few key physical properties are given. These may include, information on physical appearance, color, melting and/or boiling point, refractive index, flash point, and solubility. The editors say that when several values are found in the literature, the data are evaluated, and representative selections are made. In addition, information is included on toxicity and on general, medical, and veterinary uses. Furthermore, for most of the entries, references are given as to the first key preparative or product patents, and journal citations are also provided. Additional references, notably reviews, are provided when available, thereby offering good starting points in the literature for the reader.

The newest edition includes a section describing organic name reactions. This section was absent in the previous (11th) edition but has now been revised and reintroduced. There is a comprehensive name index for approximately 60,000 synonyms (chemical, trivial, and generic names as well as trademarks and drug codes), a formula index, a very handy CAS Registry Number index (by both chemical name and CAS number), an index of therapeutic category and biological activity, and a variety of useful tabular information such as conversion factors. Approximately 65% of the entries have structural depictions, and stereochemistry is included where applicable.

The extensive scope and low price of this book make it a purchase of unusually good value for any chemist. It is published on a nonprofit basis as a service to the scientific community. Susan Budavari is the editor. The publication is searchable online through Chemical Information System, DIALOG, Questel, Orbit, and STN. The online files are updated semiannually. A CD-ROM edition was offered in mid-1996 in cooperation with Chapman & Hall, London. Some of the electronic products offer substructure searching features in addition to full-text searching.

The volumes cited here are examples of the many kinds of handbooks of value to chemists.

Chemists and engineers should use the data in desk handbooks—although valuable, convenient, and indispensable from a practical day-to-day working standpoint—only with some appropriate cautions:

1. New data are constantly being generated and published in journals and other sources. Handbooks especially those not updated regularly, cannot (and do not) claim to provide the most recent data. Thus desk handbooks can be described as selective and sometimes out of date but not necessarily incorrect. The *Merck Index* is an exception since it is updated online.
2. Tables printed in most handbooks have not been subjected to direct and recent critical review and evaluations. Hence the data are not necessarily the *best* data—an important factor in many chemical investigations.
3. Details of conditions under which data were originally generated usually cannot be given in handbooks. Limitations are not always evident. If complete references to the original sources of the data are also missing, this further complicates attempts at data evaluation by handbook users.

4. The latest editions of some handbooks omit useful data found in earlier editions. It is worthwhile to keep on hand or consult at least one prior edition.

Even with these and other limitations, the chemist or engineer will doubtless continue to make heavy use of handbooks as convenient sources of information for investigations in which the highest degree of up-to-dateness, accuracy, or precision is not essential. Handbooks, such as those mentioned, serve as good starting points when the latest best, and most complete data are not essential.

Please see also Section 15.13 which describes the value of some chemical company catalogs in providing handbook-like data.

15.12. EXAMPLES OF OTHER REFERENCE SOURCES

Some key reference sources are noted here. Some of the descriptions are as given in NBS Special Publication 454 (8), *Compiled Thermodynamic Data Sources for Aqueous and Biochemical Systems,* the successor to which is NBS Special Publication 685 (9).

J. D. Cox, D. D. Wagman, and V. A. Medvedev, *CODATA Recommended Key Values for Thermodynamics,* Hemisphere Publishing, (now part of Taylor and Francis), Philadelphia, PA, 1989.

T. E. Daubert, R. P. Danner, H. M. Sibul, and C. C. Stebbins, *Physical and Thermodynamic Properties of Pure Compounds: Data Compilation,* extant 1994 (core with 4 supplements), Taylor & Francis, Bristol, PA.

E. S. Domalski, W. H. Evans, and E. D. Hearing, "Heat Capacities and Entropies of Organic Compounds in the Condensed Phase," *J. Phys. Chem. Ref. Data,* **13**(Suppl. 1) (1984); **19**(4), 881–1047 (1990).

L. Haar, J. S. Gallagher, and G. S. Kell, *NBS/NRC Steam Tables,* Hemisphere Publishing, New York, 1984. Accurate representation of data for thermodynamic properties of water and steam from triple point to 1000°C and for pressures up to at least 10 bars.

C. H. Horsley, Ed., *Azeotropic Data—III,* American Chemical Society, Washington, DC, 1973. Data on azeotropes, nonazeotropes, and vapor-liquid equilibria for more than 17,000 systems. No attempt has been made to evaluate accuracy of the data, most of which are from the original literature. This volume is a revision of *Azeotropic Data I* and *II* and includes new data collected since 1962.

A. S. Kertes, Ed., *Solubility Data Series,* Pergamon, Oxford, various years. Series of critically evaluated solubility data volumes planned to cover all physical and biological systems. Sponsored by IUPAC (International Union of Pure and Applied Chemistry).

D. R. Lide, *Handbook of Organic Solvents,* CRC Press, Boca Raton, FL, 1995.

D. R. Lide and H. V. Kehiaian, *CRC Handbook of Thermophysical and Thermochemical Data,* CRC Press, Boca Raton, FL, 1994. Includes computer disk that permits calculation of properties as a function of temperature.

D. R. Lide and G. W. A. Milne, Eds., *Handbook of Data on Organic Compounds,* 3rd ed., CRC Press, Boca Raton, FL, 1994. Also available as a CD-ROM database.

W. F. Linke and A. Seidell, *Solubilities: Inorganic and Metal-organic Compounds—A Compilation of Solubility Data from the Periodical Literature,* Vol. I: A–Ir, Van Nostrand, Princeton, NJ, 1958; Vol. II; K–Z, American Chemical Society, Washington, DC, 1965. Comprehensive compilation of mostly unevaluated solubility data for inorganic and metal-organic compounds. Both aqueous and nonaqueous solvent systems are included. References are given to the data sources.

W. J. Lyman, W. F. Reehl, and D. H. Rosenblatt, *Handbook of Chemical Property Estimation Methods,* American Chemical Society, Washington DC, 1990.

V. Majer and V. Svoboda, *Enthalpies of Vaporization of Organic Compounds,* Blackwell Scientific Publications, Oxford, 1985.

K. N. Marsh, Ed., *Recommended Reference Materials for the Realization of Physicochemical Properties,* Blackwell Scientific Publications, Oxford, 1987.

J. B. Pedley, R. D. Naylor, and S. P. Kirby, *Thermochemical Data of Organic Compounds,* 2nd ed., Chapman & Hall, London, 1986.

Physical Constants of Hydrocarbon and Non-Hydrocarbon Compounds, ASTM Data Series DS 4B, ASTM, Philadelphia, 1988.

R. C. Reid, J. M. Prausnitz, and B. E. Poling, *The Properties of Gases and Liquids,* 4th ed., McGraw-Hill, New York, 1987.

J. A. Riddick, W. B. Bunger, and T. K. Sakano, *Organic Solvents,* 4th ed., Wiley, New York, 1986.

H. Stephen and T. Stephen, *Solubilities of Inorganic and Organic Compounds* (5 vols.), Pergamon, London, 1963. Selection of data on the solubilities of elements, inorganic compounds, and organic compounds in binary, ternary, and multicomponent systems. References are given to sources of data in the literature. The data are unevaluated.

R. M. Stevenson and S. Malanowski, *Handbook of the Thermodynamics of Organic Compounds,* Elsevier, New York, 1987.

T. S. Storvick and S. I. Sandler, Eds., *Phase Equilibria and Fluid Properties in the Chemical Industry—Estimation and Correlation,* American Chemical Society, Washington, DC, 1977. A symposium volume containing state of the art reviews.

D. R. Stull, "Vapor Pressure of Pure Substances. Organic Compounds," *Ind. Eng. Chem.,* **39,** 517–540 (1947); "Vapor Pressure of Pure Substances. Inorganic Compounds," *Ind. Eng. Chem.,* **39,** 540–550 (1947). Evaluated vapor-pressure data on over 1200 organic and 300 inorganic compounds.

D. R. Stull, E. F. Westrum, and G. C. Sinke, *The Chemical Thermodynamics of Organic Compounds,* Wiley, New York, 1969. Monograph divided into three parts. The first gives the theoretical basis and principles of thermodynamics and thermochemistry, some experimental and computational methods used, and some applications to industrial problems. The second part gives thermal and thermochemical properties in the ideal gas state from 298 to 1000 K. In this section the sources of data are listed and discussed, and standardized tables are presented for 918 organic compounds. Values of C_p°, S°, $-(G - H_{298}^\circ)/T$, $H^\circ - H_{298}^\circ$, ΔH_f°, ΔG_f°, and log K_p are given at 100 K intervals. In the third section are listed selected values of enthalpy of formation, entropy, and consistent values of ΔG_f° and log K_p of organic compounds at 298 K. More than 4000 compounds are listed, including a few inorganic compounds. A chapter briefly discusses methods of estimating thermodynamic quantities.

D. D. Wagman, W. H. Evans, V. B. Parker, R. H. Schumm, I. Halow, S. M. Bailey, K. L. Churney, and R. L. Nutall, *The NBS Tables of Chemical Thermodynamic Properties—Selected Values for Inorganic and C_1 and C_2 Organic Substances in SI Units,* American Chemical Society, Washington, DC, 1983. Published as Supplement 2 to Volume 11 of the *Journal of*

Physical and Chemical Reference Data, this volume lists 26,000 values for chemical thermodynamic properties of over 14,000 substances, counting each substance in each phase and each concentration listed for a solution. This is a comprehensive, updated edition of the *NBS Technical Note 270* series that appeared in eight parts between 1965 and 1981 and is a modern version of the classic *NBS Circular 500* that was published in 1952. An estimated 60,000 references were used in development of the tables, which include standard state data on enthalpy, entropy, Gibbs energy, and heat capacity. Properties of aqueous solutions and pure compounds are also included. All data have been critically evaluated and checked for consistency with thermodynamic constraints by specially developed computer programs. The purpose of presenting these tables of evaluated data is to provide scientists and engineers with reliable values of chemical thermodynamic properties of the elements and their compounds, from which calculations can be made of equilibrium constants and changes in enthalpies, entropies, and heat capacities for chemical processes. These data can be used in such areas as chemical engineering design, environmental modeling, and chemical research. The tables contain values, where known, of the enthalpy and Gibbs energy of formation, entropy and heat capacity at 298.15 K (25°C), the enthalpy difference between 298.15 and 0 K, and the enthalpy of formation at 0 K, for inorganic substances and for organic substances containing one or two carbon atoms. In some instances, such as complexes with organic ligands and metal–organic compounds, data are given for substances in which each organic ligand contains one or two carbon atoms.

I. Wichterle, J. Linek, and Z. Wagner, *Vapor-Liquid Equilibrium Bibliographic Database,* ELDATA SARL, Montreuil, France, 1993.

15.13. MANUFACTURER TRADE LITERATURE

Manufacturer trade literature is another important source of physical properties, and is, in fact, a key source of many types of chemical information. The literature available on any specific chemical or series of related chemicals can be extensive and will typically include data not only on properties but also on the chemistry, applications, analytical methods, safe handling, and environmental aspects. Material Safety Data Sheets (MSDSs) are also available from all manufacturers. Some data from manufacturers may not be readily available elsewhere. However, it is crucial to note the specific grade for which the properties are given and to also remember that most properties found in trade literature will not have been subjected to critical evaluation by outside specialists. The quality of the data varies with the company.

A few pieces of chemical trade literature are, in effect, like handbooks, and this fact should be of special interest to students and others who cannot afford to purchase handbooks. For example, the extensive product catalogs published by such firms as Alrich, Sigma, Fluka, and Lancaster are in this category.

The Aldrich Chemical Company catalog, which is published every other year (PO Box 355, Milwaukee, WI 53201-9358) is over 2 in. thick and includes over 25,000 chemicals manufactured or distributed by the company. It provides a considerable amount of data about many of these chemicals, including not only selected physical properties but also references to *Beilstein* and other literature sources in some cases. The content of entries from an Aldrich catalog is shown in Figure 15.1; however, not

Key to chemical listings

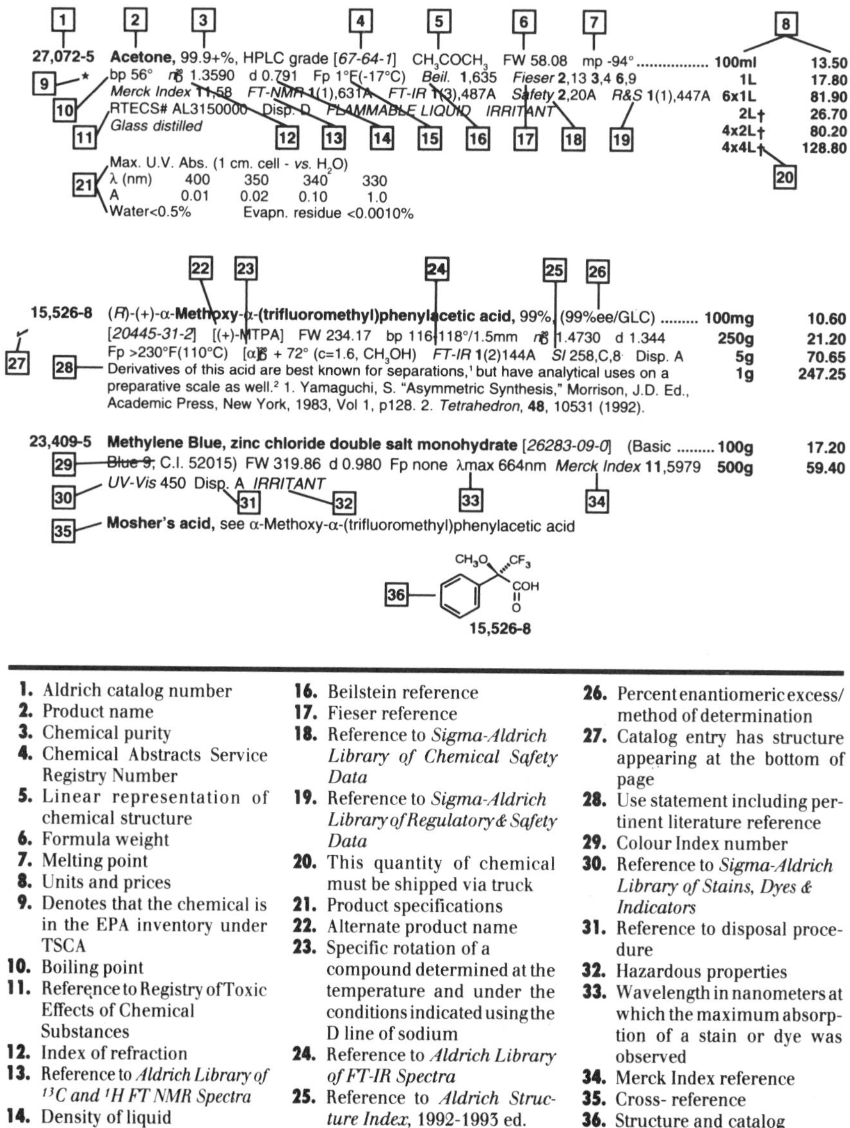

1. Aldrich catalog number
2. Product name
3. Chemical purity
4. Chemical Abstracts Service Registry Number
5. Linear representation of chemical structure
6. Formula weight
7. Melting point
8. Units and prices
9. Denotes that the chemical is in the EPA inventory under TSCA
10. Boiling point
11. Reference to Registry of Toxic Effects of Chemical Substances
12. Index of refraction
13. Reference to *Aldrich Library of ^{13}C and ^{1}H FT NMR Spectra*
14. Density of liquid
15. Flash point
16. Beilstein reference
17. Fieser reference
18. Reference to *Sigma-Aldrich Library of Chemical Safety Data*
19. Reference to *Sigma-Aldrich Library of Regulatory & Safety Data*
20. This quantity of chemical must be shipped via truck
21. Product specifications
22. Alternate product name
23. Specific rotation of a compound determined at the temperature and under the conditions indicated using the D line of sodium
24. Reference to *Aldrich Library of FT-IR Spectra*
25. Reference to *Aldrich Structure Index*, 1992-1993 ed.
26. Percent enantiomeric excess/ method of determination
27. Catalog entry has structure appearing at the bottom of page
28. Use statement including pertinent literature reference
29. Colour Index number
30. Reference to *Sigma-Aldrich Library of Stains, Dyes & Indicators*
31. Reference to disposal procedure
32. Hazardous properties
33. Wavelength in nanometers at which the maximum absorption of a stain or dye was observed
34. Merck Index reference
35. Cross- reference
36. Structure and catalog number

See Abbreviations & Acronyms for additional information

▪ F7 ▪

Figure 15.1. Contents of entries from Aldrich Catalog. (Reprinted with permission from Aldrich Chemical Company, Inc.)

all entries have all of these data. Aldrich specializes in supplying research and fine chemicals, broadly defined. Its product line, and hence its catalog, includes both organics and inorganics, as well as a wide range of laboratory supplies, books, and computer products. Aldrich chemicals in its catalog also include ACS reagents, monomers and polymers, pure elements, analytical standards, flavors and fragrances, stains and dyes, chiral compounds, and stable isotopes. The catalog is arranged alphabetically by chemical name and includes valuable indexes by molecular formula, CAS Registry Number, and product number. The company also has available a diskette version of the catalog.

The Sigma Chemical Company catalog includes over 36,000 products with a focus on biochemicals, organic compounds, and diagnostic reagents (PO Box 14508, St. Louis, MO 63178). Some inorganic chemicals are also included in the Sigma catalog. Its catalog is available on CD-ROM and includes an order entry module. Specialty catalogs are available in such areas as forensic chemistry, peptides and amino acids, and radiochemicals; these have further details beyond the main catalog and are arranged more conveniently.

Approximately 18,000 products are included in the Fluka catalog. Fluka Chemie AG is headquartered in Buchs, Switzerland, and has offices elsewhere, including 980 South 2nd Street, Ronkonkoma, NY 11779. Types of chemicals included are organic, biochemical, and analytical.

Aldrich, Sigma, and Fluka are affiliated companies. A new tool that is available is a CD-ROM that includes all the chemicals from all three companies, as well as from Supelco (140,000 chemical products and 90,000 searchable chemical structures). Structure and substructure searching is among the several search options possible with this product.

Another example of a valuable catalog is that of Lancaster Synthesis Inc. (PO Box 1000, Windham, NH 03087-9977). The focus is on organic reagents and intermediates, but inorganics and biochemicals are included. The catalog is distinguished by extensive reference to the literature, and it includes reaction schemes. Some 12,000 compounds are included.

Most major chemical manufacturers back up their trade literature with knowledgeable technical support people who can often provide additional details and more recent information about a chemical at no charge, especially if they believe that such information can help a caller make a more informed decision about potential purchase. These representatives are frequently found in the technical service department, and they often specialize in specific products or end uses. Toll-free telephone numbers are frequently available to facilitate connection with the technical support people. When calling a general customer service number, it is important to ask for a technical person (if the question is technical in nature), and to speak with a person who truly knows the product or end use. The best contact in many cases is the product manager for the specific chemical of interest.

Another example of extensive company trade literature is the *ICI Polyurethanes Book,* second edition, written by George Woods (Wiley, New York, 1990). This hardcovered book contains 364 pages, including 13 chapters, a bibliography, a glossary, and an index. The significant aspects of polyurethane chemistry and technology are covered here. Multicolored illustrations are included.

In almost all cases, company trade literature is provided at no charge to customers or potential customers. In the future, chemists can expect an increasing amount of company technical data to be made available at no cost on the Internet and, in addition, to be made available in diskette or CD-ROM form.

15.14. DATA QUALITY AND EVALUATION

Considerable attention is still being paid to data quality, in large measure because of intense concern about the environment and continued growth of online databases. It is all too easy to assume that just because data may come from a computerized database it is accurate and reliable. This may or may not be the case. The speed and ease of online access can lead to considerable misinterpretation.

In February 1982, the Chemical Manufacturers Association, jointly with U.S. government officials from such agencies as the NBS and EPA, sponsored a definitive workshop on data quality. One conclusion was that *data indicators* are needed to describe such factors as methods used to obtain the data, how the data have been evaluated, description of accuracy, and source. These data indicators, it is recommended, should accompany all data sources, especially online databases, so that users can make appropriate decisions. If there is an estimate of accuracy or certainty, this can be especially helpful.

On May 8–9, 1984, the Chemical Manufacturers Association, again jointly with U.S. government officials, held another workshop on the issue of quality under the chairmanship of Curtis Elmer, Monsanto Company. This time the focus was to develop indicators of data *documentation* quality, rather than data quality itself, because it was felt that the latter might be too ambitious at this time.

The consensus was that the following need to be documented:

1. Scope, purpose, and rationale.
2. Experimental design and strategy.
3. Conditions of test, site, and test system.
4. Substance, agent, test chemical, and test organism (by accurate name, composition, analysis, and source).
5. Sampling procedures, methods, equipment, and conditions for those areas where materials are sampled from the environment or the workplace.
6. Method descriptions, including protocol, dosage in the case of animal tests, route of administration and duration, and controls used.
7. Analytical methods, including standards, equipment type, validation, and calibration.
8. Results, tables and graphs, statistical analysis methods, and other data analyses.
9. Quality assurance (QA) statement, that is, was there a quality assurance program? Was the study signed off by QA?
10. Was there a peer review of the studies and/or publication?

Several types of indexing or rating scales were proposed, depending on how well a database, report, or paper met the criteria. Data that are important and critical to one chemist may not be to another. Thus, the significance of data quality depends on the chemist and the use to which the data will be put.

One good example of an evaluated online database is the *Hazardous Substances Data Bank* of the National Library of Medicine. Before the complete records are made available to the public, they are reviewed by a *Scientific Review Panel* to accurately convey what is known about the toxicity of the chemicals being considered.

Maizell (10) and Luckenbach et al. (11) are among those who have published important papers on methods of data assessment. Some of the highlights of Maizell's paper are summarized later in this section. The paper by Luckenbach et al. is a detailed report on screening procedures followed by editors at *Beilstein's Handbook of Organic Chemistry.* Among other things, they point out the importance, occasionally overlooked by some, of consistency with known chemical knowledge and principles. In some cases, 50% of the literature being considered can be excluded when the procedures described are employed.

What if rigorously evaluated data such as those provided by NIST's SRDP or other sources are not available? What if adequate verification of the data in the chemist's laboratory is not possible?

In such cases chemists may need to make their own relatively subjective assessment of published data on physical and chemical properties. Data from several sources may coincide or may vary significantly. Data from a single source may be suspect. In determining reliability of data found in the literature, chemists need to consider factors such as these examples:

1. Is the chemical identity and grade of the substance unambiguously specified? This is a basic starting point, because it helps avoid pitfalls due to variations in nomenclature and in the many commercial trade names and grades.
2. Is there agreement between several independent sources? When there is lack of agreement, one choice is to take the safest, most conservative value. For example, in considering different flash-point values, the chemist could select the lowest value to provide the greatest possible safety margin (*note the importance of checking in several different sources for values of the same constant*). Even if two measurements agree, however, both may have systematic errors.
3. Is the source of the data a paper written to determine specific physical constants rather than a paper in which the constant is determined only incidentally as in the course of a laboratory preparation? The former is a more reliable source.
4. How recent is the source? Recently determined values (and the more recent the better) are usually preferred over older values because of improvements in techniques and apparatus and advances in knowledge—if everything else is equal. But this does not mean that older sources of information should be totally discounted or neglected.
5. Is the source a specialized book such as a monograph on specific compounds or classes of compounds? Such books are frequently preferred as sources of physical property data over more general treatises, which are less specialized.

6. Do the sources being compared refer to the same original source for their data? In comparing secondary (i.e., unoriginal or second-hand) sources, the chemist should beware of the possibility that the secondary sources are all based on the same original source, which may be erroneous.
7. Is the author (investigator) a specialist in the chemical or property being studied, and what is that person's reputation for good work?
8. What is the reputation of the laboratory or research center where the work was done? For example, high confidence is placed in work done at the U.S. National Institute of Standards and Technology.
9. What is the reputation of the publisher or publication? For example, values appearing in a source such as the *Journal of Chemical and Engineering Data* would ordinarily be given more credence than a value appearing in a journal of unknown or questionable reputation.
10. Do the data appear in tables or graphs in which important errors are detected? If the answer is "yes," such data need to be looked on with extraordinary scrutiny.
11. How much experimental detail is given? Inclusion of full detail enhances confidence in the results. Lack of detail diminishes confidence. Some specifics for which detail is helpful and which should be looked for include
    a. Purity of material on which the determination is made
    b. Source of the material (supplier)
    c. Type of apparatus used in making the determination and precision of that apparatus
    d. Reliability and reputation of methods used
    e. Limits of error or confidence limits

    The lack of such information could reduce or even entirely negate the placing of confidence in data reliability. Inclusion of such information increases confidence and facilitates evaluation.
12. Are all pertinent facts reported, including those unfavorable to the author's position or theory? To what extent are opposing interpretations and views included?
13. Do the results appear to be overprojected and overextended?
14. Are the data internally consistent?
15. Is the work written so that it can be correlated, repeated, or verified by others?
16. Are interpretations clearly labeled as such?
17. Is there an attempt to evaluate and assess the reliability of the results critically?

15.15. SOURCES FOR PREDICTING OR ESTIMATING CHEMICAL PROPERTIES

A number of versatile computer programs or systems are available to help the chemist or engineer predict or estimate physical properties. Some representative sources are described here. An extensive listing of these can be found in the *CEP Software Directory,* which is published annually as part of the December issue of *Chemical Engineering Progress.*

CHETAH, also known as the *ASTM Computer Program for Chemical Thermodynamic and Energy Release Evauation* (NIST Special Database 16), which was first released in 1974, is now in its third edition (1994), Version 7.0. This is a powerful tool for predicting certain thermochemical properties (e.g., standard enthalpies of formation, heat capacities, entropies, Gibbs energies) and certain "reactive chemical hazards"—such as classifying materials as to their ability to explode with violence, for estimating enthalpies of reaction or combustion, and for predicting lower flammable limits. The program is suitable for pure chemicals, chemical reactions, or chemical mixtures. Radicals, liquids, and kinetics are beyond the scope of the program. Capability of the program has been expanded to include inorganics as well as organics. It is intended for use with IBM or IBM-compatible computers. An excellent, easy-to-understand 92-page user manual is provided. The product was developed by ASTM Subcommittee E27.07 on Estimation Methods of Committee E-27 on Hazard Potential of Chemicals. The basis of the program is the molecular structure of the components in question. Private datafiles may be included. The latest version of the program is said to be a considerable improvement over the previous version with regard to user-friendliness and other features.

Another well known product is the NIST *Structures and Properties Database and Estimation Program,* developed by Steven E. Stein at NIST. This includes a database of properties for several thousand chemicals, data prediction capability, and a structural drawing module. Properties that may be estimated include gas-phase heats of formation, entropies, heat capacities, vapor pressures, and boiling points.

In addition to these examples of "freestanding" programs, there are a number of important simulation systems that permit estimation or prediction of physical properties. A few are mentioned here. One is *ASPEN PLUS,* a steady-state process simulation program offered by Aspen Technology, Inc., Cambridge, MA (617-577-0100). Physical property models and data in *ASPEN PLUS* include equations of state, fugacity coefficients, molar volume; enthalpy–entropy–free energy, vapor pressure, vapor–liquid equilibrium ratio, Henry's constant, complex solids density, complex solids enthalpy, thermal conductivity, surface tension, viscosity, and diffusion coefficient. The system is said to include databanks containing property data for almost 5000 pure components, and over 37,000 pairs of binary parameters representing 4000 component pairs. There is also a databank for 800 aqueous ionic reactions that includes stoichiometry, equilibrium, and salt solubility constants. An interface to the Dortmund physical property databank is available through collaboration with DECHEMA. Proprietary databanks may be interfaced into the system as desired. Activities also include bioprocess simulation, simulation of hazardous-waste disposal; and flowsheet optimization.

PRO/II is a general purpose flowsheet simulation program offered by Simulation Sciences, Inc., 601 Valencia, Brea, CA 92623 (800-827-7999, Barbara Kolbl). This includes methods for predicting physical and thermodynamic properties and also phase equilibrium prediction, as well as general-purpose simulation and flowsheet optimization. There is an extensive (over 1700) pure-component library with all the basic data required for simulation, including physical, thermodynamic, and transport properties. Sources of data include DECHEMA and DIPPR, mentioned elsewhere in this chapter. Users may override and supplement these data is desired. The program is available worldwide with versions for PCs, UNIX workstations, and client/server environments.

Another leading source of physical property databases and simulation tools in Hyprotech, 110 Centre Street, Calgary, Alberta, Canada T2E 2R2 (800-661-8696). Their *HYPROPROP* (now expanded and improved to HYSIS Concept) software includes links to the TRC, DIPPR, DDB (Dortmund databank of DECHEMA) and PPDS electronic databanks of pure components.

In addition to the major simulation and process design systems mentioned above, certain of the general-purpose online systems have estimation and prediction capabilities.

15.16. PROPERTIES OF PLASTICS, METALS, AND OTHER MATERIALS

A number of well-established databases and other sources emphasize plastics, metals, and other materials.

The well-known and powerful materials selection database *PLASPEC* now covers over 90 properties and characteristics for over 12,800 grades of plastic materials. (*PLASPEC* is a product of D&S Data Resources, 218 East Bridge Street, Morrisville, PA, 19067, 888-752-7732.) Additionally, materials can be selected on over 200 feature attributes, as for example, weatherable, crosslinkable, self-extinguishing, transparent, UV-curable, biodegradable, and high-gloss. A separate, but associated file, developed by Plastics Design Library (PDL), covers chemical compatibility; this is known as the *Compatibility Data Base*). The PLASPEC materials selection database may be accessed either directly from PLASPEC (through a toll free number or through the STN or DIALOG networks (see Chapter 10). Internet access is also possible (www.plaspec.com). The compatibility database is available separately on STN under the file name PDLCOM.

In addition to materials properties, there are other associated databases that are available through PLASPEC. One of these, *Plaspec Machinery Selection,* provides specifications on over 5000 models of plastics processing machinery. Another in the same family of databases, *Plaspec News,* covers daily news in the plastics industry, including a subset on developments in Europe, Japan, and other East Asian nations. A related file includes resin and feedstock pricing. (A version of *Plaspec News, Plasnews,* is available on STN; this excludes the news edited by the Japanese associates of PLASPEC). *PLASPEC* also includes directories of manufacturers and suppliers of chemicals, additives, and equipment.

In addition to all the above, *PLASPEC* offers online the full text and abstracts of the SPE journals *Polymer Engineering and Science, Polymer Composites,* and *Journal of Polymer Technology.*

One of the additional advantages of using *PLASPEC* is the remarkably extensive, expert, and patient technical support that is provided to users of the system by the company.

PLASPEC was the brainchild of Malcolm Riley, former publisher of *Plastics Tech-*

nology, and it was originally established in 1985 as a service of that magazine. D&S Data Resources was formed in 1991 by two of the key executives who had managed PLASPEC for *Plastics Technology.* At the same time, the new company was licensed to market and to continue to develop PLASPEC.

Another useful plastics selection tool is offered by International Plastics Selector, D.A.T.A. Business Publishing, Engelwood, Colorado, (part of Information Handling Services). This product is available on CD-ROM, which is optionally updated semi-annually and leased under the name *Plastics DIGEST.* The file is said to include parameters on over 25,000 plastic materials from several hundred manufacturers as well as manufacturer specification sheets. Data can be accessed by properties, generic or trade names, or manufacturers. In addition to current materials, discontinued products back to 1977 are covered for use in repair and upgrading operations. The file is available online through STN International under the name *IPS.* The product is also available in the form of two printed volumes but the book is not as current as the electronic form. A related book product, *Adhesive Digest,* covers the properties of adhesives, sealants, and primers. Internet site is http://data.ihs.com.

The firm Plastics Design Library (PDL), 13 Eaton Avenue, Norwich, NY, 13815 (607-337-5080) markets a number of products that facilitate selection and use of plastics, elastomers, rubbers, and related materials. Most PDL products are available in diskette form, and some are available in online and printed forms. PDL is an imprint of William Andrew, Inc., a publisher and developer of information management and publishing that was founded in 1989. The president of PDL is William A. Wolshnis. Internet address is www.williamandrew.com.

PDL develops its own products, and many others that it markets are originally produced at the large and highly respected British rubber and plastics research and consulting firm Rapra Technology LTD, located in Shawbury, Shrewsbury, Shropshire, UK (www.rapra.net). PDL acts as the exclusive software agent and nonexclusive publications agent for Rapra in North America. PDL will also conduct custom searches of their products on a fee basis. See also Chapter 8.

Of its printed volumes relating to polymers, Plastics Design Library is perhaps best known for its PDL Handbook Series, which consists of eight books that provide testing data and other information about properties of polymers. These include *Chemical Resistance,* Volumes I and II, dealing respectively with thermoplastics and thermoplastic elastomers, thermosets, and rubbers (these books are available in more complete diskette form (DOS version) and online as PDLCOM, mentioned below); *Permeability and Other Film Properties of Plastics and Elastomers; The Effect of Creep and Other Time Related Factors on Plastics; Fatigue and Tribological Properties of Plastics and Elastomers; The Effect of UV Light and Weather on Plastics and Elastomers; The Effect of Temperature and Other Factors on Plastics;* and *The Effect of Sterilization Methods on Plastics and Elastomers.*

Each of these handbooks is published in expanded, electronic form, as part of the PDL "Rover" Electronic Handbook or Databook Series, with extensive Windows-based searching capabilities. The parts of the series are available in either full or subset form.

The PDL Rover Electronic Databook series now covers such areas as plastics and rubbers; steels and other ferrous alloys; aluminum, copper, and other nonferrous alloys; chemical and environmental resistance and corrosion; and dynamic mechanical analysis. Some specific titles include

Chemical Resistance of Plastics and Elastomers
The Effect of Creep and Other Time Related Factors on Plastics
The Effect of Temperature and Other Factors on Plastics
Effects of Sterilization Methods on Plastics and Elastomers
Permeability and Other Film Properties of Plastics and Elastomers
Effect of UV Light and Weather on Plastics and Elastomers
Fatigue and Tribological Properties of Plastics and Elastomers
Dynamic Mechanical Properties of Plastics and Elastomers
Worldwide Guide to Equivalent Alloys and Steels
Woldman's Engineering Alloys
Worldwide Guide to Equivalent Nonferrous Alloys
Structural Steels and Their Properties
Alloy Steels and Their Properties
Stainless Steels and Their Properties
Aluminum Alloys and Their Properties
Copper Alloys and Their Properties
Magnesium Alloys and Their Properties
Titanium Alloys and Their Properties
Alloy Digest 1952–1996
Chemical Resistance of Plastics and Elastomers
Corrosion of Metals Handbook
Electronic Materials and Solvents Compatibility

Data in the Rover series are developed from journals, materials supplier data, conference proceedings, independent test laboratories, and other sources. All original data sources are keyed to the appropriate references. Some of the data are said to be exclusive to PDL products and based on independent test data. Results are presented in the form of text, graphs, and tables. A utility that allows users to add their own data will also be available.

In addition, PDL markets several hundred additional books related primarily to plastics and related products. For example, these books include directories of plastics processing companies, especially in Europe, and of trade names in the rubber and plastics industry.

Among the PDL diskette products is PLASCAMS. This is a Rapra product that permits the user to select types of plastics for specific applications by specifying the general, mechanical, thermal, and electrical properties; chemical and radiation resistance; production and postprocessing methods; and cost factors desired. One or more of 72 properties and characteristics (including upper and lower limits) can be selected, and these can be compared with several hundred different kinds of plastics. Supplier and trade-name data are given so that the user can then contact the proper manufacturer. An important feature is that both advantages and disadvantages of materials are indicated. The data supplied are generic, rather than for specific trademarked materials, but also included are Rapra staff independent value judgment ratings on more than 350 variations of generic materials (such as forms of ABS). All data are based on original studies and tests at Rapra rather than on manufacturer-supplied information; this should help ensure objectivity.

A similar file is RUBACAMS, which is intended to facilitate the selection of various "rubbers."

The PDL product known as PDLCOM is an extensive diskette database on the chemical and environmental resistance of plastics. Over 100,000 specific tests are said to be included based on resistance to 4000 different exposure media. This file is based on manufacturer-supplied data. It is available online through the STN network and also by direct connection with PLASPEC.

Another product offered is MATCOMPAT (registered trademark) or Materials Compatibility System intended for use by those involved in the manufacture and/or cleaning of microelectronic parts, with emphasis on new non-ozone-depleting cleaning solutions. This system was developed by a consortium of some leading U.S. manufacturers known as the National Center for Manufacturing Sciences. The database covers materials such as polymers, composites, elastomers, adhesives, nonmetallic coatings, marking materials, tapes, and metal surface coatings. Test results are given when the materials included are treated with specific solvents. Periodic updating is planned.

Another PDL product is a diskette database known as SOLV-DB (originally produced by Syracuse Research Corporation and the National Center for Manufacturing Sciences). This is intended to assist in identification of alternative solvents by permitting the user to specify the desired characteristics. Some 321 solvents are included. For each solvent included, data are given on properties, environmental fate, health and safety, regulations, and vendors. Other information is also included.

In the field of metals (and some other materials), ASM International, Materials Park, Ohio 44073-0002 (800-336-5152), formerly known as the *American Society for Metals,* is a major source of information. ASM's principal focus is still metals, but the organization's interests have expanded to include nonmetallics, primarily structural engineering materials.

The major reference source in the field of metals and metallurgy is the 18-volume *ASM Handbook,* previously titled *Metals Handbook.* Scott Henry is the current editor of the *ASM Handbook* series. This important series is an excellent starting point for any study of metals. There are extensive discussions, numerous data compilations, and frequent references to other pertinent literature.

Large parts of the *Metals Handbook* (9th ed.), published over the time period 1978–1989, are still considered to be useful by many, although some of the information is dated and must be used with caution. However, Volumes 7–17 of this compendium are still considered by ASM to be the most current *Handbook* treatment of these topics, and have been incorporated into the current *ASM Handbook* series without changes in content.

ASM also is the publisher of the *Mat.DB* program and databases, which can be searched on a PC. The databases, which cover a broad range of properties, range from aluminum to titanium, and include engineered materials such as engineering thermoplastics, thermosets, and composites. In addition, ASM offers a CD-ROM index to the *ASM Handbook Series, Engineered Materials Handbook Series, Electronic Materials Handbook,* and all volumes of the ninth edition of the *ASM Metals Handbook.*

An example of ASM publishing activities in nonmetallics is the *ASM Engineered Materials Handbook Desk Edition* (1995). This includes textual discussion and tabular data related to plastics, composites, ceramics, and glasses.

Corrosion properties and data are of special importance because some estimates are that corrosion costs the United States alone many billions of dollars per year. Corrosion data are extensive, but are often widely scattered over many publications and an extensive period of time.

Publications and databanks are available to help chemists and engineers cope with this scatter. Use of literature data, however, can be at best a guide or first approximation toward selecting appropriate materials of construction. Engineers usually need to conduct their own experimental studies (with candidates selected from the literature) under the precise conditions under which the proposed material of construction is to be used.

One of the most valuable literature sources in this field is *Corrosion Data Survey* published by NACE International, formerly known as the National Association of Corrosion Engineers, Houston, TX. While these books (one on metals and one on plastics) are somewhat dated at this writing, they are currently being updated, and software incorporating the revision is due to become available in 1997. Subsequently, new editions of the books are to be published. The updating efforts are being accomplished in conjunction with NIST.

DECHEMA, a German organization, is another major source of corrosion data. One example is the *DECHEMA Corrosion Handbook,* 1987–1993, VCH Publishers, John Wiley, NY, 12 volumes. This multivolume source describes the corrosion behavior of important materials and outlines possible ways to combat corrosion problems. Arrangement is by type of chemical. This series has had various editors: D. Behrnes, G. Kreysa, and H. Eckermann.

As previously mentioned, ASM is a primary source of metals data, and Volume 13 of the *ASM Metals Handbook* is devoted to corrosion.

PDL offers a ROVER CD-ROM product that is packed with corrosion data. The publisher says that all the data included have been updated as of 1996. This product includes (a) Plastic Design Library's *Chemical Resistance Handbooks* (originally published in 1994) with data on how conditions influence the properties of neat and reinforced thermoplastics, thermosets, thermoplastic elastomers, and rubbers; (b) ASM International's *Handbook of Corrosion Data* (originally published in 1995, 2nd ed.), which includes data on over 1400 metals and alloys in corrosive environments with over 300 exposure media; and (c) *Material Compatibility System-MAT-COMPAT* developed by the U.S. Navy and the U.S. National Center for Manufacturing Sciences and dealing with compatibility data for substances utilized in cleaning or manufacture of microelectronic assemblies. These books are available on separate

CD-ROMs or on one CD-ROM as the product "Chemical Resistance and Corrosion Suite," 1996. See also the discussion of PDL on page 428.

NACE also publishes *Corrosion Abstracts* and many other publications. *Corrosion Abstracts* is available in both printed form and CD-ROM.

The well-known online database system Questel•Orbit includes as one of its databases the file *CORROSION*. Data included are the effects of over 600 agents on metals, carbon, glass, plastics, and rubbers. At this writing, this file had not been updated for a number of years. However, the corresponding book, *Corrosion Resistance Tables,* edited by P. A. Schweitzer, is available in a fourth edition (Marcel Dekker, New York, 1995).

Other information of value will be found in abundance in major abstracting and indexing sources such as *Chemical Abstracts, Metadex,* and *Ei CompendexPlus*. Entries pertinent to corrosion can be found in different indexes under a number of scattered headings, as indicated in Chapter 10. It is especially important to be aware of this type of scatter in corrosion literature searching efforts and to utilize all pertinent alternative index entries as required.

15.17. IMPROVING ACCESS TO, AND STANDARDIZATION OF, PHYSICAL PROPERTIES DATA

As mentioned previously, NIST has long been a leader in the development of improved access to property data. Also, as noted, CODATA has been a factor at the international level, although its performance has fallen short of original expectations of some.

ASTM has operated its Committee E49 for some years. The name of this committee is Computerization of Material and Chemical Property Data. A proposed new name for the Committee is Chemical and Material Information Technology and Systems. The committee "promotes and develops standard classifications, guides, practices, and terminology for building, accessing, and exchanging information among computerized material and chemical property databases." In addition, the committee "works in concert with other ASTM technical committees and outside organizations that develop standards related to materials, chemicals, and their properties." The committee is concerned not only with materials data but also with analytical data such as spectra, chemical structure systems, and laboratory information management systems (LIMS). It has held symposia every other year, and these are published by ASTM under the title *Computerization and Networking of Materials Databases.*

The National Research Council, Washington, DC and ASM International are among the important groups studying ways to improve access to data, including property data.

Reports concerning the status of materials databases have been published by J. H. Westbrook and others (13–22). Westbrook, formerly of General Electric Company and now a private consultant, has long been a leader in the efforts to improve the delivery of property data to the scientific and technical community.

The online availability of a number of important numeric databases on STN International is a major step forward, although some believe that a more user-friendly approach would be desirable. The concern is that present access to these databases is aimed at the information professional without enough attention to direct access by the end user. An approach like that utilized in *SciFinder* (see Section 7.3) would help solve this matter. The following section continues the discussion of sources of data in electronic form.

The *Index Guide* to *Chemical Abstracts* has provided a very convenient and useful List of Data Collection and Analysis Centers both within and outside the United States. The list was categorized by type of data such as chemical kinetics and nuclear data. Addresses and phone numbers were included. Unfortunately, the 1994 *Index Guide* is the last year for which such information will be provided by this source.

15.18. SELECTED SOURCES OF CHEMICAL DATA IN ELECTRONIC FORM

As indicated throughout this book, a vast amount of chemical information is now available in electronic form, that is, either online/Internet, or in diskette, and in CD-ROM versions, or in all three forms. Physical properties are among the many types of chemical information that are available in this way, and this offers considerably more speed, power, and flexibility over previous methods of searching for properties. Thus, data may searched for in ranges; or equal-to, greater-than, or less-than requirements can be specified. In addition, a number of the electronic sources of data offer the chemist or engineer the opportunity to generate, predict, or calculate data as required. A further development is that books dealing with physical properties now frequently appear together with accompanying diskettes, which permit the reader to use and work with the data.

This section summarizes some representative electronic sources of physical property data. Others are mentioned elsewhere in this book. So many new sources appear regularly, seemingly almost continuously, that it is difficult to keep up. An orderly solution to this matter and a good source for identifying software that covers physical properties (as well as other applications) is the *CEP Software Directory.* This is published annually as part of the December issue of *Chemical Engineering Progress* and may also be purchased separately from the American Institute of Chemical Engineers, New York; it covers all electronic media.

A. Selected Online Sources of Physical Properties

1. STN International. STN International (see also Chapter 6, 7, and 10) offers a number of online files with a wide variety of properties and other numeric data that can be accessed. Several of these are discussed further elsewhere in this book. Numeric files on STN include

AAASD—aluminum alloy data.

ALFRAC—aluminum fracture toughness data

ASMDATA—emphasis on metals data from the ASM, but also includes information on composites and plastics

BEILSTEIN—organic compounds

COPPERDATA—data on copper alloys

CRYSTMET—crystallographic data on metals and intermetallic compounds

DETHERM—500 thermophysical properties for pure compounds and mixtures; approximately 12,000 inorganic and organic substances covered

DIPPR—property constants and temperature-dependent properties for commercially important chemical substances as compiled and evaluated by Design Institute for Physical Property Data (DIPPR) of the American Institute of Chemical Engineers

GMELIN—inorganic compounds and organometallics

HODOC—*CRC Handbook of Data on Organic Compounds*

HSDB—*Hazardous Substances Data Bank*

IPS—*International Plastics Selector;* properties of commercial plastics.

JANAF (Joint Army, Navy, Air Force)—thermochemical properties of more than 1100 inorganic substances or organic substances with only one or two carbons; compiled 1959–1985

MDF—numerical properties of nearly all ferrous and nonferrous alloy systems

METALCREEP—design data on creep and rupture stress of aluminum and magnesium alloys and steels

MRCK—*Merck Index*

NEWCRYST—data on new organic and inorganic crystal structures

NISTCERAM—thermal and mechanical property data for many structural ceramics

NISTFLUIDS—thermophysical and transport properties for important industrial fluids (no searchable fields, but can be displayed)

NISTTHERMO—thermochemical property at standard state conditions for over 8000 inorganic and small organic molecules

PDLCOM—chemical and environmental compatibility data of plastics

PLASPEC—design and processing data on more than 11,900 grades of plastics

RTECS—toxicity data on many chemicals

TRCTHERMO—evaluated thermodynamic data from the Thermodynamics Research Center, Texas A&M University

2. Technical Database Services (TDS). Another key source of chemical data in electronic form is *Numerica* from Technical Database Services (TDS), 135 West 50th Street, New York, NY 10020 (212-245-0044). TDS offers ready access to a wide variety of important numerical (chemistry and chemical engineering), regulatory, and environmental databases online and/or in diskette or magnetic tape forms. In addition, the firm has staff who will do the searching for the user on a fee basis. *Numerica* specializes in factual and numerical data. In recent years, the scope of the databases offered has been significantly increased. A number of the databases are exclusive to TDS.

The TDS online system and software products are known as *Numerica.* TDS is accessible through the MCI Tymnet Network and the Internet (www.tds-tds.com), a site that is under intensive development at this writing. The files consist of two primary clusters: Chemistry/Engineering and Environmental. A file known as SYNDEX, which includes over 135,000 discrete chemicals, indicates which TDS databases contain information on the chemical in question.

The firm was founded in 1983 by Mildred Green, a science educator and executive, who is president. She describes her firm as being like a "boutique among supermarkets," the "supermarkets" being the very large information providers such as DIALOG. Many *Numerica* users, perhaps 40–50%, are believed to be end users. The system is menu-driven rather than command-driven; that is, the user is continually provided with a series of options that prompts the user during the course of every online session. TDS describes its system as "plain vanilla"—that is, the system is intended to be easy to use, and there are no frills. In contrast to bibliographic databases, the numerical data offered by TDS *Numerica* are already extracted along with the citation, and, in some cases, some indication of quality may be given. Thus, in the case of PPDS (see first list immediately below) data are accompanied by quality codes, while TRC values each have a confidence interval. In AQUIRE (see second list below), each record is evaluated in terms of the accuracy of the study. The most reliable physical and environmental property data in the Environmental Fate Database (EFD—see second list below) are labeled "recommended." The LOGKOW Database (see second list below) often provides a "recommended" value when two or more octanol–water partition coefficients are reported.

In order to utilize the system, an up-front fee of several hundred dollars per year is required, and there are connect-time charges as well as per-hit fees for some databases.

Products in the online Chemistry/Engineering cluster are listed below. If the product is also available from TDS as separately purchased software for offline use, this is so indicated. (Information regarding TDS files is supplied through the courtesy of TDS.)

DIPPR Pure Component Data Collection (see also Section 15.6)—provides pure-component data for 26 constant and 13 temperature-dependent properties of over 1500 industrially important chemicals.

PPDS-2 (Physical Property Data Service)—thermodynamic and transport properties and phase equilibrium data. A databank of over 1500 components (with specialized files for petroleum fractions, refrigerants, and steam). Part of this module is LOADER2, which is a program that predicts up to 40 physical properties for pure components on the basis of structure and other data entered by the user; results can be stored. Offline software also available. Exclusive for TDS in North America. A product of the National Engineering Laboratory (UK).

Thermodynamics Research Center/Thermophysical Datafile Service (TRC/TDS)—a wide range of thermophysical and phase equilibrium data for pure components and mixtures. Compiled and evaluated by the Thermodynamics Research Center (TRC), Texas A&M University, College Station. TDS has plans to include virtually all other TRC products in its online files. See also Section 15.5.

The Environmental cluster includes

AQUIRE—aquatic toxicity data as retrieved from the literature. Over 117,000 individual toxicity tests covering over 5200 chemicals are included. Over 160 acute, sublethal, and bioaccumulation effects are included. Offline software also available. A product of the U.S. Environmental Protection Agency's Environmental Research Laboratory Duluth, MN.

CHEMEST—methods for predicting 12 physical and environmental properties. Offline software also available. A TDS exclusive. A product of Arthur D. Little, Inc. based on the *Handbook of Chemical Property Estimation Methods* by Warren Lyman et al.

CCM—Chemical Compliance Monitor: U.S. federal and state and Canadian regulations and guidelines for chemical substances. Exclusive to TDS. Updated quarterly. A product of Logical Technology Inc.

CIDES—Carcinogenicity Information Data of Environmental Substances. Literature references to experimental data on carcinogenic and mutagenic effects. Exclusive to TDS. A product of TDS with support from National Institute of Environmental Health Sciences.

ECDIN—Environmental Chemicals Data and Information Network. This file includes the EINECS list of over 122,000 chemicals and provides extensive data for approximately 30,000 substances. Data included relate to EINECS identification, CAS Registry Numbers, other identification, basic physical and chemical properties, occupational health and safety data, concentrations and fate in the environment, hazards during transportation and storage, uses, worldwide coverage of legislation and rules, toxicity, and chemical detection methods. This is a product of the Environmental Research Program of the Joint Research Center of the Commission of the European Communities.

EFD—Environmental Fate Database. Literature references on transport, fate, persistence, and biodegradation of chemical substances. Exclusive to TDS. Offline software also available. A product of Syracuse Research Corporation.

LOGKOW—contains over 27,000 critically evaluated partition coefficients for over 18,000 organic compounds, including many recommended values. A partition or distribution coefficient is the equilibrium ratio of the concentration of a substance dissolved in an organic solvent (typically octanol) to the concentration of the same substance dissolved in water. This database was developed at, and is maintained by, Sangster Research Laboratories, Montreal.

QSAR—chemical property estimation, evaluating risk of hazardous exposure and predicting environmental fate and persistence, using quantitative structure–

activity relationships methods. Chemical structure and property values are stored in the database or may be entered by the user. Offline software also available. Exclusive to TDS. A product of Hunter Systems and the Institute for Biological and Chemical Process Analysis (IPA) at Montana State University.

TSCATS—a database of health and environmental studies submitted under the U.S. Toxic Substances Control Act. Over 7500 chemicals are included.

TDS also offers some of these online files in diskette or magnetic tape form. These include PPDS2, LOADER2, AQUIRE, ECDIN, LOGKOW, EFD, and QSAR. In addition, scheduled for implementation is the MSDS database (includes over 100,000 MSDSs) of MDL Information Systems (see Chapter 10). Another file scheduled for implementation is PredictTox. This is basically an algorithm for predicting chemical toxicity that was developed by research workers at the U.S. Department of Interior, National Biological Services Laboratories, Ann Arbor, MI; this utilizes the SMILES (simplified molecular input line entry specification) method of denoting chemical structure.

B. Selected Diskette and CD-ROM Sources of Physical Properties

Properties of Organic Compounds (*POC 4.1*). Over 27,000 organic compounds are included in this CD-ROM product. Data provided include melting point, boiling point, density, and other physical properties; solubility; and infrared, Raman, ultraviolet, mass, and NMR spectra. Data can be accessed by chemical name or synonym, molecular formula, CAS (and other) registry numbers, and values of the physical properties and spectral peaks. Structure diagrams are displayed along with the property values. (CRC Press, 1995.)

Handbook of Thermophysical and Thermochemical Data. A broad range of thermodynamic and transport properties of important chemical substances, including mixtures, are covered in this combined book and diskette. The book tabulates property values at standard conditions. The diskette permits the user to calculate the same properties at any desired temperature, pressure, or mixture composition. Thus, custom tables can be generated for printing or transfer to a computer file. (D. R. Lide and H. V. Kehiaian, CRC Press, 1994.)

Thermochemical and Physical Properties (*TAPP 3.0*). This database includes thermal properties, crystal structure, mechanical (elasticity) and electrical properties, and phase diagrams for about 31,000 substances. Inorganic materials at high temperatures are emphasized, but about 6000 organic compounds are included as well. Values can be calculated over a range of temperatures. It is distributed in diskettes or CD-ROMs. (ESM Software, 2234 Wade Court, Hamilton, OH 45013, 513-738-4773.)

ELDATA—The International Electronic Journal of Physico-Chemical Data. This new journal publishes reports of original measurements of property data, with emphasis on thermochemical and thermophysical properties of pure substances and mixtures. Tabular data are provided on a diskette and in the printed journal; the electronic records are cumulated, so that subscribers receive a database that grows with time. (ELDATA SARL, 81-83 rue Michetet, 93100 Montreuil, France.)

15.19. TRENDS

In an important 1981 article *Critical Data for Critical Needs,* Lide of NBS identified the most likely future trends (23). These include

1. Increasing need for reliable evaluated and assessed data
2. Increased use of computer-based dissemination of data
3. Increased need for coordinated development of computer systems

Years after these forecasts were made, their validity is even more apparent. No quick and easy solution to data problems is anticipated, nor is it apparent, even with enormous progress in computer capabilities. The road ahead will be difficult and complex.

REFERENCES

1. K. N. Marsh, R. C. Wilhoit, S. J. Xu, and D.-P. Yin, *TRC Databases for Chemistry and Engineering—Comprehensive Index of TRC Thermodynamic Tables, Spectral and IDS Data,* Thermodynamics Research Center, Texas A&M University, College Station, TX 77843-3111. Updated annually.
2. "Thermophysical Properties Research Literature Retrieval Guide: 1900–1980," Purdue University, Center for Information and Numerical Data Analysis and Synthesis, 1982.
3. *Landolt-Börnstein's Zahlenwerte und Funktionen aus Physik, Chemie, Astronomie, Geophysik und Technik,* 6th ed., Springer-Verlag, Berlin, Heidelberg, New York, 1950–1980. New Series volumes in progress.
4. D. R. Lide, Jr., *CRC Handbook of Chemistry and Physics,* 77th ed., CRC Press, Boca Raton, FL, 1996. 78th Edition, 1997.
5. J. A. Dean, *Lange's Handbook of Chemistry,* 14th ed., McGraw-Hill, New York, 1992. Revised at intervals.
6. R. Perry and D. W. Green, Ed., *Perry's Chemical Engineers Handbook,* 6th ed., McGraw-Hill, New York, 1984.
7. *The Merck Index: An Encyclopedia of Chemicals, Drugs, and Biologicals,* 12th ed., Susan Budavari, et al., Eds., Merck & Co., Whitehouse Station, NJ, 1996. Updated online.
8. G. T. Armstrong and R. N. Goldberg, *An Annotated Bibliography of Compiled Thermodynamic Data Sources for Biochemical and Aqueous Systems* (1930–1975), NBS Special Publication 454, U.S. Government Printing Office, Washington, DC, 1976.
9. R. N. Goldberg, *Compiled Thermodynamic Data for Aqueous and Biochemical Systems* (1930–1983), NIST Special Publication 685, U.S. Government Printing Office, 1984. Available from National Technical Information Service, Springfield, VA.
10. R. E. Maizell, "Techniques of Data Research in Chemical Libraries," *J. Chem. Educ.,* **32,** 309–311 (1955).
11. R. Luckenbach, R. Ecker, and J. Sunkel, "The Critical Screening and Assessment of Sci-

entific Results without Loss of Information—Possible or Not," *Angew. Chem., Int. Ed. Engl.*, **20,** 841–849 (1981).

12. D. L. Graver, Ed., *Corrosion Data Survey, Metals Section,* 6th ed., NACE, Houston, TX, 1985; N. E. Hammer, Ed., *Corrosion Data Survey, Nonmetals Section,* 5th ed., NACE, Houston, TX, 1975.
13. J. H. Westbrook, J. G. Kaufman, and F. Cverna, "Electronic Access to Factual Materials Information: The State of the Art," *MRS Bulletin,* pp. 40–48 (Aug. 1995).
14. J. H. Westbrook and J. G. Kaufman, "Impediments to an Elusive Dream: Computer Access to Numeric Data for Engineering Materials," in *Data and Knowledge in a Changing World, Models for Information Systems: Knowledge, Tasks and Tools.*
15. J. H. Westbrook, "Materials Databases Since Schluchsee," in *New Data Challenges in Our Information Age,* Proceedings of 13th International CODATA Conference, Beijing, P. S. Glaeser and M. T. L. Millward, Eds., CODATA, Paris, 1994, pp. A157–A172.
16. J. H. Westbrook, "Overview of Current Materials Data Systems," in *The ASTM Fifth International Symposium on the Computerization and Networking of Materials Property Data,* NRIM, Tsukuba Science City, Japan, 1995.
17. J. H. Westbrook, "Data Compilation, Analysis, and Access: The Role of the Computer," *MRS Bulletin,* pp. 44–49 (Feb. 1993).
18. J. H. Westbrook, "Materials Information Sources," in *Encyclopedia of Materials Science and Engineering,* M. B. Bever, Ed., Pergamon (now Elsevier), Oxford, 1986, pp. 527–542.
19. J. H. Westbrook, "Data Sources for Materials Economics, Policy, and Management," in *Concise Encyclopedia of Materials Economics, Policy and Management,* M. Bever, Ed., Pergamon (now Elsevier), Oxford, 1993, pp. 35–43.
20. J. H. Westbrook, "Current Activity in North America on Numerical Databases on Materials Properties," in *Materials Information for the European Communities,* N. Swindell, Ed., CEC, Luxembourg, 1990, pp. 62–73.
21. H. Wawrousek, J. H. Westbrook, and Walter Grattidge, "Data Sources of Mechanical and Physical Properties of Engineering Materials," in *ASTM Special Technical Publication 1106,* ASTM, West Conshohocken, PA, 1991.
22. J. H. Westbrook, "Review of Existing Material Properties Compilations," *AIChE Symp. Series,* **80**(237), (1984).
23. D. R. Lide, Jr., "Critical Data for Critical Needs," *Science,* **212,** 1343–1349, (1981).

List of Addresses

ASM International
(formerly American Society for Materials)
Materials Park, OH 44073

AMPTIAC
201 Mill Street
Rome, NY 13440-6916

Aspen Technology, Inc.
251 Vassar Street
Cambridge, MA 02139

Battelle Memorial Institute
505 King Avenue
Columbus, OH 43201

Chem Share Corp.
2500 Transco Tower
Houston, TX 77056
(713-414-6700)

CODATA
Secretariat, 51 Boulevard
de Montmorency
75016 Paris, France

CRC Press, Inc.
2000 Corporate Blvd, NW
Boca Raton, FL 33431

D.A.T.A. Business Publishing
15 Inverness Way East
Engelwood, CO 80012

DECHEMA e.V.
Postfach 150 104
D-60061 Frankfurt am Main
Germany

Dr. David R. Lide, Jr.
Editor-in-Chief, *CRC Handbook of Chemistry and Physics*
13901 Riding Loop Drive
Gaithersburg, MD 20878

Fachinformationszentrum Chemie GmbH (FIZ Chemie)
Postfach 12 60 50
D-10593 Berlin
Germany
(www.dgm.de/FIZ)

Institution of Chemical Engineers
165–171 Railway Terrace
Rugby, CV21 3HQ, England

Marcel Dekker, Inc.
270 Madison Avenue
New York, NY 10016

NACE
P.O. Box 218340
Houston, TX 77218
(713-492-0535)

Simulation Sciences Inc.
601 Valencia
Brea, CA 92623
(800-827-7999)

Technical Database Services
135 West 50th Street
New York, NY 10020-1201
(212-245-0044)

16 Chemical Marketing and Business Information Sources

16.1. INTRODUCTION

Chemical marketing and business information is important to almost all chemists. Particularly in industry, the chemical researcher has been brought out of the relative anonymity and alleged "ivory tower" of the laboratory and into the mainstream of the business world. Industrial researchers are, or should be, accepted as full partners by their marketing counterparts. Technological and marketing decision making have become highly interdependent.

The boundary between marketing information sources and technical or scientific information sources is blurred, and the two types of sources are complementary. Important scientific information may appear for the first time in chemical marketing or business sources. Similarly, scientific and technical literature can provide important leads to chemical marketing specialists.

This chapter summarizes what is available and how to access it. For more detail, the reader should consult books in the field, such as Giragosian's classic *Chemical Marketing Research* (1) or *Industrial Marketing Research* (2) by Donald Lee. Another useful book by Giragosian is his *Successful Product and Business Development,* Marcel Dekker, New York, 1980, currently being revised for a new edition to be published in 1998. The original Giragosian and Lee books, although published some years ago, still contain general principles of value today.

As compared to most scientific or technical data, some marketing data, especially estimates or projections into the future, are relatively "soft" or less certain. Laboratory vertification of marketing data is not possible. As chemists become more and more involved with the business aspects, they can expect to make increasing use of the tools mentioned in this chapter. In so doing, they should call on the expertise of chemical marketing specialists, especially when there are questions of interpretation or extrapolation to the future.

Many types of marketing data are highly proprietary to individual companies.

16.2. INFORMATION ABOUT MANUFACTURERS

In addition to determining who makes or supplies a product, chemists may be interested in chemical manufacturers for other reasons:

1. If a company is a possible employer
2. If a company is being considered for personal or other financial investments
3. If the company represents a present or potential customer or competitor
4. If a joint research or other business venture is being contemplated
5. If the company is a potential source for a research grant or other funding

Information about chemical manufacturers that is usually found most valuable includes

1. Products made by the company
2. Locations (principal offices and plants)
3. Names, titles, and backgrounds of officers and other key personnel
4. Historical background, present status, and future plans
5. Financial data such as sales and profits, preferably broken down by major product line
6. Patents and other publications
7. Research and development staff and facilities

In addition to company product catalogs and data sheets, a variety of information sources is available. For example

1. Reports issued by companies to stockholders, especially annual reports. These have information about present and historical financial status, new products and other major achievements, names of key officials, principal locations, and other information.

2. Reports designated as "10-K" filed by companies with the Securities and Exchange Commission (SEC) in Washington, DC. These filings are required by law for all companies that are publicly owned. Most reports filed with the SEC are readily available at SEC offices, directly from the company, on the Internet, or from private organizations that provide this kind of material, for example, Disclosure, Inc. Bethesda, MD. These reports are important because they often contain more detailed financial and marketing information than is found in the annual report issued to stockholders.

3. Reports from stockbrokers, and handbooks and services issued by the financial investment community (see Refs. 3–5). This material is useful in obtaining a description and evaluation of the financial picture of the company—past, present, and future—but it also frequently contains other information.

4. Patent documents assigned to the organization and other publications by persons employed by the organization. Patent documents can be identified by methods described in Chapter 13. A study of patents and other publications helps indicate areas of interest as well as strength of technical effort.

5. News about the company as reported in chemical news magazines, general business and trade periodicals, and newspapers.

6. Buyer's guides and other tools discussed in the following sections.

7. Biographical data about officers and other key personnel as found in *Who's Who* sources such as those published by Marquis Who's Who, for example, *Who's Who in America* (6); related sources covering various regions of the country; and related or similar sources covering the business community. Data about executives can also sometimes be found in reports to stockholders or to the government, in directories such as Dun & Bradstreet's *Reference Book of Corporate Management* (7), and in Dun & Bradstreet and other financial reports. Biographical data about scientists may be found in such sources as *American Men and Women of Science* (8a), which is available in both printed and online forms. A newer source is *Who's Who in Science and Engineering* (8b). Absence of a person from biographical directories does not necessarily mean that a person is unimportant; some persons decline to provide the data requested or shun publicity of all types. Similarly, inclusion of a person does not necessarily connote exceptional expertise since those selected write their own biographies. Additional sources of data about people include their publications, patents, and news items about them. The Industrial Research Institute, Washington, DC, publishes a directory of industrial research executives, but this is available only to appropriate employees of companies that belong to the Institute. Finally, a personal letter or phone call to the person of interest can frequently obtain the needed biographical information more accurately and quickly than any printed or other secondary source. At a minimum, a list of publications of a chemist or engineer can usually be easily obtained either directly or indirectly, and this alone can be very informative.

Many of the types of data listed here are readily available online.

16.3. BUYER'S GUIDES AND RELATED TOOLS—UNITED STATES AND OTHER COUNTRIES

Finding out what companies make specific chemicals of interest is one of the most frequent information needs of the chemist and chemical engineer. Fortunately, a number of useful tools are available. Most of these are updated annually, and some even more frequently. In addition to the popular "conventional" tools that are described below, an increasing number of chemical firms have their product catalogs on the Internet.

SRI Consulting's *Directories of Chemical Producers* are far more than just the usual buyer's guides. They have been important and unique reference tools ever since they were first published in 1961. These directories are probably the most useful guides that cover the world chemical industry. They provide current information on basic chemical manufacturers and their products.

There are three main sections in each directory: companies, products, and regions. The companies section includes addresses, plant locations, and products produced at each site. Former company names and affiliations are provided. Company phone and fax numbers are given, but in most cases only for the headquarter locations. The products section includes product names, company and plant location, cross-references to alternate names for chemicals produced, and plant-by-plant nameplate capacity for some large-volume products. Some products are organized into functional or chemical groups (such as enzymes, fibers, and plasticizers). In the regions section, all plant sites are sorted geographically, first by country, then by state, county, or province; for each town or city, companies with chemical manufacturing facilities are shown.

To expand on what has been said above, capacity tables are given company-by-company for some 180 products manufactured in the United States, and the non-U.S. directories, except for China, also include some capacity information. Unfortunately, the SRI directories no longer give capacities for petrochemicals, but leave this function to the *SRI World Petrochemicals Program,* which can be purchased separately. However, in the cases of some petrochemicals, the directories do rank the top 50% of the manufacturers by percentage. In some cases, estimated at about 15%, the processes utilized for manufacture are listed in the capacity tables.

Separate directories are available for a number of countries and regions as follows: Canada, China, East Asia (Indonesia, Japan, Korea, Malaysia, Philippines, Singapore, Taiwan, and Thailand), Europe, India, Mexico, the Middle East (Bahrain, Egypt, Iran, Israel, Oman, Pakistan, Qatar, Saudi Arabia, and UAE), South America (Argentina, Bolivia, Brazil, Chile, Columbia, Ecuador, Peru, Trinidad and Tobago, and Venezuela), and, of course, the United States.

The directories are available in both printed and CD-ROM versions, and online availability is planned for 1998 through *DIALOG.* To give additional data about the scope of the directories, the U.S. directory alone (1997 edition) includes 1,250 companies, over 3,600 plants, and over 8,100 products. The directory for Europe is even larger. Cost of the subscription includes a free inquiry service.

Most of the data included are obtained by direct questionnaire to the manufacturers. SRI is headquartered in Menlo Park, CA.

Among the other buyer's guides, the two that are probably the most widely utilized in industry are published by weekly chemical industry news magazines. *Chemical Market Reporter* (Schnell Publishing, New York, NY) publishes *OPD The Green Book—Chemical Buyers Directory.* (*OPD* stands for *Oil, Paint and Drug Reporter,* the original name of *Chemical Market Reporter,* the weekly industry news magazine.) Contents include chemical suppliers index, 800 (toll-free) phone directory, *ChemFile 96* (a brief collection of some company ads and catalogs), chemicals and suppliers, custom manufacturers, transport and storage suppliers, and transport and storage services. The book includes over 1600 chemical suppliers and over 17,000 chemicals (Schnell Publishing, 80 Broad Street, New York, NY 10004. 212-248-4177). The Internet address for this buyer's guide, some magazine text, and other features is http://www.chemexpo.com.

Chemical Week (New York, NY) publishes the *Chemical Week Buyer's Guide,* which is similar to and competitive with the *Green Book* mentioned above. It contains a very useful index of trade names. In addition to the printed version, a CD-ROM product may be purchased; this has as an optional feature, nearly 1000 MSDSs as provided by Genium Publishing Corporation. (Chemical Week Associates, 888 7th Ave., New York, NY 10106, 212-621-4900.) The Internet address is http://www.

chemweek.com. This site includes the *Buyer's Guide* as well as searchable full text from the "Executive Edition" of the weekly news magazine and links to other chemical industry sites most, notably those of chemical manufacturers.

Usually, it is a good idea to consult both of the above-mentioned buyer's guides, since each has certain listings not found in the other. Although neither of these contains all of the data found in the SRI Consulting directories, they are believed to be much more widely available and are heavily utilized. In addition, it can be very helpful to consult one or more of the other buyer's guides as noted below.

The important buyer's guide *Chemcyclopedia* is published by the American Chemical Society. The main sections of this significant product include a listing by chemical showing the producers, a custom synthesis and custom manufacturing directory, and a company directory with telephone numbers, telex codes, and fax numbers. For the chemicals listings, suppliers were asked to supply available trade names, grades and physical forms, packaging, special transportation requirements, and potential end uses. If available, CAS Registry Numbers are given. There are brief summaries of the market outlook for each of the 11 chemical markets into which the product listings are categorized. More than 10,000 chemicals are said to be included. There is a special directory with descriptions of custom synthesis firms. This product is available on the Internet without charge at www.acs.org.

Thus, as noted above, at least three of the major chemical buyers guides are available without charge on the Internet. These include the *Chemical Week* and *Chemical Market Reporter* buyers guides and *Chemcyclopedia.* In addition, many individual chemical companies maintain Web sites that list and describe their products, and other buyers guides may also be available on the Web.

Purchasing CPI Edition publishes *Chemicals Yellow Pages.* The principal sections are chemicals and raw materials, containers and packaging, and environmental services. The chemicals section includes (a) company listings with all home, branch, and district office addresses, telephone numbers, including toll-free numbers, and fax numbers; (b) distributor locator guide arranged by state; (c) product section listing the North American manufacturers, distributors, manufacturers agents, importers, and sales affiliates of foreign manufacturers; (d) custom and toll chemical manufacturers; and (e) chemical trade names. (CPI Purchasing, 275 Washington Street, Newton, MA 02158 or PO Box 497, New Town Branch, Boston, MA 02258, 617-964-3030.)

Chemical Information Services (PO Box 8344, Dallas TX 75205, 214-340-4345) publishes

- *Database and Directory of World Chemical Producers.* Said to include nearly 56,831 product titles from 7076 manufacturers in 81 countries in the 1995–1996 edition.
- *International Database and Directory of Pharmaceutical Ingredients* (active and inactive ingredients).

Both of these above are available both on diskettes and in printed form.

Directory of International Chemical Suppliers is a two-volume set that is said to include over 105,000 chemicals and synonyms and over 5000 sources worldwide. It is available from Frontier Data Publishing, 3148 Riverside Drive, Beloit WI 53511 (800-859-7925).

There are several sources on the Internet that will help companies "advertise" their search for suppliers of specific chemicals; in some cases, this service is offered on a fee basis.

A number of compilations are particularly good for locating suppliers of fine, research, and specialty chemicals. Examples include

The *Chemcats* online file on STN International, introduced in 1995, offers a collection of catalogs from many 40 worldwide suppliers of fine and specialty chemicals. Information given about chemicals listed varies with the company catalogs in question, but may include names, addresses, and phone and fax numbers of suppliers, grade, pricing, packaging and shipping information, quantities available, chemical name, CAS Registry Number, structure, basic properties, safety and regulatory information, and other data as provided by the supplier. At this point, emphasis is on fine and specialty chemicals, rather than on high-volume or commodity products. The file currently includes chemicals as well as enzymes, proteins, and other biochemicals. This file has no printed equivalent; it is available only online.

MDL Information Systems, Inc. (see Chapter 10) offers *The Available Chemicals Directory* (*ACD*), a database covering commercially available chemicals from many suppliers worldwide. Most of the vendors included are suppliers of research or fine chemicals. Pricing information is included. Updating is twice a year. This database can be searched by such entry points as chemical name or synonym, structure, substructure, molecular formula, molecular weight, or CAS Registry Number. It is configured for installation on in-house server computers.

Fine Chemicals Database, searchable on CD-ROM, is produced by Chemron, San Antonio, TX. This buyer's guide includes thousands of products from companies in North America and Europe. It focuses on laboratory and specialty chemicals, and on unusual and reagent chemicals.

Chem Sources, Clemson, SC 800-222-4531 produces several widely used buyer's guides that are especially useful for fine or specialty chemicals that may be difficult to locate elsewhere, although other types of chemicals are also included. These are *Chem Sources-International* (worldwide coverage), also available on CD-ROM; *Chem Sources-USA;* and *Chem Sources-Europe.* Chem Sources data are also searchable online through STN International (*CSCHEM* and *CSCORP* files).

In addition, some of the buyer's guides listed in previous pages (those of SRI, *Chemical Market Reporter,* and *Chemical Week)* also list manufacturers of fine and specialty chemicals.

A number of specialized guides are available to assist in purchase or identification of chemicals and formulated products intended for specific end uses. There are also special guides for laboratory and other kinds of equipment. Some examples are:

1. *Adhesive Age Directory,* Intertec Publishing, Atlanta, GA.
2. *Analytical Chemistry LabGuide,* American Chemical Society, Washington, DC. Now includes not only instruments, but also chemicals, standards, and consulting and other services. There are now over 3600 product categories, including 750 environmental and 850 biotechnology. ACS had previously published separate environmental and biotechnology directories. Internet searching is available at www.acs.org.
3. *Chemical Engineering Buyers Guide,* McGraw-Hill, New York, NY. Includes equipment; computer hardware and software; design, engineering, and construction; instrumentation; engineering materials; chemicals by type of function such as catalysts, defoamers, dispersing agents; and environmental services. There are indexes by trade name, products and services, and manufacturers.
4. *Farm Chemicals Handbook,* Meister Publishing Company, 37733 Euclid Avenue, Willoughby, OH 44094.
5. *Lockwood-Post's Directory of the Pulp, Paper, and Allied Trades,* Miller Freeman, Inc., San Francisco, CA (415-905-2200; 800-848-5594). Detailed listing of the pulp and paper mills in the United States, Canada, and Mexico, with listings also for suppliers of chemicals and other such firms.
6. *Modern Paint and Coatings Paint Red Book,* Intertec Publishing, Atlanta, GA.
7. *Modern Plastics Encyclopedia,* published annually by McGraw-Hill, New York, NY.
8. *Rubber Red Book,* Intertec Publishing, Atlanta, GA.
9. *Soap, Cosmetic and Chemical Specialties Blue Book,* PTN Publishing, Melville, NY.

A. Japan, India, and China: Market and Related Information

In addition to the buyer's guides mentioned earlier in this chapter, a major publisher of commercial and related information about chemical products and their manufacturers in Japan is Chemical Daily Co., LTD, 3-16-6 Nihinbashim Hama-cho, Chuo-ku, Tokyo 103 Japan. Most of these products are available in English and can be purchased in the United States from TEKNO-INFO Corp., PO Box 436627, Louisville, KY 40253 (502-254-5728). One such product is *CHEMINDEX,* which is a buyer's guide for chemicals manufactured in Japan.

A somewhat related product is *JCW* (*Japan Chemical Weekly*) *Chemical Products Handbook,* which includes, for over 10,000 Japanese chemicals that are marketed worldwide, not only properties and uses but also other information that can be useful in commerce. Data include manufacturers, CAS Numbers, EINECS Numbers [European Inventory of Existing Commercial Chemical Substances—an inventory from the European Commission equivalent to the U.S. Toxic Substances Control Act (TSCA) Inventory], UN Number (useful for transport of dangerous goods), basic physical properties, uses, prices, and basic Japanese laws and regulations. It also includes Class Reference Numbers of Existing Chemical Substances (list of existing chemical substances registered by Japanese Ministry of International Trade and Industry or MITI). Also available is *Japan Chemical Directory;* this is a guide to Japanese chemical companies that is organized by company. About 100 other Southeast Asia companies outside of Japan are said to be included as well.

There are also two products that relate primarily and specifically to the Japan equivalent of the U.S. TSCA Inventory. One is *Guide to Current ENCS Listed Chemicals* (the ENCS number is the Japanese TSCA number). The other is *Handbook of Existing and New Chemical Substances,* which is said to provide related information as to products that may be imported or produced in Japan.

TEKNO-INFO Corp., in addition, is a sales agent for the news magazine *Japan Chemical Weekly* (in English). They also have available *India Chemical Weekly Buyers Guide* and *India Chemical Weekly Annual.*

An excellent source of virtually all types of information about the chemical industry in China is China National Chemical Information Center, 53 Xiaoguanjie, Anwai, Beijing, 100029 China. This organization was established in 1959. An extensive range of chemical information services and products is available, including online search and retrieval. It is reported that over 300 technical professionals work here. Publications include directories, news weekly (in English), an abstracting service, statistics, and other. Once again, a sales agent for some of the products in the Americas is TEKNO-INFO Corp., Louisville, KY.

One such publication is *Directory of Chemical Products and Producers in China* (Chinese and English). Besides a section on chemical products, there are sections on producers, with basic information on 11,000 producers, including products, number of employees, and annual output; there is also a chemical names index and an appendix of enterprises engaged in import and export. This book is also available from Schnell Publishing, New York, NY.

A related product is *China Chemical Industry Yearbook* (in Chinese and English), which includes major trends and statistics. Also available is *China Chemical Reporter* (in English), bimonthly.

The China National Chemical Information Center has a number of other important publications of chemical interest that are too numerous to list here.

B. Mexico

Schnell Publishing Co., Two Rector Street, New York, NY, 10006 distributes *Mexican Chemical Industry Guide.* This includes an English-Spanish product index. A diskette version is available. Informatica Cosmos is the original publisher.

C. *Thomas Register of American Manufacturers*

On occasion, the chemist or engineer needs to identify sources of equipment or materials not necessarily related directly to chemistry. For these occasions, a comprehensive overall guide for U.S. manufacturers is *Thomas Register of American Manufacturers.* This is a buyer's guide for all types of products that also includes a trade-name index. The volumes are available in both hard-copy and online form and can be found very widely in all types of libraries, both large and small. It is available from Thomas Publishing Co., New York, NY, at a relatively low price considering the volume of information provided. The printed volumes are updated annually. This is available online through DIALOG.

16.4. OTHER SOURCES FOR LOCATING CHEMICALS

Chemists who cannot locate sources for chemicals of interest in buyer's guides need not despair. Several options remain open.

Some companies either specialize in stocking hard-to-find chemicals or will consider making these on request perhaps in small (research) quantity as well as in larger quantities. Examples include

Aldrich Chemical Co.
1001 W. Saint Paul Ave.
Milwaukee, WI 53233
(Primarily organics and biochemicals: off-the-shelf, custom synthesis service, or in larger-than-laboratory quantities.)

Alfa Aesar-Johnson Matthey
30 Bond Street
Ward Hill, MA 01835.
(Inorganics are a specialty.)

Carbolabs, Inc.
Fairwood Rd.
Bethany, CT 06524
(Specializes in "difficult" reactions.)

Lancaster Synthesis, Inc.
P.O. Box 1000
Windham, NH 03087

Sigma Chemical Company
P.O. Box 14508
St. Louis, MO 63178
(Specializes in biochemical and organic compounds for research and diagnostic and clinical reagents.)

Note that Aldrich, Carbolabs, and Sigma are now part of the same corporate entity.

The identification of companies that can manufacture chemicals on a custom basis is a matter of increasing importance, and "outsourcing" has become a major trend. There are many such companies worldwide, in addition to those mentioned above, and most have particular strengths and technologies of emphasis such as certain types of reactions or special equipment. Fortunately, there are a number of useful guides to such companies. Examples include

1. *Brandon Guide to Custom Chemical Synthesis Services—North America* and *Brandon Guide to Custom Chemical Synthesis Services—Europe,* Brandon Associates, PO Box 1244, Merrimack, NH 03054 (603-424-2035). Other related publications include *Brandon Worldwide Monomer Reference Guide and Sourcebook* (includes both suppliers directory and some basic properties),

Brandon Amino Acid and Peptide Reference Book (includes both suppliers directory and some basic properties); *Brandon Buyer's Guide to Pharmaceuticals and Pharmaceutical Manufacturers in India, Brandon Buyer's Guide to Fine Chemicals and Intermediates from India,* and *Brandon Guide to Custom Processing and Packaging Services.* Some of these are available in both printed and diskette forms.

2. Chem Sources, Clemson, SC. See Section 16.3.
3. *Chemcyclopedia.* See Section 16.3.
4. *Custom Processing Services,* produced by Custom Guide Co., PO Box 358, Closter, NJ 07624. This is said to be distinguished by in-depth indexes as to reactions, such as phosgenation, and by unit operations, such as spray drying. Another feature is full descriptions of companies. The scope is primarily United States and western Europe. Planned for future editions is an index of equipment and a matrix that facilitates location of companies with combinations of reaction capabilities, such as phosgenation and sulfonation.
5. *Directory of Custom Chemical Manufacturers,* Delphi Marketing Services, 400 East 89th Street, Suite 2J, New York, NY 10128. Edited by Newton H. Giragosian, a leader in chemical marketing research.
6. *Guide to Custom Chemical Manufacturing and Processing,* Quest Data, PO Box 210288, Nashville, TN 37221. This directory is said to include listings for more than 425 companies in the United States, Canada, the United Kingdom, France, Japan, and 14 other countries. There are indexes by services and area of specialization, chemical products, and plant locations.
7. *SOCMA Commercial Guide,* Synthetic Organic Chemical Manufacturers Association, 1330 Connecticut Ave. NW, Washington, DC 20036 (212-414-4100). This excellent guide contains detailed profiles on over 160 custom manufacturers and service providers, with an index for reactions. Coverage is limited primarily to the United States. It is believed to be the lowest cost publication of its type by a wide margin. Internet location is www.socma.com., but not all data are included.

16.5. CHEMICAL PRICING; IMPORTS AND EXPORTS

Because prices depend on quantity, grade, location of buyer of and manufacturer, and other variables, an individualized quotation from the manufacturer provides the most accurate pricing data in a specific situation. However, there are a number of published sources that can be very helpful.

Chemical Market Reporter is probably the best and least expensive source of U.S. list price data for most significant large-volume chemicals, and these data are published each week. (Excellent at this listing is, by no means are all chemicals that are available commercially included.) However, these prices do not necessarily reflect the actual prices that are negotiated in commerce; these can vary widely from list prices depending on volume, location of customer and manufacturer, and other factors. *CMR* offers a diskette version of its price pages on a quarterly basis.

A number of other trade magazines list prices. For example, *Chemical Week* offers printed and electronic versions of spot and contract pricing for major commodities in the United States and Europe.

Another source of chemical pricing data is ICIS-LOR [Independent Chemical Information Services, London Oil Reports, a division of Reed Business Publishing (Reed Elsevier); Reed also publishes *European Chemical News* and many other publications], 3730 Kirby Drive, Houston, TX 77098 (713-525-2600).

Data available from ICIS-LOR on a subscription basis include contract and spot pricing, "done deals" confirmed, market commentary, list prices, and export prices. Products covered include some 75 chemicals in the following categories: analgesics, aromatics, alcohols, base oils, crude oil, feedstocks, fertilizers, inorganics, intermediates, olefins, plastics, refined products, and solvents. Separate reports are available for the United States, Europe, and "Asia Pacific." Reports are faxed to subscribers on a weekly basis. ICIS-LOR says that their editors are on the phones daily to discover market transactions that lead to their price assessment of the market.

Chemical Data, Houston, TX (713-683-3900) is another source of pricing data. This firm provides a variety of multiclient and single-client studies dealing primarily with marketing and manufacturing issues within the petroleum, natural-gas, petrochemical, and plastics industry. Two multiclient studies are published 12 times a year: *Monthly Feedstocks and Fuels Analysis* and *Monthly Petrochemical and Plastics Analysis.* Forecasts of demands, costs, and prices are made.

Chem Systems, Tarrytown, NY, offers a subscription program known as *Petroleum and Petrochemical Economics (PPE)—United States,* and also has separate related programs covering Europe and Asia. The objectives are to monitor and evaluate current events and to forecast medium-term trends. Initiated in 1987, the U.S. program includes quarterly business analysis reports, which include margins and pricing analysis for U.S. petrochemicals and polymers; executive reports, which include, among other information, price forecasts for U.S. petrochemicals and polymers; annual reports that include energy and refined products supply–demand and pricing and U.S. petrochemical supply, demand, and economics; and limited consulting time. For example, one of the reports in the program is *U.S. and Petrochemical and Polymer Price Forecasts,* the most recent edition of which was published in 1995, which is available for separate purchase. Over 15 products are covered in this report.

The SRI PEP and Chem Systems PERP programs (see Chapter 17) are other sources of price information.

There are many other sources of pricing data, some of which are noted elsewhere in this book.

Chemical Market Reporter's chemical imports pages appear in each week's issue. The data show monthly volume and value of shipments imported into the United States. Other data given include country of origin, U.S. customs port of entry, and current and previous year volumes. Names of importers and exporters are not given. Some 400 chemicals are included. The database is "rolling"—not all the chemicals appear each week.

In addition to the import data that *CMR* presents in the weekly newspaper, it sells customized reports on imports and exports. These reports vary from single to multiple commodities, and scope ranges from worldwide to single-port and single-country, depending on customer requirements. Longer reports are available on disk as well as paper. As *CMR* has access to the entire U.S. Department of Commerce trade database, reports on nonchemical commodities are also provided.

The *Journal of Commerce PIERS* (*Port Import Export Service*) file is an excellent source of data about both imports and exports at a large number of U.S. ports. Types of data include product name, weight, point of origin, name of exporter, and U.S. importer. Prices are not given. The import and export databases are available online through DIALOG or the *Journal of Commerce,* New York, which may be contacted directly.

16.6. GENERAL CHEMICAL BUSINESS INFORMATION

To help the chemist or engineer locate the latest chemical business information, several good abstracting and indexing services and other tools are available.

One of the most complete services is PROMT. This tool is distinguished by in-depth abstracting and versatile indexing access. PROMT is available online through a number of major host systems. It is a product of Information Access Co., Foster City, CA. Mentioned earlier in this book is *Chemical Industry Notes,* a significant tool in this field published by CAS, but not as complete as PROMT.

An information service aimed primarily at covering the commercial and marketing aspects of the chemical industry in Europe, but covering other continents as well, was initiated by The Royal Society of Chemistry in January 1985. The service, *Chemical Business NewsBase,* is accessible via several online host systems and is updated weekly.

CPI Digest, published monthly, provides abstracts of recent developments in some of the "hottest" developments in the chemical industry. Its coverage includes such topics as marketing and management, adhesives and sealants, coatings, waterborne systems radiation processing, pigments and dyes, polymer research, fibers and textiles, rubber and elastomers, plastics and resins, and printing inks. Abstracts are provided on the basis of recent periodical articles, and there are also brief notes on pertinent U.S. patents in the various categories. The publisher is CPI Information Services, 2117 Cherokee Parkway, Louisville, KY 40204 (502-456-6288). The subscription cost of this eminently practical source is quite reasonable.

The *Wilson Business Abstracts—Full Text* is available online through *DIALOG.* It refers to many articles of chemical business interest and is easy to use. The corresponding print version can be found in many libraries. H. W. Wilson Co. is the publisher.

Chemical Products Synopsis is a reporting service on over 200 individual major chemical commodities. The succinct reports provide general marketing information, including present situation, short- and long-term outlook, pricing, producers, capacities, uses, and brief information on the marketing and environmental aspects. The publisher is Mannsville Chemical Products. A similar publication, but less detailed, is *Chemical Market Reporter's* service *Chemical Profiles.* This is published weekly as part of the parent magazine and can also be purchased separately.

Trends in End Use Markets for Plastics is an especially valuable indexing and abstracting source on plastics markets with indexes by markets and materials. The publisher is STR.

In addition to what is mentioned elsewhere in this chapter, there are several good weekly news sources that cover the chemical industry as a whole. For European chemical industry news, an excellent source is *European Chemical News.* For those who require more frequent updates than the weekly magazines, an especially good source is the daily *Chemical Week Newswire,* which is transmitted electronically to subscribers to this service by the publishers.

The U.S. Department of Commerce offers a huge amount of statistical data on the Internet (www.stat-usa.gov) on a fee basis.

16.7. *CHEMICAL ECONOMICS HANDBOOK*

Perhaps the largest single independent chemical market research effort in the worldwide chemical industry (outside of chemical companies themselves) is the *Chemical Economics Handbook* (*CEH*), which is a product of SRI Consulting, Menlo Park, CA. Coverage includes business, marketing, and some technical information on virtually all major commercial chemical products and products groups. Initiated in 1950, this program now comprises some 20,000 looseleaf notebook pages that cover data on over 1000 chemicals, with market research reports and product reviews on some 400 chemicals or chemical industry topics. There is an excellent index that facilitates locating information. Eric J. Linak is CEH Director.

Each market research report includes such data as the following: summary; manufacturing process description; suppliers, including producers, locations, and capacities; industry production and sales; demand, including consumption patterns and trends; prices, trends, and factors; import and export data; and bibliography. The product reviews are more concise. Coverage focuses on the United States, western Europe, and Japan, for the most part, although some reports have expanded geographical coverage.

Most reports are completely revised every 3 years, some more frequently. In addition, a quarterly *Manual of Current Indicators* updates the major statistical series for *CEH* topics. There are also a quarterly *Economic Environment of the Chemical Industry,* with graphic and tabular summaries of general economic conditions and trends, and a bimonthly *Chemical Industries Newsletter.*

The contents of *CEH* are available online through DIALOG. SRI Consulting offers limited inquiry and consulting privileges as part of the subscription price.

A related program is the *Specialty Chemicals Update Program* (*SCUP*) of SRI Consulting, Menlo Park, CA. This is under the direction of Phil J. Calderoni. The program began in 1979. Twelve completely revised reports are issued each year. Purchasers may subscribe to the entire program and receive all reports, or alternatively, purchase individual reports. Inquiry and consulting privileges with SRI Consulting staff are available to participating companies.

Geographic scope of the reports is the United States, western Europe, and Japan.

Some 32 categories of products are included. These may be either market-directed such as electronic chemicals, or functional chemicals, such as biocides.

Another product closely related to *CEH* is SRI Consulting's *World Petrochemicals Program* (*WP*). This program, initiated in 1972, provides supply–demand and producer–capacity information for over 60 petrochemical raw materials, intermediates, and end-use products in over 90 countries. The program is divided into six segments as follows: feedstocks, methanol and derivatives, ethylene and derivatives, propylene and derivatives, C4 hydrocarbons and derivatives, and aromatics and derivatives. Companies may purchase the entire program if desired, or only segments of interest. Each segment includes world product summaries, regional product summaries, and country product reviews. The supply–demand information is updated annually, and producer–capacity data are updated twice per year.

All products covered in *WP* are also covered in *CEH,* but the information in *CEH* is updated every 2–3 years, and only for the United States, western Europe, and Japan. However, the levels of analyses in *CEH* tend to be in greater depth than in *WP.*

Producers, products, processes and feedstocks, and capacities for a range of years are provided for more than 8000 petrochemical plants worldwide. Company ownerships and interrelationships are shown. Supply–demand data are also given for a range of years. Annual prices are supplied for many products in the United States, Germany, and Japan. Data for individual countries are aggregated into regional and world summaries. There is textual information that includes discussions of the product; technology/process/feedstock; capacity/company/logistics; foreign trade; derivative/use; price/cost; analysis/outlook; and feedstock trends.

The printed volumes are updated annually. Subscription includes limited inquiry and consulting privileges. An option is annual CD-ROM service, which has the advantage of being easier to search and use. The service is not available online. J. Paul Bjacek is acting director.

16.8. MULTICLIENT STUDIES

Multiclient reports are usually extensive market research reports prepared by consulting organizations that are intended for purchase by a limited number of clients, typically approximately 10–30 clients (although some reports may have a much wider client base). Such reports frequently deal with a specific chemical product or group of allied products with a common end use such as adhesives, biocides, coatings, or electronics chemicals. They will typically describe and analyze present markets, plant capacities, pricing, and future outlook, and will include discussion of the major companies in the business. The nature and impact of new technologies on the market are often discussed. Alternatively, multiclient reports may focus on investment opportunities, and in such cases, will profile and analyze specific companies. Some multiclient producers, in addition, will sponsor conferences in areas of special interest to them; for example, Business Communications Co., Norwalk, CT, sponsors periodic conferences.

Data are usually gathered by detailed field interviews (personal interview or phone calls) to market participants (producers, end users, and others), as well as in other ways such as literature and patent searches. Purchase of a study usually entitles the buyer to several copies of the report, and to a limited specified amount of free consulting time.

Multiclient reports can vary quite widely in number of pages, cost, and quality. Some reports are much more exhaustive than others because they include analysis of a much more extensive number of field contacts. These reports could ordinarily be expected to be more reliable, more useful, and more expensive, than those with only a relatively limited number of field contacts. Cost of purchase can range from a few thousand dollars up to $10–$20 thousand or even more. In some cases, individual chapters, sections, or even individual tables may be available for separate purchase. Some of the producers of multiclient studies have developed outstanding reputations and track records for quality. Some producers employ freelance researchers or consultants, while others have a full-time staff. Most multiclient producers also generate single-client studies; these are, of course, confidential to the clients.

In considering the purchase of a multiclient study, the prospective buyer will want to look at a brochure describing the study in detail, including a page or two of a nonconfidential summary, a list of tables and figures, and the table of contents. Methodology should be described, and ideally there will be a list of the field contacts made during the study. Especially in the case of the more extensive studies, many multiclient producers will make question-and-answer presentations about their reports, at a potential purchaser's office, and at this time, the purchaser will usually have some time to page through the complete study before deciding whether to purchase. The name(s) of the person(s) who wrote the study and their qualifications should be provided. Some purchasers will buy more than one study on the same topic in order to compare results from different producers in those cases involving crucial decisions. Currency and up-to-dateness are important; the value of most multiclient market reports degrades quickly with time.

Some of the best-known producers of multiclient reports include, for example, SRI International, Charles Kline and Company, Strategic Analysis, Decision Resources, Freedonia Group, Frost and Sullivan, *Scrip Reports* (represented in the United States by Pharmabooks, New York), Hull and Company, and Business Communications Co. The inclusion or exclusion of a company in this list is not intended to convey an endorsement or lack of endorsement; each market research firm should be evaluated by the potential buyer with respect to specific studies and the associated costs and benefits.

Special note should be made of *DIRASS—Strategies and Research Trends of Japanese Chemical and Pharmaceutical Companies*. This is a multiclient report series published by DIA Research Institute, Tokyo. Representation in the United States and Europe is through Technology Catalysts, Falls Church, VA. The series consists of technoeconomic appraisals of major Japanese chemical and pharmaceutical companies. In addition, there is a series of reviews on technical developments. Much of the assessment is done through analysis of published patent applications (Kokai).

Some multiclient reports can be found online through DIALOG in full-text form. The file series is known as "Marketfull." Users can choose to look at either the full report or only those parts of interest. DIALOG also offers an online file, *FINDEX,* which helps identify market research reports on a wide variety of topics from many different sources; this is a product of Cambridge Scientific Abstracts, Bethesda, MD.

The Freedonia Group, Cleveland, OH, offers the full text of its market research reports for a fee on the Internet (www.freedoniagroup.com). Individual pages and/or tables may be purchased.

16.9. PRODUCT DATA

When the manufacturer of a product has been located, the chemist can obtain extensive information about the product by requesting product bulletins and other trade literature from the manufacturer. Some of this material contains information not readily available from any other source, as mentioned earlier. Examples of information found in trade literature include (a) specifications, (b) physical and chemical properties in detail, (c) handling precautions (MSDS and other safety data), (d) analytical methods, (e) methods of appropriate use in specific applications, and (f) a list of pertinent articles and other references. Some manufacturers maintain centralized product information centers that can provide considerable data about their products over the telephone. Please see also Section 15.13 about the importance of talking with a technical person.

A convenient source of product data is the Vendor Catalog Service of Information Handling Services. A number of chemicals and other products are included in this extensive collection of product catalogs.

Several excellent tools are available to help chemists and engineers match materials with desired properties. Some of these tools are described in Chapter 15 on properties.

16.10. *CHEMICAL INTELLIGENCE SERVICES*

Useful online sources of chemical market data are *Chemplant Plus* and *Chemstats,* both published by Chemical Intelligence Services (Chem-Intell), a division of Reed Telepublishing Ltd., a member of the Reed Elsevier Group, and located in London, England.

Chemplant Plus comprises data for manufacturing plants worldwide for over 100 major organic and inorganic chemicals. For each product, the type of information is company, contractor, licensor, present capacity, expansion capacity, future/planned capacity and costs, start-up date, project status, contractor, process, and feedstock. A companion subfile is a historical database from 1980 to near the present and covering similar data for the building of production plants worldwide for several thousand chemicals or chemically related products.

Chemstats contains annual trade and production figures for over 100 organic and inorganic chemicals worldwide. Coverage is from 1978 to the present.

These databases are online through DIALOG and DataStar.

16.11. CHEMICAL INDUSTRY TRADE AND RELATED ASSOCIATIONS

Although they exist primarily to serve their members, trade associations are an excellent way to obtain leads to chemical marketing information as well as other information.

Trade associations are, of course, sources of public information as to which companies are active in the field and about broad industry trends. Most associations track proposed new regulations and legislation carefully, especially as related to the environment. They hold meetings and trade shows, often open to the general public, which are excellent places to learn about new products, new processes, and other developments. In some cases, a number of valuable publications are available to the public, including membership lists, company capabilities, trade statistics, and safety and environment data. Most of the U.S. chemically related trade associations are located in the Washington, DC area, except as noted below. A few examples of trade associations include

CMA (Chemical Manufacturers Association)—the leading chemical industry association to which almost all larger U.S. chemical companies belong

Chlorine Institute

CSMA (Chemical Specialties Manufacturers Association)

Cosmetic, Toiletries and Fragrance Association (New York, NY)

Drug, Chemical and Allied Trades Association (DCAT) (Syosset, NY)

Halogenated Solvents Industry Alliance

National Paint and Coatings Association

National Association of Chemical Distributors

Pharmaceutical Research and Manufacturers of America

SOCMA (Synthetic Organic Chemicals Manufacturers Association)—many smaller companies belong, especially those engaged in custom and fine-chemical manufacture

Soap and Detergent Association, New York, NY

Society of the Plastics Industry

At the international level, associations of chemical interest include Canadian Chemical Producers Association, Asociación Nacional de la Industria Química, Chemical Industries Association (CIA), Society of Chemical Industry, CEFIC (European Chemical Industry Council), and ICCA. Other associations will be found listed in directories of associations.

Chemists engaged in market research also belong to associations. In addition to the American Chemical Society Division of Business Development and Management, these include the Commercial Development Association, Washington, DC; and CMRA (Chemical Management and Resources Association), Staten Island, New York.

16.12. SOME OTHER SOURCES OF MARKETING INFORMATION

In previous years, the U.S. and state governments have been good sources of chemical market research information. However, with ongoing reductions in the size of government, the continued future availability of some of this information is uncertain. See also Chapter 9.

However, regulatory information that is readily publicly available remains a source. For example, if a chemical is not on EPA's TSCA (Toxic Substances Control Act) inventory of chemicals in commerce, a PMN (Premanufacturing Notice) must be submitted to EPA before the chemical can be manufactured or imported into the United States. EPA reviews and evaluates the PMN before market introduction. Some PMN data are kept confidential, but if not, this can be an excellent source of what companies plan to do. A few thousand PMN applications are filed each year, and these are a matter of public record (*Federal Register* and sources that abstract or index it).

A service that is unique in the chemical and allied industries—and, for that matter, rare in any category of manufacturing—is the statistical census of the manufacturing capability of over 17,000 U.S., Canadian, and Mexican plastics processing plants. The product is known as the *Plastics Manufacturing Census* database. It has now been further expanded in scope. The producer says that the product covers all of North America with detailed information on the activities of at least 93% of the manufacturing plants performing one or more of the processes covered.

This census was initiated in the 1970s by *Plastics Technology Magazine* and is now available through PLASPEC, D&S Data Resources, Morrisville, PA (see also Chapter 15). Plants included are those that perform one or more of the following process types: injection molding, RIM, extrusion, blow molding, thermoforming, and compounding. For each plant covered, the data provided include types and nature of equipment, types of material processed, pounds processed annually, and end uses of product. Data are available on a total censuswide basis or by individual plant. New software allows use of this information easily and effectively to establish marketing programs, prequalify sales leads, manage sales and distribution, and identify new market opportunities. **PLASPEC** has made numerous enhancements to and developed proprietary software for this database. These data are not available online, but rather only by contacting PLASPEC directly.

REFERENCES

1. N. H. Giragosian, Ed., *Chemical Marketing Research,* Reinhold, New York, 1967.
2. D. Lee, *Industrial Marketing Research,* 2nd ed., Van Nostrand Reinhold, New York, 1984.
3. *Moody's Industrial Manual,* Moody's Investors Service, New York. Published annually with supplements.
4. *Dun & Bradstreet Million Dollar Directory,* Dun & Bradstreet, New York. Published annually.
5. *Standard and Poor's Register of Corporations,* Standard and Poor's, New York. Published annually.

6. *Who's Who in America,* Marquis Who's Who, Division of Reed Elsevier, New Providence, NJ. Also available online. Published annually.
7. *Reference Book of Corporate Management,* Dun & Bradstreet, New York. Published annually.
8. (a) *American Men and Women of Science,* R. R. Bowker Co., Division of Reed Elsevier, New Providence, NJ. Updated every 3 years. Also available online; (b) *Who's Who in Science and Engineering,* Marquis Who's Who, Division of Reed Elsevier, New Providence, NJ.

List of Addresses

Disclosure, Inc.
5161 River Road
Bethesda, MD 20816

H. W. Wilson Co.
950 University Avenue
Bronx, NY 10452

Industrial Research Institute
1550 M Street
Washington, DC 20005-1712
(202) 296-8811

Information Handling Services
PO Box 1154
Englewood, CO 80110

Mannsville Chemical Products
PO Box 220
Adams, NY 13605

Marquis Who's Who
Division of Reed Elsevier
121 Chanlon Road
New Providence, NJ 07974

SRI Consulting
333 Ravenswood Avenue
Menlo Park, CA 94025
(415) 326-6200

STR (formerly Springborn Laboratories)
10 Water Street
Enfield, CT 06082

Strategic Analysis, Inc.
Box 3485, R.D. 3, Fairlain Road
Reading, PA 19606

17 Process Information

17.1. INTRODUCTION

How individual chemical products are best made in full-scale commercial practice is of utmost interest and importance. Understandably, this information can be difficult or impossible to obtain from the literature. The reason is that efficient manufacture is essential to profits and therefore regarded as proprietary.

Although the chemist or engineer may find the broad features of a process and even considerable detail described in nonproprietary literature, the fine detail and ongoing improvements that contribute to optimum production efficiency are rarely published in journals and books.

Patents are often an invaluable source of information on processes, as explained in Chapter 13, and often contain details not found anywhere else. However, patents usually give *several* examples or specify a *range* of pressures, temperatures, or other conditions. The *optimum* conditions can, therefore, be difficult to determine from patents, except by educated guessing. More details are likely to be available for industrial products and processes that are *mature* (older) such as basic heavy inorganics like sulfuric acid. Other areas in which process detail is likely to be more readily available are those that affect the public good, such as pollution abatement and energy conservation. Despite the limitations noted, the astute chemist or engineer who knows the technology can sometimes glean valuable insight by careful study of pertinent literature.

A first step is often to study what is written in the major chemical encyclopedias such as *Kirk–Othmer* (see Section 12.2.A). These often provide any basic background required. Briefer treatments of process details for the best known products can be found in such single volume works as Wittcoff and Reuben (1) and Kent (2). These are handy for regular desk use.

Accounts of processes, or related pertinent information, may appear in such journals as *AIChE Journal, Chemical Engineering, Chemical Engineering Progress, Hydrocarbon Processing* (especially the November issue), *Industrial and Engineering Chemistry Research,* and *International Chemical Engineering.*

These sources and others can be accessed directly or via such tools as *Chemical Abstracts* and *Derwent,* both of which are described earlier in this book. But the chemical engineer or chemist will frequently find the most comprehensive detail on process technology and economics for key chemical products in the services described in Section 17.2.

17.2. SPECIALIZED SERVICES

Several highly specialized services provide recent process technology and economics information, primarily for important chemicals of commerce, in considerable detail. The evaluations of processes and comparisons of competing processes are significant features. These sources are relatively expensive (typical costs range from about $30,000 to $40,000 per year), but they can be found in most major chemical companies and in some major universities. The price is well worth it. Because the information is private to clients of the services, reference to reports from these sources is not ordinarily found, unless the information is made publicly available by the source. The leading services currently available are

SRI Consulting
333 Ravenswood Avenue
Menlo Park, CA 94025—*Process Economics Program* (*PEP*)

Chem Systems, Inc.
303 S. Broadway
Tarrytown, NY 10591—*Process Evaluation/Research Planning* (*PERP*)

The two services cover some of the same products. When this is the case, the chemical engineer or chemist should consult the corresponding reports of both services, because differences in information content, evaluations, and economic estimates may be found.

The SRI *PEP* service started in 1963 and now totals hundreds of reports. Reports include an intensive technical review and analysis of each basic process. The aim is to establish commercially feasible operating conditions and to assess technical factors underlying present limitations as well as prospects for improvement. Sufficient details are presented to permit verification of design calculations and cost estimates.

PEP studies relate to polymers, specialty chemicals, refinery and commodity chemicals, and special process technology topics. Evaluations are thorough and detailed. An especially noteworthy feature is the in-depth analysis of the patent literature, which is then fully referenced as a part of each report. Analyses are conducted in-house and are then submitted to licensors and operators for review and comment.

The program includes the following components:

- A series of reports on important chemical and refinery products, based on studies of process technology and cost. This is the core of the program. There are 12 such reports each year.
- *Process Economics Reviews,* a highlighting of implications of certain new developments in the industry (about 12 issues per year).
- An annual *Yearbook* that updates cost figures presented earlier in the reports. The *Yearbook* includes economics of chemical production in the United States (Gulf Coast), as well as Germany and Japan. A newer feature is the inclusion of some data for production of selected products in the Middle East. The most recent

yearbook includes data for about 820 processes to make over 500 products, and it is available in both printed and CD-ROM formats. Most recently, the program published a comparable volume relating to the production of chemicals in China; plans call for this is to be updated every other year.

- Another newer related optional feature is an *Environmental Economics Handbook,* which was issued for the first time in 1993. This covers the economics of waste abatement and soil remediation technologies.

The Chem System *PERP* service, which is under the direction of Jeffrey S. Slotkin, is intended to provide the following

- A bulletin describing the status of current process studies issued at 4-month intervals.
- License monitoring.
- Consultation with members of program staff on reports and other matters of interest to individual clients.
- A continuing evaluation of commercial significance of both existing and emerging technological developments—specific products, product groupings, product applications, and technologies
- Realistic planning information on manufacturing economics of chemicals, polymers, and technologies
- Information on status of chemical process, product, and technology development efforts of companies throughout the world
- Identification of opportunities for purchasing or licensing partially developed technology
- Placement of technological developments in a commercial context by highlighting supply/demand trends

Both the technology and the commercial status are examined in arriving at forecasts. An analysis of company activities is important for certain reports since plant size will be a function of market concentration as well as engineering and economic factors. Geographic coverage usually includes the United States, western Europe, and East Asia. The potential commercial impact of technology developed in other geographic regions is considered in any examination of a product area.

The service is designed to be useful in new project planning, R&D, licensing evaluation, purchasing analysis, technological forecasting, and strategic planning for market entry.

It provides subscribers with

- A series of eight update reports each year, each covering current and future technology, economics, and commercial profiles for one of some 36 large-volume chemicals and polymers with respect to chemical processes, product applications, and technologies.

- Fourteen "Special Topic" reports discussing commercial impact of important recent technology developments. Each report develops and presents an economic and commercial analysis of current developments that Chem Systems staff believe to be significant. Special Topic reports cover a wide range of subjects, including refinery issues, intermediates, specialty chemicals, and engineering plastics.
- Chem Systems Annual Petrochemical conferences, which are held periodically around the world including the United States, western Europe, the Middle East, Latin America, Southeast Asia, and Japan. PERP subscribers may attend at special rates.

17.3. OTHER SOURCES OF PROCESS INFORMATION

Among the most overlooked sources of process information are published environmental studies. In the course of describing pollution problems and solutions, detailed information on processes is often provided. Government publications are excellent examples of this type of treatment, for example, some of the studies sponsored by EPA. Other examples are found in the proceedings of the annual Purdue University meetings on industrial waste pollution abatement. Accordingly, although the main thrust of pollution abatement studies is environmental, chemists and engineers seeking process information should consult these published studies.

REFERENCES

1. H. A. Wittcoff and B. G. Reuben, *Industrial Organic Chemistry,* Wiley, New York, 1996.
2. J. A. Kent, Ed., *Riegel's Handbook of Industrial Chemistry,* 9th ed., Van Nostrand-Reinhold, New York, 1992.

18 Analytical Chemistry: A Brief Review of Some of the Literature Sources

Tools of potential interest to analytical chemists are discussed in several preceding chapters. In this chapter, some of the more specialized sources that are of particular interest to analytical chemists are briefly discussed.

18.1. INTRODUCTION

Analytical chemists rely on a large number of both separately available (printed, CDs, diskettes, and online) and "other" sources of analytical data and other information. Available libraries of spectral data and of much other data of interest to analytical chemists are frequently, and preferably, available in electronic digitized formats. However, a number of printed volumes of such data also continue to be available.

The "other" sources include the leading analytical instrument manufacturers. Thus, many instruments can be purchased with libraries of spectral tables and other analytical data included by the instrument manufacturers. Such instruments are often interfaced with a CD that contains these tables and other data of interest.

For example, Nicolet Instruments, Madison, WI, recently (1997) announced the availability of a Raman database (Aldrich Raman Library) with more than 14,000 spectra developed jointly with the Sigma-Aldrich Chemical Company, Milwaukee, WI. In addition, Nicolet has a very large database of FT-IR spectra and other spectral libraries as well. Similarly, Perkin-Elmer Corp., Norwalk, CT, offers a very large number of Sadtler and other digital infrared spectral tables.

In addition, some instrument manufacturers can be very helpful in supplying application bulletins for methods of interest and even in suggesting methods to solve specific problems. One instrument manufacturer, Perkin-Elmer, Norwalk, CT, publishes a scientific peer-reviewed journal, *Atomic Spectroscopy.*

Other less formal sources of the most current information for analytical chemists include the two huge analytical chemistry meetings, the so-called Pittsburgh Conference (which, of course, may be held anywhere in the United States), and the Eastern Analytical Meeting. In addition to the technical papers, there are important exhibits. Analytical departments who want to keep with the latest are well advised to send at least one representative to at least one such meeting annually. These meetings are in addition to the usual national, regional, and special meetings that continue to be important. An example of another group of interest that meets periodically is the Directors of Industrial Research Analytical Group.

Other good sources of methods include the chemical companies that manufacture the chemical products of interest.

See also the discussion about analytical data on the Internet in Section 18.11.

18.2. MASS-SPECTRAL DATA

The largest mass-spectral reference library known is that assembled and developed by Professor Fred W. McLafferty (Cornell University, Ithaca, NY) and his associates. This is the *Wiley Registry of Mass Spectra Data,* sixth edition, and it is available commercially through John Wiley & Sons, New York, NY. It includes over 230,000 spectra for over 200,000 compounds; this represents an increase of over 90,000 new spectra as compared to the previous edition.

In addition, the firm Palisade Corp., Newfield, NY, (800-432-7475) sells the complete reference library software package known as the BenchTop/PBM MS Library Search System. The *Wiley Registry* described above is available optionally in combination with the *NIST Mass Spectral Database* (compiled by Dr. Stephen E. Stein) to yield a total of approximately 275,000 spectra. A key feature of the total package is that unknown spectra can be matched against the reference system utilizing the algorithm known as Probability Based Matching (PBM). This, it is said, can yield results in about one second with a modern desktop computer of reasonable speed and power. PBM has two unique features: reverse searching, which improves identification capability for mixture components, and weighting of the mass and abundance data, which improves retrieval performance and provides a quantitative confidence in the degree of match.

The fifth edition of the Wiley mass spectral database is available online through the Chemical Information System (see Section 14.4) It may also be obtained through instrument manufacturers when an instrument is purchased.

STN International has available online the database *SPECINFO.* This is said to include 80,000 NMR spectra for over 152,000 compounds, 17,000 IR spectra, and nearly 66,000 mass spectra. Structure searching capability is included, as is the capability to estimate NMR spectroscopic information. The producer is Chemical Concepts GmbH.

The Royal Society of Chemistry offers *Eight Peak Index of Mass Spectra* (1). This contains the eight most abundant *m/z* (mass/charge) values with intensities for over 81,000 mass spectra covering over 65,000 compounds. Data given also include molecular formula, compound name, and parent peak intensity. The CAS Registry Number is listed where readily available. The data are sorted into three separate volumes: Volume 1 (two books), by molecular weight subindexed on formula; Volume 2 (two books), by molecular weight subindexed on *m/z* values; and Volume 3 (three books), by *m/z* of the two most abundant ions.

18.3. DATA FROM SADTLER DIVISION OF BIO-RAD LABORATORIES

Sadtler, Philadelphia, PA (215-382-7800) has long been a provider of useful spectral data. Sadtler was founded in 1874, and it published its first collection of spectra in 1947. The firm is now a division of Bio-Rad Laboratories, Inc., Hercules, CA. Marie Scandone is in charge of data analysis. Sadtler says that it currently has available over 400,000 reference spectra in digital and printed form as shown in Table 18.1. Its FT-IR spectral libraries are available in a large number of categories, based largely on types of uses, as well as types of compounds, such as adhesives and sealants, automobile paint chips, dyes, flame retardants, intermediates, lubricants, unmodified monomers and polymers, polyols, priority pollutants, and many others. NMR databases are also supplied. Sadtler also offers software that permits search of the spectra by chemical structure, by peaks, and in other ways. Its products are frequently sold "bundled in" with Perkin-Elmer Instruments, or they may be purchased separately.

TABLE 18.1. Spectral Data Available from Sadtler Division of Bio-Rad

Digital IR Spectra	
Standard IR spectra	75,620
Commercial IR spectra	87,600
Vapor-Phase IR spectra	9,200
Total	172,420
Digital NMR Spectra	
Carbon-13 NMR spectra	40,000
Total	40,000
Hard-Copy Spectra	
Standard IR spectra	92,000
Commercial IR spectra	56,300
Vapor-phase IR spectra	9,200
Proton NMR spectra	64,000
Proton NMR (300-MHz) spectra	12,000
Carbon-13 NMR spectra	42,000
C-13 NMR of monomers and polymers	800
UV spectra	76,000
Fluorescence spectra	2,000
Raman spectra	4,400
Total	357,700

Data supplied by Marie Scandone of Sadtler, Division of Bio-Rad.

18.4. DATA FROM ALDRICH CHEMICAL COMPANY

Aldrich Chemical Company, Milwaukee, WI (800-231-8327) sells several spectral libraries, primarily FT-IR spectral libraries; examples are

Aldrich Vapor-phase FT-IR Library
Nicolet/Aldrich Condensed Phase Library (and Supplement)
Nicolet/Aldrich Vapor Phase Library
Nicolet/Sigma Biochemical Condensed Phase Library
Nicolet/Sigma Steroids Condensed Phase Library
Nicolet Expanded Hummel Polymer Condensed Phase Library
Nicolet Polymers, Polymer Additives and Plasticizers Condensed Phase Library
Nicolet Food Additives Condensed Phase Library
Nicolet/Aldrich Flavor and Fragrances Vapor Phase Library

Aldrich also sells software to search its databases, and it also sells printed books with NMR and FT-IR spectral data. Its spectral products may be found "bundled" with Nicolet Instrument Co. products, or they may be purchased separately.

18.5. DATA FROM NIST

NIST analytical chemistry databases include those listed below in Table 18.2. Many of these are available as part of the analytical instrument when it is purchased.

18.6. HUMMEL

Hummel has written a three-volume treatise on polymer and plastics analysis (2). Volume 1 is a collection of about 1900 infrared spectra of polymers. Volume 2 gives a gen-

TABLE 18.2. NIST Analytical Chemistry Databases

NIST/EPA/NIH Mass Spectral PC Version
The NIST/EPA/NIH Mass Spectral Database is available on floppy disk for PC use; permits the user to search for spectra of unknown compounds as well as to locate a particular spectrum; also users can enter their own spectral data; includes structural drawings for most compounds in the database and provides chemical names, including alternative names, and CAS Registry Numbers; both graphic and tabular display are possible
NIST/NIH Desktop Spectrum Analyzer Program and X-Ray Database
NIST/EPA Gas-Phase Infrared
NIST Crystal Data
NIST/Sandia/ICDD Electron Diffraction
NIST X-ray Photoelectron Spectroscopy

eral introduction to spectra and methods of identification of plastics, fibers, rubbers, and resins, and it provides spectra and other data on the most important commercial products. Volume 3 deals with additives and processing aids. A new edition is in progress.

18.7. AMERICAN SOCIETY FOR TESTING AND MATERIALS (ASTM)

Another major source of analytical methods and standards is the American Society for Testing and Materials (ASTM), West Conshocken, PA. This is a not-for-profit organization founded in 1898 that now has 35,000 members. The ASTM *Annual Book of Standards* now comprises 72 volumes containing over 10,000 standards, test methods, specifications, practices, guides, and classifications for materials, products, systems, and services. The material is available in CD-ROM form as well, and standards may also be purchased individually, and sometimes in special collections, as needed.

The work is by done by 132 main technical committees with volunteer members primarily from industry, but including also university, government, and consulting specialists from the United States and abroad. Any necessary laboratory work is done in the laboratories of the committee members. Adoption of standards or methods is based on consensus among committee members.

Companies that make new products or develop new applications may be the first to develop new methods and then bring them forward to the appropriate ASTM committee, subcommittee, or task group for ultimate publication in one of its many volumes of test methods. The ASTM approval process may sometimes take a longer time than some chemists would like (it usually takes about 2 years to develop a standard) since this distinguished group necessarily operates by a consensus of its committees. As a result, some published ASTM methods may not necessarily be the latest utilized in practice by industry. Nevertheless, the ASTM Standards represent an important and authoritative source of a large amount of widely accepted analytical and other test method information, and ASTM methods are widely quoted and referred to.

ASTM offers an *International Directory of Testing Laboratories* on the Internet at www.astm.org/labs/index.html. This may be searched full-text without charge and permits access by both types of services and geographic location. ASTM offers a similar Internet product relating to consultants and expert witnesses.

ASTM also publishes *Special Technical Publications* (STPs), which are collections of reviewed technical papers reflecting the state of the art in certain matters of interest to ASTM. There are also manuals, monographs, handbooks, reference radiographs, and *ASTM Standards InfoBriefs,* a quarterly update of new and revised standards. ASTM publishes a monthly magazine and five journals on testing and standards.

18.8. ANALYTICAL METHODS FOR OCCUPATIONAL HEALTH, ENVIRONMENTAL, AND SAFETY APPLICATIONS

Chapter 14 is a broad discussion of sources of environmental- and safety-related information that should be consulted.

Among the most widely used sources for occupational health methods are publications from the U.S. National Institute for Occupational Safety and Health (NIOSH). These include the NIOSH *Manual of Analytical Methods,* now in its fourth edition. The earlier editions are still useful; they list some older, but still useful, wet and other methods. Other related publications are published by the OSHA (U.S. Occupational Safety and Health Administration); an example is a CD-ROM popularly known as simply "OSHA Industrial Hygiene Methods."

The U.S. Environmental Protection Agency (EPA) issues a very large number of methods and protocols. One convenient way to keep up with these is through the *Federal Register* (published daily and available online) and the *U.S. Code of Federal Regulations.*

The American Public Health Association, Washington, DC, in cooperation with the American Water Works Association and the Water Environment Federation, publishes the *Standard Methods for the Examination of Water and Wastewater* (19th Edition, 1995). There is a 1996 supplement and the volume is fully updated every 3 years. Editors are: Andrew D. Eaton; Lenore S. Clesceri, and Arnold E. Greenberg.

Among the many journals in the field are the *Journal of the American Conference of Governmental and Industrial Hygienists.* The *Journal of the American Industrial Hygiene Association* is another example.

18.9. *ANALYTICAL CHEMISTRY*

As part of its June 15 issue (issue No. 7 of the year), *Analytical Chemistry* (ACS journal) publishes an extensive series of annual reviews on analytical chemistry. This series started shortly after World War II to help chemists catch up with developments during the war years.

18.10. ABSTRACTING AND INDEXING SERVICES

Analytical chemists are fortunate in having access to a wide range of excellent abstracting services and current awareness tools produced by these services. *Chemical Abstracts* covers analytical chemistry thoroughly, with two sections dedicated to analytical chemistry, while topics of analytical interest appear in many of the other sections. Other CAS products of interest include the *CA Selects* series (see Chapters 4 and 6), which contains topics that are of interest to analytical chemists and is an important current awareness tool. The Chemical Abstracts *Surveyor* Series, another CAS tool, also has a number of topics that are highly pertinent.

In addition to its journals and books in the field, The Royal Society of Chemistry provides a number of products of special interest to analytical chemists. These include

Analytical Abstracts. This important service is the only major abstracting publication dedicated to all aspects of analytical chemistry. It is available in printed form (monthly) and also online through Data-Star, DIALOG, Questel•Orbit, and STN International, and it is also available as CD-ROM. *Analytical Abstracts* provides comprehensive coverage of new techniques and applications in all branches of analytical chemistry. There are extensive abstracts. This product has recently been further improved in several ways. The number of journals fully covered as to scientific content has been increased, abstracting is now more current, there is an even fuller subject index in each monthly issue, and abstracts are more complete. Indexing of techniques has been extended so that all routine applications are covered, as well as fundamental developments. Author indexes do not appear in the monthly issues.

Chromatography Abstracts. This series is published in collaboration with the Chromatographic Society. Designed for the practicing chromatographer. As the name implies, this covers such techniques as HPLC (high-performance liquid chromatography), GC (gas chromatography), supercritical-fluid chromatography, capillary electrophoresis, and TLC (thin-layer chromatography), as well as other methods. Each issue contains subject and author indices. This is a subset of *Analytical Abstracts.*

The Royal Society of Chemistry (RSC) also offers several current awareness products of special interest to analytical chemists. For example, current awareness profiles can be established with *Analytical Abstracts.* In addition, RSC offers the *Mass Spectrometry Bulletin,* an alerting bulletin, with bibliographic citations and keywords, but not abstracts. This product, which includes over 500 abstracts per month, covers journals, patents, and books. There are monthly subject and author indexes. It is also available for the desktop computer. This product began in 1966.

Window on Chemometrics in a current awareness RSC publication that provides summaries of recently reported work on computer handling of analytical data. Specific topics included are calibration, expert systems, and applications in spectroscopy, chromatography, and other analytical techniques. It is available in printed form and as CD-ROM and is a subset of *Analytical Abstracts.*

As previously mentioned (Section 12.9), the RSC *Journal of Analytical Atomic Spectroscopy* has a diskette counterpart that contains "Atomic Spectrometry Updates" tables and references (*JAASbase*).

ISI (see Chapter 4) also has important current awareness products of interest to analytical chemists.

18.11. ANALYTICAL CHEMISTRY ON THE INTERNET

There are a number of pointers to such information. Diablo Analytical, Inc. maintains such a list at www.diab.com/analyt.com. Another is offered by Umea University, Sweden at www.anachem.umu.se/jumpstation.htm. The American Chemical Society Division of Analytical Chemistry maintains a list of databases and information resources at http://nexus.chemistry.duq.edu/analytical/databass.html."Murray's Mass Spectrometry Page," (http://tswww.cc.emory.edu/~kmurray/mslist.html) is a collection of links to mass spectrometry Internet sites. The general lists of chemistry resources on the Internet, as mentioned in Chapter 10, are also helpful.

In addition, there are analytical listservs and newsgroups such as Analysis-L and sci.chem.analytical.

The Internet is a very good source of new information about the product offerings of the companies that manufacture and sell analytical instruments. In addition, such companies as Bruker offer software on the Internet (the home page is www.-BRUKER.com/). In the case of Bruker, a principal interest is NMR. In addition, Bruker provides links to other related sites on the Web.

Stephen R. Heller, U.S. Department of Agriculture Research Service, Beltsville, MD, edits an interesting and helpful column on the Internet and chemistry for *Trends in Analytical Chemistry* (*TrAC*), an Elsevier publication. The articles are available in print form and on the Internet at the address www.elsevier.com:80/inca/homepage/saa/trac/intntcol.html.

As mentioned earlier in this book (Chapter 15), the NIST Standard Reference Data Program is in process of making a considerable amount of its data available on the Internet.

18.12. EXAMPLES OF OTHER SOURCES OF ANALYTICAL DATA

Chapter 15 discusses examples of other sources of selected analytical data. These include the Thermodynamics Research Center, College Station, TX (see Chapter 15). Another example can be found in the compilation *Physical Properties of Organic Compounds*. Other chapters also contain sources of interest to analytical chemists.

Probably the most extensive series of books in analytical chemistry is *Chemical Analysis: A Series of Monographs on Analytical Chemistry and Its Applications* (3). This series comprises 144 volumes (as of 1997) and includes the fundamental aspects, instrumentation, and applications. The editor of the series is J. D. Wineforder, University of Florida, and the publisher is John Wiley & Sons, New York. Elsevier Science Publishers, Amsterdam, offers a similar extensive series, which is entitled *Comprehensive Analytical Chemistry.*

There are a number of "official methods" compilations published by important professional organizations. Thus, the most definitive analytical standards for prescription pharmaceuticals are the *United States Pharmacopeia* (*USP*) and the *National Formulary* (*NF*). The latest versions as of 1997 are the *USP* 23rd ed. and the *NF* 18th ed., which were published together in one book in 1995 by the U.S. Pharmacopeial Convention, Inc., Rockville, MD. New editions appear every 5 years, and there are semiannual supplements with revisions. The distinction between the two books is that the *USP* contains therapeutically active materials, whereas *NF* includes excipients and other such materials that may be utilized in formulating. *USP* also includes nutritional supplements and materials that may be foodstuff ingredients. Typical data included in *USP/NF* for each compound are name, structure, molecular formula, CAS Registry Number, preparation, and legally enforceable standards as to strength, quality, purity, packaging, labeling, and storage. Other major countries also publish their own pharmacopeias.

Other examples of official methods can be found in the publications of the Association of Official Analytical Chemists, McLean, VA. Their official methods publication is *Official Methods of Analysis of the AOAC,* 16th edition, Volumes I and II. Patricia A. Cunniff, Editor (AOAC International, 1970 Chain Bridge Road, McLean, VA 22109 (800-379-2622; 703-522-3032). This classic source contains over 2300 collaborative methods covering such products as agricultural chemicals of all types, food compositions, drugs, contaminants, oils and fats, and hazardous substances. These methods have all been tested in more than one laboratory, and the standard deviation expected is given. Previously revised every 5 years, this volume is now published in looseleaf form, and there are annual updates. The books are available in printed and CD-ROM form. The AOAC was founded in 1884 with an original focus on resolving issues between federal and state analytical chemists regarding fertilizer analysis. Since then, AOAC focus has broadened to include many other types of products, and the members include analysts from industrial and academic laboratories as well as government.

AOAC also publishes *Journal of AOAC International,* which relates to state-of-the-art methodology for chemical and biological analysis. Joseph T. Tanner, FDA, is Chairman, Editorial Board.

Another official methods product is *Official Methods and Recommended Practices of the American Oil Chemists Society,* 4th ed., 1994, David Firestone, Ed. This looseleaf binder compilation covers analytical techniques for fats and oils and related products.

An annual series, *Organic Electronic Spectral Data* (4), edited by J. P. Phillips and others, is a cooperative effort to abstract and publish in formula order all the ultraviolet–visible spectra of organic compounds presented in the journal literature. Over 50 chemists have searched 100-plus titles during the course of this project to assemble over 350,000 spectra throughout these volumes.

An important source for many years is the American Chemical Society's *Reagent Chemicals* (5). The eighth edition was published in 1993, and it is updated through announcements in *Chemical and Engineering News* and a supplement on the Internet (http://www.pubs.acs.org/books/reagents/). Specifications and analytical procedures are provided.

REFERENCES

1. The Royal Society of Chemistry, *Eight Peak Index of Mass Spectra,* 3 vols., Cambridge, UK, 1991.
2. D. O. Hummel and F. Scholl, *Atlas of Polymer and Plastics Analysis,* 2nd ed., Verlag Chemie, Weinheim and Deerfield Beach, FL, 3 vols. 1978–1984; 3rd edition 1988, 1991.
3. J. D. Wineforder, Ed., *Chemical Analysis: A Series of Monographs on Analytical Chemistry and Its Applications,* Wiley, New York, 1972 to date.
4. J. P. Phillips, Ed., *Organic Electronic Spectral Data,* **31,** 1995 (1989).
5. American Chemical Society, *Reagent Chemicals,* 8th ed., Washington, DC, 1993.

APPENDIX A*
Herman Skolnik Award

In 1976, in honor of the first recipient, the American Chemical Society Division of Chemical Information established the Herman Skolnik award to recognize outstanding contributions to and achievements in the theory and practice of chemical information science. The award consists of a plaque and an honorarium of $2000.

Below are brief citations for every award recipient.

1976: Herman Skolnik—for outstanding and sustained service in the field, as one of the founders of the Division, founder of the ACS Delaware Valley Chemical Literature Group, science historian, founder and editor of the *Journal of Chemical Documentation* (*Journal of Chemical Information and Computer Sciences* since 1975), inventor of a notation system, innovator in indexing, and organizer of symposia and panel discussions at the ACS local, regional, and national level

1977: Eugene Garfield—for contributions to information science that have had considerable impact on both the academic world and the information industry, especially the successful application of scholarly work to the business of information, such as founding of the Institute for Scientific Information (ISI), publication of innovative secondary journals and indexes (*Current Contents, Index Chemicus,* and *Science Citation Index*), and enjoying acceptance of his innovations through effective educational marketing programs

1978: Fred A. Tate—for conceiving, developing, and implementing computer-based information-handling systems and procedures across the full range of Chemical Abstracts Service (CAS) operations, which had provided prototypes for other secondary services, for his leadership in the development of the CAS Chemical Registry System, and for close international cooperation between the United Kingdom, West German, French, Japanese, and United States groups in the development and use of chemical information systems and services

1980: William J. Wiswesser—for pioneering mathematical, physical, and chemical methods of punched-card and computer-stored representation of molecular structures, leading to the creation of the Wiswesser Line Notation (WLN)

*Reprinted with permission of American Chemical Society's Division of Chemical Information, from *50 Years of Chemical Information in the American Chemical Society: 1943–1993.* Courtesy of Dr. W. Val Metanomski who updated the listings.

for concise storage and retrieval of chemical structures, which was adopted by the largest chemical and pharmaceutical companies worldwide to manage their respective chemical structure files, and by a number of secondary indexes, atlases of data, and catalogs of chemical compounds

1981: Ben H. Weil—for distinguished and dedicated services to the chemical profession, particularly in definition and documentation of chemical literature. Pioneering and continuing work in chemical information systems and copyright, including one of the first punched card indexing systems placed in actual use, founding and editing of the Divisional bulletin, *Chemical Literature,* standardization of abstracts, and contribution to the creation of the Copyright Clearance Center, Inc. (CCC)

1982: Robert Fugmann—for development of the GREMAS system (Genealogical REtrieval of MAgnetic tape Storage), the first truly sophisticated computerized retrieval system, based on a faceted hierarchical fragment code for each part of a chemical molecule, and for development of the TOSAR system (TOpological representation of Synthetic and Analytical system Relations) for the retrieval of reactions and other concepts, including establishment of indexing concepts for nonstructural information and creation of theoretical basis of information systems

1983: Russell J. Rowlett, Jr.—for guiding *Chemical Abstracts'* transition from a manually produced abstracting and indexing publication to a computer-generated family of products, and for his leadership in the improvement of patent coverage, the CAS Registry System, timeliness of CA Volume and Collective Indexes, and quality control through a shift from volunteer abstractors to full-time professional document analysts and through the unified document analysis utilizing to the fullest extent human–machine interactions

1984: Montagu Hyams—for contribution to handling of patents by founding in 1951 a one-person business from his house, Derwent—which, through his vision, leadership, and business acumen, has become, as Derwent Publications Limited, the world leader in patent-based information services producing a diversified range of patent- and journal-based information services available both in printed form and as online computer-searchable databases

1986: Dale B. Baker—for leadership of Chemical Abstracts Service (CAS) in its move from the conventional abstracting and indexing service of the 1950s to the world's premier automated information storage and retrieval system through courageous embarkation on new paths and approaches, including promotion of international sharing of scientific and technical information, which provided direction for the entire information industry

1987: William Theilheimer—for pioneering a chemical reaction documentation system, embodied in 40 yearbooks of "Theilheimer's Synthetic Methods of Organic Chemistry" and paving the way to modem chemical reaction databases through codification of chemical reactions and categorization of reactions in terms of reaction type and essential bond breaking and formation

1988: David R. Lide, Jr.—for the creation of the National Standard Reference Database Series of computer-searchable numeric databases, administration of the Standard Reference Data Program of the National Bureau of Standards, founding and editing the *Journal of Physical and Chemical Reference Data,* and participation in national and international data activities of the International Union of Pure and Applied Chemistry (IUPAC) and the Committee on Data for Science and Technology (CODATA)

1989: Michael F. Lynch—for pioneering research of more than two decades on the development of methods for the storage, manipulation, and retrieval of chemical structures and reactions as well as related bibliographic information, including generic structure storage and retrieval, automatic subject indexing, articulated subject index production, document retrieval system, and database management

Stuart A. Marson—for development of innovative, user-friendly software that has allowed the bench chemist to more productively utilize chemical information as a daily resource such as the first complete commercial system for graphic input, storage, searching, and retrieval of chemical structures (MACCS) and the chemical reaction information system (REACCS)

1990: Ernst Meyer—for playing a major role in revolutionizing chemical information technology through the use of computer methodology since the late 1950s for input and searching techniques for topological and fragment representation of chemical substances, including generic or Markush structures and considering structure–activity correlations

1991: W. Todd Wipke—for pioneering work in the development of methods for representing and manipulating chemical information such as computer-assisted design of organic syntheses, simple interfaces and smart systems, methods for molecular modeling and conformational analysis, and editorial innovations in starting an electronic journal *Tetrahedron Computer Methodology*

1992: Jacques-Emile Dubois—for the development of the DARC Topological System, which led to various applications in search and retrieval of chemical substructures and structures and in artificial intelligence such as in applying sequences of substructure, structure, and hyperstructure in locating chemical entities in their structural context and in evaluation of their local or global properties according to topological or topographical information

1993: Peter Willett—for contributions to the development of chemical information science, including the identification of reaction sites and the development of maximal common subgraph algorithms in reaction retrieval systems, the introduction of similarity measures through classification and clustering in chemical substructure searching, three-dimensional searching of chemical molecules and biological macromolecules, and text searching

1994: Alexandru T. Balaban (of Bucharest, Romania)—for his pioneering contributions to chemical graph theory and its applications to chemical nomenclature, classification of chemical structures and reactions, and coding of chemical information for computer storage and retrieval

1995: Reiner Luckenbach and **Clemens Jochum** of Frankfurt am Main, Germany—for significant contributions to the design and development of the electronic version of the *Beilstein Handbook of Organic Chemistry,* thus making chemical structural information, bibliographic information, and chemical property data available and accessible online.

1996: Milan Randić—for application of mathematical chemistry to nomenclature and the characterization of chemical structures, seminal work on topological indices and orthogonalized descriptors, research on structure-property and structure-activity correlations, and building tools to model biological activities of molecules

1997: Johann Gasteiger—for advancing computerization of chemistry in areas of artificial intelligence, neural networks, chemical reactions, 3D databases, and structure-activity relationships

APPENDIX B*
The Austin M. Patterson–E. J. Crane Award

The Austin M. Patterson Award, predecessor of the Patterson–Crane Award, was established in 1949 by the Dayton Section of the American Chemical Society to acknowledge "meritorious contributions in the field of chemical literature and especially in the documentation of chemistry." Dr. Patterson was the first recipient of the award in recognition of his work as editor of *Chemical Abstracts* from 1909 through 1914 and his leadership in organic chemical nomenclature.

In 1975, the award was expanded to honor E. J. Crane, who was editor of *Chemical Abstracts* from 19915 to 1958 and the first director of Chemical Abstracts Service from 1956 to 1958. The expanded award is sponsored and administered jointly by the Columbus and Dayton Sections of ACS. The award, which is presented biennially, consists of a personalized commendation and an honorarium of $2000.

The rules of eligibility for the award state that a nominee shall have made "contributions of national notice" to the documentation of chemistry or to chemical information theory or practice. Such contributions may be in the areas of the production of books, articles, reviews, and bibliographies; editorial work; abstracting; nomenclature; construction of indexes, codes, or methods of classification; development and use of methods for searching literature and of mechanical or electronic aids to information storage and retrieval; chemical library work; or related activities.

Past recipients of the award, which is international in scope, have included research chemists who made major contributions to the information aspects of their specialties, as well as scientists who devoted full time to information activities and their management.

Below are brief citations for every winner.

I. AUSTIN M. PATTERSON AWARD (1949–1975)

1949: Austin M. Patterson—for developing of systematic nomenclature and indexing of organic compounds and work on chemical definitions and chemical dictionaries

*Data supplied courtesy of Dr. W. V. Metanomski.

1951: Arthur B. Lamb—for guiding the *Journal of the American Chemical Society* for over 32 years and for establishing its preeminence among scientific research publications

1953: E. J. Crane—for guiding *Chemical Abstracts* for 42 years and for advancing the art of scientific indexing

1955: Howard S. Nutting—for contributions to chemical nomenclature and codification, and to compilation, classification, and utilization of chemical literature

1957: Melvin G. Mellon—for teaching, promoting, and publishing on proper use of chemical literature

1959: Leonard T. Capell—for contributions to chemical nomenclature, including the coauthorship of *The Ring Index*

1961: G. Malcolm Dyson—for designing and developing a notation system for organic compounds and application of computers to chemical information storage and retrieval

1963: W. Albert Noyes, Jr.—for contributions as an editor of ACS journals; teacher and researcher

1965: Elmer Hockett—for pioneering work in standardizing and simplifying chemical editing and indexing

1967: Melville L. Wolfrom—for contributions to nomenclature of carbohydrates

1969: Herman Skolnik—for innovations in chemical notation systems and indexing, and founding of the *Journal of Chemical Documentation*

1971: Charles D. Hurd—for leadership in definition and documentation of chemical literature, and in standardizing and simplifying chemical indexing

1973: Pieter E. Verkade—for pioneering work in formulation and standardization of rules for chemical organic nomenclature

1975: William J. Wiswesser—for pioneering methods of punched-card and computer-stored representations of molecular structures, leading to the creation of Wiswesser Line Notation (WLN)

II. AUSTIN M. PATTERSON—E. J. CRANE AWARD (1977–1995)

1977: Benjamin H. Weil—for development of chemical indexing systems, abstracting standards, and copyright guidelines

1979: Dale B. Baker—for leadership in abstracting and indexing of world's chemical literature, and pioneering automated methods of chemical information processing

1981: W. Conard Fernelius—for developing and standardizing of nomenclature for inorganic and coordination compounds

1983: Eugene Garfield—for innovations in conceiving, designing, and establishing timely publications and services providing easy access to scientific and technical information

1985: Bruno J. Zwolinski—for founding the Thermodynamics Research Center, and collecting, organizing, and publishing data on thermodynamic properties of compounds

1987: Kurt L. Loening—for contributions to the development of chemical nomenclature and terminology, both nationally and internationally

1989: George E. Vladutz—for pioneering work on the theory and practice of computerized indexing and retrieval of chemical reactions

1991: David R. Lide, Jr.—for leadership in the development of data programs and databases of chemical, physical, biological, geologic, and astronomical properties

1993: Hideaki Chihara—for international contributions to chemical information and documentation, especially for making available worldwide access to the Japanese chemical literature

1995: Arthur E. Martell—for work on establishing standards for the measurement and evaluation of critical stability constants

APPENDIX C
Tabulation of Selected Representative Online Databases That are of Interest to Chemists and Chemical Engineers

INTRODUCTION

This tabulation lists selected representative online databases that are available through some of the major online vendors in the U.S. Many databases are available in other electronic forms, and a few of these are listed here as well. Most databases provide abstracts for the documents covered, except of some of those files that relate to patent concordances, citations, and changes in patent status, and except for those that are essentially structure or numeric files such as *CA Registry* and *NISTTHERMO*. Some databases are full-text. Related databases from the same or other producers are shown in this Table when feasible and appropriate. In addition to the commercial hosts, databases are increasingly available on the Internet whether or not so stated in the Table. Many files are also available in printed and/or CD-ROM format. Print counterparts may have much earlier start dates than corresponding electronic files. A number of important files, such as some of those of MDL Information and ISI, are available through in-house configurations; most of these are not covered in this Table but are discussed elsewhere in this book. Database, producer, and host names are trademarked or registered trademarks or service marks. Database names or abbreviations may vary from host to host. Hosts shown are representative (limited for the most part to DIALOG, Orbit•Questel, STN International, and MEDLARS) and are not intended to be all-inclusive; gateways to other hosts are available in some cases, and there is a strong trend toward nonexclusivity of databases. Thus, relatively few databases are found on just one host system. Beginning dates of coverage, content, and search capabilities may vary depending on host. Placement of database by category is necessarily somewhat arbitrary; all categories should be scanned. Because databases are modified, added and dropped over time, database producers and vendors should be contacted for the latest data.

A. GENERAL PURPOSE DATABASES

Chemical Abstracts

[DataStar and DIALOG, OCLC, Ovid, Questel•Orbit, STN (See Appendix D).] Outstanding file covers journal, patents, dissertations, books, reports, and other literature from 1967 to date. Very highly regarded comprehensive file that offers expert indexing and abstracting and that is usually first choice to investigate in most searches of chemical literature. If optimum completeness is desired, should always be one of the first sources consulted in any search of chemical and related literature.

Abstracts available only on STN and the OCLC file (see below), although the deeply indexed bibliographic files are on other systems such as DIALOG, Questel• Orbit, and Ovid. Also available on the Internet within *STN Easy* (easy searching) file as well as in conventional form through Telnet on the Internet. Subset version *Chemical Abstracts Student Edition,* intended for undergraduates available only on OCLC, Columbus, Ohio (www.oclc.org); covers approximately 250 journals and all dissertations in *CA* from 1967. *CASSI* (list of libraries that hold CAS-covered documents) available only on Orbit. *CHEMCATS* (buyer's guide to supplier catalogs) available only on STN. See also entries in other categories in this Table.

CA also available, in whole or part, via CD and through client server arrangements (e.g., *SciFinder*).

CA REGISTRY [DataStar and DIALOG, Questel•Orbit, STN.] Very highly regarded and exhaustive product of Chemical Abstracts Service, provides information on over 16 million substances registered by CAS since 1957. Information includes structures, CAS Registry Numbers, alternative names, and other similar basic data essential to most searches. STN, Questel•Orbit permit substructure searching. Major parts of this database are also searchable on other systems, but without structures.

Compendex [DIALOG, Questel•Orbit, STN.] Excellent coverage of engineering literature in most disciplines including chemical literature. Does not cover patents. Coverage since 1970. Well-regarded.

Current Contents Search [DIALOG, Ovid.] Multidisciplinary coverage from 1990 including author abstracts when available.

Index Chemicus [In-house only.] Excellent file that, for certain carefully selected journals, covers synthesis, isolation, or identification of new organic compounds. Includes both narrative abstracts and diagrams of reaction flows. Electronically searchable for in-house use beginning with 1993. Sister product, *Current Chemical Reactions,* provides details on new reactions from 1985. Electronically searchable for in-house use.

NTIS [DIALOG, Questel•Orbit, STN.] Abstracts of mostly U.S. government-sponsored reports and other documents since 1964.

SciSearch [DIALOG, Orbit•Questel, STN.] Unique multidisciplinary tool that permits the user to determine citation relationships between scientific papers and to conduct citation searches. Begins online coverage in 1974. Abstracts included since 1991.

B. DATABASES DEDICATED TO PATENT DOCUMENTS

Derwent World Patents Index [DIALOG, Questel•Orbit, STN.] The leading abstracting service that covers world-wide patent documents for all technologies. Beginning date ranges from 1963 to early 1970s, depending on subject matter. Deep indexing and abstracting. Very highly regarded. *Derwent World Patents Index Markush* (see also list of Markush databases below) permits Markush structure searching from 1987 (subscribers only). *Derwent World Patents Index with API Indexing* (subscribers only) offers also in-depth American Petroleum Institute indexing from 1964. *Derwent LitAlert* covers U.S. patent and trademark infringement suits filed in U.S. District Courts and reported to the Commissioner of Patents and Trademarks (from 1970's). *Derwent Patents Citation Index* covers citations from certain patent-issuing authorities since 1974 and for a few authorities from later years to date.

IFI CLAIMS [DIALOG, Questel•Orbit, STN.] Covers all issued U.S. patents since 1950. Chemical patents are deep-indexed. Highly regarded. Note that *CLAIMS* also offers a U.S. patent citation file beginning with 1947, and further, a file that tracks reassignment and reexamination of U.S. patents from 1980.

INPADOC [DIALOG, Questel•Orbit, STN.] The most comprehensive worldwide patent document concordance (to equivalents) with most coverage dating from 1968. Legal status data are also provided for some patent-issuing authorities.

MARPAT, Derwent World Patents Index Markush, PharmSearch (Mpharm) [MARPAT on STN, others Questel•Orbit.] Permit searching of Markush structures in patents from approximately the mid-later 1980's to date.

USPATFULL (or CHEMICAL PATENTS PLUS on the Internet), U.S. PATENTS FULLTEXT, QPAT-US(Internet only) [DIALOG, Internet, Orbit•Questel, STN.] Contain searchable full text of U.S. patents from 1971, usually with some gaps in the early 1970's. Chemical Abstracts Service versions (the first two above) offer the advantage of *CA* indexing. IBM offers free searching of U.S. patents on the Internet.

DIALOG offers full-text searching of European patent documents since 1978: *European Patents Fulltext.*

C. "HANDBOOK" TYPE DATABASES

Beilstein Handbook of Organic Chemistry [DIALOG, Orbit•Questel, STN.] Outstanding evaluated coverage of organic chemistry literature from 1779. Excellent treatment of preparation and properties. Very highly regarded. Over 140 chemical journals included for more recent years.

Internet and special versions for subscription, *CrossFire* and *Netfire,* offer unique searching power; very extensive coverage of chemical reactions is also offered.

Chapman and Hall Chemical Database [DIALOG.] Covers Chapman and Hall's printed *Dictionary of Organic Compounds* (formerly *Heilbron*) and several other of the publisher's Dictionary products covering a broad range of organic and inorganic compounds. The publisher's CD-ROM is more current.

Gmelin Handbook of Inorganic and Organometallic Chemistry [STN.] Outstanding coverage of inorganic chemistry since 1817, with more selective coverage for more recent years. Also covers organometallics. Very highly regarded. Also available on Beilstein *CrossFire* in-house system.

D. AGRICULTURAL, BIOLOGICAL, AND MEDICAL SCIENCES*

Agricola Good coverage of agricultural literature, including agricultural chemistry, since 1970. A similar file, which should also be consulted is *CAB Abstracts* (see below).

BIOSIS Previews Comprehensive coverage of biological and related literature since 1969. *EMBASE* and *Medline* should also be consulted. Well-regarded.

CAB Abstracts [DIALOG, STN.] Excellent coverage of agricultural and biological sciences literature since 1972. *Agricola* should also be consulted.

EMBASE [DIALOG, STN.] Excellent coverage of medical journals since 1974. *BIOSIS Previews* and *Medline* should also be consulted.

Medline [DIALOG, Internet, MEDLARS system, Orbit•Questel, STN.] Provides excellent coverage of over 4000 biomedical journals, with time coverage beginning in 1966 and to be extended back through 1962. Internet version is free to public through PubMed (http://www.ncbi.nlm.nih.gov/PubMed/); includes links to full-text of retrieved articles for certain journals. *PREMEDLINE* (Internet only) gives advance notice of new input, covers citations only prior to indexing and abstracting. *Chemline* and *CHEMID* are closely related chemical dictionary/identification files. See also entries under *EMBASE* and *BIOSIS Previews* which overlap *Medline* somewhat but also provide additional coverage.

E. ENVIRONMENTAL, SAFETY, AND REGULATORY MATTERS

See also Agricultural, Biological, and Medical Sciences. See also discussions in text re *CIS* (*Chemical Information System*) and *Technical Database Services* (*TDS*).

*See also Environmental, Safety, and Regulatory Matters.

Chemical Safety NewsBase; Occupational Safety and Health [DIALOG, STN.] Very useful to identify specific experiences, incidents, and research studies with the safety aspects of chemicals from 1981, 1973 respectively. The second database on DIALOG only.

ChemList [STN.] Information on regulatory status of chemicals with reference to TSCA and other national, international, and state lists and regulations.

DOSE (Dictionary of Substances and their Effects) [DIALOG.] Contains data on over 4000 chemicals and their biological/environmental impacts as well as some regulatory and physical properties data. References given. Product of The Royal Society of Chemistry.

HSDB (Hazardous Substances Data Bank) [DIALOG, MEDLARS, STN.] Peer-reviewed information on over 4500 chemicals as to hazards and safe handling.

MSDS-CCOHS, MSDS-OHS [Both on STN, DIALOG second only.] Collections of Material Safety Data Sheets.

RTECS (Registry of Toxic Effects of Chemical Substances) [DIALOG, MEDLARS, STN.] Extensive compilation of toxicology data, but data are not evaluated by compilers. Original references need to be consulted.

TOXLINE, TOXLIT [DIALOG, MEDLARS, STN.] Toxicology literature from 1940s. Some overlap of *Medline.*

F. TECHNOLOGY- OR INDUSTRY-SPECIALIZED INFORMATION SERVICES

APILIT, APIPAT, TULSA [DIALOG, STN.] Literature on petroleum technology from 1964. APIPAT covers patents. Excellent indexing.

PAPERCHEM [DIALOG, STN.] Outstanding coverage of pulp and paper science and technology, both journal and patents literature from 1967.

RAPRA [DIALOG, STN.] Excellent coverage of literature and some patents on plastics, rubbers, adhesives, related topics from 1972. Thoroughly indexed.

G. NUMERIC DATABASES*

See also *Beilstein* and *Gmelin* above and sections about *Technical Database Services* (*TDS*) *Numerica* and *CIS* (*Chemical Information System*) in text.

*Some of these databases contain other valuable information as well.

DETHERM [STN.] Thermophysical properties for approximately 17,000 inorganic and organic chemicals. Updated semiannually. Includes abstracts of literature covered.

DIPPR [STN, TDS.] Evaluated physical property data for approximately 1,600 commercially important chemicals. Updated annually. *Technical Database Services* (*TDS*) plans to have on their Web site in 1998 (www.tds-tds.com)

HODOC (CRC Handbook of Data on Organic Compounds) [STN.] Data on more than 25,000 organic compounds.

NISTTHERMO [STN.] Critically evaluated thermochemical data for over 8,000 inorganic and small organic molecules. See also NIST web site (www.nist.gov/srd).

PLASPEC [DIALOG, STN. Also available directly from producer.] Facilitates plastics selection by providing searchable properties data for thousands of polymer materials. Also includes data on applications, prices, CAS Registry Numbers, supplier names. Companion files include news of plastics industry.

TRCTHERMO [STN; some data also on *TDS Numerica* (calculated vapor pressures and boiling points).] Evaluated thermodynamic data from the Thermodynamics Research Center at Texas A and M University.

H. FULL-TEXT ENCYCLOPEDIAS

Kirk-Othmer Encyclopedia of Chemical Technology; Polymer Online (Encyclopedia of Polymer Science and Engineering). [DIALOG.] For *Kirk-Othmer,* both 3rd and 4th editions are available. For *Polymer,* 2nd edition. Provides full-text of printed versions for in-depth searching.

I. FULL-TEXT JOURNALS

CJACS, CJRSC, CJAOC, CJELSEVIER, CJVCH, CJWILEY [STN, Internet.] Full-text journals of the American Chemical Society, Royal Society of Chemistry, Journal of AOC International, VCH, and John Wiley and Sons, respectively. Journals have different start dates, depending on publisher. Full-text of many of these and other journals available on Internet or in other electronic forms.

J. REACTION DATABASES

CASREACT [STN.] Reactions of organic substances from 1985 as reported in the patents and journals coverage of *Chemical Abstracts.* Structure-based searching is possible.

Derwent Journal of Synthetic Methods [Questel•Orbit, CRDS (*Chemical Reactions Documentation Service*) subscribers only; STN.] Structure-searchable organic reaction information based on the journal and patent literature. Continuation of the well-known book series edited by Theilheimer. Start date is 1975 on STN, 1942 on Questel•Orbit. Also available for private in-house use through MDL Information, Inc.

K. ANALYTICAL CHEMISTRY*

Analytical Abstracts [DIALOG, Questel•Orbit, STN.] The only such product devoted solely to all aspects of analytical chemistry. Very highly regarded product of The Royal Society of Chemistry. Coverage from 1980. Deeply indexed.

SPECINFO [STN.] NMR spectra for over 152,000 organic and inorganic compounds, as well as some infrared and mass spectra.

L. CHEMICAL NEWS AND BUSINESS INFORMATION

CIN (Chemistry Industry Notes), CBNB (Chemical Business NewsBase) [CIN on STN, CBNB on DIALOG, STN.] Cover chemical industry and other news. CIN begins 1974, CBNB in 1984.

PLASNEWS [STN. Also available directly from producer.] Covers news in plastics industry from 1987. Daily updates.

*See also the section on Numeric Databases.

APPENDIX D.
CAS Online Databases Available from Major Hosts (Vendors) Used in the United States

	STN	DIALOG	DATASTAR	QUESTEL	ORBIT	OCLC	OVID
CA	X[a]	X[b]	X[b]	X[b]	X[b]		X[b]
CAPLUS	X[a]						
CAOLD	X						
CA Student Edition						X[a]	
REGISTRY	X[c]	X[d]	X[d]	X[d]	X[d]		
CHEMCATS	X						
CHEMLIST	X						
MARPAT	X[a,c]						
CASREACT	X[a,c]						
CASSI					X		
CIN	X	X	X		X		

*In addition, CAS data is available in batch services and in the *Toxline* file offered by a number of vendors.

[a]With *CA* abstracts.

[b]Without abstracts.

[c]With structures.

[d]Without structures.

Source: Chemical Abstracts Service, Jan Williams.

INDEX

*The designation *KR* has been replaced by DIALOG as of 1998.